# ADVANCED MATERIALS, STRUCTURES AND MECHANICAL ENGINEERING

PROCEEDINGS OF THE INTERNATIONAL CONFERENCE ON ADVANCED MATERIALS, STRUCTURES AND MECHANICAL ENGINEERING, INCHEON, SOUTH KOREA, 29–31 MAY, 2015

# Advanced Materials, Structures and Mechanical Engineering

*Editor*

Mosbeh Kaloop
*Department of Civil and Environmental Engineering,
Incheon National University, South Korea*

CRC Press
Taylor & Francis Group
Boca Raton  London  New York

CRC Press is an imprint of the
Taylor & Francis Group, an **informa** business

A BALKEMA BOOK

Published by:
CRC Press/Balkema
P.O. Box 447, 2300 AK Leiden, The Netherlands
e-mail: Pub.NL@taylorandfrancis.com
www.crcpress.com – www.taylorandfrancis.com

First issued in paperback 2020

© 2016 by Taylor & Francis Group, LLC
*CRC Press/Balkema is an imprint of the Taylor & Francis Group, an informa business*

No claim to original U.S. Government works

Typeset by V Publishing Solutions Pvt Ltd., Chennai, India

ISBN 13: 978-0-367-73720-7 (pbk)
ISBN 13: 978-1-138-02793-0 (hbk)

**Visit the Taylor & Francis Web site at**
**http://www.taylorandfrancis.com**

**and the CRC Press Web site at**
**http://www.crcpress.com**

*Advanced Materials, Structures and Mechanical Engineering – Kaloop (Ed.)*
© 2016 Taylor & Francis Group, London, ISBN 978-1-138-02793-0

# Table of contents

# Preface

The 2015 International Conference on Advanced Materials, Structures and Mechanical Engineering (ICAMSME 2015) took place in Incheon, at Incheon National University (INU), South-Korea, on May 29–31, 2015. This conference was sponsored by the Incheon Disaster Prevention Research Center (IDPRC) in INU. The conference program covered invited, oral, and poster presentations from scientists working in similar areas to establish platforms for collaborative research projects in this field. This conference will bring together leaders from industry and academia to exchange and share their experiences, present research results, explore collaborations and to spark new ideas, with the aim of developing new projects and exploiting new technology in this field.

The book is a collection of accepted papers. All these accepted papers were subjected to strict peer-reviewing by 2–3 expert referees, including preliminary review process conducted by conference editors and committee members before their publication by CRC Press. This book is separated into four main chapters including 1. Civil and structural engineering application and managements, 2. Hydraulic and soil engineering design and applications, 3. Industrial, information technology, advanced materials and mechanical applications, 4. Composite and textile materials design and applications. The committee of ICAMSME 2015 express their sincere thanks to all authors for their high-quality research papers and careful presentations. All reviewers are also thanked for their careful comments and advices. Thanks are finally given to CRC Press as well for producing this volume.

The Organizing Committee of ICAMSME 2015
*Committee Chair Prof. M. Kaloop,*
*Incheon National University, South Korea*

*Advanced Materials, Structures and Mechanical Engineering – Kaloop (Ed.)*
© 2016 Taylor & Francis Group, London, ISBN 978-1-138-02793-0

# Soap film Enneper model in structure engineering

H.M. Yee & N.A.H. Mohd
*University Teknologi MARA, Pulau Pinang, Malaysia*

ABSTRACT: Tensioned membrane structure in the form of Enneper minimal surface can be considered as a sustainable development for the green environment and technology, it also can be used to support the effectiveness used of energy and the structure. The objective of this study is to carry out soap film in the form of Enneper minimal surface. The combination of shape and internal forces for the purpose of stiffness and strength is an important feature of the membrane surface. Form-finding using soap film model has been carried out for Enneper minimal surface models with variables $u = v = 0.6$ and $u = v = 1.0$. Enneper soap film models with variables $u = v = 0.6$ and $u = v = 1.0$ provides an alternative choice for structural engineers to consider the tensioned membrane structure in the form of Enneper minimal surface applied in the building industry. It is expected to become an alternative building material to be considered by the designer.

## 1 INTRODUCTION

A Tensioned Membrane Structure (TMS) also known as the tensioned fabric structure is a structure where very thin doubly-curved coated fabric in tension is used as the main structural element. It has been employed throughout recorded history as in tents. Yee (2011) has stated that the materials used for the membranes generally consist of a woven fabric coated with a polymeric resin. Xu et al. (2009) have stated that in recent years, the tensile surface structure business has grown considerably and is predicted to grow further.

Soap film model is used to find the surface form that minimizes area subject to appreciate constraints. Ireland (2008) has presented some puzzling measurements of the tension between soap film and a solid surface. Moulton & Pelesko (2008) have investigated Catenoid soap film model subjected to an axially symmetric electric field. The experimental and theoretical analysis for this Field Driven Mean Curvature Surface (FDMC) surface provides a step in understanding how electric fields interact with surfaces driven by surface tension. Such interactions may be of great benefit in microscale systems such as Micro Electro Mechanical Systems (MEMS) and self-assembly. Brakke (1995) has treated the area minimization problem with boundary constraints. Koiso & Palmer (2005) have studied a variational problem whose solutions are a geometric model for thin films with gravity which is partially supported by a given contour. Boudaoud et al. (1999) have studied Helicoid soap film. The vibration equation shows that the Helicoid is the stable surface when its winding number is small. Caten-

oid is locally isometric to Helicoid so that their vibration spectra are strongly related. The normal forms of the bifurcations confirm the analysis. Huff (2006) has considered soap films spanning rectangular prisms with regular n-gon bases. As the number of edges n varies, there are significant changes in the qualitative properties of the spanning soap films as well as a change in the number of spanning soap films.

It can be seen from the previous paragraph that soap film model has been used in many different studies. Yee et al. (2013) have mentioned that the main reason for its use is due to the fact that the surface formed can be observed physically. Another main reason is that experimental study of surface form using soap film model is relatively easy and simply. In the field of a shell and spatial structures, the form-finding analysis is needed in order to determine surface form that can be obtained under certain given boundary constraint and pre-stressed pattern. One particular case where the form-finding analysis is needed is related to tensioned membrane structures. Tensioned membrane structures are structured where very thin architectural membranes in tensioned are used to span a large column-free space. This category of the using a computational or structure can be used to form a variety of surface due to the flexibility of the material used. In order to determine surface form of tensioned membrane structures, form-finding can be carried out using the computational or experimental method. Yee (2011), Yee et al. (2011), Yee et al. (2013), Yee et al. (2013), Yee et al. (2013), Mohd Noor et al. (2013), Yee et al. (2014), Yee et al. (2015), Yee et al. (2015) and Yee et al. (2014) have carried out form-finding using nonlinear analysis method and soap film models by Yee

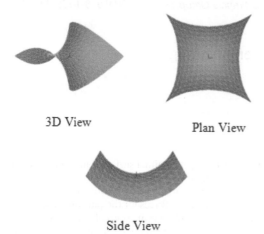

3D View

Plan View

Side View

Figure 1.   Enneper minimal surface model.

(2011). The objective of this study is to carry out soap film in the form of Enneper minimal surface. In this paper, form-finding using soap film model is carried out in order to determine surface form under boundaries corresponding to shape defined by mathematical equations for the Enneper minimal surface with variables $u = v = 0.6$ and $u = v = 1.0$ and expected to become an alternative building material shape to be considered by the designer.

## 2   ENNEPER MINIMAL SURFACE IN TENSIONED MEMBRANE STRUCTURE

Figure 1 shows Enneper minimal surface. The boundary of Enneper minimal surface in Figure 1 can be obtained by using Equation (1), Gray (1998).

$$X = u - \frac{u^3}{3} + uv^2, Y = -v + \frac{v^3}{3} - vu^2, Z = u^2 - v^2 \quad (1)$$

For $u$ and $v$ = variables

## 3   EXPERIMENTAL OF ENNEPER SOAP FILM MODEL

The experimental soap film model is based on Yee (2011) work. The material and equipment have been used in the project is steel, aluminium wires, plywood, super glue, rubber band, wire, glycerin, concentrated car detergent, distilled water, theodolite, plumb bob and camera.

Soap film model has been shown in Figure 2. The boundary frame on the surface has been built based on the coordinate from Enneper minimal

surface. The coordinates have been calculated using mathematical equations. Plywood has been used as x-coordinate and y-coordinate of the boundary, standing steel rods has been used to support the wire frame at the desired height corresponding to z coordinate of the wire frame as shown in Figure 2. Aluminum wires have been used to build the boundary of models as shown in Figure 2. Super glue has been used to fix the steel and plywood. The glycerin has been used to secure the wire to steel rods. The wire has been tied to the steel rod at specified coordinates with rubber bands in order to produce the desired boundary defined by equations Enneper minimal surface. Then, glycerin concentrated car detergent and distilled water used to prepare a soap solution. The preparation of soap solution has been contained the composition of water, detergent and glycerin. The composition used is 25.7% of glycerin, 22.8% of concentrated car detergent and 51.1% of distilled water.

Experimental form-finding using soap film model has been carried out as shown in Figure 3.

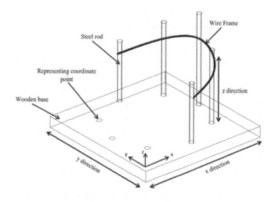

Figure 2.   Soap film model with boundary frame.

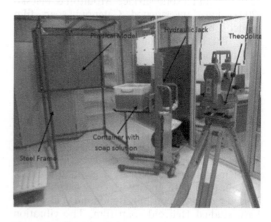

Figure 3.   Experiment setup.

2

In this experimental setup contain theodolite, hydraulic jack, the container with soap solution, steel frame and physical model. Soap film container has been placed on a hydraulic jack and raised up by using the hydraulic jack to dip the model into the soap solution. In order to measure the shape of the soap film produced, a steel frame has been prepared. Theodolite has been used to check the horizontal alignment and plan position alignment. It used the angle to balance the angle of the model. This is to make sure that the soap film model is suspended at the desired orientation and images taken using a camera can be compared with computational results. Soap film model has been used to determine the surface form corresponding to the boundary shape defined by mathematical equations for Enneper minimal surface in detail.

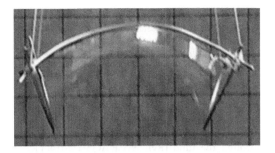

Figure 5. Enneper soap film model. ($u = v = 0.6$).

## 4 RESULTS AND DISCUSSION OF SOAP FILM MODELS

Figure 4 shows the mathematically defined Enneper surface with $u = v = 0.6$. The corresponding result obtained through soap film experiment as shown in Figure 5. Comparison of the two surfaces shows that they are in good agreement. Yee (2011) has studied that Enneper surface of the parameter $u = v \leq 0.87$ has been found to be in good agreement with the mathematically defined Enneper.

Figure 6 shows the mathematically defined Enneper surface with $u = v = 1.0$. The corresponding result obtained through soap film experiment is shown in Figure 7. Comparison of the two surfaces shows that they are in good agreement. Yee (2011) has studied that parameter $u = v \geq 0.88$, the mathematically defined Enneper surface no longer corresponds to the stable minimal surface. Nitsche (1976) have mentioned that the Enneper disk $Dr = (X(u,v); u^2 + v^2 \leq r^2)$ is stable for $r < 1$.

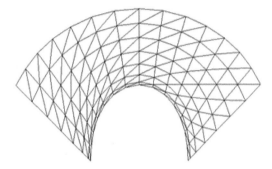

Figure 6. Mathematically defined Enneper surface. ($u = v = 1.0$).

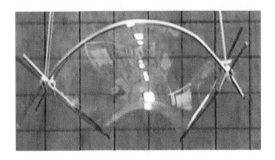

Figure 7. Enneper soap film model. ($u = v = 1.0$).

## 5 CONCLUSIONS

Form-finding with the surface in the form of Enneper tensioned membrane structure with variables $u = v = 0.6$ and $u = v = 1.0$ has been carried out successfully using the soap film model. The results from this study show that tensioned membrane structure in the form of Enneper minimal surface $u = v = 0.6$ and $u = v = 1.0$ is a structurally viable surface form to be considered by an engineer for sustainable development. Therefore, the proposed study TFS in the form of Enneper with variables

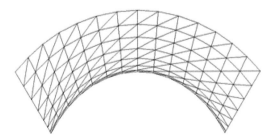

Figure 4. Mathematically defined Enneper surface. ($u = v = 0.6$).

$u = v = 0.6$ and $u = v = 1.0$ can save materials, provision of economic gains and reduce the environment impact by giving natural diffuse light with reduced heat load.

## ACKNOWLEDGEMENT

The researchers wish to thank Ministry of Education, Malaysia for funding the research project through the fund [Ref. No RAGS/1/2014/TK02/UITM//2] and the Research Management Institute (RMI), Universiti Teknologi MARA (UiTM) for the administrative support.

## REFERENCES

Boudaoud, A. Patrício, P. & Amar, M.B.. 1999. The Helicoid versus the Catenoid: Geometrically Induced Bifurcations. *Physical Review Letters,* 83(19): 3836–3839.

Brakke, K.A. 1995. Numerical solution of soap film dual problems. *Experimental Mathematics,* 4(4): 269–287.

Gray, A. 1998. *The Modern differential geometry of curves and surfaces with Mathematica,* CRC Press LLC: United States of America.

Huff, R. 2006. Soap films spanning rectangular prisms, *Geometriae Dedicata,* 123(1): 223–238.

Ireland, P.M. 2008. Some curious observations of soap film contact lines. *Chemical Engineering Science,* 63(8): 2174–2187.

Koiso, M. & Bennett, P. 2005. On a variational problem for soap films with gravity and partially free boundary. *Journal of the Mathematical Society of Japan,* 57(2): 333–355.

Mohd, N. Mohd, S. Yee, H.M. Choong, K.K. & Haslinda, A.H. 2013. Tensioned Membrane Structures in the Form of Egg Shape. *Applied Mechanics and Materials,* 405–408: 989–992.

Moulton, D.E. & Pelesko, J.A. 2008. Theory and experiment for soap-film bridge in an electric field. *Journal of colloid and interface science,* 322(1): 252–262.

Nitsche, J. 1976. Non-Uniqueness for Plateau's Problem. A Bifurcation Process. *Annales Academip Scientiarum Fennicre,* 2: 361–373.

Xu, W. Ye, J.H. & Jian, S. 2009. The application of BEM in the membrane structures interaction with simplified wind. *Structural Engineering and Mechanics,* 31(3): 349–365.

Yee, H.M. 2011. *A Computational Strategy for Form-Finding of Tensioned Fabric Structure using Nonlinear Analysis Method.* Ph.D Dissertation, School of Civil Engineering, Universiti Sains Malaysia, Pulau Pinang, Malaysia.

Yee, H.M. Abdul Hamid, H. & Abdul Hadi, N. 2015. Computer Investigation of Tensioned Fabric Structure in the Form of Enneper Minimal Surface. *Applied Mechanics and Materials,* 754–755: 743–746.

Yee, H.M. & Choong, K.K. 2013. Form-Finding of Tensioned Fabric Structure in the Shape of Möbius Strip. *Iranica Journal of Energy & Environment,* 4(3): 247–253.

Yee, H.M. Choong, K.K. & Abdul Hadi, M.N. 2015. Sustainable Development of Tensioned Fabric Green Structure in the Form of Enneper. *International Journal of Materials, Mechanics and Manufacturing,* 3(2): 125–128.

Yee, H.M. Choong, K.K. & Kim, J.Y. 2011. Form-Finding Analysis of Tensioned Fabric Structures Using Nonlinear Analysis Method. *Advanced Materials Research,* 243–249: 1429–1434.

Yee, H.M. Choong, K.K. & Kim, J.Y. 2013. Experimental Form-Finding for Möbius Strip and Enneper Minimal Surfaces Using Soap Film Models. *International Journal of Engineering Science and Innovative Technology,* 2(5): 328–335.

Yee, H.M. Kim, J.Y. & Noor, M.S. 2013. Tensioned Fabric Structures in Oval Form. *Applied Mechanics and Materials,* 405–408: 1008–1011.

Yee, H.M. & Samsudin. M.A. 2014. Development and Investigation of the Moebius Strip in Tensioned Membrane Structures. *WSEAS Transactions on Environment and Development,* 10: 145–149.

Yee, H.M. & Samsudin, M.A. 2014. Mathematical and Computational Analysis of Moebius Strip. *International Journal of Mathematics and Computers in Simulation,* 8: 197–201.

*Advanced Materials, Structures and Mechanical Engineering – Kaloop (Ed.)*
© 2016 Taylor & Francis Group, London, ISBN 978-1-138-02793-0

# Hencky bar-chain model for buckling and vibration of beams with elastically restrained ends

H. Zhang & C.M. Wang
*Engineering Science Programme and Department of Civil and Environmental Engineering,*
*National University of Singapore, Kent Ridge, Singapore*

N. Challamel
*UBS—LIMATB, Centre de Recherche, Université Européenne de Bretagne,*
*University of South Brittany, Lorient Cedex, France*

ABSTRACT: This paper presents the Hencky bar-chain model for buckling and vibration analyses of beams with elastically restrained ends. The Hencky bar-chain model is composed of rigid beam segments (of length $a = L/n$ where $L$ is the total length of beam and $n$ the number of beam segments) connected by frictionless hinges with elastic rotational springs of stiffness $EI/a$ where $EI$ is the flexural rigidity of the beam. The key contribution of this paper lies in the modelling of the elastic end restraints of the Hencky bar-chain that will simulate the same buckling and vibration results as that furnished by the first order central finite difference beam model. The establishment of such a physical discrete beam model allows one to obtain accurate solutions for beam like structure with repetitive cells (or elements).

## 1 INTRODUCTION

In 1920, Hencky (1920) proposed that in buckling of beams, the continuum structure may be replaced by a structural model comprising rigid segments connected by frictionless hinges with elastic rotational springs that are capable of resisting elastic deformation. In the literature, this structural model has now been referred to as a Hencky "bar chain". When Salvadori (1951) published a paper showing how one can compute the buckling loads of beams and plates using the finite difference method, Silverman (1951) promptly wrote a discussion note on Salvadori's paper. Silverman pointed out the interesting analogy between the first order central finite difference formulation and the physical Hencky's bar chain model if one sets the segmental bar length to be the same and the internal rotational spring constant to be equal to $EI/a$ where $EI$ is the beam flexural rigidity and $a$ is the length of the rigid segment. When it comes to an elastically restrained end, the rotational spring constant and lateral springs constant have to be calibrated to make the central finite difference formulation analogous to the Hencky bar-chain model. This paper investigates the expressions for the end spring restraints of the Hencky bar-chain model with the aid of the central finite difference beam model for beam buckling and vibration.

## 2 PROBLEM DEFINITION

Consider a uniform beam of length $L$, cross-sectional area $A$, flexural rigidity $EI$, density per unit length $\rho$, subject to an axial compressive load $P$. The beam ends are elastically restrained with a rotational spring and a lateral spring as shown in Figure 1. The stiffnesses of the rotational spring and the lateral spring are $K_{RA}$ and $K_{LA}$, respectively, for the left end of the beam while the corresponding stiffnesses for the right end are $K_{RB}$ and $K_{LB}$. The problem at hand is to determine the stiffnesses of the rotational spring $C_{RA}$, $C_{RB}$ and the lateral spring $C_{LA}$, $C_{LB}$ for the Hencky bar-chain model as shown in Figure 2.

As the central finite difference beam model is assumed to be equivalent to the physical Hencky bar-chain model, we shall match the end moments and shear forces of these two models so as to obtain the spring stiffness expressions that we are seeking.

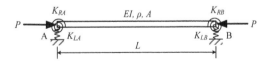

Figure 1. Beam with elastic end restraints under an axial compressive load $P$.

*Hinge with rotational spring stiffness C = EI/a and concentrated mass m = ρAa*

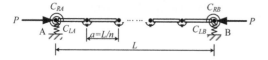

Figure 2. Hencky bar-chain with $n$ segments and equivalent elastic end restraints under an axial compressive load $P$.

## 3 FINITE DIFFERENCE BEAM MODEL BASED ON LOCAL EULER BEAM THEORY

The governing equation for the vibration problem of an Euler beam under an axial compressive load $P$ (see Fig. 1) is given by

$$EI\frac{d^4w}{dx^4} + P\frac{d^2w}{dx^2} - \rho A\omega^2 w = 0 \qquad (1)$$

where, $w$ is the transverse displacement, $x$ the longitudinal coordinate with its origin at end $A$ of the beam and $\omega$ is the angular vibration frequency of the beam.

At the two elastically restrained ends, the boundary conditions are given by (Wang & Wang 2013).

$$EI\left[\frac{d^3w}{dx^3}\right]_{x=0} + P\left[\frac{dw}{dx}\right]_{x=0} + K_{LA}w(0) = 0 \text{ at } x=0$$

$$(2a)$$

$$EI\left[\frac{d^2w}{dx^2}\right]_{x=0} - K_{RA}\left[\frac{dw}{dx}\right]_{x=0} = 0 \text{ at } x=0 \qquad (2b)$$

$$EI\left[\frac{d^3w}{dx^3}\right]_{x=L} + P\left[\frac{dw}{dx}\right]_{x=L} - K_{LB}w(L) = 0 \text{ at } x=L$$

$$(2c)$$

$$EI\left[\frac{d^2w}{dx^2}\right]_{x=L} + K_{RB}\left[\frac{dw}{dx}\right]_{x=L} = 0 \text{ at } x=L \qquad (2d)$$

In the following, we apply the central finite difference method to solve the governing Equation (1) and boundary conditions (2) of the Euler beam under an axial compressive load $P$. As shown in Figure 3, the beam is discretized into $n$ segments of length $a = L/n$. Note that four fictitious nodes (i.e., $j = -2, -1, n+1$ and $n+2$) are created with two points extending from each end to handle the Neumann boundary conditions at these two ends.

Based on the finite difference method, the governing Equation (1) is given by

$$\left(w_{j+2} - 4w_{j+1} + 6w_j - 4w_{j-1} + w_{j-2}\right)$$
$$+ \frac{\alpha}{n^2}\left(w_{j+1} - 2w_j + w_{j-1}\right) - \frac{\Omega^2}{n^4}w_j = 0 \text{ for } j=0...n$$

$$(3)$$

where, $\alpha = \frac{PL^2}{EI}$, $\Omega = \sqrt{\frac{\rho A\omega^2 L^4}{EI}}$, $C = \frac{EI}{a}$.

The boundary conditions given by (2) at the two ends can be approximated by

$$\left(w_2 - 2w_1 + 2w_{-1} - w_{-2}\right)$$
$$+ \frac{\alpha}{n^2}(w_1 - w_{-1}) + \frac{2K_{LA}a^2}{C}w_0 = 0 \text{ at } x=0 \qquad (4a)$$

$$\left(w_1 - 2w_0 + w_{-1}\right) - \frac{K_{RA}}{2C}(w_1 - w_{-1}) = 0 \text{ at } x=0 \qquad (4b)$$

$$\left(w_{n+2} - 2w_{n+1} + 2w_{n-1} - w_{n-2}\right)$$
$$+ \frac{\alpha}{n^2}(w_{n+1} - w_{n-1}) - \frac{2K_{LB}a^2}{C}w_n = 0 \text{ at } x=L \qquad (4c)$$

$$\left(w_{n+1} - 2w_n + w_{n-1}\right) + \frac{K_{RB}}{2C}(w_{n+1} - w_{n-1}) = 0 \text{ at } x=L$$

$$(4d)$$

The discretized governing Equations (3) and boundary conditions (4) may be assembled and written in a matrix form as

$$\begin{bmatrix} 0 & \gamma_{01} & -2 & \gamma_{02} & 0 & \cdots & \cdots & \cdots & \cdots & 0 \\ -1 & -e & \gamma_{03} & e & 1 & \cdots & \cdots & \cdots & \cdots & \vdots \\ 1 & g & h & g & 1 & 0 & \cdots & \cdots & \cdots & \vdots \\ 0 & 1 & g & h & g & 1 & 0 & \cdots & \cdots & \vdots \\ \vdots & \vdots & \vdots & \vdots & \vdots & \vdots & \vdots & \vdots & \vdots & \vdots \\ 0 & \cdots & 0 & 1 & g & h & g & 1 & 0 & 0 \\ \vdots & \vdots & \vdots & \vdots & \vdots & \vdots & \vdots & \vdots & \vdots & \vdots \\ \vdots & \cdots & \cdots & \cdots & 0 & 1 & g & h & g & 1 & 0 \\ \vdots & \cdots & \cdots & \cdots & \cdots & 0 & 1 & g & h & g & 1 \\ \vdots & \cdots & \cdots & \cdots & \cdots & \cdots & -1 & -e & \gamma_{13} & e & 1 \\ 0 & \cdots & \cdots & \cdots & \cdots & \cdots & 0 & \gamma_{11} & -2 & \gamma_{12} & 0 \end{bmatrix}_{(n+5)\times(n+5)}$$

$$\times\begin{Bmatrix} w_{-2} \\ w_{-1} \\ w_0 \\ \vdots \\ w_{j-1} \\ w_j \\ w_{j+1} \\ \vdots \\ w_n \\ w_{n+1} \\ w_{n+2} \end{Bmatrix} = [K]\{w\} = 0 \qquad (5)$$

6

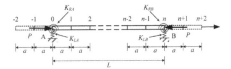

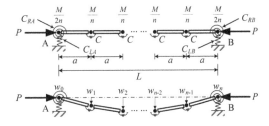

Figure 3. Finite difference discretization of a beam with elastic lateral and rotational end restraints.

where, $g = \frac{\alpha}{n^2} - 4$, $h = 6 - 2\frac{\alpha}{n^2} - \frac{\Omega^2}{n^4}$, $e = g + 2$,

$\gamma_{01} = 1 + \frac{K_{RA}}{2C}$, $\gamma_{02} = 1 - \frac{K_{RA}}{2C}$, $\gamma_{03} = \frac{2K_{LA}a^2}{C}$,

$\gamma_{11} = 1 - \frac{K_{RB}}{2C}$, $\gamma_{12} = 1 + \frac{K_{RB}}{2C}$ and $\gamma_{13} = -\frac{2K_{LB}a^2}{C}$.

By setting the determinant of the above coefficient matrix $K$ to zero and solving the characteristic equation, we obtain multiple solutions of $\omega$, each eigenvalue solution corresponds to a natural frequency of the finite difference beam model with elastic end restraints modelled by rotational springs $K_{RA}$, $K_{RB}$ and lateral springs $K_{LA}$, $K_{LB}$ under an axial compressive load $P$.

## 4 HENCKY BAR-CHAIN MODEL

Next, we consider using the Hencky bar-chain to model the foregoing beam. The model comprises $n$ rigid segments with lumped mass and internal elastic rotational springs having stiffness $C = EI/a$ where $EI$ is the bending rigidity of the beam, as shown in Figure 4. The stiffnesses of the elastic rotational springs are $C_{RA}$, $C_{RB}$ and the corresponding lateral springs are $C_{LA}$, $C_{LB}$.

The elastic potential energy $U$ of the Hencky bar-chain model is given by (Wang et al. 2013).

$$U = \sum_{j=1}^{n-1} \frac{1}{2}C\left(\frac{w_{j+1} - 2w_j + w_{j-1}}{a}\right)^2 + \frac{1}{2}C_{RA}\left(\frac{w_1 - w_0}{a}\right)^2$$
$$+ \frac{1}{2}C_{LA}w_0^2 + \frac{1}{2}C_{RB}\left(\frac{w_n - w_{n-1}}{a}\right)^2 + \frac{1}{2}C_{LB}w_n^2$$

(6)

The potential energy $V$ due to the axial compressive load $P$ can be expressed as

$$V = -\sum_{j=1}^{n} P\frac{1}{2}a\left(\frac{w_j - w_{j-1}}{a}\right)^2$$

(7)

The kinetic energy $T$ of the Hencky bar-chain model is given by

$$T = \sum_{j=1}^{n-1} \frac{1}{2}\frac{M}{n}\left(\frac{\partial w_j}{\partial t}\right)^2 + \frac{1}{2}\frac{M}{2n}\left(\frac{\partial w_0}{\partial t}\right)^2 + \frac{1}{2}\frac{M}{2n}\left(\frac{\partial w_n}{\partial t}\right)^2$$

(8)

Figure 4. Hencky bar-chain model with elastically restrained ends.

where, the total mass $M$ of the Hencky bar-chain model beam is distributed as follows: For internal nodes $j = 1, 2, ..., n-1$, the lumped mass is $M/n$ and for the two end nodes $j = 0$ and $n$, the lumped mass is $M/(2n)$.

To derive the equations of motion, we used Hamilton's principle that requires.

$$\delta\int_{t_1}^{t_2}(U + V - T)dt = 0$$

(9)

where, $t_1$ and $t_2$ are the initial and final times. By substituting Equations (6), (7) and (8) into Equation (9) and assuming a harmonic motion, i.e., $w_j(x,t) = w_j(x)e^{-i\omega t}$ where $i = \sqrt{-1}$ and $\omega$ is the angular frequency of vibration, one obtains the following matrix assembled with the governing equation and boundary conditions for the vibration problem of the Hencky bar-chain model.

$$\begin{bmatrix} h_0 & g_0 & -1 & 0 & \cdots & \cdots & \cdots & \cdots & 0 \\ g_0 & h_1 & -g & -1 & 0 & \cdots & \cdots & \cdots & \vdots \\ 1 & g & h & g & 1 & 0 & \cdots & \cdots & \vdots \\ 0 & 1 & g & h & g & 1 & 0 & \cdots & \vdots \\ \vdots & \vdots & \vdots & \vdots & \vdots & \vdots & \vdots & \vdots & \vdots \\ 0 & \cdots & 0 & 1 & g & h & g & 1 & 0 & \cdots & 0 \\ \vdots & \vdots & \vdots & \vdots & \vdots & \vdots & \vdots & \vdots & \vdots \\ \vdots & \cdots & \cdots & \cdots & 0 & 1 & g & h & g & 1 & 0 \\ \vdots & \cdots & \cdots & \cdots & 0 & 1 & g & h & g & 1 \\ \vdots & \cdots & \cdots & \cdots & \cdots & 0 & -1 & -g & h_{n-1} & g_n \\ 0 & \cdots & \cdots & \cdots & \cdots & 0 & -1 & g_n & h_n \end{bmatrix}_{(n+1)\times(n+1)}$$

$$\times \begin{Bmatrix} w_0 \\ \vdots \\ \vdots \\ \vdots \\ w_{j-1} \\ w_j \\ w_{j+1} \\ \vdots \\ \vdots \\ w_n \end{Bmatrix} = [K]\{w\} = 0$$

(10)

7

where, $h_0 = \frac{\Omega^2}{n^4} + \frac{\alpha}{n^2} - 1 - \frac{C_{RA}}{C} - \frac{C_{LA}a^2}{C}$, $g_0 = \frac{C_{RA}}{C} + 2 - \frac{\alpha}{n^2}$,

$h_1 = \frac{\Omega^2}{n^4} + 2\frac{\alpha}{n^2} - 5 - \frac{C_{RA}}{C}$, $h_{n-1} = \frac{\Omega^2}{n^4} + 2\frac{\alpha}{n^2} - 5 - \frac{C_{RB}}{C}$,

$g_n = \frac{C_{RB}}{C} + 2 - \frac{\alpha}{n^2}$, $h_n = \frac{\Omega^2}{2n^4} + \frac{\alpha}{n^2} - 1 - \frac{C_{RB}}{C} - \frac{C_{LB}a^2}{C}$,

$g$ and $h$ are defined in Equation (5).

Upon setting the determinant of the above coefficient matrix to zero and then solving the characteristic equation, one obtains the natural frequencies of the vibrating Hencky bar-chain model with elastic end restraints having constants $C_{RA}$, $C_{RB}$, $C_{LA}$ and $C_{LB}$ under an axial compressive load $P$.

## 5 EQUIVALENCE BETWEEN FINITE DIFFERENCE BEAM MODEL AND HENCKY BAR-CHAIN MODEL

The stiffness equivalence is derived based on matching the buckling loads and vibration frequencies of finite difference beam model and Hencky bar-chain model. By comparing Equation (5) to Equation (10), we find that the governing equations of these two models are exactly the same while the boundary conditions are not. Therefore, to make the boundary conditions equivalent, we have to apply the equivalence rule of deflection, bending moment and shear force of the end points. The derivation is divided into two parts, namely rotational spring stiffness equivalence and lateral spring stiffness equivalence.

In the first part, the rotational spring stiffness equivalence is derived based on the equivalence of deflection and bending moment as

$$w^H = w^F \quad \text{and} \quad M^H = M^F \tag{11}$$

where, the superscript $H$ denotes the quantities associated with the Hencky bar-chain model and the superscript $F$ denotes the quantities associated with the central finite difference Euler beam model.

First, let us consider the Hencky bar-chain model (see Fig. 4). The bending moment at $x = 0$ is given by

$$M_0^H = -C_{RA} \cdot \frac{w_1 - w_0}{a} \tag{12}$$

Next, we consider the finite difference beam model (see Fig. 3). The bending moment at $x = 0$ is given by

$$M_0^F = -EI\frac{d^2 w}{dx^2}\bigg|_{x=0} = -EI\left(\frac{w_1 - 2w_0 + w_{-1}}{a^2}\right) \tag{13}$$

According to the boundary condition Equation (4b) for elastically restrained end at $x = 0$, we have

$$w_{-1} = \frac{2w_0 - \left(1 - \dfrac{K_{RA}}{2C}\right)w_1}{1 + \dfrac{K_{RA}}{2C}} \tag{14}$$

By substituting Equation (14) into Equation (13), we have

$$M_0^F = -\frac{K_{RA}}{1 + \dfrac{K_{RA}}{2C}} \cdot \frac{w_1 - w_0}{a} \tag{15}$$

By invoking the moment equivalence $M_0^H = M_0^F$, the rotational spring stiffness relationship is derived by comparing Equation (12) to Equation (15),

$$C_{RA} = \frac{K_{RA}}{1 + \dfrac{K_{RA}}{2C}} = \frac{2C}{1 + \dfrac{2C}{K_{RA}}} \tag{16}$$

The stiffness of the right end restraint can be derived in a similar way as

$$C_{RB} = \frac{K_{RB}}{1 + \dfrac{K_{RB}}{2C}} = \frac{2C}{1 + \dfrac{2C}{K_{RB}}} \tag{17}$$

In the second part, the lateral spring stiffness equivalence is derived based on the correspondence principles of deflection and shear force,

$$w^H = w^F \quad \text{and} \quad F^H = F^F \tag{18}$$

For the Hencky bar-chain model (see Fig. 4) at $x = 0$, the shear force is given by

$$F_0^H = C_{LA} \cdot w_0 \tag{19}$$

while, for the finite difference beam model (see Fig. 3) at $x = 0$, the shear force is expressed by

$$F_0^F = K_{LA} \cdot w_0 \tag{20}$$

By invoking the shear force equivalence $F^H = F^F$, we can derive the same stiffness relationship by comparing Equation (19) to Equation (20) as

$$C_{LA} = K_{LA} \tag{21}$$

The lateral spring stiffness at the right end can be derived in a similar way as

$$C_{LB} = K_{LB} \tag{22}$$

## 6 EXAMPLE SOLUTIONS OF HENCKY BAR-CHAIN MODEL

### 6.1 Clamped-free beam

Consider the buckling and vibration problem of a clamped-free beam ($K_{RA} = K_{LA} = \infty$, $K_{RB} = K_{LB} = 0$) under an axial compressive load $P$ as shown in Figure 5.

Table 1 shows the buckling load parameter $\alpha_{cr} = (P_{cr}L^2)/EI$ and fundamental frequency parameter $\Omega_1 = \omega_1\sqrt{\frac{\rho A L^4}{EI}}$ of the clamped-free beam with $n$ segments modelled by the Finite Difference beam Model (FDM) or Hencky Bar-chain Model (HBM). Note that $\alpha = (PL^2)/EI$ is the axial load parameter applied on the beam. It is to be emphasized that the results obtained by FDM and HBM are exactly the same and they converge to the continuum results (the results in the last row of the table referring to Wang & Wang, 2013) when $n$ approaches infinity.

### 6.2 Elastically restrained beam

Consider the buckling and vibration problem of an elastically restrained beam ($K_{LA} = K_{LB} = \infty$, $K_{RA} = K_{RB} = 10EI/L$) under an axial compressive load $P$ as shown in Figure 6.

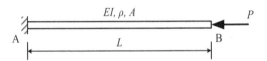

Figure 5. Buckling and vibration of clamped-free beam under an axial compressive load $P$.

Table 1. Buckling loads and fundamental frequencies of clamped-free beam with $K_{RA} = K_{LA} = \infty$ and $K_{RB} = K_{LB} = 0$ for various segmental numbers $n$.

| $n$ | Buckling load $\alpha_{cr}$ ($\Omega_1 = 0$) | Fundamental frequency $\Omega_1$ with $\alpha = \pi^2/8 \approx 1.2337$ | Fundamental frequency $\Omega_1$ with $\alpha = 0$ |
|---|---|---|---|
| 10 | 2.4623 | 2.5116 | 3.4866 |
| 20 | 2.4661 | 2.5288 | 3.5086 |
| 40 | 2.4671 | 2.5331 | 3.5142 |
| $\infty$ | $\pi^2/4 \approx 2.4674$ | 2.5346 | 3.5160 |

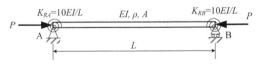

Figure 6. Buckling and vibration of elastically restrained beam under an axial compressive load $P$.

Table 2. Buckling loads and fundamental frequencies of the elastically restrained beam with $K_{LA} = K_{LB} = \infty$ and $K_{RA} = K_{RB} = 10EI/L$ for various segmental numbers $n$.

| $n$ | Buckling load $\alpha_{cr}$ ($\Omega_1 = 0$) | Fundamental frequency $\Omega_1$ with $\alpha = 14.0838$ | Fundamental frequency $\Omega_1$ with $\alpha = 0$ |
|---|---|---|---|
| 10 | 27.3402 | 11.7744 | 16.8483 |
| 20 | 27.9585 | 12.1321 | 17.1622 |
| 40 | 28.1153 | 12.2228 | 17.2426 |
| $\infty$ | 28.1677 | 12.2531 | 17.2695 |

Table 2 presents the buckling loads and fundamental frequencies of the beam as shown in Figure 6.

## 7 CONCLUSIONS

It is shown herein that the finite difference Euler beam model is equivalent to the Hencky bar-chain model provided that the rotational spring stiffnesses $C_{RA}$, $C_{RB}$ of the latter model are related to the rotational spring stiffnesses $K_{RA}$, $K_{RB}$ of the former model by the relations given in Equations (16) and (17); the lateral spring stiffnesses $C_{LA}$, $C_{LB}$ of the latter model are equal to the lateral spring stiffnesses $K_{LA}$, $K_{LB}$ of the former model. The advantage of the discrete beam model (i.e. the central finite difference beam model and the Hencky bar-chain model) is the ease in determining the buckling loads and natural frequencies of beams without the need to solve a differential equation. Instead, one deals with only a set of homogenous equations for solution. Moreover, the discrete beam model gives accurate solutions for beam-like structures with repetitive cells (or elements).

## REFERENCES

Hencky, H. 1920. Über die angenäherte lösung von stabilitätsproblemen im raum mittels der elastischen gelenkkette, *Der Eisenbau* 11: 437–452.

Salvadori M.G. 1951. Numerical computation of buckling loads by finite differences, *Transactions of the ASCE*, 116: 590–636.

Silverman, I.K. 1951. Discussion on the paper of "Salvadori M.G., "Numerical computation of buckling loads by finite differences", *Transactions of the ASCE*, 116: 625–626.

Wang, C.M. Zhang, Z. Challamel, N. & Duan, W.H. 2013. Calibration of Eringen's small length scale coefficient for initially stressed vibrating nonlocal Euler beams based on microstructured beam model, *Journal of Physics D: Applied Physics*, 46(34): 345–501.

Wang, C.Y. & Wang, C.M. 2013. *Structural Vibration: Exact Solutions for Strings, Membranes, Beams and Plates, Boca Raton, Florida*, USA: CRC Press.

# Application of the ball indentation technique to study the tensile properties across the P92 steel weld joint

Dipika R. Barbadikar, A.R. Ballal & D.R. Peshwe
*Department of Metallurgical and Materials Engineering, VNIT Nagpur, Nagpur, India*

T. Sakthivel & M.D. Mathew
*Mechanical Metallurgy Division, IGCAR, Kalpakkam, Chennai, India*

ABSTRACT: Ball indentation technique is used to study the tensile properties such as yield strength, ultimate tensile strength, strain hardening exponent, and strength coefficient. This technique is very useful to study the gradient in mechanical properties across the weld joint. In the present study, the microstructural and mechanical properties of the heat-affected zone in the P92 steel weld joint fabricated by the Shielded Metal Arc Welding process (SMAW) are determined at room temperature and at 350°C. The P92 weld joint consists of the weld metal, coarse grain region, fine grain region, critical region and base metal. The strength gradually decreases from the weld metal to enter the critical region and then increases towards the base metal. The trough enters the critical region because of the coarsening of the $M_{23}C_6$ precipitate and sub-grain formation with a decrease in dislocation density.

## 1 INTRODUCTION

### 1.1 *Material*

P92 steel is a modified version of P91 steel, which is obtained after the addition of 1.5–2.0 wt.% tungsten and decreasing the content of molybdenum from 1 wt.% to 0.5 wt.%, with the addition of 0.001–0.002 wt.% boron. The addition of tungsten offers solid solution strengthening and a reduction in the coarsening rate of $M_{23}C_6$ precipitates, at high surface temperatures. P92 steel is widely used in the steam generator application in fossil-fired and nuclear power plants due to its excellent mechanical and physical properties. High temperature strength and thermal conductivity, low thermal expansion coefficient, resistance to stress corrosion cracking and oxidation favors it to be used as a material in power plants to increase the thermal efficiency (Ennis & Czyrska-Filemonowicz 2003, Carl et al. 2000, Wang et al. 2013, Kouichi et al. 2001, Shen et al. 2009, Klueh 2005, Giroux et al. 2010, Ennis et al. 1997). The steel offers good weldability and hence can be used in different weld joints. Thus, the P92 steel is a good choice over the austenitic stainless steel. P92 steel derives its strength from the tempered martensitic lath microstructure, precipitation and solid solution strengthening offered by $M_{23}C_6$ and MX precipitates and a high dislocation density obtained after normalizing the steel (Kouichi et al. 2001).

### 1.2 *Ball Indentation (BI) technique*

A lot of research is being focused on the miniaturized sample techniques such as Ball Indentation (BI), shear punch, small punch creep and impression creep. The BI technique has been used for the evaluation of tensile properties such as Yield Strength (YS), Ultimate Tensile Strength (UTS), strength coefficient (K), strain hardening exponent (n),% elongation and hardness of the material at room temperature as well as at high temperatures. The technique requires a minimal material for testing, and hence can be used to study the remaining life assessment of in-service components of nuclear and thermal power plants. The system has to be standardized for a particular material to evaluate the mechanical properties. Efforts have been taken to convert the load-depth data obtained from the BI test into true stress-true plastic strain data using empirical relations (Mathew et al. 1999, Murty et al. 1998, Mathew & Murty 1999, Murty, & Mathew 2004). The mechanical properties across the weld joint, as well as various regions in heat-affected zones, have been studied using this technique (Miraglia 1997, Murty et al. 1999).

In the present investigation, tensile properties of the heat-treated P92 steel weld joint at room temperature were obtained using the BI technique. The results obtained were compared with that of the hardness data obtained by carrying out the Vickers hardness test. The conventional tensile test

Table 1.   Welding process parameters.

| Welding of Grade 92 steel plates using P92 steel SMAW electrode (S80) | |
| --- | --- |
| Base material | P92 Ferritic steel |
| Dimensions of plate | 250 (L) × 125 (W) × 12 (T) approx. |
| Welding procedure | SMAW |
| Position | The welds are in the IG (Flat) position |
| Joint geometry | Single V-groove<br>Root gap: 1.5 mm<br>Angle: 35° (half-angle on each plate) |
| Pre-heating and inter-pass temperature | Maintained at 200–250°C during welding. Pre-heating: 200°C min. Inter-pass temperature: 300°C max. |
| Electrode details | 3.2 mm φ P92 steel SMAW electrode |
| Electrical supply | Type: Direct current<br>Current: 115 amps<br>Voltage: 20–22 volts<br>Welding speed: ~ 100 mm/min |
| Polarity | Straight polarity (electrode negative) Root-pass by TIG welding using Grade-92 steel filler wire |
| PWHT and radiography examination | PWHT will be carried out at 780°C for 2 hours followed by air cooling (weld pad should be loaded into the furnace below 150°C). X-ray examination of the weld pad shall be carried out after the weld |

was carried out on the base metal, and the result was compared with that obtained from the BI technique.

## 2   EXPERIMENTAL PROCEDURE

The P92 steel plate was procured from Mishra Dhatu Nigam, Hyderabad. The plate was initially normalized at 1050°C for 30 minutes and then tempered at 780°C for 2 hours. The P92 steel (9 Cr 0.5 Mo 1.8 W) weld joint was fabricated by the Shielded Metal Arc Welding process (SMAW). The welding process parameters are listed in Table 1. The weld pad was subsequently post weld heat treated at 780°C for 2 hours followed by air cooling. The weld joint was X-ray radiographed for its soundness. The weld joint was cut into a block and well polished for the microstructural study and ball indentation test. Care was taken to ensure that both the surfaces of the block were perfectly flat and parallel. The optical microscopy of the different zones present in the weld, especially in the heat-affected zones, was carried out. Mechanical properties across the weld joint were investigated using the ball indentation technique. The multiple indentations with partial unloading were carried out at the single test location using a tungsten carbide ball indenter of diameter 0.254 mm. The indenter was penetrated up to 30% of the indenter

radius with a velocity of 0.00635 mm/s. This depth was kept constant for all the tests and 6 cycles (loading-unloading) were been fixed to reach the defined depth. The BI tests were carried out across the weld joint with the interval of 0.762 mm. This process was continuously monitored and the load depth data for all the cycles were recorded using the data logger. Young's modulus of WC and P92 steel used was 627.42 GPa and 213.7 GPa, respectively. The load depth data were then converted to true stress-true plastic strain data by using the existing empirical relations. The constraint factor and yield slope values used for the conversion were 1.32 and 0.24, respectively. The variation of strengths (YS, UTS) and hardness across the joint were studied by using the BI technique. The tensile test of the base metal was then carried out.

## 3   RESULTS AND DISCUSSION

### 3.1   Microstructure

#### 3.1.1   Actual weld joint
Optical micrography was performed on a cut from the actual weld structure. The specimen was ground, polished to mirror finish and etched by Villella's reagent. The optical microstructure was produced using the Olympus model optical microscope at 50× magnification. Approximately 70 pictures

were taken and coupled together to produce the image, as shown in Figure 1, which had a sufficient contrast between major weld regions. Only a portion of the weld and base metal of the entire weld joint was considered. The different passes are clearly visible inside the weld region. The different zones including the weld metal, fusion boundary, heat affected zone, and base metal are marked in the micrograph.

The weld thermal cycle generates an inhomogeneous microstructure across the Heat-Affected Zone (HAZ) in the weld joint. The weld joint consists of different microstructural zones such as weld metal, HAZ, and base metal due to the thermal cycle during the welding process. In the weld region, there is a peak temperature above 1500°C, which is the melting temperature of the steel. A small amount of delta ferrite was observed in the fusion boundary.

In the HAZ, the temperature of the base metal near the weld reaches above the $Ac_3$ temperature, whereas the zone beyond the HAZ, the base metal is unaffected since the temperature in this region is below the $Ac_1$ temperature. Furthermore, the HAZ can be distinguished into a coarse prior Austenite Grain (CGHAZ) and Fine Prior Austenite Grain (FGHAZ) and enters the critical region depending on the temperature reached during welding.

The microstructures of the different zones across the weld joint are shown in Figure 2.

The microstructure consists of the tempered martensitic lath structure across the weld joint. The grain boundaries are very well decorated with the $M_{23}C_6$ and MX precipitates. These precipitates are responsible for the precipitation and solid solution strengthening in the P92 Steel. The average value of 25 measurements was taken to calculate the prior austenite grain size, which was found to be 14 μm, 32 μm, 24 μm and 11 μm for the base metal, weld metal, CGHAZ, and FGHAZ, respectively. The grains during the entering of the critical region were not well defined. The weld metal was found to have the largest grain size since during welding, the weld metal was heated up for a longer period of time, after recrystallization, the grain growth occurs. The CGHAZ is the region next to

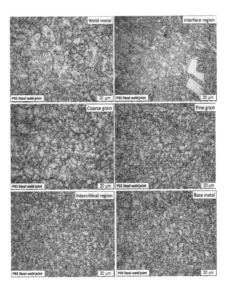

Figure 2. Optical microstructures of the different zones across the weld joint.

the weld metal that exhibits the long-term exposure at a temperature of 1100 to 1500°C, and hence a partial grain growth takes place in this region as well. The FGHAZ is the region having a heat input below 1100°C and above $Ac_3$. This region is far from the actual weld, where the heat input tends to recrystallization and hence new grains of small dimensions are formed. When entering the critical region, the heat input is between $Ac_3$ and $Ac_1$, and could not initiate the recrystallization, but it increases the size of the precipitates already present, leading to partial transformation.

### 3.2 Study of mechanical properties

#### 3.2.1 The hardness measurement
Hardness measurement was made under the conditions of 300 g load and 15 seconds dwell time. The variation of the hardness with distance across the weld joint is shown in Figure 3. The decrease in the hardness from the fusion boundary (302 $H_{v0.3}$) to the base metal (252 $H_{v0.3}$) with a trough in the ICHAZ region (238 $H_{v0.3}$) of the weld joint was observed.

#### 3.2.2 Effect of the test temperature on YS/UTS calculation
A gradient in the YS and UTS across the different zones of the weld joint was studied at 25°C and 350°C by using the BI technique. The variation in strength occurs, as shown in Figure 4.

It was found that the strength decreased by nearly 100 MPa with the increase in the test

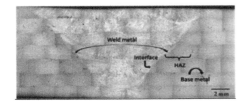

Figure 1. The macrostructure of the P92 steel weld joint.

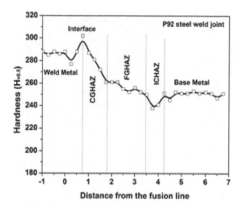

Figure 3. Variation of the hardness of the actual weld and the simulated HAZ.

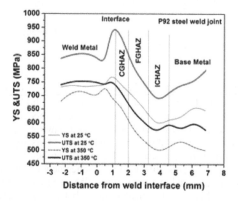

Figure 4. Variation in strength by using the BI tests carried out at 25°C and 350°C.

temperature. The strength was maximum at the interface and continuously decreased up to the entering of the critical region. The same trend in strength was also observed at high temperatures. The YS and UTS showed the same trend as that of the hardness across the weld joint. Among all the heat-affected zones, the CGHAZ showed the highest strength. This is due to a very high temperature exposure in this region, contributing to the dissolution of the precipitates in the matrix. Thus, the CGHAZ region exhibited the maximum strength due to solid solution strengthening. On the other hand, in the FGHAZ, the precipitate partly dissolved and hence showed a comparatively lower strength than the CGHAZ. When entering the critical region, the heat input was not sufficient to dissolve the precipitates, but it was used for precipitate growth and tempering, thereby reducing the dislocation density, and hence showed the minimum strength.

## 4 CONCLUSIONS

1. The YS and UTS values of the different regions of the weld joint were efficiently measured using the ball indentation technique.
2. The prior austenite grain size was found to be 14 μm, 32 μm, 24 μm and 11 μm for the base metal, weld metal, CGHAZ, and FGHAZ, respectively.
3. The strength and hardness were found to be highest in the weld region, while it was lowest in the inter-critical region. This is because the grain structure in this region was not well defined, and it was supposed to have a minimum dislocation density. The coarsening of the precipitate is the main reason for the lowering of strength, which attributed to the type IV cracking in the P92 steel weld joints.

## ACKNOWLEDGMENTS

The authors are grateful to UGC-DAE-CSR for financial support. The first author thanks Dr. G. Amarendra, Scientist-Incharge UGC-DAE center, Kalpakkam node for extending the facilities and Mr. J. Ganeshkumar and Mr. N.S. Thampi for technical discussion and support.

## REFERENCES

Ennis, P.J. & Czyrska-Filemonowicz, A. 2003. Recent advances in creep resistant steels for power plant applications. Sadhana—*Academy Proceedings in Engineering Sciences*, 28: 709–730.

Ennis, P.J. Zielinska-Lipiec, A. Wachter, O. & Czyrska-Filemonowicz, A. 1997. Microstructural stability and creep rupture strength of the martensitic steel P92 for the advanced power plant, *Actamaterialia*, 45(12): 4901–4907.

Giroux, P.F. Dalle, F. Sauzay, M. Malaplate, J. Fournier, B. & Gourgues-Lorenzon, A.F. 2010. Mechanical and microstructural stability of P92 steel under uniaxial tension at high temperature. *Materials Science and Engineering A*, 527 (16–17): 3984–3993.

Klueh, R.L. 2005. Elevated temperature ferritic and martensitic steels and their application to future nuclear reactors, *International Materials Reviews*, 50(5): 287–310.

Kouichi, M. Kota, S. & Junichi K. 2001. Strengthening Mechanisms of Creep Resistant Tempered Martensitic Steel, *ISIJ International*, 41(6): 641–653.

Lundin, C.D. Liv, P. Cui, Y. 2000. A Literature review on Characteristics of High Temperature Ferritic Cr- Mo Steels and weldments, *WRC Bulletin*, 454: 1–36.

Mathew, M.D. & Murty, K.L. 1999. Non-destructive studies on tensile and fracture properties of molybdenum at low temperatures (148 to 423 K), *Journal of Materials Science*, 34: 1497–1503.

Mathew, M.D. Murty, K.L. Rao, K.B.S. & Mannan, S.L. 1999. Ball indentation studies on the effect of aging on mechanical behavior of alloy 625, *Materials Science and Engineering A*, 264: 159–166.

Miraglia, P.Q. 1997. *Characterization of mechanical and fracture properties of a reactor pressure vessel steel weldment using automated ball indentation,* thesis masters of science.`

Murty, K.L. & Mathew, M.D. 2004. Nondestructive monitoring of structural materials using Automated Ball Indentation (ABI) technique, *Nuclear Engineering and design,* 228: 81–96.

Murty, K.L. Mathew, M.D. Wang, Y. Shah, V.N. & Haggag, F.M. 1998. Nondestructive determination of tensile properties and fracture toughness of cold worked A36 steel, *International Journal Pressure Vessels and Piping*, 75: 831–840.

Murty, K.L. Miraglia, P.Q. Mathew, M.D. Shah, V.N. & Haggag, F.M. 1999. Characterization of gradients in mechanical properties of SA-533B steel welds using ball indentation, *International Journal for Pressure Vessels and Piping*, 76(6): 361–369.

Shen, Y.Z. Kim, S.H. Han, C.H. Cho, H.D. & Ryu, W.S. 2009. TEM investigations of MN nitride phases in a 9% chromium ferritic/martensitic steel with normalization conditions for nuclear reactors, *Journal of Nuclear Materials*, 384: 48–55.

Wang, S.S. Peng, D.L. Chang, L. & Hui, X.D. 2013. Enhanced mechanical properties induced by refined heat treatment for 9Cr–0.5Mo–1.8 W martensitic heat resistant steel, *Materials and Design*, 50: 174–180.

*Advanced Materials, Structures and Mechanical Engineering – Kaloop (Ed.)*
© 2016 Taylor & Francis Group, London, ISBN 978-1-138-02793-0

# The continuum model considering the drivers' anticipation effect

C.Y. Yu & R.J. Cheng
*Department of Fundamental Course, Ningbo Institute of Technology, Zhejiang University, Ningbo, China*

H.Q. Liu
*Faculty of Maritime and Transportation, Ningbo University, Ningbo, China*
*National Traffic Management Engineering and Technology Research Centre, Ningbo University Sub-Centre, Ningbo, China*
*Jiangsu Key Laboratory of Urban ITS, Southeast University, Nanjing, China*

ABSTRACT: The new continuum model mentioned in this paper is developed based on optimal velocity car-following model, which takes the drivers' anticipation effect into account. After the critical condition of the steady traffic flow is derived, the nonlinear analysis shows density waves occur in traffic flow because of the density fluctuation. Near the neutral stability line, the KdV-Burgers equation is derived and one of the solutions is given. Last some simulation is carried out to show the evolution of the traffic flow.

## 1 INTRODUCTION

In 1955, Lighthill and Whitham made the first contribution to continuum models and later Richards drew the similar conclusion independently (for short, the LWR model). In this model, the relationship of the three basic parameters of the fluid is built by,

$$\frac{\partial \rho}{\partial t} + \frac{\partial q}{\partial x} = 0$$

where, $\rho$, $q$, $t$, $x$, represent the density, flow, time, and space, respectively.

In 1971, adding a dynamic equation to the continuity one, Payne established a high-order continuum traffic flow model. The dynamic equation is:

$$\frac{\partial v}{\partial t} + v\frac{\partial v}{\partial x} = -\frac{\alpha}{\rho T}\frac{\partial \rho}{\partial x} + \frac{v_e - v}{T}$$

where, $T$ is relaxation time, and $\alpha = -0.5\partial v_e/\partial \rho$ is the anticipation coefficient.

In 1995, Bando et al. (1995) proposed the Optimal Velocity Model (for short, OVM). The model is:

$$\dot{v}_n(t) = a[V(\Delta x_n(t)) - v_n(t)]$$

in which the item $V(\Delta x_n(t))$ means an optimal velocity depended on space headway $\Delta x_n(t)$.

The rest of this paper is organized as follows. In Section 2, based on OVM a new model is derived and then the stability analysis is used. Through nonlinear analysis, the KdV-Burgers equation is derived in Section 3 and the simulation is carried

out in Section 4. Finally, the conclusion is given in Section 5.

## 2 MODEL AND STABILITY ANALYSIS

As we know, the time lag produced by the driver's reflection and the transmission of automotive crankshaft system exists in the real traffic. But this influence is ignored in most previous investigations. With the Bando model we consider the drivers' expected effect and get:

$$\frac{dv_n(t)}{dt} = a[V(\Delta x_n(t) + T\Delta v_n) - v_n(t)] \quad (1)$$

in which $T$ means the expected time and the term $T\Delta v_n$ represent distance influenced by the expected time.

Through Taylor expansion and the following relation to rewrite the above micro variables into macro ones:

$$v_n(t) \rightarrow v(x,t), v_{n+1}(t) \rightarrow v(x+\Delta,t),$$
$$V(\Delta x_n(t)) \rightarrow V_e(\rho), V'(\Delta x_n(t)) \rightarrow \overline{V}'(h) \quad (2)$$

where, $\Delta$ represents the distance between two adjacent vehicles. Through the density $\rho$ and the mean headway $h = 1/\rho$, we define the equilibrium speed $V_e(\rho)$ and the equality $\overline{V}'(h) = -\rho^2 V_e'(\sigma)$ can be deduced. Considering the continuous conservation equation, we get:

$$\frac{\partial \rho}{\partial t} + \frac{\partial(\rho v)}{\partial x} = 0 \quad (3)$$

$$\frac{\partial v}{\partial t}+[v+aT\rho^2V_e'(\rho)\Delta]\frac{\partial v}{\partial x}$$
$$=a[V_e(\rho)-v]-aT\rho^2V_e'(\rho)\frac{\Delta^2}{2}v_{xx} \qquad (4)$$

We rewrite the system (3)–(4) as follow:

$$\frac{\partial \overline{U}}{\partial t}+\overline{A}\frac{\partial \overline{U}}{\partial x}=\overline{E} \qquad (5)$$

and the eigenvalues of $\overline{A}$ are

$$\lambda_1=v,\ \lambda_2=v+aT\rho^2V_e'(\rho)\Delta \qquad (6)$$

We can easily find that the macroscopic traffic flow speed $v$ is bigger than the characteristic speed because the equilibrium speed $V_e(\rho)$ declines along with the density, i.e., $V_e'(\sigma)<0$, which means our model possesses the anisotropic property of traffic flow.

Considering that the steady state is a uniform flow, we initially apply an infinitesimally perturbation to the homogenous flow.

$$\begin{pmatrix}\rho(x,t)\\v(x,t)\end{pmatrix}=\begin{pmatrix}\rho_0\\v_0\end{pmatrix}+\sum\begin{pmatrix}\hat{\rho}_k\\\hat{v}_k\end{pmatrix}\exp(ikx+\sigma_k t) \qquad (7)$$

Combining Equations (3)–(4) with Equation (7) and then neglecting the nonlinear higher-order terms, we have,

$$\begin{cases}(\sigma_k+ikv_0)\hat{\rho}_k+ik\rho_0\hat{v}_k=0\\\sigma_k\hat{v}_k+\left[v_0+aT\rho_0^2V_e'(\rho_0)\Delta\right]\hat{v}_k ik\\\quad=a\left[\hat{\rho}_kV_e'(\rho_0)-\hat{v}_k\right]+aT\rho_0^2V_e'(\rho_0)\frac{\Delta^2}{2}k^2\hat{v}_k\end{cases} \qquad (8)$$

Taking $\hat{\rho}_k$ and $\hat{V}_k$ as the unknown quantities of the equations, we can obtain,

$$(\sigma_k+ikv_0)^2+\left(a+ikaT\rho_0^2V_e'(\rho_0)\Delta\right.$$
$$\left.-\frac{aT\rho_0^2V_e'(\rho_0)\Delta^2}{2}k^2\right)(\sigma_k+ikv_0)$$
$$+ika\rho_0V_e'(\rho_0)=0 \qquad (9)$$

So the neutral stability condition for this steady state is given by;

$$a_s=\frac{1}{T\rho_0\Delta-\dfrac{T^2\rho_0^3V_e'(\rho_0)k^2\Delta^3}{2}} \qquad (10)$$

$$\mathrm{Im}(\sigma_k)=-kv_0-\frac{k\rho_0V_e'(\rho_0)}{1-\dfrac{\Delta^2}{2}T\rho_0^3V_e'(\rho_0)k^2} \qquad (11)$$

Considering the Taylor expansion, we can get

$$\mathrm{Im}(\sigma_k)\approx-k(v_0+\rho_0V_e'(\rho_0))+O(k^3) \qquad (12)$$

Based on Equation (12), we deduce the critical speed of disturbance propagation

$$c(\rho_0)=v_0+\rho_0V_e'(\rho_0) \qquad (13)$$

which, is similar to those mentioned in Ref. (Helbing & Tilch 1998).

## 3 NONLINEAR ANALYSIS

We are interested in the system behavior near the neutral stability condition determined by Equation (13). Then a new coordinate is introduced to transform the reference system in Reference (Berg & Woods 2001).

$$z=x-ct \qquad (14)$$

Then we get

$$-c\rho_z+q_z=0 \qquad (15)$$

$$-cv_z+\left[v+aT\rho^2V_e'(\rho)\Delta\right]v_z$$
$$=a\left[V_e(\rho)-v\right]-aT\rho^2V_e'(\rho)\frac{\Delta^2}{2}v_{xx} \qquad (16)$$

where, $q$ is defined as the product of density and velocity. From Equation (15), we get,

$$v_z=\frac{c\rho_z}{\rho}-\frac{q\rho_z}{\rho^2} \qquad (17)$$

The flow $q$ can be expanded as,

$$q=\rho V_e(\rho)+b_1\rho_z+b_2\rho_{zz} \qquad (18)$$

Two parameters $b_1$ and $b_2$ can get easily solved because the flow $q$ is homogeneous, stable and stability. Substituting Equations (17)–(18) into Equation (16), we have,

$$-c\left(\frac{c\rho_z}{\rho}-\frac{q\rho_z}{\rho^2}\right)+\left[\frac{q}{\rho}+aT\rho^2V_e'(\rho)\Delta\right]\left(\frac{c\rho_z}{\rho}-\frac{q\rho_z}{\rho^2}\right)$$
$$=a\left[V_e(\rho)-\frac{q}{\rho}\right]-aT\rho^2V_e'(\rho)\frac{\Delta^2}{2}$$
$$\left[\frac{c\rho_{zz}}{\rho}-\frac{2c\rho_z^2}{\rho^2}-\frac{q\rho_{zz}}{\rho^2}+\frac{2q\rho_z^2}{\rho^3}\right]$$
$$ \qquad (19)$$

18

So we obtain:

$$
\begin{cases}
b_1 = \dfrac{1}{a}\left[aTV_e'(\rho)\rho^2\Delta + V_e(\rho) - c\right]\left[V_e(\rho) - c\right] \\
b_2 = \dfrac{1}{a}aTV_e'(\rho)\rho^2\dfrac{\Delta^2}{2}\left[V_e(\rho) - c\right]
\end{cases}
\tag{20}
$$

Under the neutral stability condition and using the Taylor expansions to rewrite Equation (18), we have

$$
\rho V_e(\rho) \approx \rho_h V_e(\rho_h) + (\rho V_e)_\rho\big|_{\rho=\rho_h}\hat{\rho} + \frac{1}{2}(\rho V_e)_{\rho\rho}\big|_{\rho=\rho_h}\hat{\rho}^2
\tag{21}
$$

Combining Equation (16) with Equation (19) and turning the $\hat{\rho}$ to $\rho$, we can get

$$
-c\rho_z + \left[(\rho V_e)_\rho + (\rho V_e)_{\rho\rho}\rho\right]\rho_z + b_1\rho_{zz} + b_2\rho_{zzz} = 0
\tag{22}
$$

Performing the following transformations:

$$
U = -\left[(\rho V_e)_\rho + (\rho V_e)_{\rho\rho}\rho\right],\ X = ms,\ T = -mt
$$

Then considering the Equation (22), we can derive the KdV-Burgers equation

$$
U_T + UU_X - mb_1U_{XX} - m^2b_2U_{XXX} = 0
\tag{23}
$$

And one of the solutions is

$$
U = -\frac{3(-mb_1)^2}{25(-m^2b_2)}
$$

$$
\left[
\begin{aligned}
&1 + 2\tanh\left(\pm\frac{-mb_1}{10m^2}\right)\left(X + \frac{6(-mb_1)^2}{25(-m^2b_2)}T + \varsigma_0\right) \\
&+ \tanh^2\left(\pm\frac{mb_1}{10m^2}\right)\left(X + \frac{6(-mb_1)^2}{25(-m^2b_2)}T + \varsigma_0\right)
\end{aligned}
\right]
\tag{24}
$$

in which $\varsigma_0$ is an arbitrary constant.

## 4 SIMULATION

For the simulation, some assumptions are needed as follows:

1. The initial condition of flow is the local equilibrium everywhere;

2. The parameter values;

$$
v_f = 30\ \text{m/s},\ \rho_m = 0.2\ \text{veh/m},\ \Delta x = 100\ \text{m},
$$
$$
a = 0.1,\ \Delta t = 1\text{s},\ L_m = 32.2\ \text{km}
$$

And the anticipation time $T$ is set as 1 s in the whole traffic flow.

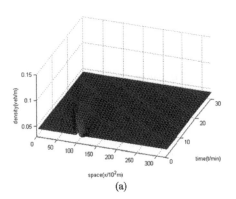

(a)

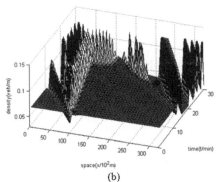

(b)

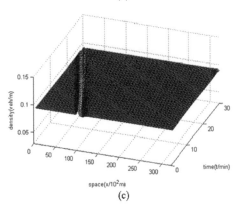

(c)

Figure 1. With a uniform initial traffic and a localized perturbation of amplitude $\Delta\rho_0 = 0.01$ veh/m the time of traffic flow evolves on range of 32.2 km circumference, for: (a) $\rho_0 = 0.04$ veh/m; (b) $\rho_0 = 0.065$ veh/m; (c) $\rho_0 = 0.090$ veh/m.

19

At the density 0.042(a), the perturbation of the traffic flow dissipates with the evolution of traffic flow. When the density goes up to 0.065(b), a quite complex wave pattern including two or more clusters forms appears, which shows the evolutionary process of the stop-and-go traffic phenomenon. Last at the 0.090(c), the traffic flow becomes a stable state again which shows that in the heavy traffic flow density, the apparent density fluctuations is not easy to appear.

## 5 SUMMARY

In this paper, a new continuum model has been proposed based on optimal velocity model, in which the drivers' anticipation effect is considered. The neutral stability line has been obtained and the KdV–Burgers equation and its solution have been derived to describe the evolution of density wave happened in the traffic congestion. Local cluster phenomena can be found by use of our model with certain conditions.

## ACKNOWLEDGEMENT

Project supported by the Science Foundation of Ningbo (Grant Nos. 2012A610162, 2014A610022 and 2014A6111015), Humanities and Social Science Project of Ministry of Education of China (Grant No. 14 jyc630171), Soft Science Project of Zhejiang Province (Grant No. 2014C35076) the Open Research Fund of Jiangsu Key laboratory of Urban ITS, Southeast University and the K.C. Wong Magna Fund in Ningbo University, China.

## REFERENCES

Bando, M. Hasebe, K. Shibata, A. & Sugiyama, Y. 1995. The dynamical model of traffic congestion and numerical simulation, *Physical Review E*, 51(2): 1035–1042.

Berg, P. & Woods, A. 2001. Traveling waves in an optimal velocity model of freeway traffic, *Physical Review E*, 64(3): 035602.

Ge, H.X. Lai, L.L. & Zheng, P.J. 2013. The KdV-Burgers equation in a new continuum model with consideration of driver's forecast effect and numerical tests. *Physics Letters A*, 377(44): 3193–3198.

Helbing, D. & Tilch, B. 1998. Generalized force model of traffic dynamics, *Physical Review E*, 58: 133.

Lai, L.L. Cheng, R.J. & Li, Z.P. 2013. The KdV-Burgers equation in a modified speed gradient continuum model. *Chinese Physics B*, 22(6): 060511.

*Advanced Materials, Structures and Mechanical Engineering – Kaloop (Ed.)*
© *2016 Taylor & Francis Group, London, ISBN 978-1-138-02793-0*

# Fire resistance of fiber reinforced Concrete Filled Box Columns

C.W. Tang
*Cheng Shiu University, Kaohsiung, Taiwan*

H.W. Liao
*Ling Tung University, Taichung, Taiwan*

H.J. Chen & K.C. Jane
*National Chung-Hsing University, Taichung, Taiwan*

ABSTRACT: Two groups of full-size experiments were carried out to consider the effect of type of concrete infilling on the fire resistance of Concrete Filled Box Columns (CFBC). The control group was a steel box filled with plain concrete while the experimental group consisted of a steel box filled with fiber concrete and two steel boxes filled with fiber reinforced concrete. Due to the experimental group filled with fiber concrete, or configured with longitudinal reinforcements and transverse stirrups, their fire resistances are quite better than that of the control group. Test results show that the configuration of longitudinal reinforcements and transverse stirrups can significantly improve the fire resistance of CFBC.

## 1 INTRODUCTION

Concrete is a relatively heat-resistant materials of construction. However, explosive spalling had been considered a common phenomenon occurring in-consistently when high strength concrete was subjected to high temperature. In view of this, the mechanical properties of fiber reinforced concrete after the effects of high temperatures have received considerable attention in the last two decades (Lie 1994, Chen & Liu 2004, Poon et al. 2004, Lau & Anson 2006, Sideris et al. 2009, Ozawa & Morimoto 2014, Yan et al. 2015, Ezziane et al. 2015).

Concrete Filled Box Columns (CFBC) typically consist of rectangular or square hollow structural sections filled with concrete. The fire behavior of circular and square hollow sections filled with concrete has been extensively investigated (Lie & Kodur 1996, Kodur & Sultan 1998, Kodur et al. 2004, Eurocode 4, 2005, Hong & Varma 2009, Schaumanna et al. 2009, Song et al. 2010, Tokgoz & Dundar 2010, Tang & Jiang 2014). More recently, Espinos et al. (2015) studied the fire behavior of rectangular and elliptical slender concrete-filled tubular columns. The influence of the cross-section shape, load eccentricity and percentage of reinforcement on the fire behavior of these columns was investigated. This investigation showed that the eccentricity had a detrimental effect on the fire resistance time while the presence of reinforcing bars contributed to increase the load-bearing capacity of the columns in the fire situation.

Compared with bare steel or reinforced concrete columns, CFBC has several structural and constructional advantages. As a result, CFBC have been widely used as primary axial load carrying members in high rise buildings.

Against the above background, this study aims to explore the effect of type of concrete infilling on the fire resistance of CFBC. The experiments were also intended to provide useful experimental data for the development of CFBC in Taiwan.

## 2 EXPERIMENTAL PROCEDURE

### 2.1 Column specimens

Details of test specimens are given in Table 1. Two groups of full-size experiments were carried out to considering the effect of type of concrete infilling on the fire resistance of CFBC. The control group (i.e., Group A) was a steel box filled with plain concrete while the experimental group (i.e., Group B) consisted of a steel box filled with polypropylene fiber concrete and two steel boxes filled with polypropylene fiber reinforced concrete.

The columns have square cross-sections. The width of the square columns was 400 mm and the wall thickness was 12 mm, as shown in Figure 1. The length of the columns was 3060 mm, as shown in Figure 2. No external fire-proofing was provided for the steel. As shown in Figure 3, each of the CFBC had end plates welded to them in order

Table 1. Details of test specimens.

| Group | Specimen no. | $\eta_{fi,t}$ | F | R | $u_s$ (mm) | S (mm) | $f_c'$ (MPa) | Fire resistance (min.) |
|-------|--------------|---------------|---|---|-----------|--------|--------------|------------------------|
| A | TA1 | 0.28 | – | 0 | – | – | 48.5 | 40 |
| B | TB1 | 0.28 | 1% | 0 | – | – | 44 | 46 |
| | TB2 | 0.28 | 1% | 6% | 60 | 150 | 44 | 105 |
| | TB3 | 0.28 | 1% | 6% | 60 | 300 | 44 | 68 |

Notes: $\eta_{fi,t}$ = Load level for fire design; F = Fiber content (Volume%); R = Reinforcement ratio = $A_s/(A_s+A_c)$, $A_c$ = Cross-sectional area of the concrete, $A_s$ = Cross-sectional area of the reinforcing bars; $u_s$ = Minimum axial distance of reinforcing bars; S = Stirrup spacing; $f_c'$ = Concrete compressive strength.

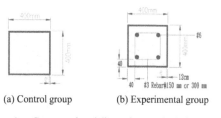

(a) Control group  (b) Experimental group

Figure 1. Cross-sectional dimensions and reinforcement arrangement of the tested columns.

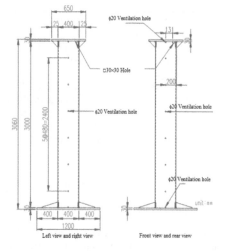

Figure 2. Column specimen size chart.

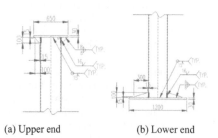

(a) Upper end  (b) Lower end

Figure 3. End plates and stiffeners of the tested columns.

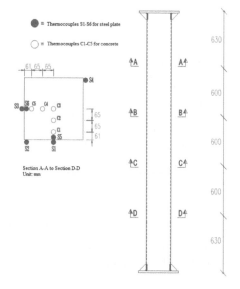

Figure 4. Scheme and thermocouples layout of CFBC.

to transfer the load, and end stiffeners were also introduced to ensure that end conditions did not affect the failure resistance of thermal load.

## 2.2 Test set up and procedure

The furnace, concrete, and steel temperatures as well as the axial deformations were recorded until a failure of the columns. The temperature from the specimen's surface to the inner central core was measured with type K thermocouples located at different depths in four sections of the columns as represented in Figure 4. Thermocouples S1–S6 was welded to the steel plate surface while thermocouples C1–C5 were embedded in the concrete core.

During the whole test, the columns were subjected to a constant compressive load. This load was controlled by a load cell of 19.6 MN, located on the head of the piston of a jack. The applied load corresponded to 28% of the design value of buckling resistance of the columns at room temperature. The Thermal load was applied on the columns in form of CNS 12514 time-temperature curve in a natural gas-fired large-scale laboratory furnace until the set experiment termination condition was reached. The current failure criterion specified in CNS 12514 is adopted in this study, which is based on the amount of contraction and the rate of contraction. For the columns under consideration, these criteria correspond to a maximum contraction of 30.6 mm and a rate of contraction of 9.18 mm/min.

## 3 TEST RESULTS

### 3.1 Axial deformation

Four CFBC were tested to failure by exposing the loaded columns to fire. The temperature followed nearly the CNS 12514 fire curve and was very uniform in all tests, indicating a precise control of furnace temperature. For example, Figure 5 presents the furnace temperature in the fire resistance tests for Specimen TA1.

The axial deformation of the columns was measured by LVDTs and displacement meters located outside the furnace. Figure 6 presents the axial deformation (y-axis) versus the fire exposure time (x-axis) curve registered during the fire test. Basically, the deformation in the tested columns results from several factors such as load, thermal expansion, and creep. The effect of load and thermal expansion is significant in the early stages while the effect of creep becomes pronounced in the later stages.

The axial displacement versus the fire exposure time curves for the tested columns have three similar stages: expansion, stable contraction, and rapid contraction, as shown in Figure 6. Owing to thermal

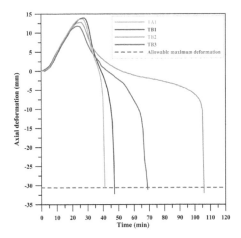

Figure 6. Axial deformation versus fire exposure time.

expansion behavior of the material, the columns experienced an expansion phase before being compressed to failure. In addition, because of its higher thermal conductivity and its direct exposure to fire, the steel plate heated up more rapidly and subsequently expanded faster than the core concrete. Moreover, the axial deformations of the specimens increase with the increased fire exposure time until they reach their maximum elongations. At those points, the compressive deformation overcomes the expansion and gradually increases. The reason is the deterioration in the mechanical properties of the steel and concrete at elevated temperatures. Then the compressive deformation sharply increases in a very short time interval, as shown in Figure 6. At this moment, the specimens can no longer withhold the load and attain the fire endurance because of the successive and rapid axial deformation.

The axial deformations of the tested columns are compared in Table 2, in which a positive value means expansion. Taking Specimen TA1 as an example, it is shown in Table 2 that during the first 20 minutes in the fire damage test, the axial elongation of Specimen TA1 was greater than that of Specimen TB1. In other words, Specimen TB1 with polypropylene fiber concrete had a smaller axial elongation. On the whole, Specimen TA1 behaved in a very ductile manner. By comparison, Specimens TB2 and TB3 failed by gradual contraction. As a result, it can be concluded that the presence of re-bars not only avoids a sudden loss of strength but also contributes to the load-carrying capacity of the concrete core.

### 3.2 Fire resistance

Provision of fire resistance has been one of the major fire safety requirements in building design.

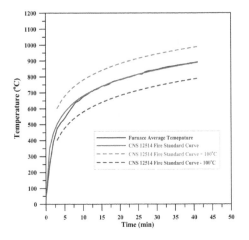

Figure 5. The temperature in the furnace for Specimen TA1.

Table 2. Axial deformation for the tested columns.

| Fire damage time (min.) | Axial deformation (mm) | | | |
|---|---|---|---|---|
| | Specimen TA1 | Specimen TB1 | Specimen TB2 | Specimen TB3 |
| 5 | 0.9 | 0.8 | 1.1 | 1.2 |
| 10 | 4.8 | 4.2 | 4.6 | 4.9 |
| 15 | 8.5 | 7.9 | 8.2 | 8.1 |
| 20 | 11.7 | 11.3 | 11.3 | 11.0 |
| 25 | 13.5 | 13.6 | 12.7 | 11.6 |
| 30 | 10.6 | 12.5 | 9.2 | 7.1 |
| 35 | 2.9 | 4 | 5.2 | 3.7 |
| 40 | −7 | −1 | 3.1 | 1. |
| 45 | – | −9.5 | 1.5 | −0.6 |
| 50 | – | – | 0.4 | −2.2 |
| 55 | – | – | −0.4 | −4.2 |
| 60 | – | – | −1.1 | −7.7 |
| 65 | – | – | −1.6 | −12.5 |
| 70 | – | – | −1.9 | – |
| 75 | – | – | −2.3 | – |
| 80 | – | – | −2.8 | – |
| 85 | – | – | −3.3 | – |
| 90 | – | – | −4.0 | – |
| 95 | – | – | −4.9 | – |
| 100 | – | – | −6.2 | – |
| 105 | – | – | −10.3 | – |

Therefore, several codes contain specifications for fire resistance rating requirements for structural members. Table 1 presents the resulting fire resistance rating, obtained according to CNS 12514. Owing to the experimental group filled with fiber concrete, or configured with longitudinal reinforcements and transverse stirrups, their fire resistances are quite better than that of the control group.

Results from the fire tests indicate that the fire resistance of Specimen TB1 was 46 minutes, as compared to 40 minutes for Specimen TA1. This result shows that the use of polypropylene fiber concrete can improve the fire resistance of CFBC. But its effect is not significant. By comparison, the fire resistance of Specimen TB2 was 105 minutes, as compared to 46 minutes for Specimen TB1. This result shows that the configuration of longitudinal reinforcements and transverse stirrups can significantly improve the fire resistance of CFBC.

On the other hand, under the same load conditions, the fire resistance of Specimen TB2 with transverse stirrup spacing of 15 cm was 105 minutes, while the fire resistance of Specimen TB3 with stirrup spacing of 30 cm was reduced to 68 minutes. This shows that the stirrup spacing has a close relationship with the fire resistance of CFBC.

## 4 CONCLUSIONS

In this study, a series of four fire tests on CFBCs filled with different types of concrete was presented. Based on the experimental results, the following conclusions can be drawn:

1. The use of polypropylene fiber concrete can improve the fire resistance of CFBC. But its effect is not significant.
2. In the bar-reinforced fiber concrete-filled box columns, the presence of rebar contributes to the load-carrying capacity of the concrete core. The fire resistances of these columns were improved significantly compared to the plain concrete-filled box column.
3. Under the same load conditions, reducing the spacing of stirrups can significantly improve the fire resistance of concrete filled box columns.

## REFERENCES

Chen, B. & Liu, J. 2004. Residual strength of hybrid-fiber-reinforced high-strength concrete after exposure to high temperatures. *Cement and Concrete Research*, 34(6): 1065–1069.

CNS 12514. *The Method of fire resistance test for structural parts of building*. Bureau of Standards, Metrology and Inspection (BSMI), MOEA, R.O.C

Espinos, A., Romero, M.L., Serra, E. & Hospitaler, A. 2015. Experimental investigation on the fire behavior of rectangular and elliptical slender concrete-filled tubular columns. *Thin-Walled Structures*, 93: 137–148.

Eurocode 4 2005. *The design of composite steel and concrete structures–Part 1–2*: General–Structural fire design.

Ezziane, M., Kadri, T., Molez, L. Jauberthie, R. & Belhacen, A. 2015. High temperature behavior of polypropylene fibres reinforced mortars. *Fire Safety Journal*, 71: 324–331.

Hong, S. & Varma, A.H. 2009. Analytical modeling of the standard fire behavior of loaded CFT columns. *Journal of Constructional Steel Research*, 65: 54–69.

Kodur, V.K.R. Wang, T.C. & Cheng, F.P. 2004. Predicting the fire resistance behavior of high strength concrete columns. *Cement & Concrete Composites*, 26(2): 141–153.

Lau, A. & Anson, M. 2006. Effect of high temperatures on high performance steel fiber reinforced concrete. *Cement and Concrete Research*, 36: 1698–170.

Lie, T.T. 1994. Fire resistance of circular steel columns filled with bar-reinforced concrete. *Journal of Structural Engineering*, ASCE, 120(5): 1489–509.

Lie, T.T. & Kodur, V.K.R. 1996. Fire resistance of steel columns filled with bar-reinforced concrete. *ASCE Journal of Structural Engineering*, 122(1): 30–36.

Ozawa, M. & Morimoto, M. 2014. Effects of various fibres on high-temperature spalling in high-performance concrete. *Construction and Building Materials*, 71: 83–92.

Poon, C.S. Shui, Z.H. & Lam, L. 2004. The compressive behavior of fiber reinforced high performance concrete subjected to elevated temperature. *Cement and Concrete Research*, 34(12): 2215–2222.

Schaumanna, P. Kodur, V. & Bahr, O. 2009. Fire behavior of hollow structural section steel columns filled with high strength concrete. *Journal of Constructional Steel Research*, 65: 1794–1802.

Sideris, K.K. Manita, P. & Chaniotakis, E. 2009. The performance of thermally damaged fiber reinforced concretes. *Construction and Building Materials*, 23(3): 1232–1239.

Song, T.Y. Han, L.H. & Yu, H.X. 2010. Concrete filled steel tube stub columns under combined temperature and loading. *Journal of Constructional Steel Research*, 66: 369–384.

Tang, C.W. & Jiang, Y.J. 2014. *Concrete filled steel tubular columns under combined temperature and loading*. Proceedings of 9th International Symposium in Science and Technology at Cheng Shiu University 2014, 90–93, 18–20 August, Kaohsiung, Taiwan.

Tokgoz, S. & Dundar, C. 2010. Experimental study on steel tubular columns in-filled with plain and steel. *Thin-Walled Structures*, 48(6): 414–422.

Yan, Z., Shen, Y., Zhu, H., Li, X. & Lu, Y. 2015. Experimental investigation of reinforced concrete and hybrid fibre reinforced concrete shield tunnel segments subjected to elevated temperature. *Fire Safety Journal*, 71: 86–99.

*Advanced Materials, Structures and Mechanical Engineering – Kaloop (Ed.)*
© 2016 Taylor & Francis Group, London, ISBN 978-1-138-02793-0

# Theoretical and experimental comparative study of strains acting on the hub fork of a cardan joint

I. Bondrea & E. Avrigean
*Lucian Blaga University, Sibiu, Romania*

ABSTRACT: The present research aims to analyze a component of the cardanic transmission, the hub fork, from a theoretical point of view, by using the analytical calculation, and also from an experimental perspective, trying to simulate the functioning of the specific part within the assembly of the cardanic transmission. The real strains are applied (similar to those in practice) for the operation of the cardanic transmission of a Dacia vehicle.

## 1 INTRODUCTION

The cardanic transmission refers to a set of machine parts (joints, shafts, intermediate bearings etc.) used for the remote transmission of mechanical energy by rotation, without torque gain, between units with variable or invariable position in space. The judicious design of these machine parts and the execution technology ensures an increase in the operational reliability and a low metal consumption.

In general, in the technical field, a joint means a kinematic coupling (the kinematic coupling is the mobile and direct link between two elements, which restricts the possibilities for relative motion) where only rotating relative motions are possible.

The classification of the cardan joints is the same as that of cardanic transmissions and therefore the mechanisms (spatial) which serve to transmit the rotating motion between two concurrent shafts, having a usually variable angle between axes and whose transmission ratio is a periodic quantity, with a mean value of one, are called asynchronous universal joints. These can be obtained by successive kinematic, constructive and structural transformations of the spherical quadrilateral. In the technical literature, all the constructive versions of the spherical quadrilateral mechanism with quadratic mobile sides (the corresponding center angles are right) and the hyper-quadrant base (the corresponding center angle is obtuse) are called cardan joints. Therefore, all the universal joints deriving from the spherical quadrilateral are called cardan joints (more precisely, monocardanic) and these joints are asynchronous.

## 2 THE ANALYTICAL CALCULATION APPLIED TO THE HUB FORK OF THE CARDAN JOINT OF A DACIA 1307 VEHICLE

### 2.1 *Determining the calculation moment*

The torque of the cardanic transmission is calculated according to the vehicle type and to the operating conditions, and for a vehicle with a single drive axle for which the computing moment of the cardanic transmission $M_c$ is determined taking into account the situation when the engine reaches maximum torque $M_M$, and the gearbox is in gear 1, and it is expressed by the relation:

$$M_c = M_M \cdot i_{cv_1} \qquad (1)$$

where, $M_M = 300 \text{ N} \cdot \text{m}$, and $i_{cv_1}$ is the transmission ratio of the first gear and comes under the value 1.

$$M_c = 300 \cdot 1 = 300 \text{ N} \cdot \text{m} \qquad (2)$$

### 2.2 *The calculation of the cardan hub fork*

The cardan fork is strained by force F (is perpendicular to the fork—Fig. 1).

The dangerous section A-A is strained at bending and twisting. The force F which strains each arm of the cardan hub fork is given by the relation:.

$$F = \frac{M_c}{2 \cdot R} = \frac{300 \cdot 10^3}{2 \cdot 27,7} = 5,42 \text{ kN}, \qquad (3)$$

where, $M_c$ is the calculating moment of the hubbed cardanic transmission; $R$—the average radius where the force F acts.

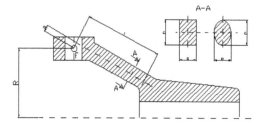

Figure 1. Scheme for calculating the cardan joint fork.

The normal bending stress in section A-A is:

$$\sigma_i = \frac{M_i}{w_i} = \frac{F \cdot l}{w_i} = \frac{5,42 \cdot 10^3 \cdot 32}{2312} = 75,02 \text{ MPa}, \quad (4)$$

and for the elliptical section is:

$$w_i = \frac{b \cdot h^2}{10} = \frac{20 \cdot 34^2}{10} = 2312 \text{ mm}^3.$$

Under the force F, the arm of the hub fork in section A-A is subject to the torsion strength:

$$\tau_t = \frac{M_t}{w_t} = \frac{5,42 \cdot 10^3 \cdot 7}{3360} = 11,28 \text{ MPa}, \quad (5)$$

where, $w_t = \frac{\pi b^2 h}{16} \approx 0,2 \cdot 20^2 \cdot 42 = 3360$ mm$^3$ for the elliptical cross section.

For the materials used in the cardan forks, the allowable bending stress is $\sigma_{ai} = 100\text{--}180$ N/mm$^2$.

The cardan hub forks are made of a medium carbon steel, 0,35–0,45%, or of low alloy improvement steels. After quenching and tempering, the hardness of the fork ranges between 197 and 300 HB depending on the type of vehicle.

## 3 STATIC ANALYSIS APPLIED TO THE CARDAN JOINTS

The issue of the geometrical modeling can be currently approached by using computer aided design software packages or modules incorporated in the finite element analysis programs for computer aided design. Such a program of a finite element analysis, is the Cosmos program, a product incorporated into the Solidworks program, which, by the facilities and the accuracy of the results it provides, is often used in studying the static and dynamic behavior of the components of technological systems. In fact, the Solidworks program is used for geometrically modeling the cardanic transmissions components and for assembling them, and the Cosmos program, based on the geometry

taken from Solidworks, generates the finite elements network, idealizes the contacts between the component parts and allows the application of strains and stresses. The Cosmos program has a set of modules dedicated to specific areas, such as: the analysis of structures, in general, the fluid mechanics, the thermal analysis. Each module, in its turn, possesses a complete set of analyses for linear or nonlinear, static or dynamic problems, which can help perform a complete research.

One of the modules of this program is the structural module which, due to the extended facilities it enables, has been chosen for researching the static and dynamic behavior of the cardanic transmission assemblies.

For approaching the present research, we chose the cardanic transmission of the Dacia 1307 vehicle, manufactured in Romania, being able to transfer the conducted researches on other different size vehicles of the same class.

As we mentioned before, the geometrical modeling was done with the help of the Solidworks program. We chose this solution due to the problems arising when transferring a model saved under IGS format from another computer aided program in the Cosmos program.

The cardanic transmission assembly of the Dacia 1307 vehicle was modeled on the following components: the flange from the end of the drive, the Φ 630 tube, the splined arbor, the splined hub, the cardan cross 1, the hub fork 1, the Φ 730 tube, the hub fork 2, the cardan cross 2 and the flanged fork towards the differentiator and then the entire model was assembled. The bearings which are mounted on the ends of the cardan crosses were also modeled. Figure 2 presents the model of the cardanic transmission in closed loop variant.

In order to obtain a model with a behavior as similar to reality as possible and at the same time

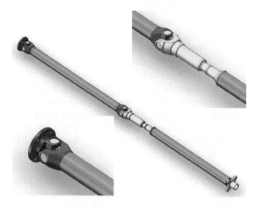

Figure 2. The assembly of the cardanic transmission modeled with Solidworks (general view).

trying to obtain the shortest functioning time, the insignificant details have been eliminated (small radius connections or niches) and the inhomogeneous areas on the structure have been approximated with homogenous finite elements.

The static analysis has the purpose of determining the strains and stresses which act on the model in a static regimen. This analysis has been applied on the parameterized model of the cardanic transmission assembly of a Dacia 1307 vehicle. The static behavior of the cardanic transmission components was studied in order to compare the results with those resulted from the finite elements analysis.

In order to carry out the static analysis on the end towards the differential, we apply strains on the flanged fork, annulling all degrees of freedom of movement, so that this end is fixed. At the end located towards the drive, we apply a constant moment of 300 Nm. The position of the strains and stresses can be seen in Figure 3.

The components of the cardanic transmission are: 1—flanged fork, STAS 880-80; 2—hub fork, STAS 880-80; 3—hollow shaft encoder, STR 302-88; 4—cardan cross, 18 Mn Cr 10 – STAS 791-88; 5—safety ring, 6—needle roller bearing, 7—boss fork, OLC 45, STAS 880-80; 8—intermediate shaft, 40 Cr 10 – STAS 791-88; 9—hollow shaft encoder, 10—flange, STAS 880-80.

The sequence of Figures 4 and 5 show the variation graphs of the von Mises stress and of the main strains of the cardanic transmission assembly.

By analyzing the graphs of the von Mises stress for the entire cardanic transmission assembly, we notice that its maximum value is 77, 85 MPa, a value achieved on the hub fork. The von Mises

Figure 4. Variation graph of von Mises stress ($\sigma_{VM}$ = 269,15 MPa).

Figure 5. Variation graph of the unit strain. ($\varepsilon_{VM}$ = 9, 03 · 10⁻⁴).

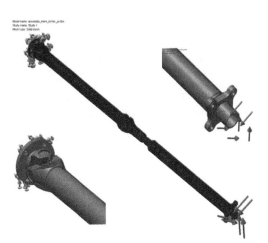

Figure 3. The meshing and the application of stress and strain on the cardanic transmission (general view).

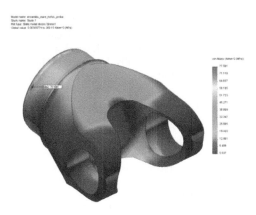

Figure 6. Variation graph of von Mises stress. ($\sigma_{VM}$ = 77, 58 MPa).

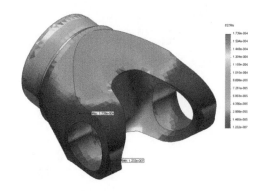

Model name: ansamblu_mare_inchis_proba
Study name: Study 1
Plot type: Static strain Strain1
Deformation scale: 1

ESTRN

1.739e-004
1.594e-004
1.449e-004
1.304e-004
1.159e-004
1.015e-004
8.899e-005
7.251e-005
5.803e-005
4.356e-005
2.898e-005
1.460e-005
1.232e-007

Max: 1.739e-004

Min: 1.232e-007

Figure 7. Variation graph of unit strain. ($\varepsilon_{VM} = 1,739 \cdot 10^{-4}$).

stress in the other areas of the hub fork is below the allowable value of the material (180 MPa).

The subsequent Figures 6 and 7 present the von Mises stress and the unit strain variation graphs.

## 4 CONDUCT OF THE MECHANICAL TESTS

### 4.1 Generalities of the testing machine

In order to achieve these tests, we will use the testing machine Instron 4303. The universal machine testing the tension, compression and buckling Instron 4303 is a universal testing tool.

The machine has a maximum load capacity of 25 kN, controlled via the IEEE-488 interface and the specialized software Material Testing System series IX. This universal testing tool allows the control of the speed of the mobile beam with an accuracy of 0.5% and the recording of the force with an accuracy corresponding to ASME 4-E Class or DIN 51221 Class 1. The obtained data can be plotted directly in the coordinates: force-displacement. However, they are generally converted in the coordinates: stress-strain.

### 4.2 Checking unit strain of the hub fork with electric resistance tensiometry

Determining the state of stress and strain on a point from the surface of a structure with tensiometric techniques is typically based on transforming the unit strain variation in that point in the variation of an electrical measurement (voltage) by means of a circuit element which is called a transducer.

The tensiometric techniques using capacitive transducers and semiconductors allow the determination of very small deformations with high precision. These transducers are however very expensive and the equipment necessary for this technique is sophisticated, that is why it is rarely used. The inductive electrical tensiometry is typically used to measure the displacements of the resistance elements.

The resistive electrical tensiometry is the most used technique to determine the strain in a certain point. This technique allows the measurement of specific strains of level $10^{-6}$ μm/m, with high precision. The transducers used in this technique are simple and pretty cheap compared to the other types of transducers mentioned above, and the measuring equipment is not very complicated.

### 4.3 The data acquisition and dissemination system. Conduct of tests

Wishing to complete the knowledge base related to the static and dynamic behavior of the cardanic transmission assemblies in general and of the hub forks in particular, in the present chapter we present a series of experimental researches conducted by the author and their results. The researches were targeted on the experimental validation of the theoretical research performed on the finite element models of the cardanic transmission assembly of the Dacia 1307 vehicle.

Setting the goals of the experimental research was based on highlighting the current state of knowledge in the field and on the basis of practical considerations related to the applicability of the research results. Based on these, the main objective of the experimental research is the analysis of the static behavior of the hub forks included in the cardanic transmission assembly.

The method of investigation used in the experimental research was the resistive electrical tensiometry.

The experiments were performed in the laboratories of the Faculty of Engineering in Sibiu, with the help of a stand mounted on the tension, compression and buckling test machine INSTRON 4303 (Fig. 8). The tension, compression and buckling test machine INSTRON 4303 allows both the control of the charge applied on the hub fork and the control of the displacement (maximum arrow).

Planning and conducting the experimental research were done in accordance with the standing standards. Also, the methods of acquiring and statistical processing of the experimental data are those established by the Romanian and international norms.

The developed experimental research methodology considered the validation of the theoretical results obtained by the numerical analysis through

Figure 8. General view and detail of the stand mounted on the INSTRON 4303 machine.

Table 1. Experimenting plan no. 1: Analysis of the static behavior of the hub fork.

| | |
|---|---|
| Objective: | Determining the strains of the press structure in static stress in the extent 0 to 20 kN |
| Input parameters: | Stress force |
| Measured quantities: | Unit strain of the structure at the level of the hub fork in two measurement points |
| Number of parallel measurements: | 3 |

the finite element method, both the results of the static and of the dynamic analyses.

Thus, an experimenting plan has been developed, which includes Table 1.

The stand used in the experimental research, using the tensiometry, consists of: the employed stand, the system of applying the stress, measurement transmitters and the system of data acquisition (Fig. 8).

The data acquisition system is composed of four modules: the transmitters mounted on the hub fork

analyzed lines, the signal conditioning modules (MB-38 produced by Keithley Instruments Inc.), the analog-digital converter board (KPCI 3108, Keithley Instruments Inc.) and a software package that controls the acquisition system and processes the collected data. The two transmitters included in the stand are connected and powered by stabilized electric current from the analog-digital converter board through the conditioning modules. These modules provide the selection of the signal amplification factor, of the measuring range and of the filtering of the received signal.

In order to measure the unit strains of the hub fork, two tensiometric stamps HBM 350XY11 were applied on each side of the hub fork.

The place where the tensiometric stamps were applied was established after the conducted observations, based on the results of the static and dynamic analyses, through the finite element method. The position of the stamps was determined in such a way as to match the position of two components of the finite element model.

The experimental determination of the unit strain $\varepsilon$, according to the two lines, was done by applying the Formulas 6 or 7, depending on the used installation, namely:

– for a half-bridge assembly:

$$\varepsilon = \frac{-4V_r}{GF\left[(1+v)-2V_r(v-1)\right]} \cdot \left(1+\frac{R_L}{R_g}\right); \qquad (6)$$

– for a full-bridge assembly:

$$\varepsilon = \frac{-4V_r}{GF\left[(1+v)-V_r(v-1)\right]}, \qquad (7)$$

where, $GF$ is the tensiometric transmitter factor; $v$—Poisson's ratio; $R_L$—the resistance of the

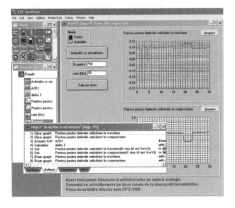

Figure 9. Virtual tool used for acquiring unit strains.

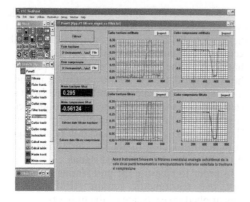

Figure 10. Virtual tool used for filtering the electric signal.

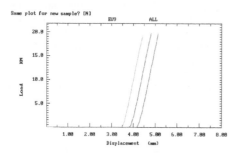

Figure 11. The loading curve of the hub fork for 20 kN strains.

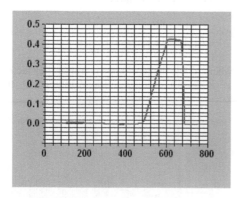

Figure 12. Variation graph of the tensile train of the hub fork for a 20 kN [mV/V] strain.

connecting wires; $R_g$—the resistance of the tensiometric transmitter; $V_r$—the value of the received signal, calculated with the relation:

$$V_r = k_c \cdot \frac{\Delta V}{V_{ex}}, \qquad (8)$$

where, $k_c$ is the amplification factor, selected through the conditioning module; $\Delta V$—the variation of the received signal; $V_{ex}$—voltage of the tensiometric transmitter.

The signal acquired from the analog-digital board KPCI 3108 is processed, filtered and saved by two

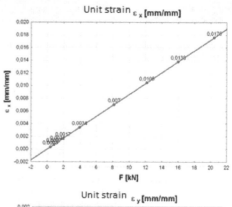

Figure 13. Variation graph of the compressive train of the hub fork for a 20 kN [mV/V] strain.

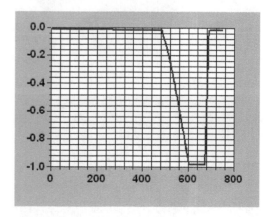

Figure 14. Variation curve between Unit strain $\varepsilon_x$ and the force F in tension-compression of the hub fork arm.

virtual tools created by the author in the Test Point program, a software which accompanies the acquisition board and which is designed for data acquisition. The tools contain blocks made available to the user by the program, that allow the modification of the number of channels by which the acquisition is made, the acquisition rate, the total duration of the acquisition, the data filtering in order to eliminate "the noise" inherent in any acquisition and the saving of the data in the form of text files.

A tool allows the data acquisition and visualization and the other allows the filtration and saving of the data in the form of text files (ASCII).

The loading curve of the testing machine for tension, bending and buckling for the hub fork strains are presented in Figure 11.

In order to determine the values of the two considered characteristics, in each point of the experimental program were used the readings of the values of the maximum stresses developed during the active running (Fig. 14).

## 5 CONCLUSIONS

From the graphs of the variation of the strains in the two directions we can observe, on the one hand, that the linearity of the experimental results is respected and, on the other hand, there is a good conformity with the finite element simulation results. Thus, in the worst case, the error percentage compared to the numerical simulations is around 9%.

The Von Mises stress experimentally obtained after applying a torque of 300 Nm is 79,64 MPa, and the Von Mises stress obtained with the finite element method in the same area of study is 77,58 MPa, the analytical calculation providing the value of 75,02 MPa. The obtained tensions come below the allowable values under normal functioning conditions, what happens in choke starting conditions remaining to be studied.

The unit strain experimentally obtained after applying a torque of 300 Nm is 0,0066 mm, and the unit strain obtained with the finite element method in the same area of study is 0,0073 mm.

The presented model may be taken as an example for other components of the cardanic transmission, a fact which will be considered in the future.

The cardanic transmission of a Dacia 1304/1307 vehicle was chosen in order to illustrate the theoretical and experimental research on a real physical model;

The optimization of the hub fork, a cardanic transmission component, leads to the development of optimal elements, but with relatively small differences in the newly obtained stresses (8%).

After studying the hub fork of the cardanic transmission through the finite element method, the variation graphs of the deformations on the two directions show, on the one hand, that the experimental results are linear, and, on the other hand, that there is a great similarity with the simulation results by the finite element method. Thus, in the worst case, the error percentage compared to the numerical simulations is around 7%.

## REFERENCES

Avrigean, E. 2009. *Optimizing transmission shafts under conditions of strength and stiffness.* Lucian Blaga, University of Sibiu Publishing House. 276.

Avrigean, E. 2014. *Comparative Study of the Loads Acting on the Operating Cardanic Transmission in the Closed and Open Loop Configurations,* 2014 4th International Conference on Civil Engineering and Transportation ICCET 2014, December 24–25, Xiamen, China. 2014.

Avrigean, E. 2014. *Veryfing the strength on the cardan transmission joint through the finite element method.* International Conference on Civil, Materials and Computing Engineering ICCMC 2014 Taiwan, December 6–7.

Duse, D.M. Bondrea, I. & Avrigean, E. 2003. *Fabricatia integrata de calculator CIM a transmisiilor cardanice.* Lucian Blaga. University of Sibiu Publishing House.

Oleksik, V. & Pascu, A.M. 2007. *Proiectarea optimala a maşinilor si utilajelor, Lucian Blaga,* University of Sibiu Publishing House.

∗Internal research conducted by the personnel of SC COMPA SA on the functioning of the cardanic transmissions, 142.

# Experimental investigation of residual stress in the micro-grinding of maraging steel

Z. Ding, B. Li, P. Zou & J. Yang
*Advanced Manufacturing Center of Donghua University, Shanghai, China*

S.Y. Liang
*Georgia Institute of Technology, Atlanta, USA*

ABSTRACT: This study investigated the effect of process parameters and thermo-mechanical load-ings on the residual stress during the micro-grinding of maraging steel by experimental validation. Both the degrees of the effects of process parameters and thermo-mechanical loadings on residual stress were evaluated using the partial relation coefficient analysis. In order to avoid the drawback of the corre-lation between single variable-controlled parameters and residual stress, effective wheel revolution was calculated using the coefficient analysis. Invalidation of the correlation, a series of maraging steel micro-grinding experiments and X-ray diffraction measurements analyses were carried out. To investigate the residual stress caused on the grinding surface, micro-grinding force and temperature were examined experimentally. Finally, this study offers suggestions on the combination range of relevant parameters based on the correlation analysis. It is practical to alter the part material properties through the reduction of residual stress with thermo-mechanical optimal control.

## 1 INTRODUCTION

Over the past 40 years, generic classes of ultra-high strength maraging steels have been developed mainly for aircraft, aerospace, and tooling applica-tions (Guo et al. 2004). The diameter of the hole from 1 mm to 3 mm is defined as the small hole (Sen & Shan 2005). Small holes are produced in various aircraft, aerospace, and tooling applications of components made of maraging steel. Grinding at the micro-scale is typically the essential process; it provides a competitive edge over other proc-esses in the fabrication of small-sized features and parts (Park & Liang 2008, 2009). It is also known that during the machining process, plastics defor-mation and non-uniform temperature gradient are generated by grinding, which affects the material's surface integrity (Shao & Liang 2014, Fergani & Liang 2015, Ding et al. 2014, 2015a). Residual stress-induced deformations on thin manufacturing parts are still one of the most challenging and relevant problems in engineering material applications.

As a consequence, enhancement of a part func-tionality is needed in aerospace, energy and medical applications, which becomes possible as reported in the literature (Brockman et al. 2012, Pu et al. 2012). Finite element methods (Özel & Ulutan 2012, Jiang et al. 2013) and numerical methods (Laamouri et al. 2013, Moussa et al. 2012) have

been used to predict and evaluate the residual stress distribution. All these investigations have laid the basis for understanding residual stress generation processing and improved the subse-quent optimization on controlling residual stress (Li et al. 2015). Analytical models have shown their capacity to predict residual stress accurately in a very short computational time (Chason et al. 2012, Fergani et al. 2014a). They also offer the possibil-ity to investigate the physics of the phenomena in contrast to FEA methods, known for their very large computational time and their small physics-based interaction (Fergani et al. 2014b). Unfortu-nately, these analytical models are based on some simplifying assumptions that reduce the complex interaction between abrasives and workpiece to a single-pass operation. By using several methods discussed above, many scholars (Mohamed et al. 2007, Mohammadpour et al. 2010) have optimized the generation of residual stress by analyzing dif-ferent processing parameters. However, there is no comprehensive investigation to discuss the relation-ship between process parameters and the residual stress induced by mechanical and thermal loadings during micro-grinding.

The quality of the parts produced by the micro-scale grinding process can be influenced by vari-ous factors related to the mechanical and thermal loadings. Therefore, the mechanical and thermal loadings of micro-grinding are useful to provide guidance for further development and optimization of the process. In this study, a new analysis method to address the relationship between process parameters and residual stress was developed. This developed method integrates the effects of process parameters on residual stress under given machining conditions to estimate the mechanical and thermal effects in the micro-grinding process. In order to verify these effects, the experiments based on a micro-grinding setup were performed for changing effective wheel revolution by combining the wheel speed, feed rate and depths of cut. The temperature of the workpiece surface was experimentally calibrated using embedded thermocouple measurements followed by analytical calculations. In addition to this, a certain linear correlation analysis was conducted to identify the main effective factors. The understanding of the effect of grinding parameters on the residual stress profile will contribute to the enhancement of surface integrity in grinding.

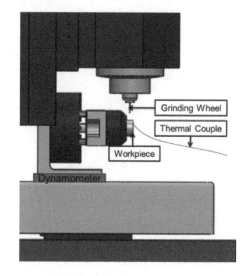

Figure 1. Experimental setup.

## 2 EXPERIMENTS

Grinding of 2.5 mm-diameter holes is a complex interrelated process. All the conditions are presented in Table 1. Internal micro-grinding experiments were carried out on a Moore G18-CNC jig grinder. The Kistler dynamometer, thermocouple temperature measurement devices and Data Acquisition card (DAQ card) were used to measure the micro-grinding force and temperature (Fig. 1). Temperature and force signal was acquired by data acquisition systems (Fig. 2). After micro-grinding, in order to validate residual stress, the hole feature was cut into two semicircle surfaces by EDM to load in the sample holder of the attachment provided in the high-power X-Ray Diffraction (XRD) meter.

For grinding of small holes, three process parameters are usually applied: wheel rotating

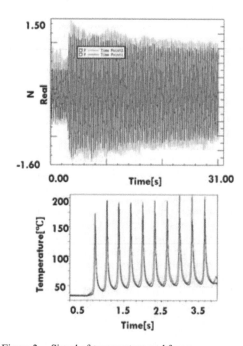

Figure 2. Signal of temperature and forces.

speed, workpiece rotating speed and depth of cut. Given the grinding machine and wheel specifications, the parameters can be planed by using a single variable control: (a) wheel rotating speed (ns)—6w, 9w and 12w rpm; (b) workpiece rotating speed (vw)—100, 150 and 200 rpm; (c) depth of cut (ap)—5, 10 and 15 μm.

Table 1. Experimental conditions.

| Items | Conditions |
| --- | --- |
| Type | Micro-grinding; dry |
| Spindle | Maximum rotating speed 150000 rpm |
| Grinding wheel | Φ2 mm, 230 grits CBN |
| Workpiece | Maraging steel |
| Forces measurement | Kistler dynamometer and NI USB-9213 DAQ Card |
| Temperature measurement | Nickel chromium and nickel silicon thermocouple |
| Residual stress | Proto (LXRD-Standard) XRD meter |

## 3 CALCULATION

### 3.1 Forces

With the relationship between the tangential and normal force in hole micro-grinding (Fig. 3) and the part of force signal (Fig. 4), the tangential and normal force can be expressed as follows:

$$F_t = \left( \left| F_{t(1)} \right| + \left| F_{t(2)} \right| + \cdots + \left| F_{t(n)} \right| + \left| F_{t(n+1)} \right| \right) / (n+1) \quad (1)$$

$$F_n = \left( \left| F_{n(1)} \right| + \left| F_{n(2)} \right| + \cdots + \left| F_{n(n)} \right| + \left| F_{n(n+1)} \right| \right) / (n+1) \quad (2)$$

### 3.2 Effective wheel revolution

The specific case of contact between a larger number of abrasive grains and workpiece determines the behaviors related to deformation,

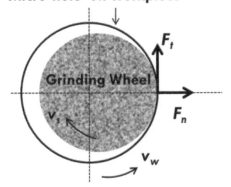

**Micro-hole on Workpiece**

Figure 3. The geometrical relationship of forces in hole micro-grinding.

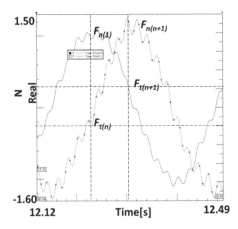

Figure 4. Part of tangential and normal force signal.

wear, temperature and integrity on the ground surface. Effective wheel revolution was proposed (Ding et al. 2015b) as the wheel revolution per contact length, which can be expressed as follows:

$$n_{eff} = n_s t \quad (3)$$

where $n_s$ is the wheel rotating speed; $t$ is the time needed for grinding a contact length $l$, which can be defined as follows:

$$t = l/v_w = \sqrt{a_p d_e}/v_w = \sqrt{a_p \frac{d_s d_w}{d_s + d_w}} \left/ \frac{n_w \pi d_w}{60} \right. \quad (4)$$

where $v_w$ is the workpiece feed rate; $a_p$ is the depth of cut; $d_e$ is the equivalent diameter for cylindrical grinding (Shah 2011); $d_s$ is the wheel diameter; and $d_w$ is the hole diameter. Effective wheel revolution can be estimated by using the following relation:

$$n_{eff} = 60 n_s \sqrt{a_p \frac{d_s d_w}{d_s + d_w}} \left/ \left( n_w \pi d_w \right) \right. \quad (5)$$

## 4 RESULTS AND ANALYSIS

The relationship between single variable-controlled parameters and residual stress cannot reflect the real correlation because there is no orthogonal experimental design for parameter planning. Three variable parameters are combined into effective wheel revolution to fill the gaps. The calculated effective wheel revolution and the resulting experimental measurement force, temperature and residual stress are given in Table 2.

To minimize the interconnectivity between all the parameters, partial relation coefficient

Table 2. Measurement results under various effective wheel revolutions.

| $n_{eff}$ [wr] | $T$ [°C] | $F_t$ [N] | $F_n$ [N] | $\sigma_t$ [MPa] | $\sigma_n$ [MPa] |
|---|---|---|---|---|---|
| 26.83 | 49 | 1.63 | 0.45 | −122.44 | −398.53 |
| 28.46 | 99 | 1.59 | 0.39 | −67.04 | −331.48 |
| 30.98 | 126 | 1.32 | 0.27 | −19.25 | −328.99 |
| 32.86 | 62 | 1.78 | 0.61 | −140.06 | −407.11 |
| 34.86 | 100 | 1.61 | 0.48 | −100.42 | −381.03 |
| 37.95 | 95 | 1.98 | 0.83 | −170.59 | −425.96 |
| 37.95 | 148 | 1.53 | 0.38 | −40.71 | −339.48 |
| 40.25 | 157 | 1.73 | 0.71 | −147.69 | −409.21 |
| 43.82 | 157 | 1.66 | 0.46 | −73.35 | −345.54 |

$n_{eff}$, effective wheel revolution; $T$, surface temperature; $F_t$, tangential force; $F_n$, normal force; $\sigma_t$, parallel residual stress; $\sigma_n$, vertical residual stress.

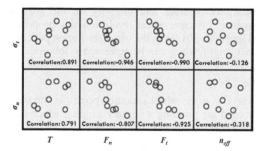

Figure 5.  Correlation analysis diagram.

analysis method is used in the following correlation analysis. According to the correlation analysis diagram shown in Figure 5, tangential and normal forces and effective wheel revolution have a negative correlation with both parallel and vertical wheel speed direction residual stresses, while the temperature has a positive correlation with them. The absolute value of residual stress will increase with the thermal expansion and larger plastic deformation by the increasing thermo-mechanical loadings. There are three major causes for residual stress on the ground surface: thermal expansion and contraction, phase transformations, and plastic deformation. It is found that mechanical loadings can be the most important factors in the generation of residual stresses during small hole micro-grinding. Besides, the magnitude of the partial relation coefficient has also shown that thermo-mechanical loadings have a much higher influence than process parameters on residual stress. It will be more effective to reduce residual stress by controlling thermo-mechanical loadings than process parameters.

## 5  SUMMARY

In this paper, in order to investigate the effect of process parameters and thermo-mechanical loadings on residual stress during micro-grinding, experiments and correlation analyses were conducted. In order to avoid the drawback of the correlation between single variable-controlled parameters and residual stress, effective wheel revolution was calculated using the coefficient analysis. Both the degrees of the effects of effective wheel revolutions and thermo-mechanical loadings on residual stress were evaluated using the partial relation coefficient analysis. This paper indicates thermo-mechanical loading as part of important controllable factors for the variation of residual stress on the grinded part surface. Thermo-mechanical loadings have a much higher influence than process parameters on residual stress. It will be more

immediate and flexible to reduce residual stress by thermo-mechanical loadings via controlling material properties, wheel topography characteristics, and some other processing conditions. In order to achieve a high surface mechanical property in the future, more details about the effects of thermo-mechanical loadings on residual stress should be analyzed based on this study.

## ACKNOWLEDGMENT

The authors would like to gratefully acknowledge the support of the National High-tech Research and Development Program of China (NO. 2012AA041309) and the Fundamental Research Funds for the Central Universities (CUSF-DH-D-2015070).

## REFERENCES

Brockman, R.A. Braisted, W.R. Olson, S.E. Tenaglia, R.D. Clauer, A.H. Langer, K. & Shepard, M.J. 2012. Prediction and characterization of residual stresses from laser shock peening. *International Journal of Fatigue*, 36(1): 96–108.

Chason, E. Shin, J.W. Hearne, S.J. & Freund, L.B. 2012. Kinetic model for the dependence of thin film stress on growth rate, temperature, and microstructure. *Journal of Applied Physics*, 111(8): 083520.

Ding, Z. Li, B. & Liang, S.Y. 2015. Material Phase Transformation at High Strain Rate during Grinding. *International Journal of Advanced Manufacturing Technology*. Doi: 10.1007/s00170-015-7014-5.

Ding, Z. Li, B. & Liang, S.Y. 2015. Maraging C250 steel phase transformation and residual stress during grinding. Materials Letters. Doi:10.1016/j.matlet.2015.04.040.

Ding, Z. Li, B. Zou, P. & Liang, S.Y. 2014. Material Phase Trans-formation during Grinding. *In Advanced Materials Research*, 1052: 503–508.

Fergani, O. & Liang, S.Y. 2015. The effect of machining process thermo-mechanical loading on workpiece average grain size. *The International Journal of Advanced Manufacturing Technology*, 1–9.

Fergani, O. Mamedov, A. Lazoglu, I. Yang, J.G. & Liang, S.Y. 2014. Prediction of residual stress induced distortions in micromilling of Al7050 thin plate. *Applied Mechanics and Materials*, 472: 677–681.

Fergani, O. Shao, Y. Lazoglu, I. & Liang, S.Y. 2014. Temperature Effects on Grinding Residual Stress. *Procedia CIRP*, 14: 2–6.

Guo, Z. Sha, W. & Li, D. 2004. Quantification of phase trans-formation kinetics of 18 wt.% Ni C250 maraging steel. *Materials Science and Engineering A*, 373: 10–20.

Jiang, X. Li, B. Yang, J. & Zuo, X.Y. 2013. Effects of tool diameters on the residual stress and distortion induced by milling of the thin-walled part. *The International Journal of Advanced Manufacturing Technology*, 68(1–4): 175–186.

Laamouri A. Sidhom H. & Braham C. 2013. Evaluation of residual stress relaxation and its effect on fatigue strength of AISI 316 L stainless steel ground surfaces: Experimental and numerical approaches. *International Journal of Fatigue*, 48: 109–121.

Li, B. Jiang, X. Yang, J. & Liang, S.Y. 2015. Effects of a depth of cut on the redistribution of residual stress and distortion during the milling of the thin-walled part. *Journal of Materials Processing Technology*, 216: 223–233.

Mohammadpour, M. Razfar, M.R. & Jalili Saffar, R. 2010. Numerical investigating the effect of machining parameters on residual stresses in orthogonal cutting. *Simulation Modelling Practice and Theory* 18: 378–389.

Mohamed, N.A. Ng, E.G. & Elbestawi, M.A. 2007. Modelling the effects of tool-edge radius on residual stresses when orthogonal cutting AISI 316L. *The International Journal of Advanced Manufacturing Technology,* 47(2): 401–411.

Moussa, N.B. Sidhom, H. & Braham, C. 2012. Numerical and experimental analysis of residual stress and plastic strain distributions in machined stainless steel. *International Journal of Mechanical Sciences,* 64(1): 82–93.

Özel T. & Ulutan D. 2012. Prediction of machining induced residual stresses in turning of titanium and nickel based alloys with experiments and finite element simulations. *CIRP Annals-Manufacturing Technology*, 61(1): 547–550.

Park, H.W. & Liang, S.Y. 2008. Force modeling of micro-grinding incorporating crystallographic effects. *International Journal of Machine Tools and Manufacture,* 48(15): 1658–1667.

Park, H.W. & Liang, S.Y. 2009. Force modeling of micro-scale grinding process incorporating thermal effects. *The International Journal of Advanced Manufacturing Technology*, 44(5–6): 476–486.

Pu, Z. Outeiro, J.C. Batista, A.C. Dillon, O.W. Puleo, D.A. & Jawahir, I.S. 2012. The enhanced surface integrity of AZ31B Mg alloy by cryogenic machining towards improved functional performance of machined components. *International journal of machine tools and manufacture*, 56: 17–27.

Shah S.M. 2011. *Prediction of residual stresses due to grinding with phase transformation.* Dissertations and Theses. INSA DE LYON. France.

Sen, M. & Shan, H.S. 2005. A review of electrochemical macro-to micro-hole drilling processes. *International Journal of Machine Tools and Manufacture*, 45(2): 137–152.

Shao, Y. & Liang, S.Y. 2014. *Predictive force modeling in MQL (Minimum Quantity Lubrication) grinding*, Proceedings of MSEC 2014, Jun 9–14. Detroit, Michigan.

*Advanced Materials, Structures and Mechanical Engineering – Kaloop (Ed.)*
© 2016 Taylor & Francis Group, London, ISBN 978-1-138-02793-0

# Flutter analysis of rotating missile's variable cross-section empennage by differential quadrature method

D. Horak

*Faculty of Civil Engineering, Brno University of Technology, Brno, Czech Republic*

ABSTRACT: Several years ago a new FRP reinforcement system was developed at Brno University of Technology. During the following years this system was refined to maximise its benefits and to lower the initial costs related to FRP materials. There was a special focus on anchoring and the development of a pre-stressed reinforcement length. There was a special focus on anchoring and the development of a pre-stressed reinforcement length. A large number of surfacing types were tested and the gained results were gathered and analysed using a bond-slip relationship. As most of today recommended calculations of the minimal anchoring length do not cover the influences of either the surfacing of the bar or the thickness of the concrete cover, a new extension was proposed. The equation was confirmed by a number of real-world tests.

## 1 INTRODUCTION

A new system for the reinforcement of structures has been developed within the framework of research projects carried out at the Faculty of Civil Engineering. This system uses non-metallic materials based on carbon (CFRP) and Glass Fibre Reinforced Polymers (GFRP).

A study was prepared to determine the influence of the surface or additional jigs on the cohesion between reinforcement and surrounding concrete. A special focus was paid to the calculation of anchoring length in such a way that allows for the influence of different FRP reinforcement surfacings.

However, the surfacing is not the only influence that determines the anchoring of the reinforcement. The thickness of the concrete cover is also a critical factor when anchoring reinforcement particularly when using only thin concrete cover. This is usually one of the main points when presenting the advantages of FRP materials—thanks to their resistibility against aggressive environments it is possible to reduce cover and thus save some material.

## 2 BOND PROPERTIES WITHOUT THE INFLUENCE OF CONCRETE COVER

For the basic theoretical approach, a bond-slip model published by Eligehausen et al. (1983) and Cosenza (2002) was used. This theory assumes constant stress along the anchored length of the reinforcement when the limit bond stress is reached. Also, the behaviour of the reinforcement material is assumed to be linear. The combination of these two conditions makes it possible to obtain a differential equation. This basis was used in the past to create several analytical models that describe the progression of bond stress $\tau(x)$ along the anchoring length.

It was decided to leave out the descending branch of the bond-slip diagram as it comes into effect after reaching the maximal bond stress, i.e. after damaging the interface between reinforcement and concrete. This results in the following single equation (also graphically expressed in Fig. 1):

$$\frac{\tau}{\tau_m} = \left( \frac{s}{s_m} \right)^{\alpha}$$

(1)

where, $\tau$ = bond stress [MPa]; $\tau_m$ = maximal bond stress [MPa]; $s$ = the slip of the bar for the actual bond stress [mm]; $s_m$ = the slip of the bar for the maximal bond stress [mm]; $\alpha$ = a curve parameter for the ascending branch.

It has to be noted that this simplification is suitable for design purposes only. When a detailed

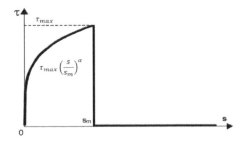

Figure 1. Modified bond-slip diagram used for calculation.

description of the behaviour of the anchoring length is required, the descending branch of the diagram cannot be neglected.

To obtain the real bond properties, a set of standard pull-out tests was performed. The length of the anchoring (cohesive) length was five times the recommended diameter (i.e. 70 mm for a bar with diameter 14 mm as shown in Fig. 2).

Basic material properties of the concrete and reinforcement were measured using standardised tests. The surface of the bar was modified by additional sanding and one-side wrapping.

All test specimens failed due to the reinforcement bar being pulled out of the concrete. The failure was mostly caused in the interface between concrete and reinforcement by the tearing off of the thin surface layer of reinforcement. A few specimens demonstrated a failure similar to that seen with steel reinforcement bars; a thin concrete layer was torn off around the reinforcement. This failure mode is caused by the crushing of the concrete in front of the surface bumps on the FRP reinforcement. Given the high compressive strength of the concrete and the relatively "soft" surface of the FRP bars such a failure mode is expected only in cases when there is significant local weakening of the concrete. Not one single case of cone-shaped concrete tearing was recorded.

The bond stress was calculated for four stages occurring during the test—for slippages of the bottom (not loaded) end of the reinforcement of 0.05 mm, 0.10 mm, 0.25 mm and for the maximal tensile force in the reinforcement. The average bond stress was then calculated using the relationship.

$$\tau = \frac{F}{C_b \cdot l_e} \tag{2}$$

where, $\tau$ = bond stress [MPa]; $F$ = tensile force [N]; $C_b$ = the circumference of the reinforcement bar

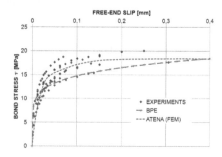

Figure 3. Comparison of the test results, the analytical solution (BPE model) and the numerical FEM model of the test.

[mm] and $l_e$ = the length of the embedded part of the reinforcement in the concrete [mm].

The derivation of variables $\alpha$ and $p$ was accomplished using the least squares method. The resulting value was used in the analytical model to describe the behaviour of the GFRP reinforcement. The accuracy of this result can be observed in Figure 3. The comparison shows a good match between the test results and the theoretical solution.

Using Equation (1), it is possible to express the tensile stress $\sigma_1$ in the reinforcement that is reached during maximal slippage $s_m$ of the bar, which is also the moment when maximal bond-stress $\tau_m$ is reached. Using a similar procedure, it is possible to express the ultimate anchoring length $l_m$ that is needed to reach maximal bond stress. The detailed procedure can be found in Cosenza (2002).

## 3 THE INFLUENCE OF CONCRETE COVER

Although accurate, the relationships described above do not include the influence of the concrete cover and are only usable in situations when the concrete cover is sufficient. For this reason, a set of numerical models was prepared to study the influence of the proximity of the reinforcement to the concrete surface.

### 3.1 Numerical models of pull-out tests with thin concrete cover

The numerical models were created using ATENA non-linear FEM software. The materials were defined as follows:

– To simulate the concrete material, a numerical model with brittle-plastic behaviour was used.
– The composite reinforcement was modelled as linear reinforcement.

The bond behaviour between reinforcement and concrete was specified using a bond material model.

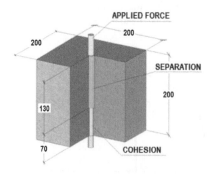

Figure 2. A schematic diagram of a pull-out test used to determine the bond properties and variables used in the analytical solution (corresponds to the ACI standard).

The bond-slip diagram used in this model was defined according to the real pull-out test described above.

The overall geometry of the model was created to fit best the real test. First, standard pull-out tests were calculated to verify the reliability of the model, as well as the calculated results of ultimate stress $\sigma_l$ and ultimate anchoring length $l_m$.

The second part of the FEM analysis was to observe the influence of various thicknesses of concrete cover. Results show the obvious influence of thin concrete cover (Fig. 5) on the bearing capacity of the anchoring. If the surrounding concrete is insufficient, the bearing capacity drops sharply. However, the thickness of concrete cover has to be observed as a relationship between the diameter of the bar and the cover thickness itself. As is shown, the decrease starts approximately when the concrete cover thickness is equal to the diameter of the reinforcement.

The zone of concentrated stresses along the reinforcement bars is almost the same in all observed specimens and is located in front of the anchorage

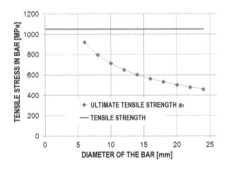

Figure 4.    Verification of the ultimate tensile stress $\sigma_l$ in the reinforcement bar.

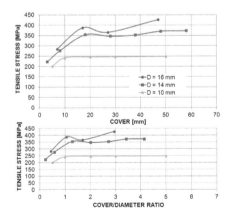

Figure 5.    Results of the numerical analysis of the influence of concrete cover (anchoring length 70 mm, concrete C30/37).

length. The dimension of the zone is approximately two times the diameter of the reinforcement. In the thin cover, the stress quickly reaches the surface of the concrete and causes the propagation of a longitudinal crack. This crack negatively affects the ability of the reinforcement to transfer the tensile stress into the concrete, which manifests itself in uneven stress distribution along the reinforcement.

### 3.2    Analytical solution

In most structures, the reinforcement is placed 15–50 mm from the surface. For such reinforcement, the failure of cohesion is demonstrated by the longitudinal cracking of the concrete cover. This cracking is caused by radial stresses around the reinforcement bar as a result of the forces acting along the anchoring zone. A theoretical description of this phenomenon can be found in fib Bulletin 40 (2007).

This distribution can be transformed into a simple strut and tie analogy to obtain the compressed diagonals (the compressed side of the cone) and stretched ties perpendicular to the axis of the bar (the transverse tensile stress along the circumference of the bottom of the cone, see Fig. 6).

If equality is presumed between the ultimate bond stress and the ultimate tensile stress in the surrounding concrete, the minimal concrete cover $c$ can be calculated using the equation.

$$c = \frac{\tau_m d \tan \alpha}{2 f_{ct}} - \frac{d}{2} \tag{3}$$

where, $\alpha$ = the angle of the distribution of the stress into the surrounding concrete (recommended value 45°–30°); $d$ = the diameter of the reinforcement bar [mm]; $f_{ct}$ = the tensile strength of the concrete [MPa]; $\tau_m$ = the maximal cohesion for the given reinforcement surfacing [MPa].

Using a similar approach, it is possible to calculate the maximal tensile force that can be anchored if the concrete cover $c$ is given, or in other words, to determine the tensile force that can act in the reinforcement bar without the risk of longitudinal crack propagation. In the case that the bottom of the cone meets the surface of the concrete, the circumference transferring tensile forces in

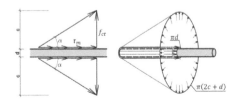

Figure 6.    Verification of the ultimate tensile stress $\sigma_l$ in the reinforcement bar.

concrete is smaller, thus increasing the stress in the concrete.

Using the theory concerning concentrated load (EC2, 2006), the condition $2c_1 < 3d$ can be used. After calculating the central angle $\theta$, it is possible to derive an equation to calculate the maximal $\tau_m$ that can be reached before the longitudinal crack opens (7), and subsequently the tensile force in the reinforcement bar.

### 3.3 Experimental verification of the models

The theoretical relationships and FEM results were tested in a real-world situation. The loading scheme, as well as measuring setup, was the same as in a standard pull-out test. The only difference was the position of the embedded reinforcement. It was placed in such a way that the two concrete cover depths tested were 10 mm and 20 mm. These values were chosen accordingly to theoretical expectations that the influence of thin concrete cover would only manifest itself in a thickness of 25 mm or less.

The results prove that the analytical approach is relatively accurate. In the light of the obtained values, it seems the angle $\alpha$ of the inner stress distribution is somewhere around 40°. The failure mode corresponds very well with the FEM analysis results. The first crack occurred along the reinforcement in the embedded part. Later, the cracks spread further from the reinforcement and the failure was completed by the complete layer of cover falling off.

## 4 CONCLUSIONS

The experimental results prove (via the differences between the displacements of the loaded and free ends of the anchored bar during the pull-out tests) the linearity of the progress of the reinforcement's displacement. It follows that the stress propagation progresses in the same manner along the whole anchoring length and therefore it is possible to use the above-stated analytical relationships. However, it has to be mentioned that this is only valid in cases when there is a sufficient concrete cover.

To determine a minimal concrete cover, a number of different design approaches (ACI (2006), JSCE (1997)), analytical solutions and numerical models have been used and studied as well as compared with the real experiments. The results show that optimal concrete cover cmin should be at least as thick as one diameter of the anchored reinforcement. This recommendation is valid in general for standard quality concrete of grade C25/30 and better. When using concrete grades with lesser strengths, it is possible to determine the minimal concrete cover using the relationship (2).

When compared to several design codes (ACI (2006), JSCE (1997) and EC2 (2006)), the calculated minimal concrete cover providing sufficient anchoring corresponds very well with design values recommended by the ACI standard (i.e. a thickness of one diameter of reinforcement). The other two codes provide thicker cover (from 1.6 up to 3.1 diameters), causing higher material usage. The difference is higher for small diameters while for large diameters the difference stabilizes.

The relationships stated above were verified by means of real-world tests. They show good correspondence with theoretical estimations. The angle of the inner distribution of stresses in concrete along the anchoring zone seems to be around 40° as this value approaches those found in test results.

As a successive work, new research was started (Zlamal et al. 2013) with the aim on the influence of the effects of fire and increased temperatures. The results and theories described above are going to be the groundwork for the theories related to this problem.

## ACKNOWLEDGEMENT

This paper was supported by the project CZ.1.07/2.3.00/30.0039 of Brno University of Technology and a research project TA03030851 "Rein-stating of tunnels—technology, materials and methodology".

## REFERENCES

ACI 440.1R-06 2006. *Guide for the Design and Construction of Structural Concrete Reinforced with FRP Bars.*

Cosenza, E. Manfredi, G. & Realfonzo, R. 2002. Development length of FRP straight rebars. *Composites: Part B.* 33(7): 493–504. ISSN 1359–8368.

CSN EN 1992-1-1 *Eurocode 2: Design of concrete structures Part 1-1: General rules and rules for buildings*, Czech Re-public, 2006.

Eligehausen, R. Popov, E.P. & Bertero, V.V. 1983. *Local bond stress-slip relationships of deformed bars under generalized excitations.* Report No. 83/23 Earthquake Engineering Research Center. University of California, Berkeley.

Fib Bulletin 40. 2007. *FRP reinforcement in RC structures: Technical report*, Lausanne, Switzerland.

Japan Society of Civil Engineers (JSCE). 1997. *Recommendation for Design and Construction of Concrete Structures using Continuous Fiber Reinforced Materials.* Concrete Engineering Series 23, ed. by A. Machida. Research Committee on Continuous Fiber Reinforcing Materials. Tokyo.

Stepanek, P. Fojtl, J. & Horak, D. 2006. *Pull-out test of non-metallic FRP reinforcement for concrete structures.* In Proceedings of the 4th International Specialty Conference on Fibre Reinforced Materials. Hong Kong, Singapore, CI-PREMIER PTE LTD.

Zlamal, M. Kucerova, A. & Stepanek, P. 2013. *Effect of fire on FRP reinforced concrete structures*, Proc. 'CESB 2013 Central Europe Towards Sustainable Building', Prague, Czech Republic.

*Advanced Materials, Structures and Mechanical Engineering – Kaloop (Ed.)*
© 2016 Taylor & Francis Group, London, ISBN 978-1-138-02793-0

# Influence of polypropylene waste carpet fiber on deformation characteristics of concrete

M.H. Hossein, A.S.M. Abdul Awal & A.A.K. Mariyana
*Faculty of Civil Engineering, University Teknologi Malaysia, Johor, Malaysia*

ABSTRACT: The utilization of waste fibers from textile industries for reinforcement of concrete is a very attractive approach because of their technical, economic, and ecological benefits. This article highlights some laboratory test results of the influence of waste carpet fibers on the deformation characteristics of concrete. Concrete mixtures containing 0.5% to 2.0% fibers were made and tested for compressive strength, flexural strength, modulus of elasticity and shrinkage. It has been observed that carpet fiber reinforced concrete had relatively lower compressive strength compared to plain concrete. Along with the lower strength development, the modulus of elasticity of carpet fiber reinforced concrete was found to be lower although a higher value in flexural strength was observed. In this study, a relatively higher drying shrinkage was also recorded in the concrete containing carpet fiber.

## 1 INTRODUCTION

The recycling of solid wastes is a significant issue all over the world. Reusing and utilization of the waste materials are expanding significantly, particularly in the development of infrastructures (Schmidt & Cieslak, 2008; Wang, 2010). Since the advent of fiber reinforcing of concrete, a great deal of work has been led on different fibrous materials to determine the real characteristics and benefits for each product. The most common fibers used in construction are steel fibers, synthetic fibers such as nylon and polypropylene, glass fibers, natural fibers and fibers from pre and post-consumer wastes. Over the decades different types of synthetic fibers have been successfully used to reinforce concrete. A large portion of the fiber waste is made out of natural and engineered polymeric materials like cotton, silk, wool, polyester, and so forth. Synthetic fibers are developed mostly to supply the high demand for carpet and textile products. Nylon and Polypropylene are the most synthetic fibers used in these industries.

The rate of carpet disposal in the USA, for example, is about 2–3 million tons in a year, and about 4–6 million tons for every year around the world (Wang, 2013). Most of the fibrous waste is composed of natural and synthetic polymeric materials such as cotton, polyester, nylon, and polypropylene. The tufted structure is the most common type of industrial carpet in the market. It mainly comprises of two layers of facing and baking yarns majority of which are polypropylene and nylon fibers. The typical industrial carpet wastes are shown in Figure 1.

In the last decades, researchers have investigated the effects of utilization of different types of fibers in plain concrete. Wang (1999), Miraftab (1999) and Vilkner et al. (2004) investigated the physical and mechanical properties of waste carpet fiber reinforced concrete. Schmidt & Cieslak (2008) studied the evaluation of surface energy of concrete with recycled carpet fibers; Reis (2009) analyzed the mechanical properties of polymer concrete containing textile fibers. The industrial waste carpet fibers after being prepared and cut into desired length were also utilized for concrete reinforcement to enhance toughness and ductility of concrete (Wang, 1999; Song et al., 2005). Indeed, the inclusion of waste carpet fibers in the concrete mix is a new area of research which needs significant attention. It is in general that a few amounts of short fibers has been recommended for the enhancement of the properties of concrete like impact resistance and tensile strength, and it paves the way to use polypropylene (PP) carpet fibers to investigate the deformation behaviour

Figure 1. Typical industrial carpet wastes.

of concrete. The main objective of this study is to investigate in detail the effect of the addition of waste PP carpet fibers on compressive strength and flexural strength, modulus of elasticity and shrinkage of concrete at different volume fractions ranging from 0.5 to 2.0%.

## 2 MATERIALS AND METHODS

### 2.1 *Material and mix proportion*

Waste carpet fibers used in this study was obtained from ENTEX Carpet Industries Snd. Bhd, Malaysia. The fibers were collected at different lengths where all the fibers are multi-filament polypropylene from the face yarn wastes. The fibers were 0.45 mm in diameter and cut into the length of 30 mm (Fig. 1), the engineering properties of the fibers being shown in Table 1.

Ordinary Portland cement conforming ASTM C150 (2005) was utilized in this study. A saturated surface dry river sand with fineness modulus of 2.9 passing through 4.75 mm ASTM sieve with water absorption and specific gravity of 0.7% and 2.6 respectively was used as fine aggregate. Crushed granite of 10 mm maximum size with the specific gravity of 2.7 and water absorption of 0.5% was used as coarse aggregate. Following DOE (1992) mix design method, concrete of grade 30 MPa with the w/c ratio of 0.50 and slump value range of 30–60 mm was designed. A control mix of Plain Concrete (PC) without carpet fibers was prepared for comparison, and four fiber volume fractions $V_f$ of 0.5%, 1.0%, 1.5% and 2.0% were investigated. It should be noted that no mineral or chemical admixtures were added in the mixes. The mix proportions of Carpet Fiber Reinforced Concrete (CFRC) are presented in Table 2.

### 2.2 *Preparation of test specimen*

Concrete cubes, 100 mm in size were cast for compressive strength in accordance with BS EN 12390-3 (2009). Shrinkage of concrete was measured using 100 mm × 200 mm cylindrical molds in accordance with ASTM C157-8 and ASTM C490-11. Prism specimens of 100 mm × 100 mm × 500 mm in size were prepared for testing the flexural strength in accordance with BS EN 12390-5 (2009). The number of specimens for each test was three and shrinkage strains was measured on four vertical gauge lines spaced uniformly around the specimen by a demountable mechanical strain gauge having a gauge length of 100 mm. After casting, the specimen was covered with a plastic sheet and was demolded after 24 hours. The specimens were submerged into water curing tank until testing, and all these tests were performed at an average room temperature of 27 °C with the relative humidity, RH of 80 ± 5%.

## 3 RESULTS AND DISCUSSIONS

### 3.1 *Compressive strength*

The results of compressive strength of Plain Concrete (PC) and Concrete Containing Carpet Fiber (CFRC) at the age of 1, 7, 28 and 91 days are presented in Figure 2. The results showed that PC exhibited higher strength development, while a relatively lower strength was observed in carpet fiber reinforced concrete at all ages. It can be seen that at the age of 28 days, the plain concrete developed 46.55 MPa, while a strength development of 42.7 MPa was found in 0.5% CFRC, which is about 8% lower than that of the PC. It has also been found that additional increase in fiber volume fraction reduced the compressive strength further.

Table 1. Properties of polypropylene carpet fiber.

| Length (mm) | Diameter (μm) | Density (kg/m³) | Tensile strength (MPa) | Melting point (°C) | Reaction with water |
|---|---|---|---|---|---|
| 30 | 450 | 940 | 324 | 170 | Hydrophobic |

Table 2. Concrete mix proportion (kg/m³).

| Mix | Carpet fiber | Water | Cement | Fine aggregate | Coarse aggregate |
|---|---|---|---|---|---|
| PC | 0.0 | 215 | 430 | 840 | 910 |
| 0.5% CFRC | 4.7 | 215 | 430 | 840 | 910 |
| 1.0% CFRC | 9.4 | 215 | 430 | 840 | 910 |
| 1.5% CFRC | 14.1 | 215 | 430 | 840 | 910 |
| 2.0% CFRC | 18.8 | 215 | 430 | 840 | 910 |

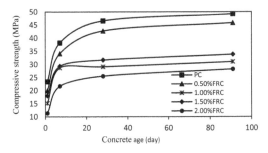

Figure 2. Effect of fiber content on compressive strength.

Table 3. The modulus of elasticity of concrete.

| Mix | Modulus of elasticity (GPa) at 28 days |
|---|---|
| PC | 29.3 |
| 0.5% CFRC | 28.5 |
| 1.0% CFRC | 25.8 |
| 1.5% CFRC | 26.1 |
| 2.0% CFRC | 25.1 |

The rate of reduction of compressive strength for concrete containing carpet fibers, for example, were 37.48%, 34.13% and 45.15% for 1.0% CFRC, 1.5% CFRC and 2.0% CFRC respectively. This decrease in compressive strength of CFRC might be attributed to the presence of pores in the composite in contrast with the plain concrete due to the addition of carpet fibers in the matrix.

A similar trend has been stated by Wang et al. (2000) and Qian & Stroeven (2000), who observed a lesser value in concrete incorporating carpet fibers than in the plain concrete. This has also been observed by Vilkner et al. (2004) where the addition of carpet fibers into the concrete was found to achieve lower compressive strength.

### 3.2 Modulus of elasticity

The modulus of elasticity of plain concrete and carpet fiber reinforced concrete was calculated with the corresponding compressive strength at the age of 28 days and are shown in Table 3. The obtained values of modulus of elasticity are calculated from the expression given Neville & Brooks (2010).

$$E_{c,28} = 20 + 0.2 \, f_{cu,28} \tag{1}$$

where, $E_c$ is the modulus of elasticity and $f_{cu,28}$ is the cube compressive strength of concrete at the age of 28 days. The results shown in Table 3 reveal

the modulus of elasticity of carpet fiber reinforced concrete was lower than that of plain concrete, but the difference was not significant. For example, the 28 days modulus of elasticity of plain concrete with the corresponding compressive strength value of 46.55 MPa was found to be 29.3 GPa, while a value of 28.5 GPa was obtained for 0.5% CFRC having a 28 days compressive strength of 42.7 MPa. From the results, it is clear that, like that in normal concrete, the modulus of elasticity increases with the increase in compressive strength of fiber reinforced concrete.

### 3.3 Flexural strength

The values obtained in the flexural test are presented in Figure 3. Compared to plain concrete without carpet fiber, the 0.5% CFRC showed an appreciable increase in flexural strength. The flexural strengths of carpet fiber reinforced concrete were 18.5%, and 1.71% higher for 0.5% and 1.0% of fiber content respectively. A reduction in the flexural strength of CFRC, however, was observed with the inclusion of additional fiber. On average, the flexural strength decreased by 2.37% and 17.34% for 1.5% and 2.0% carpet fibers as compared to the plain concrete.

Figure 4 reveals the failure mode and crack patterns as observed for plain and fiber reinforced

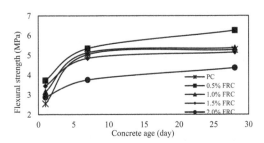

Figure 3. The effect of fiber content on flexural strength.

Figure 4. The failure mode and crack patterns: a) PC, b) 0.5% CFRC, c) 1.5% CFRC.

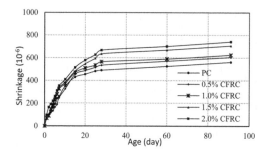

Figure 5. Drying shrinkage of carpet fiber reinforced concrete and plain concrete.

concrete with 0.5% and 1.5% of fiber content. Polypropylene fibers have been shown to be highly effective in controlling cracking in concrete through the crack bridging action. In general, carpet fibers reduced the number of cracks, total crack area, and maximum crack width. The fractured concrete illustrates that the plain concrete specimen immediately failed into two pieces without any primary cracks (Fig. 4). Similar trends have been demonstrated by Reis (2009), Altun et al. (2013) and Mazaheripour et al. (2011) in concrete containing polypropylene fiber.

### 3.4 Drying shrinkage

Experimental data on shrinkage of both types of concrete are presented in Figure 5. It can be seen that at early ages, there has been no significant difference in shrinkage development. However, with time a clear distinction can be observed in the development of shrinkage strain in concrete with carpet fiber. After 28 days the shrinkage values, for instance, were $490.44 \times 10^{-6}$, $534.66 \times 10^{-6}$, $565 \times 10^{-6}$, $635.16 \times 10^{-6}$ and $667.32 \times 10^{-6}$ for PC, 0.5% CFRC, 1.0% CFRC, 1.5% CFRC and 2.0% CFRC respectively. The higher shrinkage value in CFRC may be attributed to the higher porosity in the concrete mixtures as compared to that in plain concrete due to the balling effect for the addition of carpet fibers and voids produced by segregation of aggregates (Hsie et al. 2008, Zhou & Xiang 2011). Similar observations have been made by Wang, et al. (2000) using carpet fiber in concrete.

## 4 CONCLUSIONS

In this study, waste carpet fibers were incorporated at various volume fractions to study strength, modulus of elasticity and shrinkage behavior of concrete. It has been found that fiber, in general, reduced the compressive strength of concrete by 8% for 0.5% fiber content. However, this fiber has been

shown to enhance the flexural strength of concrete for a certain volume fraction of 0.5% by 18.5%. Although a relatively higher shrinkage strain has been recorded, the addition of waste carpet fiber has the potentials in converting the brittle nature of the concrete into a ductile material through improving the flexural strength of concrete.

## ACKNOWLEDGEMENT

The authors wish to acknowledge the technical support received from the staff of Structure and Materials laboratory of the Universiti Teknologi Malaysia (UTM) in conducting the experimental work. Special thanks are due to Mr. Henry Yee, the managing director of ENTEX Carpet Industries SND. BHD, Malaysia for making this research possible by providing fibers.

## REFERENCES

Altun, F. Tanrıöven, F. & Dirikgil, T. 2013. Experimental investigation of mechanical properties of hybrid fiber reinforced concrete samples and prediction of energy absorption capacity of beams by fuzzy-genetic model. *Construction and Building Materials*, 44: 565–574.

ASTM C150. 2005. *Standard Specification for Portland Cement.* ASTM International.

ASTM C157. 2008. *Standard Test Method for Length Change of Hardened Hydraulic-Cement Mortar and Concrete.* ASTM International.

ASTM C490. 2011. *Standard Practice for Use of Apparatus for the Determination of Length Change of Hardened Cement Paste, Mortar, and Concrete.* ASTM International.

BS EN 12390-3. 2009. *Testing hardened concrete. Compressive strength of test specimens.* British Standards Institution.

BS EN 12390-5. 2009. *Testing hardened concrete. Flexural strength of test specimens.* British Standards Institution.

Department of the Environment (DOE). 1992. The *design of normal concrete mixes.* Building Research Establishment, UK.

Hsie, M. Tu, C. & Song, P. 2008. Mechanical properties of polypropylene hybrid fiber-reinforced concrete. *Materials Science and Engineering A*, 494: 153–157.

Mazaheripour, H. Ghanbarpour, S. Mirmoradi, S. & Hosseinpour, I. 2011. The effect of polypropylene fibers on the properties of fresh and hardened lightweight self-compacting concrete. *Construction and Building Materials*, 25: 351–358.

Miraftab, M. 1999. *A novel application of pre-and post-consumer carpet wastes in concrete.* In CMRI Conference Papers, 1.

Neville, A.M. & Brooks, J.J. 2010. *Concrete Technology.* 2nd ed. Pearson.

Qian, C. & Stroeven, P. 2000. Development of hybrid polypropylene-steel fibre-reinforced concrete. *Cement and Concrete Research*, 30: 63–69.

Reis, J. 2009. Effect of textile on the mechanical properties of polymer concrete. *Materials Research*, 12: 63–67.

Schmidt, H. & Cieslak, M. 2008. Concrete with carpet recyclates: Suitability assessment by surface energy evaluation. *Waste Management*, 28: 1182–1187.

Song, P. Hwang, S. & Sheu, B. 2005. Strength properties of nylon-and polypropylene-fiber-reinforced concrete. *Cement and Concrete Research*, 35: 1546–1550.

Suna, Z. & Xu, Q. 2009. Microscopic, physical and mechanical analysis of polypropylene fiber reinforced concrete. *Materials Science and Engineering A*, 527: 198–204.

Vilkner, G. Meyer, C. & Shimanovich, S. 2004. *Properties of glass concrete containing recycled carpet fibers*. In 6th International RILEM Symposium on Fibre-Reinforced concretes. Varenna, Italy.

Wang, Y. 2013. *Fiber Recycling in the United States. School of Materials Science & Engineering,* Georgia Institute of Technology, Atlanta, Georgia.

Wang, Y. 1999. Utilization of recycled carpet waste fibers for reinforcement of concrete and soil. *Polymer-Plastics Technology and Engineering*, 38(3): 533–546.

Wang, Y. 2010. Fiber and Textile Waste Utilization. *Waste Biomass*, 1: 135–143.

Wang, Y. Wu, H. & Li, C. 2000. Concrete Reinforcement with Recycled Fibers. *Materials in Civil Engineering*, 12: 314–319.

Zhou, J. & Xiang, H. 2011. Research on Mechanical Properties of Recycled Fibers Concrete. *Applied Mechanics and Materials*, 94: 1184–1187.

*Advanced Materials, Structures and Mechanical Engineering – Kaloop (Ed.)*
© 2016 Taylor & Francis Group, London, ISBN 978-1-138-02793-0

# Constructions from wood for a power-effective and eco-friendly real estate

A.M. Krygina
*South-West State University, Kursk, Russia*

N.M. Krygina
*Moscow State University, Moscow, Russia*

ABSTRACT: In this article, conceptual questions of sustainable development of housing construction in Russia with a use of modern frame systems of wooden housing construction are considered. It is shown that under the conditions of strengthening of anthropogenous influence and increase in an imbalance between a production activity of the enterprises of an investment and construction complex and assimilatory opportunities of environment transition to construction of facilities of eco-residential real estate, technologies of "green" construction are necessary, assuming the creation of the comfortable and safe inhabited environment, rational use of natural resources and minimization of the negative impact on the nature at all stages of the life cycle, including a stage of exploitation of which it is the share to 80% of cumulative expenses. The main advantages of timber construction materials in comparison with other traditional materials and systems are considered.

## 1 INTRODUCTION

Providing an innovative sustainable development of housing construction is a key strategic task, which generally contributes to the social level and quality of life of the population, safety and well-being of citizens, and also effective functioning of economy of the state. In the Russian Federation, the problem of providing citizens with affordable and comfortable housing is still particularly acute; thus, security with living space on 1 people in the Russian Federation makes only 23 to 24 sq·m/1 person (for comparison in the countries of Europe and USA, this indicator is 40 to 70 sq·m/1 person) (Krygina 2013b).

Today, we observe universal tendencies of an aggravation of the contradictions connected with exhaustion of natural resources (and according to continuous increase in prices for energy carriers) and environmental pollution, including from activity of the construction enterprises. Therefore, the most important principle of a sustainable development of society, economy, in general, and construction branch is the implementation of the concept of the balanced development of social-and-ecological and social-and-economic system on the basis of the balance of requirements and opportunities of the enterprises builders and contract organizations of an investment and construction complex, balance of requirements and opportunities of the enterprises of building industry and reproduction opportunities of environment.

Therefore, the purpose of this article is to analyze perspectives of using wood structures for solving problems of a resource, energy and environmental performance of residential properties.

Development of housing construction has to proceed from co-measurement of androgenic impact on the nature and its resistance to this influence, ensuring first the minimization of consumption of natural resources followed by thermal and power.

Construction of buildings of the new type is represented. This forms the objects of eco-housing real estate, which is characterized by energy efficiency and environmental friendliness, meeting the following requirements:

– rational use of renewable resources (energy of the earth, water, and wind);
– minimization of the negative impact of objects of eco-real estate on the environment, as in the course of construction, and exploitation (since it is the share of this stage of the life cycle of the building to 80% of the general expenses) (Krygina & Grabovy 2014);
– safe utilization of a real estate object and providing comfortable conditions for the accommodation of people.

The total area of the buildings in the Russian Federation makes about 5 bln sq·m. About 40–45% of the developed thermal energy is spent for heating of residential buildings in Russia (Krygina 2013a). About 30% of non-renewable fuel and energy resources are spent for the development of thermal energy. The Russian system of housing consumes about 20% of electric energy and 45% of thermal energy. Comparatively, in the European Union countries, cumulative expenses of buildings on heating, ventilation, conditioning and lighting make no more than 25% of the developed energy. In European countries (e.g. Finland, whose climatic conditions are similar to those in Russia) the cost of heat sold in a year makes about 200 US dollars ($) per capita; in Russia, this indicator is in limits of 600–850$ (Krygina 2013a). Despite toughening of standards after the construction heating engineer, the specific expense of thermal energy in Russia, in particular, on residential buildings by 1.5–2 times exceeds similar standards of EU countries with similar climatic conditions.

The indicator of specific energy consumption of buildings in the states of the EU does not exceed 140 kW·h/sq·m in a year. In Russia, the actual average power consumption of heating systems and hot water supply of buildings makes 229 kW·h/sq·m in a year (350–400 kW·h/sq·m in a year in the first standard buildings of mass building achieved).

Timber constructions are eco-friendly and have high heat-insulating properties. Therefore, the technical and economic analysis of prospects of their use in housing construction is an important task.

## 2 ACTUAL DIRECTIONS OF ECO-FRIENDLY REAL ESTATE DEVELOPMENT

### 2.1 Analysis of the main materials and technologies

Until the end of the 20th century, the heat-saving problem in buildings in Russia was not much sharp because of the low cost of energy carriers and the use of the protecting designs of stone materials with low heat-insulating properties. Issues of heat-shielding of buildings were resolved at the expense of the large thickness of external walls (Krygina et al. 2013).

"Green" construction, "green" buildings are a practice of the construction and exploitation of buildings, the purpose of which is to decrease the level of consumption of energy and material (natural) resources simultaneously or to improve the quality of buildings, the comfort of their internal environment and the preservation of ecological safety (Krygina 2013a, Krygina & Grabovy 2014).

The main objectives are supposed to be solved by means of "green" construction:

– reduction cumulative negative impact of the construction activity on human health and the environment reached due to the application of innovative technologies and organizational decisions;
– creation of the market of affordable housing for various segments of the population; the increase in productivity of organizational and production construction system and the creation of new workplaces in the sphere of construction production;
– achievement of the efficiency of investments in eco-steady construction and decrease in charges of objects of housing real estate;
– the decrease in the loads of regional power networks and increase in the reliability of their work.

At the present stage of development of low housing construction in the Russian Federation, the main materials used for construction are as follows: brick and stone materials (53% and 15%, respectively); timber materials (22%); and other materials (10%) (Krygina & Grabovy 2014, Krygina 2013a).

At the same time, traditional construction technologies are the most expensive (Table 1).

Thus, from the perspective of eco-friendliness and profitability of the construction, the construction material of most concern is the use of wood and materials for its processing (Zwerger 2012, Weber et al. 2004).

The application of new technologies (low housing construction with the use of materials from cellular concrete, technologies of thermostructural panels, and the modernized technologies of large-panel housing construction with the application of elements of a fixed timbering from expanded polystyrene) allow to carry out the construction work at all seasons of the year. Thus, costs of production, transportation, installation of the bearing and

Table 1. The prime cost of 1 sq·m of premises.

| No. | Construction material | The average prime cost of 1 sq·m of premises (without finishing and external networks), $ |
|-----|-----------------------|-----------------------------------------------------------------------------------------|
| 1 | Brick | 560 |
| 2 | Timber | 440 |
| 3 | Construction materials from wood for deep processing | 280 |
| 4 | Expanded polystyrene | 248 |
| 5 | Cellular concrete | 224 |

protecting designs and their subsequent operation constructions are lower in comparison with technologies involving the use of traditional materials.

We describe the following classification of groups of projects of low construction by technology and prime cost.

1. Stone houses: brick; blocks; fixed timbering; monolithic and modular designs. The range of prime cost is 1 sq·m. of the house: the average and minimum value is 417$/sq·m and the maximum value is 583$/sq·m.
2. Frame houses: wooden, metal framework; volume and modular buildings; SIP panels (380...700$/sq·m).
3. Wooden houses: glued, profiled bar; the rounded log (270...614$/sq·m).

Thus, heat losses on 1 sq·m of the panel and brick house in terms of conditional fuel in Russia are 6 times higher in comparison with EU countries (12 kg/year and 2 kg/year, respectively).

Comparative characteristics of the construction of various types of low houses are provided in Table 2.

When designing buildings in Russia, it is necessary to consider severe climatic conditions of a low temperature for nearly 8 months in a year. For ensuring thermal efficiency of stone walls, their minimum thickness (with a heater) for the European territory of Russia has needs to be not less than 510 mm.

Timber as a constructional material also has shortcomings, including a low fire resistance and susceptibility of biological corrosion. However, the main advantage of designs and materials based on timber is its high ecological qualities. Modern technologies of fire protection and bioprotection of wooden designs allow to level successfully the wood to remain as a construction material (Table 3).

## 2.2 Comparative technical analysis of the main construction materials

In summary, we give the main merits and demerits of stone and timber construction materials for comparison.

### 2.2.1 Brick walls
#### 2.2.1.1 Advantages
Walls from a brick are very strong, refractory, durable; allow to apply ferroconcrete plates of overlapping; allow to build walls of difficult configurations, to spread decorative elements of a facade.

#### 2.2.1.2 Shortcomings
Walls possess high heat conductivity; absorb moisture at the expense of a capillary suction and freeze through in the winter that leads (at seasonal exploitation) to destruction (Fig. 1); though rather heavy, walls do not suffer deformations. In this case, the powerful base is required. For providing thermal insulation, brick walls have big sizes; after laying of walls prior to their finishing, they have to pass a year, walls before finishing have to "settle"; the main shortcoming is the high cost.

Table 2. Comparative characteristics of different types of low housing construction with a total area of 200 sq·m.

| House type | Substructure type | Laboriousness at plant/ building site | Installation term at an existence of the base | Season for constructing | Shrinkage of walls | Terms finishings of the house | The comfortable accommodation since the beginning of constructions |
|---|---|---|---|---|---|---|---|
| Brick | Tape Monolithic | 0/100% | 3–4 months | Spring, autumn | 2 years | 6 months | 1–2 years |
| Timbered | Columnar Monolithic Tape | 60%/40% | 1–2 months | Within a year | 2 years | 3 months | 1 year |
| Glued bar | Columnar Monolithic Tape | 70%/30% | 2–3 weeks | Within a year | No | 1 month | 2–3 month |
| Gas-concrete | Tape Monolithic | 0/100% | 2–3 months | Spring, autumn | 1 year | 6 months | 1–2 years |
| Wooden panel and frame | Columnar Monolithic Tape (facilitated) | 80%/20% | 1 week | Within a year | No | 1 month | 1–1.5 month |
| Wooden modular | Columnar Monolithic Tape (facilitated) | 95%/5% | 2 days | Within a year | No | Factory finishing | 1–2 weeks |

Table 3. The advantages of modern wooden structures.

| Shortcomings of wooden houses of the past | Advantages of modern wooden houses |
|---|---|
| Limited architectural concepts | Wide choice of standard projects. Development individual taking into account the desire of the customer |
| Fragility | Durability. Thanks to unique technologies of the construction the term of operation of wooden houses makes 100–150 years |
| The possibility of fire | The fire safety and moisture resistance reached as a result of the use of new materials for wood finishing |
| A lot of waste materials | Almost waste-free production brought to perfection by long centuries of the construction of wooden houses |
| High labor costs, low operational properties at the high level of humidity | The use of new equipment which allows to lower labor costs significantly |

Figure 1. Destruction of a bricklaying during the soaking and influence of negative temperatures (so-called aeration).

### 2.2.2 Light concrete walls (foam concrete, polystyrene concrete)

#### 2.2.2.1 Advantages

Concerning refractory, walls are durable; rather small sizes of blocks and ease of their processing allow to build of them walls of difficult configurations; thickness of such walls can be twice less than bricks; the laying of walls from blocks is much simpler and cheaper than bricklaying; because of the small density of cellular concrete, all design of walls turns out 2–3 times easier that simplifies a base design.

#### 2.2.2.2 Shortcomings

Owing to high porosity of a product, there is increased moisture absorption; therefore, the building facade after the end of the construction of walls needs to be covered with the structures, creating a moisture protective vapor-permeable film on a surface; walls do not suffer deformations; prior to their finishing, walls have to "settle"; at a deposit, cracks can be formed; and the costs are rather expensive.

### 2.2.3 Wooden walls (bar, log)

#### 2.2.3.1 Advantages

Walls from a tree possess low heat conductivity; therefore, if in the winter, the house was not heated, it is possible to warm it up to comfortable conditions in some hours; create a healthy microclimate in the house; bring excess humidity out of the room; is rather easy and steady against deformations; it is possible to build on the simple column base; maintain a large number of "freezing-defrosting" cycles; and the term of their service is about 100 years.

#### 2.2.3.2 Shortcomings

Walls easily ignite and are subject to the action of insect wreckers and rotting; after the end of the cabin of wooden walls prior to their finishing, walls have to pass not less than a year (a deposit to 10%); during drying, they are deformed and burst. A caulking iron of bar-shaped walls is a difficult and expensive procedure.

### 2.2.4 Frame walls

#### 2.2.4.1 Advantages

Walls possess low heat conductivity; the lungs from all aspects are also steady against deformations; it is possible to build "floating columns" on the column base or the base; expenses of means, forces and time for a construction of frame walls are minimum; before finishing it is not necessary to wait for house "rainfall".

#### 2.2.4.2 Shortcomings

Walls easily ignite and are subject to the action of insect wreckers and rotting; the design of walls does not give confidence of capital construction; the increase in the sizes of the house leads to considerable complication of a framework and decrease in reliability; it is expedient to apply at construction of the dachas intended for seasonal or year-round operation.

## 3 CONCLUSIONS

1. The use of wooden frame-panel and modular systems allows to reduce terms of input of residential buildings in exploitation at the expense of the possibility of construction at all seasons and reduction of terms of finishing work.
2. Cost reduction of 1 sq·m of housing with use of timber constructions makes 13–28% in comparison with traditional materials (e.g. stone, concrete).
3. Thus, from a technological and economic perspective, the use of frame systems of residential buildings on the basis of materials from wood, as the most available, technologically effective and ecologically safe is considered to be the most effective at present.

## REFERENCES

Krygina, A.M. 2013a. *Innovative housing construction: organizational-technical solutions.* Kusrk: SWSU.

Krygina, A.M. 2013b. Prospects of development of regional social housing policy. *Fundamental research,* 4(4): 812–817.

Krygina, A.M. & Grabovy, P.G. 2014. *Innovative development of low housing real estate.* Moscow: ASV.

Krygina, A.M. Kretova V.M. & Savenkova A.V. 2013. On the determination of the initial composition of the interior finishing of the Voskresenskii Temple of the Znamensky male Monastery in Kursk (Russian Federation). *Applied Mechanics and Materials.* 420: 281–287.

Weber, J. Hugues, T. & Steiger L. 2004. Timber *Construction: Details, Products, Case Studies (Detail Praxis).* Institute fur Internationale Architektur-Dokumentation GmbH 8 Co. KG.

Zwerger, K. 2012. *Wood and Wood Joints: Building Traditions of Europe,* Japan and China. Basel: Birkhäuser/.

*Advanced Materials, Structures and Mechanical Engineering – Kaloop (Ed.)*
© 2016 Taylor & Francis Group, London, ISBN 978-1-138-02793-0

# Study of the numerical calculation model of the sporting rifle

C. Xu & G.Q. Liu
*School of Mechanical Engineering, Nanjing University of Science and Technology, Nanjing, China*

ABSTRACT: The numerical model of the interaction between the sporting rifle bullet and the barrel was established to study the vibration rule of the muzzle, which is influenced by the change of tube structure parameters and bullet weight deviation. In order to obtain powder impetus and provide accurate payload data for the model, the bore pressure measurement experiment was conducted using the pressure test system. The muzzle vibration rule at the vertical direction during the launching process was captured using high-speed photography equipment, and the correctness of model is verified by comparing measured data and numerical results. The establishment of the numerical model provides a theoretical basis for researching the effect of parameters on muzzle vibration and numerically predicted results have a certain significance to engineering practices.

## 1 INTRODUCTION

The sporting rifle is designed for shooting competition. Compared with an ordinary rifle, it has the advantage of lower muzzle velocity, smaller caliber and shorter range, but the appearance and launching theory does not make much difference. Usually, it adopts to a single-short firing structure, and all lead bullets are designed with a raised belt. But their design principle is totally different. Sporting rifle mostly follows the principle of Gun Adapt to Person, which improves the mutual performance between the human and the rifle. Consequently, the universality of a sporting rifle is worse than an ordinary rifle, whose price is high and production capacity is low. The sporting rifle has a very high firing accuracy so that stricter requirements are needed for the bullet's design and manufacturing.

Currently, the research results of the sporting rifle are very few at home and abroad. Relevant works have been carried out by drawing on other barrel weapons' design experience; however, the systematic method of design and analysis still has not been achieved. The interaction between the artillery shell and the barrel was studied in the literature (Ge 2007), pointing out that forcing cone has a great influence on interior ballistic performance and firing accuracy, which is an important structure to adjust shell location and extrude bearing band forming score. The influence of structural parameters on launching a process of terminal guidance artillery shell has been studied in the literature (Kang & Wu 1999) and some significant conclusions were obtained. The radial magnitude of the interference of the bearing band plays a necessary role in closed powder gas and rotating artillery shell. When the radial magnitude of interference increases, the rotating ratio and rotational speed of the artillery shell will increase; meanwhile, the barrel life will be shortened because the abrasion gets worse. The influence of barrel configuration on muzzle vibration of a sniper rifle has been studied in the literature (Qi & Shen 2013). That study has shown that it is an efficient way to increase natural frequency by changing the configuration of a barrel.

In conclusion, the above research has a limited applicability; therefore, it cannot be directly applied to a sporting rifle, as there are many differences in performance parameter between the sporting rifle and other barrel weapons. Besides, some performance parameter has not been comprehensively verified through experiments, so research is urgently needed to be carried out in both theory and experiment.

## 2 CHAMBER PRESSURE EXPERIMENT

### 2.1 Experiment principle

The electrical measuring method was adopted to test chamber pressure in the internal ballistic period, which is very commonly used in the ballistic performance experiment. The characteristics of this method are of high precision, which can be continuously measured and has excellent dynamic features. This electrical measurement pressure system is composed of four parts: pressure sensor, charge amplifier, NI data processing system and peripheral. The flow diagram is shown in Figure 1.

In order to ensure the signal without distortion and reduce the pressure of storage, the equipment model and test parameters were chosen as follows. The model of a piezoelectric pressure sensor is

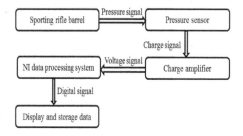

Figure 1. Test system diagram of bore pressure.

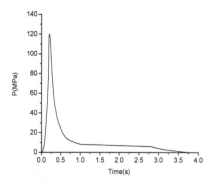

Figure 2. Bore pressure curve of a sporting rifle.

Kistler 6215, whose maximum range is 600 MPa and sensitivity rate is 14900 pC/MPa. While the model of the charge amplifier was chosen as Kistler 5018 A0001, the sample frequency of the NI data processing system was set to 100 KHz.

Under the chamber pressure, the sensor produces and outputs the charge that will be amplified and converted into a voltage signal by the charge amplifier according to the proportional relation theory. Finally, the NI data processing system converts the voltage signal into a digital signal for display and storage.

## 2.2 Experimental result and analysis

Sporting rifle's consistency of the chamber pressure is better than that of ordinary rifles. Typical pressure changing feature obtained from the experiment is shown in Figure 2. There are many differences between the sporting rifle and ordinary rifles: the prominent change feature, as shown in the figure, is the sharp curve, which rises sharply but declines slowly.

As Figure 2 shows, the internal ballistic period continues for 2.8 ms, which can be divided into three stages. During the first stage, chamber pressure rises from zero to a maximum of 119.6 MPa rapidly within 0.21 ms, which only takes half time compared with a regular rifle. Then, in the second stage, chamber pressure drops to 8 Mpa within

only 0.81 ms. After a sharp drop, chamber pressure slows down its steps at the end. While the third stage is the stage of quasi-static balance, a decline of only 2 Mpa is observed at the end of the internal ballistic period. The chamber pressure is twentieth of the maximum level when the bullet leaves the barrel. Change regularity of chamber pressure in this stage is different from that of regular rifles, which makes the sporting rifle distinctive.

## 3 MUZZLE VIBRATION EXPERIMENT

### 3.1 Experiment principle

Muzzle vibration is an important influence factor on firing accuracy; its formation and development is influenced by many factors. It is difficult to accurately obtain by the conventional test method because the vibration frequency is very high and the amplitude is very small. At present, the most direct and effective method is the use of a high-speed camera, which takes pictures of muzzle vibration and then the use of special software to extract vibration displacement in one direction from the picture. But displacement accuracy is bound up with a performance of high-speed camera.

In order to obtain muzzle vibration regularity of the sporting rifle during the internal ballistic period, a muzzle vibration experiment was carried out in this paper. Sporting rifle launching system is composed of barrel, holding device, the launcher and fixed base, as shown in Figure 3. High-speed photography system is composed of a high-speed camera, the external light source and PC terminal. The experiment object diagram is shown in Figure 4.

### 3.2 Test result and analysis

Muzzle vibration has the characteristics of regularity and repeatability under the same launch conditions. Displacement of muzzle vibration was obtained by a high-speed photography system, as shown in Figure 5, where dark spots indicate the symbol of the end of the internal ballistic period.

As Figure 5 shows, the internal ballistic period continues for 2.8 ms and the muzzle is in a rising trend at the end. At the time of 4.22 ms, the muzzle rises to a maximum value, i.e. thirty-four times of the first maximum value. In the mass, the maximum and the cycle of muzzle vibration both are in the increasing trend.

Muzzle vibration is the macroscopic behavior of forced response, which is motivated by time-varying dynamic load. Stress waves motivate by dynamic load such as high-velocity motion of the bullet and chamber pressure will propagate in the barrel. Stress wave reflection in the muzzle as the unconstrained face of the cantilever barrel will be strengthened or weakened due to the interaction

Figure 3.  Sporting rifle launching system.

Figure 4.  Test materials picture of muzzle vibration.

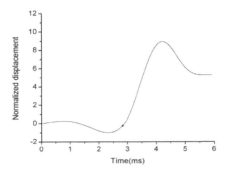

Figure 5.  Displacement curve of muzzle center.

of reflection waves with incident waves, eventually forming macroscopic vibration. The regularity of forced response is bound with natural frequency, holding condition and dynamic load variation.

# 4  NUMERICAL CALCULATION MODEL

## 4.1  *Finite element model and material parameters*

The barrel is meshed into 368352 elements and the bullet is meshed into 258140 elements. Bearing

band will be extruded and forming score during an engraving period, as a result the stress concentration will also be very serious; therefore, it needs to increase the density of the element both axially and radially, as shown in Figure 6.

Preload shaft sleeve is meshed into 8880 elements and fixed floor is meshed into 1836 elements. Most of the elements are hexahedral reduced integral elements. The finite element model of the launching system is shown in Figure 7. The Z axis is the direction of the barrel axial, the Y axis is the vertical direction and according to the right-hand rule, the X axis is determined. The related mechanical performance of the parameters is presented in Table 1.

## 4.2  *Contact*

We define the outer surface of a bullet with an inner surface of the barrel as contact pairs, similarly the outer surface of a bullet with the inner surface of the preload shaft sleeve. Then, we take the surface-to-surface form as the contact property for contact pairs, and define the outer surface of a bullet with the front and back surface of the bearing band as a self-contact. Contact property is effective from the initial step. Contact control algorithm is developed by using the penalty function method.

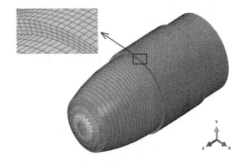

Figure 6.  Finite element model of a bullet.

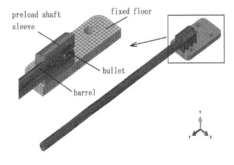

Figure 7.  Finite element model of the sporting rifle launching system.

Table 1. Parameter of mechanical properties.

| Material | Density (kg/m³) | Young's modulus | Poisson's ratio | σs (Mpa) | σb (Mpa) |
|---|---|---|---|---|---|
| Steel | $7.85 \times 10^3$ | 206000 | 0.29 | 810 | – |
| Lead | $11.34 \times 10^3$ | 17000 | 0.42 | 11.3 | 16.5 |

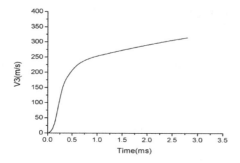

Figure 8. Velocity curve of the bullet.

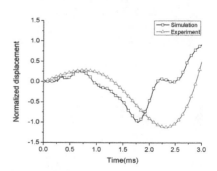

Figure 9. Displacement curve of muzzle vibration.

### 4.3 Mixed operation

In this paper, the advantages of two algorithms have combined to solve the launch state of the sniper rifle under the action of gravity. We import the stress and strain state results calculated by the implicit solver to the explicit solver and then reset the initial conditions, such as constraint situation, contact property and initial field. The interaction state of the bullet and bend barrel was obtained through the mixed calculation.

### 4.4 The result of numerical calculation

The velocity of the bullet and muzzle vibration result of numerical calculation is shown in Figure 8 and Figure 9, where the center of the muzzle is used as the zero point.

Muzzle velocity was designed to 320 m/s for this type of the sporting rifle, and the designed value was verified by the experiment of velocity measurement. As shown in Figure 8, muzzle velocity was found to be 317.6 m/s, as calculated by the numerical model. The error was about 0.75%, compared with the experiment. We prove that the numerical calculation model is valid. Displacement curve of muzzle vibration calculated by the numerical model is shown in Figure 9. It can be considered as sinusoid with changing amplitude and cycle. The time of internal ballistic was 2.8 ms and the error was about 0.7%. The error of the maximum was 11.9% and that of the minimum was 1.4%.

## 5 CONCLUSIONS

In order to research the regularity of muzzle vibration during the launching process, the numerical model of the interaction between the bullet and the barrel was established, which is based on the nonlinear finite element method. Chamber pressure experiment and muzzle vibration experiment was carried out to obtain changing regularity of chamber pressure for this type of the sporting rifle. There is about 0.75% error between the velocity of the bullet obtained from the simulation and experimental results, while vibration displacement of muzzle is almost equal to the experimental results. The numerical model is proved to be valid and correct.

## REFERENCES

Ge, J.L. 2007. *Study on Nonlinear Dynamics Simulation for Vehicle-Mounted Howitzers by FEM.* Nanjing: Nanjing University of Science and Technology.

Kang, X.Z. & Wu, S.L. 1999. *Gun System Dynamics.* Beijing: National Defense Industry Press.

Qi, X. & Shen, W.J. 2013. Analysis of Dynamic Characteristic of Sniping Rifle Barrel. *Ordnance Industry Automation*, 32(11): 27–30.

Sun, H.Y. & Ma, J.S. 2012. Study on Influence of Bore Structure on Gun's Interior Ballistic Performances. *Acta Armamentarii*, 33(6): 669–675.

Wang, C.M. & Gao, N.T. 1990. *Automatic Weapons Ballistics.* Beijing: National Defense Industry Press.

Zhang, Z.H. 2013. *Adaptive Technology and Coupling Mechanism between Tube and Terminal Guided Projectile.* Nanjing: Nanjing University of Science and Technology.

*Advanced Materials, Structures and Mechanical Engineering – Kaloop (Ed.)*
© *2016 Taylor & Francis Group, London, ISBN 978-1-138-02793-0*

# Finite element parameterized modeling of projectiles penetrating gelatin and the development of a simulation system

S.S. Liu & C. Xu
*School of Mechanical Engineering, NUST, Nanjing, Jiangsu, China*

A.J. Chen
*School of Science, NUST, Nanjing, Jiangsu, China*

ABSTRACT: For completing the simulation analysis of projectiles penetrating gelatin conveniently and efficiently, methods of finite element parametric modeling were studied. A simulation system was designed based on the parametric language APDL provided by ANSYS and the general programming tools VB. The system can be used to implement automatic modeling without the need to consider the details of the modeling process. The accuracy of simulated results and practicability of this system is proved by the contrast between the simulation results and texting data of different kinds of projectiles penetrating gelatin. In addition, for researching the effect of the initial angle of incidence on interaction, the system was used to simulate the process of a bullet penetrating gelatin with different initial angles of incidence. The realization of the finite element parametric modeling of the projectiles and gelatin improves the efficiency of the simulation, and the system reduces the difficulty in the simulation greatly.

## 1 INTRODUCTION

In recent years, numerical simulation has development enormously and has been applied in the study of the penetrating problem. The existing commercial software can be used to represent the penetration process, provide comprehensive data and reveal the mechanism of penetration. Among them, AN-SYS/LSDYNA is a three-dimensional nonlinear finite element software in the explicit dynamic analysis, which can be used to simulate various kinds of complex nonlinear dynamic process. And it has become the main software in the analysis of penetration problems. Compared with experiments, domestic and foreign scholars have achieved remarkable results on the penetration of projectiles into gelatin used in the method of numerical simulation. And some scholars have used the ANSYS/LSDYNA to simulate the process of different projectiles penetrating gelatin, such as rifle bullet and spherical fragment (Luo et al. 2012, Wen et al. 2013, Wen et al. 2012, Wen et al. 2013).

In order to understand the effect of the parameters of the projectiles, such as impact velocity, mass, angle of incidence and geometries, finite element models need to be established or modified over and over again, which is inefficient and has a poor performance of engineering appliance, and there is a high need for researchers. From this, it is obvious that the numerical simulation method of artificial operation cannot meet the demands of damage effect and evaluation of projectiles.

For completing the simulation analysis of projectiles penetrating gelatin conveniently and efficiently, methods for studying the finite element parametric modeling were investigated. A simulation system was designed based on the parametric language APDL in ANSYS software and the general programming tools VB. The system can be used to implement automatic modeling without the need to consider the details of the modeling process. The accuracy and practicability of this system have been proved. For researching the effect of the initial angle of incidence on interaction, the system was used to simulate the process of a bullet penetrating gelatin with different initial angles of incidence.

## 2 STUDY OF THE FINITE ELEMENT PARAMETERIZED MODELING

Nowadays, the parameterized technology has become an efficient means in the initial phases of product design and modification. It is required to store the whole established process of finite element models in the parametric modeling method, so that the FEM can be established all at once without considering the details. It is convenient for researchers to update the FEM repeatedly by modifying required parameters rather than to establish CAD models, select material models, plot the grid, and set the parameters step by step, which lightens workload and improves the working efficiency considerably.

Though ANSYS is a powerful software, its versatility makes it impossible to analyze specific problems in the specific technical method. In the face of the distinction between different disciplines, three conventional secondary development tools, namely UIDL, APDL and UPFs, are provided by ANSYS, which are the base of the finite element parametric modeling for the design and analysis of specific products. APDL is a kind of scripting language used to automatically complete the regular and parameterized finite element modeling, but its interaction is so weak that it is difficult to control the process of analysis and develop a friendly interface. The method of combining another programming language with APDL is widely used to develop a simulation system for the analysis of the specific problem. The system works by using interfaces to input or modify the parameters relating to geometry and initial condition at the impact point, compiling analysis files with parameterized language (APDL), and calling ANSYS to implement automatic modeling and analysis.

## 3 PARAMETERIZED MODELING OF PROJECTILES AND GELATIN

There are two main kinds of projectiles used in the experiment in wound ballistic, fragment and bullet. In the simulation, spherical balls are used to replace fragments for the implementation of simulations. A 7.62 mm rifle bullet and gelatin are selected as an example to explain the process of finite element parameterized modeling and simulation of projectiles penetrating gelatin by the simulation system, including the selection of the material model, establishment of 3-D geometries and contact definition.

### 3.1 Selection of the material model

Based on the results of the current research and experiments (Johnson & Cook 1983, Jeng et al. 2007, Ensen et al. 2008, Nestor 2008), we adopt the MAT_JOHNSON_COOK model for the bullet and MAT_RIGID model for the spherical projectiles. The former is used for problems where the strain varies over a large range and adiabatic temperature increases due to plastic heating that leads to material softening, and can be used as an equation-of-state (EOS_GRUNEISEN) to describe its mechanical response. The latter provides a convenient way of turning one or more parts comprised of beams, shells, or solid elements into a rigid body. And what's more, gelatin is modeled as an elastic_plastic material (MAT_ELASTIC_

PLASTIC_HYDRO) with a polynomial equation of state (EOS_ELASTIC_PLASTIC_HYDRO) to describe its hydrodynamic response. The stress-strain behavior of the gelatin is defined by the material model in the low pressure and pressure-volume behavior by the equation of state under the high pressure. The MP command is used to define a linear material property as a constant, and the TB command is used to activate a data table for nonlinear material properties in ANSYS.

### 3.2 Establishment of 3-D geometries and contact definition

There are many dimensions used in the modeling of projectiles, and some of them depend on an empirical formula or are not interrelated, especially the dimensions of a bullet. In order to realize the implementation of automatic parameterized modeling conveniently, the mathematical relationship among various dimensions of a bullet should be summarized and key parameters of concern need to be extracted based on the analysis of the structure of bullets, which are helpful for the modification of bullet geometries during the numerical simulation with different values of the parameter. In order to achieve good accuracy of numerical simulation, gelatin has small elements in the region encompassing the impact area where it forms the temporary cavity, and to reduce the amount of mesh, the element size increases as one move away from this region. The finite element model of a 7.62 mm bullet consisting of three parts (jacket, lead set and steel core) and a 300 mm × 300 mm × 300 mm gelatin (10% at 4°C) block is established by APDL command flow, as shown in Figure 1.

Contact definitions of the interaction between the bullet and the gelatin, among the parts of

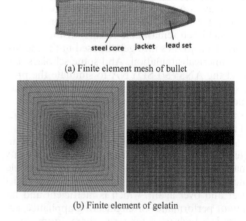

steel core    jacket    lead set

(a) Finite element mesh of bullet

(b) Finite element of gelatin

Figure 1. Finite element mesh of a bullet and gelatin.

the bullet, are ERODING_SURFACE_TO_SURFACE and CONTACT_AUTOMATIC_SINGLE_SURFACE, respectively, and both can be defined by the APDL command EDCGEN. The viscous hourglass control algorithm with the hourglass coefficient of 0.01 is employed to control the influence of hourglass deformation caused by the large deformation of elements on the reliability of numerical results. Part of the APDL command flow used for contact definition is shown in Figure 2.

### 3.3 Design and development of the parameterized simulation system

Based on the study of the method of finite element parametric modeling, a simulation system is developed by using VB and APDL. Part of the interface is shown in Figure 3.

The VB code used to make calls to ANSYS and LSPREPOST is shown below. Table 1 lists the meaning of VB code number.

```
!!Defien contact surfaces

EDPART,CREATE      !!part: 1-jacket: 2-lead set: 3-steel core: 4-gelatin
EDASMP,ADD,9,1,2,3          !!parts
EDCGEN,ESTS,9,4,0,0,0,0,0,0,1,0,,,,,6
EDCGEN,ASSC,9,,
EDCONTACT,2,
```

Figure 2. Part of the APDL command flow used for contact definition.

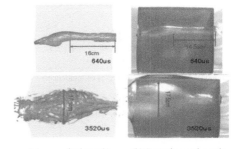

Figure 3. The interface of the simulation system for the analysis of spherical projectiles penetrating gelatin.

Table 1. Meaning of some VB code number.

| VB code | Meaning |
| --- | --- |
| ansyspath | Path of ANSYS |
| Lspath | Path of Lsprepost.exe |
| ane3flds | Product feature code of ANSYS |
| Lsprepost | Product feature code of LS-PrePost |
| outpath | Working path |
| outputfile | The output file |

Dim X

X = Shell(ansyspath + " -b -p ane3flds -i " + outpath + "\in.txt -o outputfile", 1)

Dim p

p = Shell(Lspath + " -b -p dynapp -i " + outpath + "\d3plot ", 1)

## 4  APPLICATION INSTANCES

### 4.1 Verification of the numerical results simulated by the system

In order to verify the accuracy of the system, the system was used to simulate the penetration of the 7.62 mm rifle bullet into gelatin, and the numerical results were compared with the corresponding experimental data.

In the verifying experiment, gelatin was impacted by the 7.62 mm bullet at 625 m/s and incidence angle of 1° using a rifle from the front face of the gelatin, and a high-speed camera was used to visualize the process of penetration. As shown in Figure 4 and Figure 5, the size and length of the

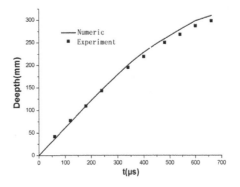

(a) numerical results　　(b) Experimental results

Figure 4. Numerical and experimental temporary cavity, shown after the moment of impact at 640us and 3520us penetration depths.

Figure 5. Comparison of computed and experimental time histories of the penetration depths.

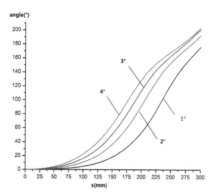

(a) Yaw-depth diagram for different angles of incidence. (1°, 2°, 3°, 4°)

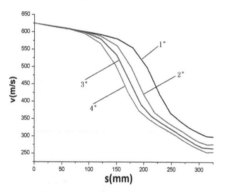

(b) Velocity-depth diagram for angles of incidence. (1°, 2°, 3°, 4°)

Figure 6. Numerical results with different angles of incidence (1°, 2°, 3°, 4°).

temporary cavities at different moments and the histories of the penetration depths between the numerical results and the corresponding experimental results are found to agree well, which proves that the results obtained by the system are reliable.

### 4.2 *Effects of incidence angle*

In order to study the effect of the initial angle of incidence on the interaction of the rifle bullet penetrating gelatin, the system was used to adopt different angles of incidence (1°, 2°, 3°, 4°) and the same impact speed (625 m/s) at the impact point to simulate the penetration. In combination with the numerical results plotted in Figure 6, there is little influence for the angle of incidence on the rule of rolling and velocity attenuation of rifle bullet, but with the increase in the incidence angle, the instable moment is ahead and the length of the narrow wound channel decreases.

## 5 CONCLUSIONS

We studied the method of the finite element parametric modeling and completed the finite element parametric modeling of kinds of projectiles penetrating gelatin. A simulation system was designed and developed based on the parametric language APDL, provided by ANSYS and the general programming tools VB. By comparing the numerical results simulated by the system with the corresponding experiment data, the system was proved to be accurate and practicable. In addition, the system was used to simulate the process of the rifle bullet penetrating gelatin at different initial angles of incidence to study the effect of the initial angle of incidence on interaction.

## REFERENCES

Ensen, M.N. Zimcik, D.G. & Lahoubi, M. et al. 2008. 07-CSME-66-Soft body impact simulation on composite structures. *Transactions of the Canadian Society for Mechanical Engineering*, 32(2): 283–296.

Jenq, S. Hsiao, F. & Lin, I. et al. 2007. Simulation of a rigid plate hit by a cylindrical hemi-spherical tip-ended soft impactor. *Computational Mater Science*, 39(3): 518–526.

Johnson, G.R. & Cook, W.H. 1983. *A constitutive model and data for metals subjected to large strains, high strain rates and high temperature*, Proceedings of the 7th International Symposium on Ballistics. Hague, Netherlands: IBC, 541–547.

Luo, S.M. & Huang, G.W. et al. 2012. Numerical Simulation Analysis of Spherical Projectiles Penetrating Gelatin. *Computer Simulation*, 11: 79–82.

Nestor, N. 2008. *Theoretical study of the motion of a rigid gyro-stabilized projectile into homogeneous dense media*, Proceedings of the 24th International Symposium of Ballistics. New Orleans.

Wen, Y. & Xu, C. et al. 2012. Numerical Simulation of Spherical Fragments Penetrating into Ballistic Gelatin at High Velocity. *Journal of Ballistics*, 24(3): 25–30.

Wen, Y. & Xu, C. et al. 2013. Numerical Simulation for the Penetration of Bullet on Gelatin Target, *Acta Armamentar-ii*, 1: 14–19.

Wen, Y. & Xu, C. et al. 2013. The impact of steel spheres on ballistic gelatin at moderate velocities. *International Journal of Impact Engineering*, 62: 142–151.

*Advanced Materials, Structures and Mechanical Engineering – Kaloop (Ed.)*
© *2016 Taylor & Francis Group, London, ISBN 978-1-138-02793-0*

# Dynamic responses of long-span cable-stayed bridges under multi-line railway trains

P. Yang, X.T. Si & K.J. Chen
*China Railway Eryuan Engineering Group Co. Ltd., Chengdu, Sichuan, P.R. China*

W.H. Guo
*School of Civil Engineering, Central South University, Changsha, Hunan, P.R. China*

ABSTRACT: This paper investigates the dynamic responses of a cable-stayed bridge under multi-line railway trains. A spatial grillage model is built for a long-span dual-deck cable-stayed railway bridge by use of finite element method with proper boundary conditions. The railway train is modeled as a mass-spring-damper system with 35 DOFs. Then a fully computerized method is employed to establish the integrated bridge-train interaction system taking into account of surface irregularities and creep phenomenon between rail tracks and wheel sets. Finally dynamic responses of the bridge are investigated by direct integration method under various moving trains. Results show that the first modes of both the bridge and train are in the lateral direction which may easily induce train-bridge coupling vibration. It is also learned that the vertical displacement of the bridge increases with the moving trains on the bridge and a bigger lateral displacement would appear when there is an unbalanced arrangement of moving trains running on the bridge.

## 1 INTRODUCTION

With the sustainable development of national economy, more and more railway networks are planned in the next decades. Long span bridges always play an important role in the railway network due to the geological locations. Besides, in order to fully make use of spaces and resources for long-term development, more and more long-span bridges are de-signed as multi-line bridges. Many studies have shown that with an increase of bridge span and moving loads on the bridge, the interaction between the bridge and moving trains become more of concern. Although many investigations have been done in the field, most of them focus on the single-line or double-line railway bridges (Au et al. 2001, Zeng & Guo 1999, Zhai & Xia 2011, Zhang & Xia 2013). Few studies have been published in terms of the multi-line railway train-bridge interaction system. Thus, it is desirable to carry out dynamic responses of long-span bridges under the multi-line railway trains.

This paper adopts a fully computerized method to form an integrated long-span bridge and moving trains interaction system. Based on this system, de-tailed investigations are carried out on the dynamic responses of the cable-stayed bridge under multi-line moving railway trains.

## 2 ESTABLISHMENT OF COUPLED LONG-SPAN BRIDGE-MOVING TRAIN INTERACTION SYSTEM

### 2.1 Modelling of a long-span bridge

A long-span bridge can be modelled by finite element method using proper elements based on a global coordinate system comprising X-, Y- and Z-axes respectively along the longitudinal, lateral and vertical directions of the bridge following the right-hand rule. The mass distribution and structural damping should also be properly introduced into the interaction system for dynamic analysis. Usually, consistent mass matrices are used for proper representation of the mass distribution. Rayleigh damping is adapted to account for the damping effect. Finally, the transient response of the bridge is governed essentially by (Bath 1996, Si et al. 2010, Zeng & Guo 1999, Zhai & Xia 2011):

$$[M_b]\{\ddot{v}_b\} + [C_b]\{\dot{v}_b\} + [K_b]\{v_b\} = \{P_{be}\} \qquad (1)$$

where, $[M_b]$, $[C_b]$ and $[K_b]$ are mass, damping and stiffness matrices of a bridge respectively, where $\{\ddot{v}_b\}$, $\{\dot{v}_b\}$ and $\{v\}$ are nodal dynamic acceleration, velocity and displacement vectors of the bridge respectively and $\{P_{be}\}$ is the corresponding force vector.

## 2.2 Modeling of a railway train

Usually, a railway train consists of a motor car and several trailer cars. Each car has four bogies and four sets of wheels. It can be modeled by a mass-spring-damper system with 35 DOFs as shown in Figure 1.

Bouncing $z$, yawing $y$, swaying $\varphi$, pitching $\theta$ and rolling $\phi$ motion of the car body, each bogie and each wheel set are incorporated into the model (Au et al. 2001, Iwnicki 2006, Zeng & Guo 1999, Zhai & Xia 2011, Zhang & Xia 2013).

Based on the idealized mass-spring-damper system, the equation for free vibration of the train can be easily written as follows (Iwnicki 2006, Zhai & Xia 2001):

$$[M_t]\{\ddot{v}_t\} + [C_t]\{\dot{v}_t\} + [K_t]\{v_t\} = \{P_t\} \qquad (2)$$

where, $[M_t]$, $[C_t]$ and $[K_t]$ are mass, damping and stiffness matrices of a bridge respectively, and $\{\ddot{v}_t\}$, $\{\dot{v}_t\}$ and $\{v_t\}$ are the acceleration, velocity and displacement vector of the train.

## 2.3 Governing equation of motion for holistic bridge-vehicle interaction system

Assuming that all relevant displacements remain small, the principle of virtual work can be used to derive the governing equation of the coupled train-bridge system. The equilibrium condition of the bridge under its self-weight without any vehicle is taken as the initial condition. Then the virtual work $\delta W$ done by inertial forces, damping forces, elastic forces, external loading, creep effect between wheel sets and rails, the gravitational and swaying effects of wheel sets can be obtained, thereby giving:

$$\delta W_I^t + \delta W_I^b + \delta W_D^t + \delta W_D^b + \delta W_S^t + \delta W_S^b$$
$$+ \delta W_{Ex}^b + \delta W_C^t + \delta W_G^t + \delta W_\varphi^t = 0 \qquad (3)$$

where, the subscripts $I, D, S, Ex, C, G$ and $\varphi$ stand for inertial, damping, elastic, external, creep, gravity and swaying effects respectively.

Finally, the governing equation of the coupled system can be obtained by means of the fully computerized approach (Iwnicki 2006) as

$$[M_{bt}]\{\ddot{v}_{bt}\} + [C_{bt}]\{\dot{v}_{bt}\} + [K_{bt}]\{v_{bt}\} = \{P_{bt}\} \qquad (4)$$

where, the subscript $[M_{bt}]$, $[C_{bt}]$, $[K_{bt}]$ and $\{P_{bt}\}$ stand for mass, damping, stiffness and interaction loading due to external load, creep effect between wheel sets and rails, the gravitational and swaying effects of wheel sets of the whole bridge-train interaction system respectively.

## 3 CASE STUDY

The New Baishatuo Yangtze River Bridge is a key project of the railway line between Chongqing and Guiyang in China. It is a steel truss cable-stayed bridge with a main span of 432 m and a total length of 918 m (81 + 162 + 432 + 162 + 81 m) as shown in Figure 2. It is the first dual-deck steel-truss cable-stayed railway bridge in the world. It has four lines of passenger trains running with a design speed of 200 km/h on the top level and two lines of freight trains running with a design speed of 120 km/h on the bottom level. This bridge is

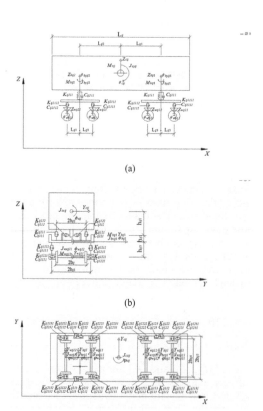

(a)

(b)

(c)

Figure 1. Car idealization of a railway train:(a) Elevation; (b) Front view; and (c) Plan.

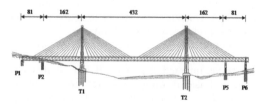

Figure 2. Elevation of New Baishatuo Yangtze River Bridge.

also the first steel-truss cable-stayed bridge with six railway lines and the longest cable-stayed railway bridge with the heaviest railway loading in the world. An orthotropic deck is used with a width of 24.5 m and a height of 15.2 m. The length of one segment is 13.5 m. The height of two towers is 171.2 m and 188.2 m respectively. There are total 64 pairs of stay cables.

### 3.1 Modelling of New Baishatuo Yangtze River Bridge

The spatial beam-plate model has a lot of finite elements especially plate elements, which increases the computational time greatly when it comes to dynamic analysis of bridge and railway trains coupled system. Therefore, it is desirable to simplify the spatial beam-plate model for dynamic analysis without losing the essential characteristics of coupled system. Grillage method (Hambly 1990) can be adopted to cater for such simplification.

The principle of grillage method is to use grillage beams with the equivalent area and stiffness to model the orthotropic deck. After proper modeling, the top level is simplified as four longitudinal beams and three transverse beams in one segment. The bottom level is simplified as two longitudinal beams and three transverse beams in one segment. The superimposed dead loads are only applied to longitudinal beams. Finally, the spatial grillage model as shown in Figure 3 is built by use of self-developed program MULIBRITIN.

### 3.2 Dynamic properties of New Baishatuo Yangtze River Bridge

By use of subspace iteration method, the dynamic properties of New Baishatuo Yangtze River Bridge were calculated, and their results are shown in Table 1 and the first two modes are shown in Figures 4–5.

It is learned from Table 1 that motor train has lower frequencies than that of trailers because they have the same stiffness while the motor train is heavier than the trailer. It also shows that the first modes of both the bridge and train are in a lateral direction. Therefore, attention should be paid to the lateral running stability and safety of the train

Figure 3. Grillage model of New Baishatuo Yangtze River Bridge.

Table 1. Dynamic properties of the trains and long-span cable stayed bridge.

| Mode | Cable-stayed bridge (Hz) | Motor car (Hz) | Trailer car (Hz) |
| --- | --- | --- | --- |
| 1 | 0.2610 | 0.3817 | 0.4096 |
| 2 | 0.4109 | 0.6816 | 0.7313 |
| 3 | 0.5494 | 0.7978 | 0.8560 |
| 4 | 0.5850 | 0.8517 | 0.9138 |
| 5 | 0.6458 | 0.8999 | 0.9656 |

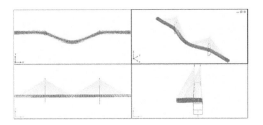

Figure 4. The first mode of New Baishatuo Yangtze River Bridge ($f_z = 0.2610$ Hz).

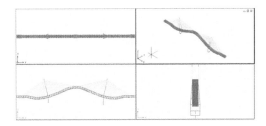

Figure 5. The second mode of New Baishatuo Yangtze River Bridge ($f_z = 0.4109$ Hz).

due to the coupling effect between the train and the bridge.

### 3.3 Load cases

The type of the moving train is assumed to be CRH2. Each line have a motorcade comprising four motor cars and four trailer cars with the running speed of 200 km/h and the running cases are: (1) Load Case (LC) 1: one motorcade runs on the Left Rail Line 1; (2) LC2: two motorcade run on the Left Rail Line 1 and the Right Rail Line 1 respectively and simultaneously; (3) LC3: two motorcade run on the Left Rail Lines 1 and 2 respectively and simultaneously; (4) LC4: two motorcades run on the Left Rail Lines 1 and 2 and one motorcades runs on the Right Rail Line 1, respectively and simultaneously; and (5) two motorcades run on the Left Rail Lines 1 and 2 and two motorcades run on the Right Rail Lines 1 and 2, respectively and simultaneously.

Table 2. Maximum acceleration of mid-span point.

| Load case | Lateral (cm/s²) | Vertical (cm/s²) |
|---|---|---|
| LC1 | 3.70 | 2.80 |
| LC2 | 5.50 | 7.70 |
| LC3 | 7.70 | 6.10 |
| LC4 | 9.80 | 11.90 |
| LC5 | 11.80 | 24.00 |

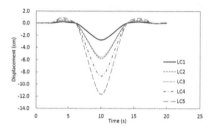

Figure 6. Vertical displacement of mid-span point under various Load Cases. (LCs).

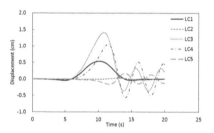

Figure 7. Lateral displacement of mid-span point under various Load Cases. (LCs).

### 3.4 Dynamic responses

Based on the aforementioned model and framework, the dynamic responses of the New Baishatuo Bridge were investigated by use of self-developed program MULIBRITIN. The maximum lateral and vertical accelerations at the mid-span of this bridge are shown in Table 2, and the lateral and vertical displacements at the mid-span of this bridge are shown in Figures 6–7. It is learned from Figure 6 that the vertical displacement increases with moving trains on the bridge. It is also indicated from Figure 7 that a bigger lateral displacement appears when there is an unbalanced arrangement of moving trains running on the bridge compared to other cases. The results also show that this bridge has good vertical and lateral stiffness.

## 4 CONCLUSIONS

The dynamic responses of long-span cable-stayed bridge under multi-line moving trains are investigated based a framework of multi-line railway bridge-train interaction system taking into account of surface irregularities and creep phenomenon between the rail tracks and wheel sets and the findings are as follows:

1. It shows that the first modes of both the bridge and train are in the lateral direction and the first mode has the most important contribution to structure vibration. Therefore, attention should be paid to the lateral running stability and safety of the train due to the coupling effect between the train and the bridge.
2. The vertical displacement of the bridge increases with the increase of moving trains on the bridge and a bigger lateral displacement would appear when there is an unbalanced arrangement of moving trains running on the bridge.
3. The study shows that the New Baishatuo Bridge has a good stiffness in both lateral and vertical direction.

## ACKNOWLEDGEMENT

This work was financially supported by Postdoctoral Research Fund (No. 14126191) of Faculty of Scientific Research of China Railway Eryuan Engineering Group Co. Ltd and China Postdoctoral Science Foundation Grant (No. 2013M540640).

## REFERENCES

Au, F.T.K. Wang, J.J. & Cheung, Y.K. 2001. Impact study of cable-stayed bridge under railway traffic using various models. *Journal of Sound and Vibration*, 240(3): 447–465.

Bathe, K.J. 1996. *Finite Element Procedures (2nd)*. New Jersey: Prentice Hall.

Guo, W.H. & Xu, Y.L. 2001. Fully computerized approach to study cable-stayed bridge-vehicle interaction. *Journal of Sound and Vibration*, 248: 745–761.

Hambly, E.C. 1990. *Bridge Deck Behaviour*. London: Spon Press.

Iwnicki, S. 2006. *Handbook of railway vehicle dynamics*. London: Taylor & Francis Group.

Si, X.T. Au, F.T.K. & Guo, W.H. 2010. *Riding comfort of a double-deck long-span bridge under both road vehicles and monorail trains*, In Second International Postgraduate Conference on Infrastructure and Environment. Hong Kong.

Zeng, Q.Y. & Guo, X.R. 1999. *Vibration Analysis of Train-Bridge Interaction System: Theory and Application*. Beijing: China Railway Publishing House.

Zhai, W.M. & Xia, H. 2011. *Train-Track-Bridge Dynamic Interaction: Theory and Engineering Application*. Beijing: Science Press.

Zhang, N. & Xia, H. 2013. Dynamic analysis of coupled vehicle-bridge system based on inter-system interaction method. *Computers and Structures*, 114: 26–34.

*Advanced Materials, Structures and Mechanical Engineering – Kaloop (Ed.)*
© *2016 Taylor & Francis Group, London, ISBN 978-1-138-02793-0*

# Multiple Tuned Mass Dampers for wind–excited tall building

Y.M. Kim & K.P. You
*Department of Architecture Engineering, Chonbuk National University, Jeonju, Korea*
*Long-Span Steel Frame System Research Center, Jeonju, Korea*

S.Y. Paek & B.H. Nam
*Department of Architecture Engineering, Chonbuk National University, Jeonju, Korea*

ABSTRACT: Multiple Tuned Mass Dampers (MTMD) is used for suppressing along-wind responses of a tall building. The performance of MTMD is compared with that of a single Tuned Mass Damper system (TMD). MTMD are located at the top floor of tall building and optimum parameters of TMD/MTMD for minimizing the variance response of the damped primary structure derived by Krenk, Igusa, and Jangid were used. Fluctuating along-wind load treated as a stationary Gaussian white noise process was simulated numerically using the along-wind load spectrum by Solari. The equation of motion is represented by state space formulation. The optimally designed MTMD system is more effective than that of a single TMD system.

## 1 INTRODUCTION

Tuned Mass Damper (TMD) is a classical vibration control device consisting of a mass, a spring and a damper supported at the primary vibrating system (Patil 2011). The original idea of TMD is from Frahm in 1909, who invented vibration control device called a vibration absorber using a spring supported mass without damper (Frahm,1909). That was effective when absorber's natural frequency was close to the excitation frequency. This shortcoming was improved by introducing damper in the spring supported mass by Ormondroyd & Den Hartog (Ormondroyd 1928). Later Den Hartog derived optimum tuning frequency and damping ratio for the undamped primary structure under harmonic load (Den Hartog 1985). While Den Hartog considered harmonic loading only Warburton & Ayroinde derived optimum parameters of TMD for the undamped primary system under harmonic and white noise random excitations (Ayroinde 1980). If the mass ratio is small and the primary structure's damping ratio is less than that value of TMD, Krenk derived the optimum parameters of TMD for the damped primary structure (Krenk 2008). And a number of TMDs have been installed in tall buildings to suppress wind-induced excitations of tall building (McNamara 1977, Housner et al. 1997).

The main disadvantage of a single TMD is its sensitivity to the error in identifying the natural frequency of the primary system and the damping ratio of TMD (Patil 2011). Therefore, MTMD having distributed natural frequencies around the natural frequency of system has been utilized (Xu 1992, Patil 2011). Iwanami & Seto had shown that two TMDs are more effective than single TMD (Iwanami 1984). Xu & Igusa carried out a parametric study of closely spaced natural frequencies of MTMD installed at a Single Degree of Freedom (SDOF) structure under a wide-band random excitation (Xu 1992). Yamaguchi & Harnpornchai performed a parametic studies on the effectiveness of MTMD compared to a single TMD under harmonic load (Yamaguchi, 1993). Kareem & Kline studied SDOF-MTMD structure under random loading and pointed out that MTMD have more advantages of portability and easy installation compared to a single TMD (Kareem 1995). Patil & Jangid studied the effectiveness of MTMD for the wind excited tall building (Patil 2011).

In this study, the performance of MTMD for suppressing wind-excitations of a tall building under along wind load is investigated. Fluctuating along-wind load was simulated numerically using the along-wind load spectra proposed by Solari (1993). Simulation procedure used in this study is taken from Shinozuka (1987). And using this simulated along-wind load, estimated along-wind responses of a tall building with MTMD/TMD and found out the effectiveness of MTMD compared to a single TMD for suppressing along-wind excitation of a tall building.

## 2   EQUATIONS OF MOTION

The dynamic analysis procedure can be simplified if the contribution of higher modes of the primary structure is ignored (Kareem 1995). Therefore, tall building-MTMD system can be modeled as the first-mode generalized SDOF/MTMD system as shown in Figure 1.

Dynamic equations of motion of the system can be written in matrix form of Equation (1).

$$M\ddot{X} + C\dot{X} + KX = f(t) \tag{1}$$

where,

$$X = \begin{bmatrix} x_s & x_1 & x_2 & \cdots & x_n \end{bmatrix} \tag{2}$$

$$M = \begin{bmatrix} m_s & 0 & 0 & 0 & \cdot & 0 \\ 0 & m_1 & 0 & 0 & \cdot & 0 \\ 0 & 0 & m_2 & 0 & \cdot & 0 \\ 0 & 0 & 0 & m_3 & \cdot & 0 \\ \cdot & \cdot & \cdot & \cdot & \cdot & 0 \\ 0 & 0 & 0 & 0 & 0 & m_n \end{bmatrix} \tag{3}$$

$$C = \begin{bmatrix} C_s \sum_{j=1}^{n} c_j & -c_1 & -c_2 & -c_3 & \cdot & -c_n \\ -c_1 & c_1 & 0 & 0 & \cdot & 0 \\ -c_2 & 0 & c_2 & 0 & \cdot & 0 \\ -c_3 & 0 & 0 & c_3 & \cdot & 0 \\ \cdot & \cdot & \cdot & \cdot & \cdot & 0 \\ -c_n & 0 & 0 & 0 & 0 & c_n \end{bmatrix} \tag{4}$$

$$K = \begin{bmatrix} K_s \sum_{j=1}^{n} k_j & -k_1 & -k_2 & -k_3 & \cdot & -k_n \\ -k_1 & k_1 & 0 & 0 & \cdot & 0 \\ -k_2 & 0 & k_2 & 0 & \cdot & 0 \\ -k_3 & 0 & 0 & k_3 & \cdot & 0 \\ \cdot & \cdot & \cdot & \cdot & \cdot & 0 \\ -k_n & 0 & 0 & 0 & 0 & k_n \end{bmatrix} \tag{5}$$

where, $M$ = the mass matrix of tall building with MTMD; $C$ = the dumping matrix of tall building with MTMD; $K$ = the stiffness matrix of tall building with MTMD; $f(t)$ = along-wind load; $m_s$ = the first-mode modal mass; $c_s$ = the first-mode modal damping constant; $k_s$ = the first-mode modal stiffness constant; $m_{i_,}$ = mass constant of each TMD of MTMD; $c_i$ = damping constant of each TMD of MTMD; $k_i$ = stiffness constant of each TMD of MTMD; $x_{s_,}$ = displacement of primary structure; $\dot{x}_s$ = velocity of primary structure; $\ddot{x}_s$ = acceleration of primary structure; $x_i$ = displacement of

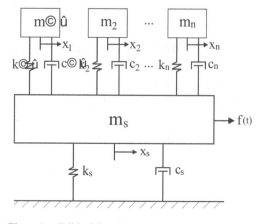

Figure 1.   Tall building-MTMD model.

TMD; $\dot{x}_i$ = velocity of TMD; and $\ddot{x}_i$ = acceleration of TMD.

Equation (1) can be written as a state-space representation as:

$$\dot{Z} = AZ + Hf(t) \tag{6}$$

where,

$$Z = \begin{bmatrix} x_s & x_1 & x_2 & \cdots & x_n & \dot{x}_s & \dot{x}_1 & \dot{x}_2 & \cdots & \dot{x}_n \end{bmatrix} \tag{7}$$

$$A = \begin{bmatrix} 0 & I \\ -M^{-1}k & -M^{-1}C \end{bmatrix} \tag{8}$$

$$H = \begin{bmatrix} 0 & M^{-1} \end{bmatrix}^T \tag{9}$$

$Z$ = state-space vector of primary structure with MTMD system; $A$ = system matrix; $H$ = the location vector of $f(t)$.

## 3   OPTIMUM PARAMETERS OF TMD

Krenk derived the optimum parameters of TMD for the damped primary structure when the mass ratio is small and the primary system's damping ratio is less than that value of TMD (Krenk 2008).

$$f_{opt} = \frac{1}{1+\mu} \tag{10}$$

$$\xi_{opt} = \frac{\sqrt{\mu}}{2} \tag{11}$$

where, $f_{opt}$ = optimum tuning frequency; $\xi_{opt}$ = optimum damping ratio; and $\mu$ = mass ratio.

## 4 OPTIMUM PARAMETERS OF MTMD

Optimum parameters of MTMD for minimizing rms responses are the number of MTMD, the frequency band width, tuning frequency ratio, mass ratio and damping ratio (Patil 2011). It was known that tuning the frequencies of every TMD to the fundamental mode in the primary structure is more effective than tuning it to different modes (Kareem 1995). Accordingly, the natural frequencies of MTMD are tuned to the fundamental natural frequency of the primary structure so MTMD have distributed natural frequencies around the fundamental natural frequency of the primary structure. Xu & Igusa proposed MTMD model which have the same stiffness and damping of each TMD for easy manufacturing of MTMD (Xu 1992).

### 4.1 *Optimal number of MTMD*

Patil & Jangid reported that the performance of MTMD is enhanced when 5 MTMD are used compared to that of a single TMD (Patil 2011). And even increasing MTMD over 5 cannot obtain a response reduction effect. So they proposed the optimal number of MTMD is 5. 5 MTMD are used in this study and the sum of all MTMD's masses is the same as the mass of a single TMD.

### 4.2 *Optimal frequency band width*

The natural frequencies of MTMD are distributed uniformly around their average natural frequency which is the same value of the fundamental natural frequency of the primary structure (Xu 1992). The natural frequency of the $j$th TMD $w_j$ is expressed as

$$\omega_j = \omega_T \left[ 1 + \left( j - \frac{n+1}{2} \right) \frac{\omega_s}{n-1} \right] \quad (12)$$

where, $\omega_T = \sum_{j=1}^{n} \omega_j / n$ is the average natural frequency of all MTMD; $\omega_s = \frac{\omega_n - \omega_1}{\omega_T}$; the non-dimensional frequency spacing of the MTMD; $\omega_j$ = the natural frequency of $j$th TMD; and $n$ = the total number of MTMD.

And the ratio of the total MTMD's mass to the primary structure's mass is defined as the mass ratio $\mu$;

$$\mu = \frac{\sum_{j=1}^{n} m_j}{m_s} = \frac{m_t}{m_s} \quad (13)$$

where, $m_t$ = total mass of MTMD; $m_s$ = mass of the primary structure; and $m_j$ = mass of the $j$th TMD.

Patil & Jangid proposed the optimal frequency band width is 0.3 when the number of MTMD is 5 and the mass ratio is 1.0% (Patil 2011).

### 4.3 *Optimal tuning frequency ratio*

Optimal tuning frequency ratio of MTMD is similar to a single TMD's, which is in Equation (10) as $1/1 + \mu$. It shows that optimal tuning frequency ratio is close to 1 when the mass ratio is small. Patil & Jangid proposed that optimal tuning frequency ratio is 1.0 when the number of MTMD is 5 and the mass ratio is 1.0% (Patil 2011).

### 4.4 *Optimal damping ratio*

If each TMD of MTMD has the same stiffness and damping constant for easy manufacturing MTMD, the stiffness constant $K_T$ and damping constant $C_T$ of each TMD can be evaluated as

$$K_T = \mu m_s \bigg/ \left( \sum_{j=1}^{n} \frac{1}{\omega_j^2} \right) \quad (14)$$

$$C_T = 2 K_T \xi_T / \omega_T \quad (15)$$

where, $\xi_T$ = average damping ratio of the MTMD expressed as

$$\xi_T = \sum_{j=1}^{n} \frac{\xi_j}{n} = \frac{\omega_T c_T}{2 k_\tau} \quad (16)$$

where, $\xi_j$ = damping ratio of the $j$th TMD.

And the mass of the $j$th TMD is expressed as

$$m_j = k_T / \omega_j^2 \quad (17)$$

Patil & Jangid reported that the optimal damping ratio is 0.03 when the other parameters of number of MTMD are 5, optimal tuning frequency is 1.0, optimal frequency bandwidth is 0.3 and the same constant stiffness and constant damping of each TMD are maintained (Patil 2011).

## 5 NUMERICAL SIMULATION OF ALONG-WIND LOAD

Fluctuating along-wind load which can be treated as stationary Gaussian white noise process is simulated numerically using the along-wind load power spectrum $S_F(n)$ proposed by Solari (1993).

$$S_F(n) = \left[ \rho B H C_D \overline{V}(h) \sigma_v(h) K_b \right]^2 S_{veq} * (n) \quad (18)$$

71

where,

$$S_{veq}*(n) = \frac{Sv(h;n)}{\sigma_v^2(h)} L\left[0.4\frac{nC_xB}{\overline{V}(h)}\right]$$

$$\frac{1}{C_D^2}\left[C_W^2 + 2C_WC_l L\left[\frac{nC_yD}{\overline{V}(h)}\right] + C_l^2\right]L\left[0.4\frac{nC_zH}{\overline{V}(h)}\right];$$

$$L(\eta) = \frac{1}{\eta} - \frac{1}{2\eta^2}(1-e^{-2\eta});$$

$$K_b = \frac{1}{H\overline{V}(h)\sigma_V(h)}\int_0^H \overline{V}(z)\sigma_V(z)\psi_1(z)dz$$

$$\frac{nS_v(z;n)}{\sigma_v^2(z)} = \frac{6.868\frac{fL_v}{z}}{\left(1+10.302\frac{fL_v}{z}\right)^{5/3}}; \text{and } f = \frac{nz}{\overline{V}(z)} \quad (19)$$

$S_F(n)$ = power spectrum of first fluctuating modal force; $C_x$, $C_z$ = lateral and vertical exponential decay coefficients; $C_y$ = cross-correlation coefficient of pressure acting on the windward and leeward face; $L_v(h)$ = integral length scale of turbulence at height; $h,\rho$ = air density; $B$ = width of bluff surface; $H$ = height of structure; $h$ = reference height of structure; $C_D$ = drag coefficient; $C_m$, $C_e$ = absolute values of mean pressure coefficients on windward and leeward face; $\overline{V}$ = mean wind velocity; $\sigma_v$ = standard deviation of longitudinal turbulence; and $n$ = frequency,

Numerical simulation procedure of fluctuating along-wind load was provided by Shinozuka as follows (Shinozuka 1987);

$$f(t) = \sum_{h=1}^{N}\sqrt{2S_F(f_1)\Delta\omega}\cos\left(\omega_k t + \phi_t\right) \quad (20)$$

where, $S_F(f_1)$ = the value of the spectral density of along-wind load corresponding to the first modal resonant frequency; $\Delta\omega = (\omega_u - \omega_1)/N$; $\omega_k = \omega_1 + (k - \frac{1}{2})/N$; $\omega_k = \omega_1 + (k - \frac{1}{2})\Delta\omega$; $\omega_u$ = upper frequency of $S_F(f)$; $w_l$ = lower frequency of $S(\omega)$; $\Phi_t$ = uniformly distributed random numbers between 0~2$\pi$; and $N$ = number of random numbers.

## 6 NUMERICAL EXAMPLE

This numerical example is from "Numerical Examples" in the reference (Solari 1993). The building's height $H$ = 180 m, width $B$ = 60 m, and depth $D$ = 30 m, first modal natural frequency $f_1$ = 0.27 Hz, critical damping ratio = 0.015, first modal mass = $2.4*10^7$ kg, $h$ = 120 m, $\overline{v}(h)$ = 40.96 m/s, $\sigma_v(h)$ = 5.39 m/s, $L_v(h)$ = 582.48 m, $C_x$ = 16, $C_z$ = 10, $C_w$ = 0.8, $C_l$ = 0.5, $K_b$ = 0.5, and $\mu$ = 0.01, and number of MTMD = 5.

Numerically simulated along-wind load and response without TMD were shown in Figure 2 & 3. Along-wind rms response using numerically simulated along-wind loading without TMD is 0.0274 and that of Solari's closed form response is 0.027 (Solari, 1993). Which shows that along-wind response using numerically simulated along-wind load is good approximation to that of closed form response.

### 6.1 Along-wind response with TMD

Along-wind response with TMD which have $f_{opt}$ = 1.0, $\xi_{opt}$ = 0.05 and $\mu$ = 1%, is shown in Figure 4. And the relative displacement of TMD with respect to tall building is shown in Figure 5.

Rms response with TMD is 0.0171, which is 37% reduction of along-wind response without TMD. And the relative displacement of TMD is about 7 times larger than that of tall building.

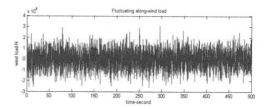

Figure 2. Fluctuating along-wind load.

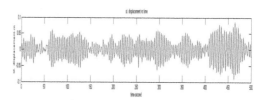

Figure 3. Along-wind response without TMD (rms = 0.0274 m).

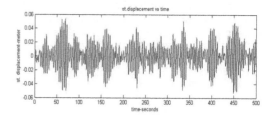

Figure 4. Along-wind response with TMD (rms = 0.0171 m).

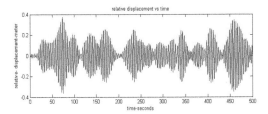

Figure 5. Relative displacement of TMD to building (rms = 0.1179 m).

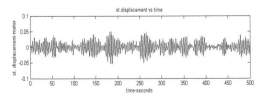

Figure 6. Along-wind response with MTMD (rms = 0.0157 m).

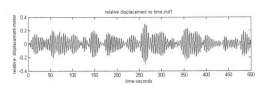

Figure 7. Relative displacement of the 1st TMD to building (rms = 0.0864 m).

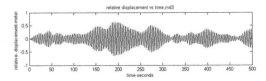

Figure 8. Relative displacement of the 3rd TMD to building (rms = 0.211 m).

### 6.2 Along-wind response with MTMD

Along-wind response with 5 MTMD which have optimal parameters of frequency bandwidth, tuning frequency, damping ratio and a mass ratio was shown in Figure 6.

Relative displacements of the 1st, 3th and 5th TMD with respect to tall building are shown in Figure 7, 8 & 9.

Reduced rms response with MTMD is about 10% reduction compared with that of a single TMD. However, maximum relative displacement

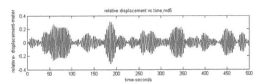

Figure 9. Relative displacement of the 5th TMD to building (rms = 0.0906 m).

of MTMD is increased about 80% compared with that of a single TMD. Threrefore, MTMD is more effective than a single TMD to suppress along-wind response of a main structure.

## 7 CONCLUSIONS

The performance of Multiple Tuned Mass Dampers (MTMD) and a single TMD (TMD) for suppressing along-wind responses of a tall building is investigated. Optimum parameters of tuning frequency ratio, applied damping ratio, frequency band width and number of MTMD for minimizing the variance response of the damped main were used. The fluctuating along-wind load acting on a tall building can be simulated numerically using the along-wind load spectra. Suppressed rms response with MTMD is smaller than that of a TMD. And the maximum relative displacement of MTMD is larger than that of TMD. Therefore, MTMD is more effective than a single TMD to suppress the along-wind response of a tall building.

## ACKNOWLEDGEMENT

This work supported by the National Research Foundation of Korea (NRF) grant funded by the Korea government (MEST) (NO. 2011-0028567).

## REFERENCES

Ayoringde, E.O. & Warburton, G.B. 1980. Minimizing structural vibrations with absorbers, *Earthquake Engineering and Structural Dynamics*, 8: 219–236.

Den Hartog, J.P. 1956. *Mechanical Vibration, 4th edn.* McGraw-Hill, New York, 1956. (Reprinted by Dover, NewYork, 1985).

Frahm, H. 1909. *Device for damped vibration of bodies.* U.S. Patent No. 989958. October 30.

Housner, G.W. Bergman, L.A. Caughey, T.K. Chassiakos, A.G. Claus, R.O. Masri, S.F. Skelton, R.E. Soong, T.T. Spencer, B.F. & Yao, J.T.P. 1997. Structural control: past, present, and future, *Journal of Engineering Mechanics (ASCE)*, 123(9): 897–971.

Iwanami, K. & Seto, K. 1984. Optimum design of dual tuned mass dampers and their effectiveness, *In Proceedings of the Japan Society of Mechanical Engineering,* 50: 44–52.

Kareem, A. & Kline, S. 1995. Performance of multiple mass dampers under random loading, *Journal of Structural Engineering ASCE,* 121(2): 348–361.

Krenk, S. & Hogsberg, J. 2008. Tuned mass damped structures under random load, *Probabilistic Engineering Mechanics,* 23: 408–415.

McNamara, R.J. 1977. Tuned mass dampers for buildings, *Journal of the Structural Division.* 103: 1785–1798.

Ormondroyd, J. & Den Hartog, J.P. 1928. The theory of the dynamic vibration absorber, *Transactions of ASME,* 50 (APM-50-7): 9–22.

Patil, V.B. & Jangid, R.S. 2011. Optimum multiple tuned mass damper for the wind excited benchmark building, *Journal of Civil Engineering and Management,* 17(4): 540–557.

Shinozuka, M. 1987. *Stochastic fields and their digital simulation,* Martinus Nijhoff Publishers, Stochastic Methods in Structural Dynamics: 93–133.

Solari, G. 1993. Gust buffeting, *Journal of Structural Engineering ASCE,* 119(2): 383–398.

Xu, K. & Igusa, T. 1992. Dynamic characteristics of multiple sub-structures with closely spaced frequencies, *Earthquake Engineering and Structural Dynamics,* 21(12): 1059–1070.

Yamaguchi, H. & Harnpornchai, N. 1993. Fundamental characteristics of multiple tuned mass dampers for suppressing harmonically forced oscillations, *Earthquake Engineering and Structural Dynamics,* 22(1): 51–62.

*Advanced Materials, Structures and Mechanical Engineering – Kaloop (Ed.)*
© *2016 Taylor & Francis Group, London, ISBN 978-1-138-02793-0*

# Influence of laser radiation on structure and properties of steel

O.V. Lobankova, I.Y. Zykov, A.G. Melnikov & S.B. Turanov
*Tomsk Polytechnic University, Tomsk, Russia*

ABSTRACT: An investigation is carried out to study the influence of laser radiation of Nd: YAG-laser on the structure and properties of annealed and quenched steel with different carbon content. The pulsed laser with the power density up to 30 kW/cm$^2$ is used for surface hardening. The characteristics of laser and the properties of steel are obtained.

## 1 INTRODUCTION

### 1.1 *Thematic justification*

Surface hardening of metal becomes an important factor in conditions of progressive mechanization and automation of manufacturing processes of machine parts. Methods of modifying the structure and properties of the material must be carefully worked out and have a predictable results according to the requirements of service parts.

Tool steels are widespread. The hardness and strength of steel can effectively resist the shock, so tool steel is used for the manufacture of cutting, stamping and measuring tool. However, carbon tool steel requires preliminary and final treatments. Pre-treatment is to obtain a structure with homogeneous properties and to facilitate mechanical processing of tool. Final treatment is to harden the steel. Tools with large size are quenched to martensite only in a thin surface layer. Material inside the tool has less brittle structure and the product works better under dynamic loads in comparison with fully hardened details.

When parts are manufactured of tool steel and parts are exposed to wearing process, the main task is to obtain a high surface hardness, and it is necessary to achieve the strength to resist the frictional force applied to the sample during operation. Another important requirement of tool steels is keeping of elasticity allowing resist loads without breaking.

### 1.2 *Treatment of material*

Tool steel undergoes surface treatment to implement the requirements of using details. Traditionally it is used such surface treatment processes as flame hardening, induction hardening, carburizing, nitriding, carbonitriding and various hardfacing. However, conventional surface treatments have several disadvantages: the high energy and time consumptions, complicated heat treatment conditions, wide heat affected zone, the absence of the solubility limit of the solids, sometimes it is harmful to the environment. Using a laser as a heat source for the surface treatment can cope with a large number of disadvantages of the above methods of treatment (El-Batahgy et al. 2013).

Thus, the laser surface hardening is one of the effective methods for surface hardening without loss of material viscosity inside the product, providing improved tribological properties and prolonging the service life of the components made of the tool and structural steels (Babu et al. 2011).

A distinctive feature of laser radiation on metals is fast heating of metal volume and rapid cooling due to heat dissipation into the metal without the need for additional cooling medium. The use of laser radiation can influence the metal briefly, localized, it can be used to handle specific areas of material and inaccessible places, providing minimal distortion. Presented number of advantages of laser irradiation shows the appropriateness and prospects of application of lasers for surface hardening (Mazumder 1983).

Nowadays two types of lasers (Nd:YAG and $CO_2$) are mainly used for the implementation of surface hardening. However, the emission wavelength of $CO_2$ laser is 10.6 μm and it's coefficient of absorption by metals is low. Therefore, the use of solid-state lasers with a wavelength of 1.06 μm is preferred because of its good absorption of metals, so there is no need to use absorbing coatings (Grigoriants et al. 2006).

Laser surface hardening has been studied in certain areas earlier (Babu et al. 2012), but on the whole sphere it is far from complete study and requires further research, it is becoming evident in the formulation of certain tasks. The aim of this work is to reveal the influence of laser surface treatment on the structure and properties of the hardened and annealed carbon steel.

## 2 THE EXPERIMENTAL PART

### 2.1 Material

The following grades of steel were selected for the study: C45 is structural carbon steel with 0.45% carbon content and CT70, CT120 are instrumental carbon steel with 0.7% and 1.2% carbon content respectively. This set of steels allows us to trace the influence of carbon content on the properties of steel after laser treatment. We used steel samples with dimensions 30 × 20 × 10 mm in the annealed and quenched condition. The heat treatment was carried out by standard methods for the selected steels (Koloskov 2001).

### 2.2 Laser complex

Laser radiation source is a solid-state Nd: YAG-laser operating in a pulsed mode with a wavelength of 1.064 μm and a pulse duration of 12 μs. Rectangular pulse is used for treatment (software of laser system allows changing the shape of pulse), the laser spot diameter is 0.35 mm. In the experiment the laser power was varied from 15.6 to 31.2 kW/cm².

### 2.3 Research methodology

As a result of experiments samples were obtained with the different areas of laser action, surface hardness, hardened depth and microstructures. Laser treated surface of the specimens and their cross-sectional view was prepared for the micro-hardness measurements and investigation of the microstructure using grinding and polishing machines, then etching was carried out with 4% solution of $HNO_3$ in ethyl alcohol. The temperature in the heat affected area is calculated by the usual method (Wang et al. 2006). The laser power is measured, power density is calculated (Losev 2011). The microhardness of the structure is measured with the PMT-3 under the load 150 g.

## 3 RESULTS AND DISCUSSION

### 3.1 Laser parameters

According to the statements of the manufacturer Nd: YAG-laser, the power of its radiation depends linearly on the duration of the radiation. At the same time, the manufacturer guarantees the average maximum power Pmax 50 W at a pulse duration τmax 20 ms. However, the calculated and measured data may not be agreed. Therefore, laser radiation power of the complex was measured and calculated by standard techniques. According to calculations by the standard technique the output power will be 31.2 W at a pulse duration of 12 ms

(single pulse). Further calculation was done with decreasing power in percent. Figure 1 shows the calculated and experimental curves of the change of laser power. It is shown that the calculated data are in good agreement with the measured values.

Changing the laser power will lead to changing of temperature of the surface processed steel and heat affected region that will change the surface structure and mechanical properties. Therefore, the temperature in the treatment area of laser radiation was calculated:

$$T = \frac{q_s \sqrt{\alpha \tau_i}}{\lambda_T} \qquad (1)$$

where, $q_s$ = flux density falling on the surface, W/cm²; $\tau_i$ = laser pulse duration, s; $\alpha$ = thermal diffusivity of material cm²/s; $\lambda_T$ = thermal conductivity W/(cm K).

### 3.2 Material structure and properties

Figure 2 shows that temperature on the surface of used steel in the experiment increases by increasing the power density, and it reaches a maximum—2450–2650 °C when the power density is 31.2 kW/cm². Such a high temperature should lead to melting of the metal corresponding to the spot of the light

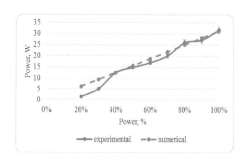

Figure 1. Comparison of the calculated and experimental values of the laser power.

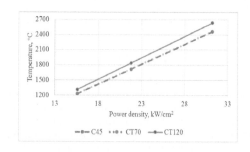

Figure 2. The attained temperature during laser action.

beam. Increasing the temperature of various steels ambiguous. Steels with 0.45% and 0.7% carbon content has the same increase in temperature dependency, while a steel with a 1.2% carbon content retains the trend dependence, but it is considerably higher. Such increase in temperature is due to the change of the microstructure of steel. In the CT120 steel there is a large quantity of the secondary cementite which leads to lower thermal conductivity of steel. Therefore, introduced into the surface layer of the steel sample heat will be distributed much more slowly which leads to an increase in temperature.

Figure 3 shows photographs obtained during a single pulse of radiation on different steels in the hardened and annealed condition under laser radiation with the power density of 31.2 kW/cm². The photographs show clearly the laser irradiation zone and heat affected zone.

Figure 4 shows the dependence of the diameter of the laser irradiation zone on the power density for steels of different composition in the annealed and quenched condition. The graphs show that the increase in power density leads to an increase in the diameter of the laser irradiation zone. As can be seen from the curves, an increase in carbon content increases the laser irradiation zone. In addition, diameter of laser irradiation zone in quenched steels is bigger than in annealed steels. Figure 5 shows the dependence of the depth of the laser irradiation on power density of the radiation and pretreatment of steel. As in the case with the laser irradiation diameter (Fig. 4), the depth increases with increasing of power density. And the depth of the laser action in quenched steel is much higher than in annealed steel.

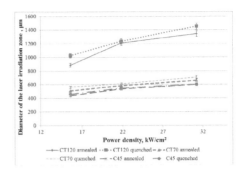

Figure 4. Laser irradiated zone diameter of different steel grades, its pretreatment and power density of laser radiation.

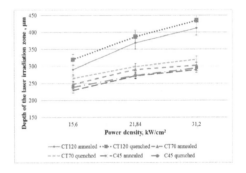

Figure 5. Laser irradiated zone depth of different steel grades, its pretreatment and power density of laser radiation.

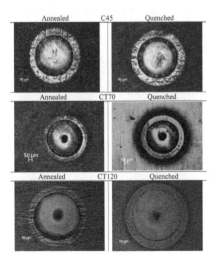

Figure 3. The macrostructure of laser treated samples at power density 31,2 kW/cm², top view.

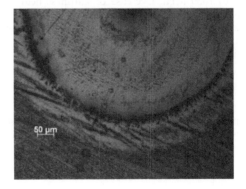

Figure 6. Photograph of microhardness measurements.

In the zone of the laser influence properties of steel (microhardness) vary depending on the condition of the steel before laser treatment. Figure 6 shows the dependence of microhardness change on the initial state in direction to the center of the treated area for CT120 steel in the

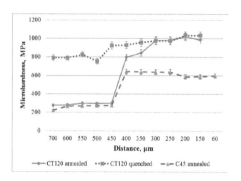

Figure 7. The microhardness of the surface area after laser treatment with a power density of 21.8 kW/cm².

annealed and quenched conditions. The change in microhardness is also shown for C45 steel in the annealed condition for comparison.

The microhardness condition increases immediately after crossing the border between the treated and untreated areas in the annealed steel. The microhardness increases from 250 to 650 MPa in C45 steel, and from 300 to 1050 MPa in the CT120 steel. Such increase of microhardness is connected with quenching of steel from the liquid state and the generation of fine structures, which is poorly etched.

Change of microhardness occurs in another way in quenched steel. The microhardness of quenched steel is high and is about 800 MPa. As it approaches the laser irradiation zone microhardness firstly decreases and then there is an increase up to 1050 MPa. Decrease of hardness near the laser irradiation zone caused by tempering. And quenching also occurs in the laser irradiation zone from the liquid state, which causes an increase of microhardness.

## 4 CONCLUSIONS

Laser treatment of carbon steels by a pulsed laser showed that the zone of laser influence in quenched steel is higher than in annealed.

Microstructure of the steel is vary under laser irradiation—fine structure with high hardness is formed in the treated area. Which causes an increase of microhardness.

ACKNOWLEDGMENT

This study was carried out under the state assignment of the Ministry of Education and Science of Russian Federation for 2014–2016 (research work № 8.2500.2015/K).

REFERENCES

Babu, P.D. Buvanashekaran, G. & Balasubramanian, K.R. 2011. Laser surface hardening: a review. *International Journal of Surface Science and Engineering,* 5–2: 131–151.
Babu, P.D. Buvanashekaran, G. & Balasubramanian, K.R. 2012. Experimental studies on the microstructure and hardness of laser transformation hardening of low alloy steel. *Transactions of the Canadian Society for Mechanical Engineering,* 36–3: 241–257.
El-Batahgy, A.M. Ramadan, R.A. & Moussa, A.R. 2013. Laser surface hardening of tool steels—experimental and numerical analysis. *Journal of Surface Engineered Materials and Advanced Technology,* 3: 146–153.
Grigoriants, A.G. Shiganov, I.N. & Misurov, A.I. 2006. *Physical foundations laser material processing.* Moskow: Moscow State Technical University.
Koloskov, M.M. 2001. *Steel and alloy grade guide.* Moskow: Mashinostroenie.
Losev, V.F. Morozova, E.Y. & Tsipilev, V.P. 2011. *Technological processes of laser treatment.* Tomsk: Tomsk Polytechnic University.
Mazumder, J. 1983. Laser heat treatment: The state of the art. *Journal of Metals,* 35–5: 18–26.
Wang, X.F. Lu, X.D. Chen, G.N. Hu, Sh.G. & Su, Y.P. 2006. Research on the temperature field in laser hardening. *Journal of Optics and Laser Technology,* 38: 8–15.

# Comparison of concentration profile prediction for slurry transportation based on the diffusion theory

L.J. Zhao

*Engineering Research Center of Dredging Technology of Ministry of Education, Hohai University, Changzhou, China*

ABSTRACT: The transportation of slurries in a pipeline is common in engineering application. The diffusion theory for the calculation of a solid-liquid two-phase flow was introduced. Three models, namely the Rouse model, Lane-Kalinske model, and Tanaka-Sugimoto model, were compared for sand with a mean particle diameter ($d_{50}$) of 0.123 mm and 0.372 mm hydraulic transportation in a pipeline. The volume concentration varied by 10% to 35%. The calculated models based on diffusion theory were compared with the experimental data. The result shows that the diffusion theory is valid for studying the concentration profiles of sand hydraulic transport in a pipeline, especially for high velocity or lower concentration.

## 1 INTRODUCTION

Hydraulic transport through the pipeline was first introduced in the mid-19th century. It has become a well-known technique in the dredging chemical and mining industries (Matousek 2005). Due to the inherent complexity of the multiphase flow, solid-liquid mixtures in horizontal pipes are quite different from the single-phase flow. The requirements for economic transport, optimization of procedure conditions and valuation of safety factors create the need for quantitative information about such flow. In the literature, lots of models have been studied for the prediction of solid-liquid flow through the pipeline, focusing on the settling slurry (Durand 1953, Matousek 1997, Kaushal & Tomita 2002, 2013, Assefa & Kaushal 2015). The performance of the solid-liquid flow is determined by the particle properties and transport circumstances. The shape of the concentration profile is associated with transportation resistance, the delivered velocity, and even transport stability. The investigation of solid distribution along the pipe cross-section is essential for the optimum operation conditions for a slurry pipeline.

## 2 DIFFUSION THEORY

Diffusion is the net movement of molecules or atoms from a region of high concentration to a region of low concentration. This is also referred to as the movement of a substance down a concentration gradient. An elementary theory of diffusion, proposed many years ago (O'Brien 1933), for the vertical sediment distribution in the open channels has been widely applied for the calculation of particle concentration distribution. And it is usually practical to a steady flow in the pipe. The sand diffusion equation is given in Equation (1) as follows:

$$\varepsilon_s \frac{dC}{dy} + \omega C = 0 \qquad (1)$$

where $\varepsilon_s$ is the sediment diffusion coefficient; $\omega$ is the group settling velocity of sands; and $C$ is the concentration at the vertical position of $y$.

In the literature, a lot of models are found to predict the concentration profile based on diffusion theory. It is a simple and highly accurate way to calculate the concentration profile of the open channel. The diffusion theory can also be used to study the sand hydraulic transport in pipelines. The boundary condition is the main difference. First, the maximum velocity point is on the free surface of the open channel, but the pipeline's boundary is circle-shaped with no free surface, and the maximum velocity point is located near the center of the cross-section. Besides, the driven force is gravity for the open channel flow, whereas the differential pressure is the power of slurry flow in pipelines. These two different situations have an influence on sand diffusion.

Through trial and error, three models of diffusion were studied to predict the concentration distribution along the pipe section. These models are recommended by Rouse, Lane-Kalinske and Tanaka-Sugimoto, respectively.

## 2.1 Rouse model

The slurry diffusion factor $\varepsilon_s$ is considered to be the same as that of water in the Rouse model (Rouse 1937). It has been suggested that the concentration distribution can be achieved in open channels as follows:

$$\frac{C}{C_a} = \left( \frac{a}{y} \cdot \frac{H-y}{H-a} \right)^z \tag{2}$$

where $H$ is the depth of a river, but in the present study, the diameter $D$ is used instead of $H$ for the slurry flow in a pipe, and $C_a$ is the concentration at a position of $y = a$. The constant $a$ is typically $0.5D$ instead for the concentration profile in the pipeline:

$$Z = \omega / \kappa u_* \tag{3}$$

where $Z$ is the suspension factor; $\kappa$ is the Karman constant; $u_*$ is the friction velocity and $\omega$ is the group settling velocity of particles.

## 2.2 Lane-Kalinske model

An exponent type is elaborated in the Lane-Kalinske model (Lane & Kalinske 1941). The suspension factor is involved in the exponent as follows:

$$\frac{C}{C_a} = \exp\left[ -\frac{6\omega}{\kappa u_*} \left( \frac{y}{H} - \frac{a}{H} \right) \right] \tag{4}$$

For slurry flow in pipelines, the depth $H$ is used instead as the diameter of the pipeline.

## 2.3 Tanaka-Sugimoto model

By implementing the logarithm velocity profile for clear water, the Tanaka-Sugimoto model (Tanaka & Sugimoto 1958) is given as follows.

$$\frac{C}{C_a} = \left( \frac{1+\sqrt{1-y/H}}{1+\sqrt{1-a/H}} \cdot \frac{1-\sqrt{1-a/H}}{1-\sqrt{1-y/H}} \right)^{\omega/\kappa u_*} \tag{5}$$

These three models are the typical examples of diffusion theory for the open channel.

## 3 SETTLING VELOCITY

### 3.1 Settling velocity of a single particle

Particles tend to fall to the bottom of the pipe if the particle's density is greater than water. The settling velocity of the single particle $\omega_0$ is due to the gravity and the strength of viscous forces at the surface of the particle, providing the majority of the retarding force.

### 3.2 Group settling velocity

Group settling velocity $\omega$ is a function of concentration and single particle settling velocity. As suggested in Mori (1956), it can be calculated by

$$\omega / \omega_0 = (1 - C_v)^2 / \left[ 1 + 3C_v / (1 - C_v/0.52) \right] \tag{6}$$

where $C_v$ is the average concentration in the pipe.

## 4 PARTICLE

Two kinds of sand particles with the density of 2650 kg/m³ were studied. The mean particle diameter is 0.123 mm (Sand 1) and 0.372 mm (Sand 2), respectively. The Particle Size Distribution (PSD) is shown in Figure 1. The PSD is the narrow width. The volume concentration varied by 10% to 35%. The test result can be obtained from Fusheng et al. (2007).

## 5 CALCULATION RESULTS

Figure 2 shows the concentration profiles comparisons of sand 1 ($d_{50} = 0.123$ mm) between the experimental data and calculated values based on the three models.

Figure 2 shows the results obtained from the method of diffusion theory. It is can be seen that the calculated profile with the measured data shows a good agreement with respect to the center of the pipe. The predicted results of Rouse, Lane-Kalinske Tanaka-Sugimoto are very similar. All of the three models failed to predict the volume concentration at the top and bottom of the pipe.

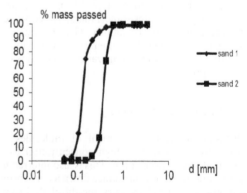

Figure 1. Particle size distribution.

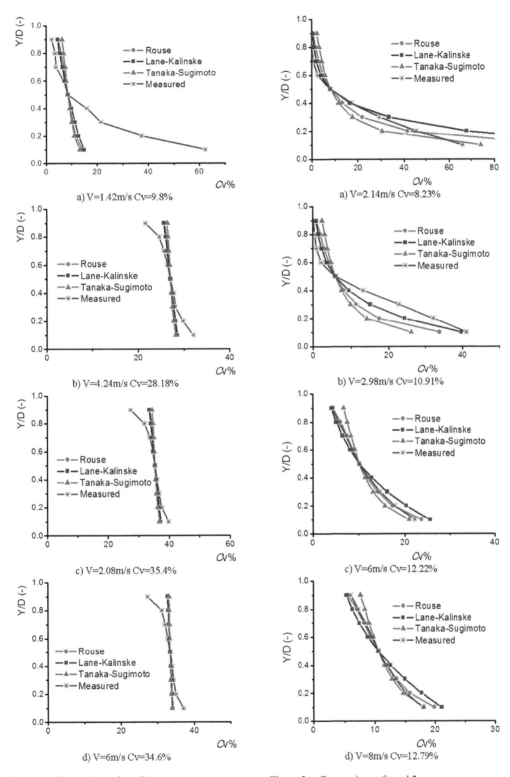

Figure 2.   Comparison of sand 1.

Figure 3.   Comparison of sand 2.

Figure 3 shows the data obtained from the studies of sand 2. Clearly, the distinction between the calculated concentration and the measured data tends to be smaller with the increasing velocity. The calculated profiles using the three models are compatible with those available in the test. The calculated concentration at the bottom is more than 100% (Fig. 3-a) for low velocity, which is impossible in reality. The reason is that the delivered velocity is under the deposit critical velocity, most particles move as a contact load and the particles are able to suspend in the slurry without sufficient energy supplied by the slurry. It is not suggested to predict the concentration profile at a low velocity or concentration.

Each concentration column has similar profile, in that the calculated concentration is smaller than the measured concentration at the center of the pipeline. This discrepancy may be explained by the fact that the velocity difference between the top and bottom of the pipe is considerable, i.e. a velocity sheer layer exits in the center of the pipe section as well as the concentration sheer layer. In the sheer layer, besides the pressure drop, turbulent eddy crash with each other. Sand particles come across the sheer force between the top layer and the bottom layer, which makes sand particles suspend easily. More and particles suspend in the center of the pipe that leads to the increasing concentration.

## 6 CONCLUSIONS

By trial calculation, the concentration profiles were predicted based on the diffusion theory. The following conclusions were drawn on the basis of the results of the present study:

1. The validation of diffusion theory depends on the situation of slurry and is influenced by the particle size, transport velocity and volume concentration.
2. It is not suggested to predict the concentration profile at either a low concentration or small flow velocity.

3. There are small differences in the model of Rouse, Lane-Kalinske and Tanaka-Sugimoto for the prediction of the concentration profile.

## REFERENCES

Assefa, K. & Kaushal, D. 2015. A comparative study of friction factor correlations for high concentrate slurry flow in smooth pipes. *Journal of Hydrol Hydromech.* 63(1): 13–20.

Durand, R. 1953. *Basic Relationships of the Transportation Of Solids In Pipes—Experimental Research.* Proceedings: 5th Minnesota International Hydraulic Convention, 15.

Fusheng, N. et al. 2007. *A model calculation for flow resistance in the hydraulic transport of sand.* Florida, USA: Proceedings of the 18th World Dredging Congress.

Kaushal, D. & Tomita, Y. 2013. Prediction of the concentration distribution in a pipeline flow of highly concentrated slurry. *Particulate Science and Technology.* 31(1): 28–34.

Kaushal, D.R. & Tomita, Y. 2002. Solids concentration profiles and pressure drop in a pipeline flow of multisized particulate slurries. *International Journal of Multiphase Flow.* 28(10): 1697–1717.

Lane, E.W. & Kalinske, A.A. 1941. Engineering calculations of suspended sediment. *Transactions, American Geophysical Union.* 22(3): 603–607.

Matousek, V. 1997. *Flow mechanism of sand-water mixtures in a pipelines.* Delft, the Netherlands: Delft University Press.

Matoušek, V. 2005. Research developments in a pipeline transport of settling slurries. *Powder Technology.* 156(1): 43–51.

Mori, Y. 1956. On the viscosity of suspensions. *Chemical Engineering,* 20: 488–493.

O'Brien, M.P. 1933. Review of the theory of turbulent flow and its relation to sediment-transportation. *Eos, Transactions American Geophysical Union.* 14(1): 487–491.

Rouse, H. 1937. Modern conceptions of the mechanics or fluid turbulence. *Transactions of the American Society of Civil Engineers.* 102(1): 463–505.

Tanaka, S. & Sugimoto, S. 1958. *On the Distribution of Suspended Sediment in Experimental Flume Flow,* ed. N.M.o.F.o. Engineering. Kyoto Japan.

*Advanced Materials, Structures and Mechanical Engineering – Kaloop (Ed.)*
© *2016 Taylor & Francis Group, London, ISBN 978-1-138-02793-0*

# CFG compound foundation in high-rise building design and application

J.Z. Li & G.X. Fang
*Structure Engineering, College of Engineering, Yanbian University, Yanji, China*

Q. Quan
*Zhongjian Group Co. Ltd., China*

ABSTRACT: At present, a lot of CFG compound foundation is applied in foundation works. Combining the Xi'an Poly International Plaza project, this paper introduces the design of CFG compound foundation and takes the bearing capacity and settlement of the compound foundation as a case, while the calculated value and the measured value are compared. As the characteristic value of the bearing capacity is lower than the measured value, formula calculation of the settlement higher than the measured value, the origin software is applied to fit the measured value. The fitted value is used to revise the bearing capacity and settlement of the compound foundation. At the same time, ANSYS simulation analysis of the compound foundation is carried out and contrast with the measured value. The result shows agree with the measured bearing capacity-settlement test, and in the single pile composite foundation settlement, pile mainly bears top load. With the number of the pile grows, the function of the soil between piles is increasing.

## 1 INTRODUCTION

As economical and practical methods of foundation reinforcement, CFG compound foundation shares the feature of big bearing capacity, strong flexibility, shortening the construction period, reducing the engineering cost, convenient installation (Zhao, H.H. 2009). With the continuous development of China's construction project, CFG compound foundation is well used in many kinds of high-rise buildings. As the building in itself has to be available for the strengthen requirement as well as the normal use, the bearing capacity of compound foundation and the foundation deformation are both needed to be considered

in the design (Su & Chen 2012). Combining the Xi'an Poly International Plaza project, this paper introduces the design of CFG compound foundation and the origin software is applied to revise the characteristic value and settlement of the compound foundation. At the same time, ANSYS simulation analysis of the compound foundation is carried out and contrast with the measured value.

## 2 HYDROGEOLOGICAL CONDITIONS

Poly International Plaza is in the New Area of Xi'an. It covers the area east to Armed Police East Road, south to the Apang 4 Street, west to the

Table 1. Soil information.

| No. | Name | Moisture w/% | Void ratio e | Liquidity index IL | Modulus of compression Es/Mpa | Coefficient of compressibility $\alpha$1-2/Mpa-1 | Characteristic value of subsoil fak |
|-----|------|--------------|--------------|--------------------|-------------------------------|-------------------------------------------------|-------------------------------------|
| 1 | Lime soil | 19.1 | 0.877 | 0.2 | 5.48 | 0.35 | |
| 2 | Loess | 22.4 | 0.877 | 0.39 | 7.32 | 0.27 | 150 |
| 3 | Medium-coarse sand | | | | | | 170 |
| 4 | Loess | 24.9 | 0.782 | 0.53 | 7.42 | 0.26 | 160 |
| 5 | Silty clay | 22.2 | 0.685 | 0.36 | 9.2 | 0.2 | 170 |
| 6 | Medium-coarse sand | | | | | | 260 |
| 7 | Silty clay | 24.1 | 0.704 | 0.37 | 8.38 | 0.22 | 180 |
| 8 | Medium-coarse sand | | | | | | 280 |
| 9 | Silty clay | 23 | 0.688 | 0.33 | 8.53 | 0.2 | 180 |

Armed Police Road and north to the Three Bridge New Street. It is consists of building number 1,4 and 7, 30 F, 30 F and 33 F in order, 2 F underground. The total construction area of the project is 161,000 m², 104.3 m overall height. The design requires the compound subgrade bearing capacity be equal or greater than 620 kPa. The soil physical properties are shown in Table 1.

## 3 DESIGN THEORY OF THE CFG, S, RA, FSK, S

According to the Building foundation treatment technology specification (Zi & Feng 2000, JGJ79 2002, GB50007 2012), for the first time of design, the following method can be used. Ra—Single pile bearing capacity measurement. FSK—Bearing capacity of the compound foundation. S—CFG pile composite foundation measurement.

$$Ra = u_p \sum_{i=1}^{n} q_{si} l_i + q_p A_p \qquad (1)$$

$$f_{spk} = m \frac{Ra}{A_p} + \beta(1-m) f_{sk} \qquad (2)$$

$$s = \omega_s s' = \omega_s \sum_{i=1}^{n} \frac{p_0}{\xi E_{si}} (Z_i \bar{\alpha}_i - z_{i-1} \bar{\alpha}_{i-1}) \qquad (3)$$

$Ra = 1038$ kN, $fsk = 813$ kPa, $S = 58.55$ mm.

## 4 CFG FIELD STATIC LOAD TEST AND APPLICABLE ORIGIN FITTING AND CORRECTION COEFFICIENTS COMPOSITE FOUNDATION BEARING CAPACITY TEST

Test method: Before the test, putting the 200 mm cushion thickness down the top of composite foundation, then the cushion thickness places the rigid bearing plate, and the rigid bearing plate uses the rigid circular plate (diameters:1.47 m; acreage: A = 1.7 m²) with the sustained loading. Loaded with nine, the first stage of 248 kpa, each increment 124 kpa, termination load for 124 kpa, each measuring more than 2.5 hours, reading every once loading is 10.10.10.15 minutes until the settlement stability. The composite foundation P-S curve shown in Figure 1.

Figure P-S by subsidence curves Figure 1, Group 4 points in the termination of the test pressure of the load 1240 kpa, and the final settlement respectively 19.242, 20.870, 23.024 and 19.811 mm, its PS settlement curve slow degeneration, corresponding load as the load bearing capacity of foundation, respectively, 917 kN,

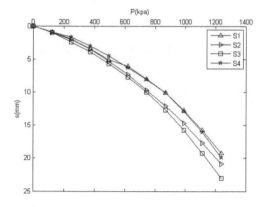

Figure 1. Composite foundation static load test.

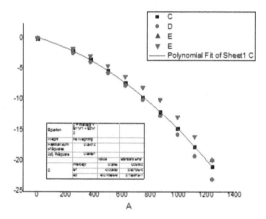

Figure 2. Polynomial fitting.

953 kN, 1033 kN, 1040 kN, then composite foundation load test method in accordance with the relevant regulatory requirements and take the s/d = 0.01, s = 14.7 mm.

CFG pile through stress analysis, selection of design parameters from ground test results consistent with is safe. Composite Foundation with CFG is a better approach.

This article uses origin tool to predict load-settlement value with polynomial fitting. The result shows in Figure 2. The results R² = 0.9981, the fitting is good. Polynomial fitting well; Polynomial fitting equation y = ax² + bx + c, a = −8.075E-6, b = −0.006931, c = 0.0965.

## 5 CORRECTION COEFFICIENT

According to the results, the composite foundation bearing capacity higher than the formula 2 value,

Table 2. Inverse coefficient results.

| $S_1$ (mm) | $\Psi_1$ | S (mm) | $\Psi_s$ |
|---|---|---|---|
| 19.242 | 0.328923 | 58.5 | 0.55 |
| 20.8 | 0.355556 | 58.5 | 0.55 |
| 23.024 | 0.393573 | 58.5 | 0.55 |
| 19.811 | 0.33865 | 58.5 | 0.55 |

and more than the theoretical value 11%, 15%, 21.2%, 21.8%. There are three points.

1. Through the soil between piles of piles of additional stress. Soil pile composite foundation to withstand the upper load, the increase in the pile side resistance. The formula for calculating the value of 1 is small in Ra.
2. Composite foundation bearing capacity of the soil leads to larger carrying capacity than the original. Because the pile inserted into soil, so that the lateral soil variations restrained and consequent reduction of vertical deformation.
3. When the pile into the hole process, there are piles of soil compaction pile and disturbance effects.

In this paper, fitting formula y = 14.7 mm, corresponding to the value of 990 × 813 and the ratio of composite foundation theory, as a composite foundation bearing capacity increase factor. Its value is 1.22. By static load curve s1 and stratified sedimentation resulting s total legal theory (GB50007 2012) value as shown in Table 2. $\Psi_1$ measured values calculated from the results of anti, $\Psi_s$ calculate deformation within the depth range of compression modulus when the value of the correction factor corresponding settlement.

Fitting formula x = 1240 kpa corresponding y = 21 mm and theoretical comparison of the settlement formula S1 = 58.5 mm as settlement correction factor. Its value is 0.36, and the average of the same.

## 6 ANALYSIS OF CFG PILE COMPOSITE FOUNDATION SETTLEMENT BY ANSYS

### 6.1 The element type and constitute relation with the soil and pile

This paper use the three-dimensional finite element to facilitate the calculation; Pile diameter value (400 mm), assuming that the pile is homogeneous elastomer, soil Drucker-Prager elastic-plastic material model to simulate a single homogeneous soil between piles of soil, pile SOLID45 unit block using soil cushion layer selection SOLID65 unit.

### 6.2 Pile composite foundation settlement is simulated by ANSYS

Pile numerical analysis and calculation should be in the range of radial over the pile diameter of 5 to 6 times. In the depth, the direction beyond the pile bottom of pile length of 0.5 to 0.6 times (Xiao 2008). This article takes a wide radial lateral soil 2.4 m, thick pile bottom soil taken following 9 m to minimize the influence of the boundary conditions. Figure 3 is a network cell division shape. Boundary conditions around the front and rear sides are bound by the horizontal displacement of the bottom surface of the fixed constraint.

### 6.3 ANSYS analysis calculation result

For example, the composite foundation static load test load 9, the first stage is 248 kpa, each increment 124 kpa, termination load for 1240 kpa. When loaded into 1240 kpa settlement amount 21.125 mm, the settlement displacement distortion cloud shown in Figure 4. Each level is loaded after pile displacement curve comparison with the measured QS shown in Figure 5.

ANSYS simulation results with the measured load-settlement curves are consistent predictable figures corresponding to the bearing capacity of CFG pile settlement value of 4. But there are some differences, because the pile, the soil is difficult to determine the parameters, on the other hand is assumed to be a homogeneous single soil. In fact, it's not a single soil layer of soil, it is easy to calculate the failure to consider the soil stratification, further research is needed.

In the figure, the pile composite foundation pile group effect is not reflected.

Pile composite foundation settlement process, the pile is mainly the upper load to bear, but the role of a small pile of soil. As the load increases, the

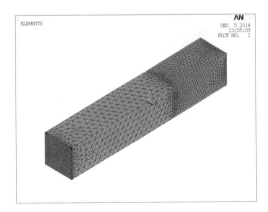

Figure 3. The cell network division.

85

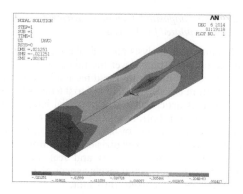

Figure 4. Settlement analog displacement distortion cloud.

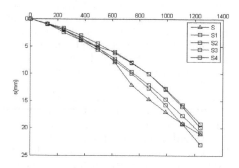

Figure 5. Composite foundation static load test and ANSYS analysis.

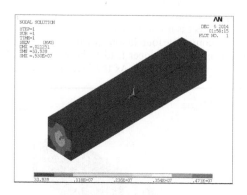

Figure 6. Load—settlement equivalent stress cloud.

number of piles and soil between files function is increased (Tong, J.X & Hu 2005). By stress concentration from the top of the pile portion of the phenomenon, Stress concentration based on the upper part of the pile load (basic). Prevent pile on the basis of punching, punching basis to prevent damage.

Cushion thickness to make piles and piles of soil between the upper loads to bear. Figure 6 is a simulated load—settlement equivalent stress cloud.

## 7 CONCLUSIONS

It shows the following conclusion by the test CFG and the analyzing ANSYS in the Xi'an Poly international plaza project.

1. Combining the engineering practice experience, this paper introduces the preliminary design of CFG compound and takes the bearing capacity higher than theoretical value in site. This paper suggests the engineering increasing coefficient of carrying capacity is 1.22.
2. The formula calculation of the settlement and the measured settlement volume are different. This paper advises the project that the settlement correction factor is 0.36.
3. The calculation result and the single pile of CFG composite foundation with the static load test value are similar. Combining the CFG pile composite foundation of the vertical load and the analysis of the ANSYS, the settlement of the CFG pile and the bearing capacity is predicted by the ANSYS software.

## ACKNOWLEDGEMENT

The corresponding author of this paper is Guangxiu Fang. This paper is supported by Yanbian university civil engineering construction technology teaching resource database construction fund projects (project number: 802014010).

## REFERENCES

GB50007 2011. 2012. *Code for design of building foundation*. BeiJing: China-abp.
JGJ79 2002, 2002. *Technical code for ground treatment of buildings*. BeiJing: China-abp.
Su, Y.S. & Chen, L. 2012. Application of compaction pile and CFG composite foundation in collapsible loess area. *Construction technology*, 2012: 52–54.
Tong, J.X. & Hu, Z.J. 2005. Definition of CFG composite foundation carrying capacity. The Civil Engineering paper, 2005.
Xiao, W. Yi, T.B. & Zhang, S.Q. 2008. Analysis of CFG composite foundation settlement calculation by ANSYS. *Subgrade engineering*. 2008: 109–110.
Zhao, H.H. 2009. *Application and design of CFG composite foundation of one engineering in Xi'an*. Xi'an University of architecture and technology thesis. 2009: 35–36.
Zi, G.C. & Feng, T. 2000. Analysis of CFG composite foundation by ANSYS. *Subgrade engineering*. 2000: 6–9.

# Design and construction of pile tip and pile lateral bored grouting

X. Zhou & G.X. Fang
*Structure Engineering, College of Engineering, Yanbian University, Yanji, China*

P. Zhang & Y.Q. Leng
*China Construction No. 3 Bureau (Beijing) Company, Beijing, China*

ABSTRACT: The commercial and financial projects of Block Z14 in the core area of Beijing CBD used post grouting bored pile compressive piles, the auxiliary building and underground garage employed raft foundation. Post grouting bored piles construction was adopted in some areas that had anti-floating problems. In order to improve the bearing capacity of single pile in pile foundation, and to reduce the settlement of pile foundation, the pile, and the pile end and side post casting were formed. In this paper, the engineering of post grouting construction is introduced briefly, and the principle of the post grouting technology is analyzed.

## 1 INTRODUCTION

With the increase of building height and foundation load in cities, in order to increase the bearing capacity and stability of foundation construction, foundation piles are used more and more widely, for the super high buildings with a large diameter, and long reinforced concrete piles are often used. Pile foundation and the pile end portion, pile lateral grouting form to improve the bearing capacity of pile body, the safety and quality of building foundation piles and stability plays a vital role. The Post grouting bored pile compressive piles are used for the main building of this project, and raft foundation is employed in the auxiliary building and the underground garage. Post grouting bored piles construction was adopted in some areas that had anti-floating problems.

## 2 THE GENERAL SITUATION OF THE ENGINEERING AND GEOLOGICAL AND HYDROLOGICAL CONDITIONS

### 2.1 Construction survey

The project is located in the central business district of Beijing city core area, which is located the northeast of World Trade Center, Chaoyang District, Beijing City, near the Guanghua Road, east of the Customs Building, south of Jinghui Jie Dong pipe gallery, on the west side of East Street and gold pipe gallery. The total construction area is about 336,000 square meters, including the building area of land, which is about 220,000 square meters, totally the building has 45 floors with the height of 238 m, function of the building for office and commercial podium Twin Towers. The underground construction area is about 96,200 square meters, with a total of 6 layers, and the excavation depth is about 34 m. The building is mainly used as commercial, logistics, parking and equipment ones.

The basic structure of this project is the pile raft foundation. The frame core-tube system is applied in the basement of the main tower; concrete frame shear wall structure is applied in the basement of the podium building; and the concrete frame structure is employed in the rest basements. To the south, the North Tower is used for the steel reinforced concrete frame and reinforced concrete core wall cantilever truss + + belt truss structure, foundation frame structure.

### 2.2 Hydrogeological conditions

#### 2.2.1 Ground water

According to the hydrogeological investigation and regional hydrogeological data, the depth of about 60 m under the natural ground, the engineering field is mainly distributed within 3 groups relative to aquifer.

According to the hydrogeological investigation and regional hydrogeological data, at the depth of about 50 m under the natural ground, the engineering field is mainly distributed within 4 groups relative to aquifer. The geotechnical engineering during drilling is measured to 3 layers of groundwater. The groundwater level is given in Table 1.

The elevation of −26.000 m is found in the foundation pit under the retention water curtain wall project of groundwater. Water sealing curtain is inside the unwatering well out of retention water curtain.

Table 1. List of groundwater.

| | The geotechnical investigation, layer number and lithology stable water level depth (m) | Stable water level depth (m) | Stable water level (m) |
|---|---|---|---|
| Interlayer water aquifer | ④ Pebbles, gravel; ④ 1 fine sand, medium sand | 2.10 (the bottom of the groove starting) | 19.16 |
| The first confined aquifer | ⑥ Pebbles, gravel; ⑥ 1 fine sand, medium sand | 11.30 (the bottom of the groove starting) | 9.96 |
| The second confined aquifer | ⑧ Pebbles, gravel; ⑧ 1 medium sand, fine sand; ⑧ 2 silty clay | 21.80 (the bottom of the groove starting) | −0.54 |

Table 2. The relevant parameters of soil pile foundation design.

| Stratigraphic sequence | Properties of rock and soil | The elevation of the top layer (m) | Remarks |
|---|---|---|---|
| ⑥ | Pebbles | 11.49~14.60 | Pile foundation work layer |
| ⑥ 1 | Fine sand | | |
| ⑦ | Heavy clay | 0.31~3.91 | The excavation surface substrate |
| ⑦ 1 | Silty clay | | |
| ⑦ 2 | Clayey silt | | |
| ⑧ | Pebbles | −2.79~0.26 | ZH-2 pile bearing stratum |
| ⑧ 1 | Medium sand | | |
| ⑧ 2 | Silty clay | | |
| ⑫ | Pebbles | −37.42~−32.51 | ZH-1 pile bearing stratum |
| ⑫ 1 | Medium sand | | |
| ⑫ 2 | Sandy silt | | |

### 2.2.2 The geological conditions

The engineering of the ZH-1 pile bearing layer is the layer of pebble gravel layer, ZH-2 type pile bearing layer is the layer of pebble and gravel layer, the two good engineering properties of soil, and the pile end bearing layer.

The relevant parameters of soil pile foundation design are listed in Table 2.

### 3 SLURRY DESIGN PRESSURE

All engineering piles are characterized by the pile end post grouting and pile side length. ZH-1 type pile configuration 3 & 25 pile shaft grouting pipe and 2 Phi 25 pile end grouting pipe along the pile, pile bottom, respectively, from above 15 m, 22.5 m, 30 m set the pile shaft grouting loop, loop tube is equipped with 3 pressure slurry valve, and arranged evenly, each pile end grouting pipe the end of the 1 set of grouting valve; ZH-2 pile base configuration 1 root diameter of pile shaft grouting pipe 25 and a 2 Phi 25 pile end grouting pipe along the pile spacing, pile bottom post grouting 4 m loop, loop is equipped with 3 pressure slurry valves, and arranged evenly, each pile end pressure grouting pipe is arranged on the end of the 1 pressure slurry valve.

Grouting valve reverse compressive strength is not less than 1.0 Mpa. Post grouting cement slurry is calculated by using PSA32.5, with the water cement ratio of 0.5~0.6. Pile post grouting technology of the main indicators is shown in Table 2 and Figure 1.

The following conditions are to be considered when meeting the needs of the moment: (1) stop grouting and grouting pressure grouting amount should reach the design requirements; (2) the grouting volume should reach the design value of 75%, and the grouting pressure should exceed the design requirements; and (3) according to the actual situation of the engineering, grouting pile group should first be lateral and internal, in order to double grouting in saturated soil to pile lateral pile end. For the unsaturated soil to pile after pile, multi-section pile side grouting should be after the first. Pile side and pile end grouting interval should not be less than 2 hours. Pile end grouting should deal with the same pile of the grouting pipe grouting for the equal implementation.

### 4 CONSTRUCTION DIFFICULTIES AND COUNTERMEASURES

1. Pile grouting is more difficult after a long lead in this project pile length of 10–60 m, in order

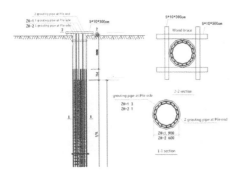

Figure 1.  The design of post grouting technology.

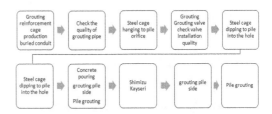

Figure 2.  Post grouting construction process.

to ensure the quality of pile construction, grouting required after construction needs reasonable deployment. The solution is to use sub-grouting, top to bottom is buried catheter grouting, grouting pressure for grouting volume and strict control.

2. Because post grouting has a greater impact on the scope of the sand layer string pulp, after taking into account the subsequent construction of the adjacent pile grouting pile into the hole on the increased difficulty factor, we should pile 3 to 30 days (not more than 30 day) begin grouting, grouting pile sequence using the first lateral pile end, multi-grouting should be on the next plane after grouting pile end and side interval should not be less than 2 h.

## 5   GROUTING

### 5.1   *Post grouting construction process is shown in Figure 2*

### 5.2   *Post grouting construction method*

1. Post grouting grouting device and conduit are installed using GB low pressure liquid delivery welded pipe. Each pile set 3 (ZH-2 piles set 2) post grouting catheter. Grouting uniform diameter catheter along the steel cage set, the lower end of the one-way to pile grouting valve connection. Grouting pipe pile side elevation determined by the design drawings, floor sealing plugs with more than 0.2 m from the longitudinal and circumferential guide tube squeezing a slurry hoses. Grouting vertical conduit through which the catheter main role is to import the slurry to a slurry hose ring, squeezing through the catheter to the ground. Central to the proposed slurry pipe made of a high-quality PVC pipe and protection devices, namely a set of 3~5 cm pulp every orifice on the PVC hose to the outside of the pile, and then sealed with transparent tape wrapped around the jacket as a protective rubber sleeve devices. Longitudinal and

circumferential catheter grouting hose from the tee after a slurry composed of the grouting device is connected, as shown in Figure 3.

Post pressure grouting device is installed in the reinforcement cage, and the general situation of post grouting device installation location to ring to grout hose at the pile end is first set, and then according to the design requirements of setting multi lateral pile grouting pipe on the pile spacing, pile side grouting pipe according to the drawings to determine design elevation, but the pressure grouting device the top from the top of the pile of not less than 200 mm, to prevent the slurry spillover of pile top. Pressure grouting device after installation, the vertical grouting pipe length according to the ring to the pulp out the position of the hoses, in consideration of the pile top elevation and ground elevation calculation, vertical grouting pipe is higher than the ground 200 mm. The ring to the inner diameter of the tube and the grout should be equal to the outer diameter of steel reinforcement cage, in consideration of the production conditions, to calculate the diameter of the steel reinforcement cage. Grouting steel tube along the longitudinal direction of the length of reinforcement cage to cage, ring to grout hose around the reinforced lateral was also tied to the steel cage.

2. The ratio of slurry and preparation of the cement slurry, adding water reducing agent and expansion agent, in order to improve the properties of slurry, grouting efficiency. Slurry configuration of 7-day strength should not be less than 5 Mpa, the grouting cement by PSA32.5. Configuration: water cement ratio of cement paste to w/c = 0.5 0.6, initial water cement ratio to prepare 0.7, grout initial setting time is >3 hours. Pressure to normal gradually after the match 0.5. Strictly controlling the water cement ratio, mixing time should not be less than 2 min, the slurry into the slurry storage barrel with 16 mesh gauze filtration, to prevent blockage of the grouting holes. Admixture: u expansion agent 6%, bentonite 2%, effects of water reducing agent 0.6%. Slurry preparation should be placed after the use of 5 min, in order to eliminate the slurry in the air. The two mixing cement slurries and stirring time

Figure 3.   After grouting equipment.

Figure 4.   Grouting equipment.

Figure 5.   Grouting process.

should not be less than 10 minutes each time, mixing the mixing barrel. The first complete mixing is made into another mixing barrel. The two objectives are to make the slurry mixing and more good, conducive to the grouting effect.

3. The grouting pressure control of grouting pressure, including the opening pressure of injection water pressure, injection pressure and injection pressure slurry. According to the engineering practice, the opening pressure is relatively large, sometimes up to 10 MPa, but because the opening pressure is to overcome the protective device of sealing grouting hole, the opening pressure is without control. In general, the opening pressure is only momentary pressure grouting device, once opened, the pressure can be decreased. Pressure pump used in the glycerine pressure should not be less than 20 MPa. In order to prevent water penetration into the soil around the pile in high pressure water injection, the pressure is about 2.0 MPa. Diluted slurry pressure when the pressure is about 2–5 MPa, injecting slurry is to make grout along the pile penetration distance farther, filling more fully, at the same time should be to prevent the slurry diffusion at. Pressure thickness should be noted when the pressure may be appropriate to increase the control in about 5.0 MPa, injecting slurry is to make further fully dense infiltration. Injection water and grouting pressure pump type selection BW-250 are included. The pressure grouting device is shown below.

4. Grouting process.
Glycerine is installed in a pile of post grouting device opened, which connected with the soil around the pile, the grouting slurry can flow smoothly through the device into the soil around the pile and pile side pore. A time in pile concrete pouring completed 8~12 hours. After the injection of water, post pressure grouting device will continue to maintain a certain pressure. And the injection pressure should not be less than 5 minutes by water injection, water, clear and open mud cake and gap, to ensure the smooth flow of slurry along the pile, to ensure the effect of slurry pressure. The practice has proved that

the effect is either good or bad, the key of slurry pressure injection water is after. Grouting in pile concrete pouring is completed within 3 to 30 days. The slurry after the pressure, pressure and slurry in 60 min. Pressure grouting termination criteria. Grouting termination criterion with grouting quantity control based, supplemented by pressure control. First, to ensure the grouting quantity estimate of the distribution of pressure grouting device is all. The following measures can be taken to deal with the problems. A. when a grouting device due to too high pressure hydraulic into slurry, pressure grouting device by its proximity to the remaining slurry hydraulic injection. B. when a grouting device grouting amount does not reach the design requirements of grouting amount, ground or on grout slurry, can temporarily stop grouting, grouting and flushing device, stop 60 min, and then press again repeatedly until it reaches the design of grouting quantity. C. when a grouting device to achieve the design of grouting amount. But it still should stop grouting and pressure is low, with thick to grouting.

5. One of the following conditions may be terminated when grouting and grouting pressure amount: A. meets the design requirements; B. the grouting volume has reached the design value of 75%, and the grouting pressure exceeds the design value 3.0 MPa.

6   DETECTION OF PILE FOUNDATION

6.1   *Vertical compressive static load test of single pile*

1. After the completion of the construction engineering pile in 28 d test platform and test equipment, preparation before the test approach 1 D.

Each pile test cycle for 3~4 D (including equipment installation, commissioning, project pile unloading equipment removed, and shift).
2. Each static load test site for an area of 10 m * 10 m, the ground should be basically flat and solid.
3. To ensure the static load test equipment successfully in place, crane lifting capacity should be greater than or equal to 30 t.
4. From the test area within 50 m, with power supply test (220~380 V).
5. Connected with the static load test of anchor pile beam straight pull system for anchor type and non-welding type, all the anchor piles shall be set before the test, chisel flat to the design elevation, and anchor pile reinforcement should be straight. The reinforcement length of anchor pile is 1.20 m.

### 6.2 *Vertical single pile static load test*

1. Project in the construction of pile cap after 28 d began to test, and test equipment to test the approach in 1 D. Each pile test cycle for 2~3 D (including equipment installation, commissioning, project pile unloading equipment removed, and shift).
2. Each static load test site for area of 3 M * 10 m, supporting the ground should be reinforced according to the requirement, advance the production of supporting the concrete platform (300 mm thick).
3. Uplift pile reinforcement should be straight, and the reserved reinforcement length is greater than or equal to 1.0 m.

### 6.3 *Sound wave transmission method*

1. Detection of the acoustic transmission method in pile foundation construction was seized after the completion of 7 d, before and after the grouting construction.
2. As acoustic pipe shallow, easy to damage, the construction side to strengthen the protection of the acoustic tube.
3. Acoustic pipe joints should be used in the welding method, in order to ensure the reliability of the acoustic pipe.
4. The low strain test conditions of low strain dynamic testing of the pile head.

## 7  CONCLUSIONS

Foundation excavation of pile, the pile position examination and testing, borehole verticality deviation is less than or equal to L/150, with most being less than or equal to L/120 (L is the length of pile).

A total of 130 root piles exist, with the pile of 3, detection of the uplift of 3, detection of acoustic transmission method 26, and the low strain detection 97. After testing, 3 root compression test of pile vertical compressive bearing capacity characteristic value is greater than or equal to 9000 kN, 3 root pulling resistance test of single pile vertical uplift bearing capacity characteristic value is greater than or equal to 1500 kN, which can meet the design requirements. A total of 26 ultrasonic testing, 20 class I compressive piles in 19, accounting for 95% of the total number of sampling; class II pile 1, accounting for 5% of the total number of sampling; acoustic transmission method not found III, IV pile, 6 pile for all class I. Moreover, 97 during the low strain test, 60 class I compressive piles in 55, accounting for 92% of the total number of sampling; class II pile 5, accounting for 8% of the total number of sampling; acoustic transmission method not found III, IV pile. The 37 type of uplift pile in pile 33, accounting for 89% of the total number of sampling; class II pile 4, accounting for 11% of the total number of sampling; acoustic transmission method not found III, IV pile.

The test data proved that, for deep foundation engineering of the project, the post grouting construction technology is reasonable and effective, to achieve the expected goal and to meet the requirements of various engineering technology, and the effect is good.

## ACKNOWLEDGMENT

The corresponding author of this paper is Guangxiu Fang. This paper was supported by the Civil Engineering Construction Technology Teaching Resource Database Construction Fund Projects (project number: 802014010), Yanbian University.

## REFERENCES

China construction science research institute of JGJ94-2008, 2008. *Building pile foundation technical specification*. Beijing: China building industry press.
China construction science research institute. Q/JY14-1999, 1999. *After filling pile grouting technology discipline*. Beijing.
GB 50007-2011, 2012. *Building foundation design specification*. Beijing: China building industry press.
JGJ 94-2008, 2008. *Building pile foundation technical specification*. Beijing: China building industry press.
Jiang, Z.R. 2001. *Construction calculation handbook*. Beijing: China building industry press.
Omega, Y. Chen, M.X. & He, Y.N. 2001. Grouting theory research status and development direction. *Journal of geotechnical mechanics and engineering*, 20(6): 839–841.

*Advanced Materials, Structures and Mechanical Engineering – Kaloop (Ed.)*
© 2016 Taylor & Francis Group, London, ISBN 978-1-138-02793-0

# An approach for road region detection

H. Fang, G.Q. Yang, J. Li, G.T. Shao, C.Y. Zhang & C.G. Fu
*State Grid Shandong Electric Power Research Institute, Jinan, China*
*Shandong Luneng Intelligence Technology Co. Ltd., Jinan, China*

ABSTRACT: In this paper, a substation environment of the road region detection method is used to realize the autonomous navigation for a substation inspection robot. First, the image acquisition equipment was the road image; the information of the road is obtained. Second, the road image region of interest is selected, the digital image processing algorithm for de-noising and its canny operator are used to extract the edge of the road. At the end of the road extraction algorithm using the Hough transform boundary, the boundary between the regions is about the road area. The experimental results show that the method can detect the road area in real time, and the algorithm used is of high accuracy and stable performance.

## 1 INTRODUCTION

Navigation technology for a substation inspection robot is the key technology of the robot system and completes the independent operation (Xiao et al. 2012, Zuo et al. 2011). Visual sensor has prominent features such as information-rich and non-contact collection vision navigation, which is used for the CCD camera shooting pavement image, using machine vision technology-related recognition path, which realized a new navigation method for automatic navigation (Huang et al. 2010).

In recent years, robot navigation system based on vision has become a research focus, which consists of the road detection and obstacle detection. The key of road detection is the positioning of the vehicle that can exercise area (Xia et al. 2013). The key of realizing substation robot navigation is fast and accurately identifies the boundary lane (Liu & Cai 2010).

In this paper, digital image processing technology is used, and visual navigation technology is used to complete the independent operation of the substation inspection robot. The boundary of the road is obtained after the image processing algorithm.

## 2 ROAD IMAGE PROCESSING

Vision is an important feel and experience in human life, which in the human brain is reflected in the image. Humans perceive about 80% information from the visual navigation in the outside. The image contains rich information, which makes communication and transmission of information become clearer (Yang & Cui 2010).

Road can be divided into a structured and unstructured road (Xie 2008). The structured road edge is regular, flat road, road lane, and other artificial markers. In this article, the path of the substation inspection robot driving is a structured road, as shown in Figure 1.

The lane edge or shoulder can be used as the boundary line road edge in the road image, and the boundary line between the left and the right part of the road is the road area of the substation inspection robot.

There are lots of noise in the road image, but the road information is mainly concentrated in the lower part of the image. In order to increase the detection accuracy, this paper selects the image below 1/3 as the image Region Of Interest (ROI), as shown in Figure 2. The ROI is considered by

Figure 1. The road image.

Figure 2. The road image of ROI.

using the digital image processing and the image segmentation method. It not only rules out the interference of background information, but also reduces the amount of calculation and improves the operation speed of the algorithm.

Image filtering algorithm is divided into two categories: spatial filtering and frequency domain filtering. The spatial filtering method is directly on the gray value of image processing in the spatial domain, such as the gray transformation, histogram modification, domain average, and median filter. The frequency domain filtering mainly involve Fourier transform and inverse transform, various wavelets transform, and inverse transform. But frequency domain filtering requires a large amount of calculation, thus it does not sufficiently meet the real-time requirements.

Median filter is a nonlinear filtering signal processing method, which, under certain conditions, can overcome the details of image blurring caused by the linear filtering such as minimum mean square filtering and average filtering (Liu & Guo 2010). Median filter in filtering noise can protect the edges of the signal, so as not to be blurred. Furthermore, median filtering algorithm is simple and easy to be used in hardware implementation.

Edge detection is the basic problem in image processing and computer vision, the purpose of which is to identify the digital image brightness change obvious points. Edge detection operator mainly includes the first-order operator and the second-order operator. Among them, the first-order operators include the Roberts operator, Sobel operator, and Prewitt Cross operator. The second-order operator includes the Canny operator and the Laplacian operator.

Due to the presence of noise, the boundary extraction would appear as pseudo edge. This brings difficulties for the road boundary of accurate positioning, as shown in Figure 3 and Figure 4.

In the edge detection operator, the Canny operator is the most commonly used method of edge detection. The Canny edge detector is the first derivative of Gaussian function, is a signal-to-noise ratio and optimizes the product positioning approximation operator. The Canny algorithm can use the dual-threshold edge detection, and the role

Figure 3. Road edge detection (1).

Figure 4. Road edge detection (2).

of the two thresholds is to reduce the fake edge. So, in this article, the Canny operator edge detector was choosen. The Canny edge detection algorithm steps are as follows:

Step 1: The image with Gaussian filter;
Step 2: To calculate the gradient magnitude and direction by means of the finite first-order partial derivative;
Step 3: Non-maxima suppression on the gradient magnitude;
Step 4: Using dual threshold detection algorithm and the edge connection.

## 3 THE ROAD EDGE DETECTION

Based on the detection of road boundary, there are mainly two kinds of detection technology: overall pavement and road edge detection method. The first important method is the inspection of the grayscale consistency for the whole region; the second method clearly identifies the path or road edge line of the road, a road image captured by a CCD camera, through digital image processing to identify road signs or road edge lines. This method has a faster processing speed and real time, with a wide range of applications.

The use of the Hough transform for line detection is an important part of image analysis and computer vision. Also, the advantage of the Hough transform is that its anti-noise performance is better, and can connect collinear short line (Zhang & Dong 2010, Ju et al. 2013).

The essence of line detection in image problem is to find all the pixels that make up the straight line. All collinear points, when the slope of the straight line is present, can be expressed by using Equation (1):

$$y = kx + b \qquad (1)$$

where $k$ is the slope of the line and $b$ is the intercept.

## 4 EXPERIMENTAL RESULTS AND ANALYSIS

In the experiments, the road on the edge of the straight line detection algorithm is designed. Straight edge detection can be outlined in the procedure shown in Figure 5.

The collected the road image through the process of processing and the experimental results are summarized in Table 1.

As can be seen from Table 1, in a variety of circumstances, this algorithm is very good to mark the edge of the road. Under the condition of the light being strong, the shadow will also be able to accurately detect the edge of the road, as shown in Figure 6. The green line on both sides of the tag to the edge of the road, in the image position, the center of the road is the dotted line and the blue line is used to detect the edge of the road angle bisector. As shown in the figure, the algorithm of the center line (blue line), which is the center of the road, also proves the correctness of the road edge.

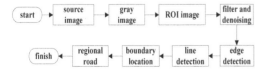

Figure 5. Detection of the flow chart.

Table 1. Experimental results of the road image.

| | Edge detection | Detection result |
| --- | --- | --- |
| 1 | | |
| 2 | | |
| 3 | | |
| 4 | | |
| 5 | | |
| 6 | | |
| 7 | | No edge |

Figure 6. Detection results under high light intensity.

## 5 CONCLUSIONS

The proposed method is used for selecting a region of interest, filtering and other related digital image processing algorithms, with the Hough transform for road boundary detection. The experimental results show that the algorithm has a good robustness in a different environment, and a well-detected edge of the road. The algorithm model is simple and fast, and has a high accuracy to meet the substation inspection robot visual navigation.

The implementation of this algorithm enhances the substation inspection robot automation and intelligent ability, and promotes the process of an unmanned substation.

# REFERENCES

Huang, X.L. Jiang, X.N. & Lu, H.Q. 2010. Survey of vision for autonomous navigation. *Journal of Jilin University* (*Information Science Edition*), 28(2): 158–162.

Ju, Q.A. Ying, R.D. & Jiang, L.T. 2013. Computer vision based fast lane detection. *Application Research of Computers*, 30(5): 1544–1546.

Liu, G.A. & Guo, W.M. 2010. Application of improved arithmetic of median filtering de-noising. *Computer Engineering and Applications*, 46(10): 187–188.

Liu, X.R. & Cai, Z.X. 2010. Robust lane detection and tracking for the structured road. *Journal of Optoelectronics Laser*, 21(12): 1834–1836.

Xia, T. Jin, C. & Liu, Y. 2013. Research on edge feature extracting for visual navigation marking the line of AGV. *Machinery Design and Manufacture*, 2013(3): 199–201.

Xiao, P. Luan, Y.Q. & Guo, R. 2012. Research of the Laser Navigation System for the Intelligent Patrol Robot. *Automation & Instrumentation*, 27(5): 5–6.

Xie, Y. 2008. Character extraction and road classification of non-structured road navigation image. *Nanjing University of Science and Technology*. 1–22.

Yang, G.Q. & Cui, R.Y. 2010. Method for natural image discrimination based on texture feature. *Application Research of Computers*, 27(7): 2783–2784.

Zhang, F.Z. & Dong, Z.T. 2010. Road Edge Detection Based on Mathematical Morphology and the Hough Transform. *Journal of Taiyuan University of Science And Technology*, 31(3): 193–195.

Zuo, M. Zeng, G.P. & Tu, X.Y. 2011. Research on visual navigation of untended substation patrol robot. *Acta Electronica Sinica*, 39(10): 2464–2468.

*Advanced Materials, Structures and Mechanical Engineering – Kaloop (Ed.)*
© *2016 Taylor & Francis Group, London, ISBN 978-1-138-02793-0*

# A university e-learning system to enhance higher education teaching environments: A model-based innovative service-oriented clouds intensive evolution approach

L. Wang
*Shanghai University of Political Science and Law, Shanghai, China*

Z. Wang
*Edinburgh Napier University, Edinburgh, UK*

ABSTRACT: This paper proposed an insight on how the e-learning system evolution can enhance higher education teaching methodology, which is an innovative approach for improving the current university's teaching environments for teachers and students. The core methodology used to enhance the higher education teaching environment is the service evolution approach in clouds proposed by Edinburgh Napier University, UK, which has been used as the foundation stone and platform for further analyzing the positive effects on higher education teaching based on the e-learning system evolution. This paper balanced software engineering and teaching methodology, which provides a new approach on the technology-enhanced modern University teaching reform, so that it shows the value for service-based e-learning system constructions, evolution and modern university teaching environment enhancement.

## 1 INTRODUCTION

This paper uses the original e-learning system from the Shanghai University of Political Science and Law (SPSL) as the object for analyzing the e-learning system evolution and its effects on enhancing higher education methodology. The current e-learning systems are mostly build based on service and deployed in the e-learning clouds. The e-learning systems contain functions and databases for storing information. As a practical example from SPSL, the e-learning system contains the following functions that potentially need evolution for improving the teaching environment. These functions include teaching course website constructions, choosing website templates, students' management, knowledge management, information interaction between teachers and students, and teaching effect analysis. All these functions are provided through customizable services in clouds. These customizable services can be further configured in order to adapt with the demands that are proposed by teachers and students, for the reason that the configuring of an existing service is limited to its predefined rules and functions, so the adapting scale of the service function is also limited. Under such context, the service evolution for these customizable services is quite needed in order to provide more flexible, adequate, robust and QoS (Quality of Service) enhanced function

to teachers and students. This can greatly enhance the higher education teaching environment (Wang 2015). This paper is organized into seven sections, section I is the introduction, which gives the overview background and interesting research object of the paper, section II is e-learning system evolution analysis in SPSL, which gives the detailed analysis of the current information system in SPSL. Section III is evolution that enhance higher education teaching methodology, which analyzes how evolution can enhance the higher education teaching methodology. Section IV is the case study in SPSL, which shows a case study based on the approach proposed by the paper and further analyzes the results of service evolution that can promote higher education methodology. Section V is the conclusion, which gives an overall summary for the paper and proposes several aspects that can be achieved in the further work.

## 2 CASE STUDY IN SPSL AND EVOLUTION THAT ENHANCES HIGHER EDUCATION TEACHING METHODOLOGY

### 2.1 *Evolution design and evolution feature*

The evolution for the knowledge management service can be further enhanced on its function and quality of service. Such as new function service can be weaved into the current service working flow, and online testing service can be enhanced by improving its database retrieving efficiency.

The 1 online testing service of evolution can be further evolved for higher efficiency in database retrieving, which is quite needed in the testing context. The evolution on the online testing service is driven by the evolution pattern; the pattern contains the evolution requirements captured from the analysis on the current subservice.

The service evolution in the campus clouds service mainly focuses on the evolved system to overcome the bottleneck of accessing the database system in the campus so as to reduce the burden of the service database accessing in clouds. It can be realized by the evolution pattern with multi-threads of accessing the system at the same time with the support of virtualization ability of the clouds. Such a service accessing ability can largely enhance the system accessing efficiency as sometimes the students' accessing peak time always occurs during the coursework time (Wang et al. 2014). A description of the entity of a single database accessing features model can be further enlarged into several topology tree-based models for complex feature model construction.

## 2.2 *The database access thread evolution*

Figure 1 shows the database access thread evolution pattern architecture. The multi-database accessing threads are used for reducing the burden of single service accessing the database, which can maximum the service accessing ability for data retrieving and database retrieving function including database connection and further dataset operation.

The case study solves the service evolution problem in campus service by using the database threads accessing evolution pattern and the approach.

The evolution pattern selected for this case study is the database access thread evolution pattern, which can enhance the data accessing by using multi-threads at the same time so as to reduce the burden of the data retrieving in the database in clouds. The feedbacks from people are excellent.

## 2.3 *The database access thread evolution for synchronously or asynchronously data accessing*

The data stored in the database can be classified into tables based on the optimized database design, such as each table represents the information acquired from one specific knowledge field. In the university teaching domain, this information is mainly focused on teaching materials, examination statistics, students' information, teachers' information and other related valuable statistics. The database threads can be applied to access these tables directly and particularly for some of these tables' data retrieving.

Figure 2 shows the data processing relation between each table, such relation is very important for database table evolution. The capture of database evolution requirements should be done based on the analysis of the data processing relation. The database accessing threads can be weaved into the current system and particularly used for accessing the specific tables in the database. The information in the databases can be triggered and acquired synchronously or asynchronously based on these implemented accessing threads. These accessing threads can be further orchestrated as a sequential working process to access the tables in the database.

As shown in Figure 3, these threads are all working under the instruction of Accessing

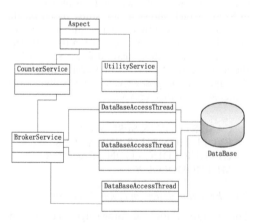

Figure 1. Database access thread evolution pattern architecture.

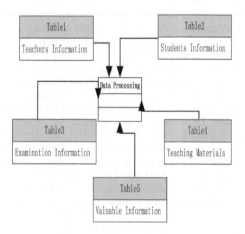

Figure 2. Data relation between tables.

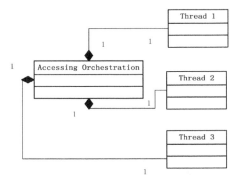

Figure 3.   Database accessing threads.

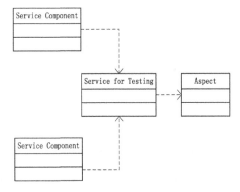

Figure 4.   Evolved service component testing.

Orchestration components so as to retrieve the data in the database in a pre-defined sequence and order so as to work properly with the data process logic that belongs to other Service Logic.

### 2.4   *Testing*

The testing of an evolved system is sometimes very complex based on the evolution status of the current system. The testing of an evolved system includes evolution aspect testing, evolved component testing and evolved systematic testing. The testing of the evolution aspect needs to be taken place on the service aspect running platform; the evolution aspect can be encapsulated by web service; and the evolved component testing should take place based on component testing through its input and output parameter. The evolved systematic testing should take place based on integrated components testing through the global input and output parameters.

The aspect testing is based on the context of evolution pattern aspect weaving; service is used for creating such context for testing the aspect. The aspect is weaved into the service environment for testing its usability and functionality.

Figure 4 shows that the evolved service component testing is based on the previous steps of Aspect testing. The testing range is enlarged as it includes the service components into the testing system. The whole testing case is based on the combination of a service component and Aspects together, so as to test their integration ability and adaptability.

Figure 5 shows that the evolved systematic testing is based on the testing of overall integrated service components, Aspects and Service Inventories, so such testing can verify the system usability and stability after evolution. The service inventories are integrated into the current testing cases so as to verify the adaptability of the service inventory after evolution.

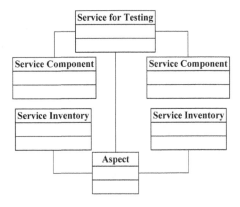

Figure 5.   Evolved systematic testing.

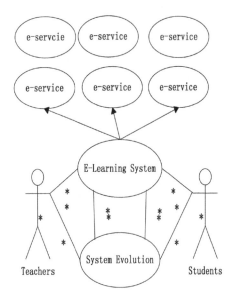

Figure 6.   E-learning system evolution.

99

## 2.5 The e-learning SOA system evolution and the modern higher education.

Figure 6 shows that the e-learning system evolution will both affect teachers and students; furthermore, it can affect the higher education pattern. The e-learning system and its evolution are based on the system evolution requirement proposed from both teachers and students. It evolves the e-learning service in the service inventory corresponding to the teachers and students evolution requirements. The above diagram shows the relation between e-learning system, evolution service, service inventory, teachers and students in the modern higher education model.

## 3 CONCLUSIONS

This paper deeply analyzes the current service-based evolution approach in clouds that can enhance the e-learning system, which makes great effects on modern university teaching environment construction. The paper further analyzes the service-based system at the component level so as to explicitly discuss the service evolution's practical value and new insights on higher education technology enhancement.

Further verification of the evolution and its evolved system should be based on formal logic system verification and service-oriented system testing methodology. The formal logic system verification for evolution can be started from system logical formalization, which includes evolution logical formalization, evolution process formalization and evolved system formalization. The testing activities of the evolved system include component-based system testing, systematical testing and evolved system integration testing. The formal system construction and testing are essential for a successful system evolution.

The guarantee of the successful system evolution is quite needed, especially in an e-learning system. The e-learning system evolution will affect not only the system itself, but also the university teaching environment, such as online coursework reviewing. The online coursework reviewing data must be protected and backed-up seriously.

The system evolution must be verified through formal logic and system testing. The evolution life cycle should include the verification activity as part of its working process, so that it can enhance the quality of further evolution.

The evolution pattern structure can also contain the verification mechanisms and XML schema as part of the evolution pattern for university e-learning system evolution.

## ACKNOWLEDGMENT

This paper was co-sponsored by the British Royal Society of Edinburgh (RSE-Napier E4161) and the Natural Science Foundation of China (Ref.: 61070030), and the case study was further sponsored by the Shanghai Education Committee University Education Enhancement Project.

## REFERENCES

Wang, L. 2015. *Building the environment that benefit for students' learning: break through the orientation of exploring the education teaching research.* The 2015 International conference on Social Science and Contemporary Humanity Development, Wuhan, China.

Wang, Z. & Chalmers, K. 2013. *Evolution Feature Oriented Model Driven Product Line Engineering Approach for Synergistic and Dynamic Service Evolution in Clouds: AO4BPEL3.0 Proposal.* In: International Conference on Information Society (i-Society 2013). University of Toronto, Canada. June 24–26, 2013. Toronto, Canada: IEEE.

Wang, Z. & Chalmers, K. 2013. *Evolution Feature Oriented Model Driven Product Line Engineering Approach for Synergistic and Dynamic Service Evolution in Clouds: Four Kinds of Schema.* In: Acadia University, Canada, The 4th International Conference on Ambient Systems, Networks and Technologies, June 25–28, 2013. Halifax, Canada: Procedia Computer Science.

Wang, Z. & Chalmers, K. 2013. *Evolution Feature Oriented Model Driven Product Line Engineering Approach for Synergistic and Dynamic Service Evolution in Clouds: Pattern Data Structure.* In: The 7th International Conference for Internet Technology and Secured Transactions (ICITST-2012), 10–12, December 2012. London, UK: Copyright © ICITST-2012 Published IEEE UK Computer Chapter.

Wang, Z. Chalmers, K. & Cheng, G.J. 2015. An approach to Synergistic and Dynamic Service Evolution in Clouds. *International Journal of Cloud Computing, Inderscience Publishers, Switzerland.* 4(2): 177–198.

Wang, Z. Chalmers, K. & Liu, X. 2013. *Evolution Feature Oriented Model Driven Product Line Engineering Approach for Synergistic and Dynamic Service Evolution in Clouds.* Journal of Industrial and Intelligent Information, 1(1) March 2013, The 2013 2rd International Conference on Database and Data Mining, Seoul, Korean, May 11–12, 2013.

Wang, Z. Chalmers, K. & Liu, X. 2014. Evolution Pattern Verification for Services Evolution in Clouds with Model Driven Architecture. *The International Journal for e-Learning Security (IJeLS), Infonomics Society, UK,* 3(3/4): ISSN 2046-4568 (Online).

*Advanced Materials, Structures and Mechanical Engineering – Kaloop (Ed.)*
© 2016 Taylor & Francis Group, London, ISBN 978-1-138-02793-0

# Coupled multiscale nonlinear dynamic analyses of Reinforced Concrete Shear Wall buildings

I. Cho, S. Yemmaleni & I. Song
*Department of Civil, Construction and Environmental Engineering, Iowa State University, Ames, Iowa, USA*

ABSTRACT: Recent extreme earthquake in Chile raises significant questions regarding unexpected dynamic behavior and failure mechanisms of Reinforced Concrete Shear Wall (RCSW) buildings. As an effort to understand the unexpected dynamic responses, this research performs "coupled" dynamic analysis of a typical RCSW building, and compares the responses against those from "uncoupled" analysis. Here, "coupled" analysis means a multiscale dynamic analysis that links millimeter length scale's microphysical damage phenomena to the building-level nonlinear dynamic responses on parallel computers. In contrast, simple macroscopic dynamic analysis using multilinear shear wall models is denoted "uncoupled." Results show that the inelastic dynamic responses from the uncoupled analysis tend to substantially deviate from those of the coupled analysis. Physical rationales are twofold: (1) Bidirectional damage interaction in multiple directions engenders "premature" development of inelasticity in the unexcited direction; (2) Path-dependence of nonlinear responses of RCSW becomes aggravated as more microscopic mechanisms are incorporated into coupled analyses.

## 1 INTRODUCTION

Recent devastating earthquakes in Japan, Chile and New Zealand raise significant questions regarding Reinforced Concrete Shear Wall (RCSW) buildings' unexpected dynamic behaviors, and call for in-depth researches on the underlying complexity behind nonlinear dynamic responses of multifaceted levels—from a primary shear wall to an entire building.

Computational investigation of the dynamic behavior of RCSW building is of significant importance for optimal design and hazard mitigation. Traditional and widespread approaches are the use of macroscopic models for a complex shear wall system (Panagiotou & Restrepo 2009, Yang et al. 2012). Although the macroscopic models provide valuable understanding of the qualitative dynamic behavior of RCSW structures, intractable challenges remain when the target RCSW buildings involve complex geometry and interrelated microscopic damage phenomena. We denote the aforementioned approaches that use macroscopic shear wall models as "uncoupled" dynamic analysis.

The principal goal of the present work is to seek answers regarding "unexpected" inelastic dynamic responses of RCSW buildings by performing comparative studies on novel "coupled" dynamic analyses. By the "coupled" dynamic analysis, we mean the fully interrelated nonlinear dynamic analysis of a RCSW building in which several millimeter scale's damage phenomena (e.g., cracks, crushing, interlocking, etc.) directly affect the building-level dynamic responses and vice versa,

through the multiscale parallel analysis technique (Cho & Porter, 2014).

For the uncoupled dynamic analysis, we derived nonlinear hysteretic models for all structural elements including rectangular walls, box-shaped walls, L-shaped walls and RC columns, by using the microphysical mechanism-based parallel finite element analysis platform, called VEEL (Virtual Earthquake Engineering Laboratory) (Cho 2013).

In what follows, we will explain how we construct the 4-story RCSW building by adopting "coupled" and "uncoupled" analysis approaches. We shall briefly touch upon the central notions of three-tiered multiscale dynamic analysis platform and advanced parallel computing algorithms, all of which are essential to realizing the novel "coupled" dynamic analysis. A detailed comparison of dynamic responses from coupled and uncoupled simulations shall be examined. Physically sound rationales behind the discrepancy between the coupled and uncoupled dynamic analyses will draw practically meaningful insight into the dynamic responses of RCSW building.

## 2 MULTISCALE MODELING OF RCSW BUILDING: COUPLED AND UNCOUPLED APPROACHES

The specimen to be studied is a representative 4-story RCSW building with identical structural layout at each story (see Fig. 1). As depicted in Figure 1, each floor contains 27 RC columns, two L-shaped walls, one rectangular wall, and one box-shaped wall with

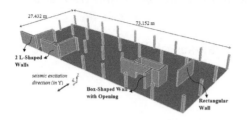

Figure 1.  Floor plan and the direction of seismic loading of the 4-story RCSW building.

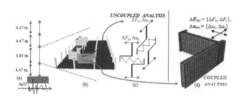

Figure 2.  Schematic illustration of modeling a 4-story RCSW building with the "coupled" and "uncoupled" approaches: (a) Reduced MDOF system for building-level dynamic analysis; (b) Floor-level modeling of columns and walls; (c) "Uncoupled" approach to a wall using two shear springs; (d) "Coupled" approach using parallel multiscale FEA engine (VEEL).

opening. The 4-story RCSW is similar to an actual building located at California region and designed according to 1970's building code.

We assumed rigid-diaphragm and imposed one-directional seismic loading along the Y direction (Fig. 1) to facilitate the comparative investigation. It should be stressed that the asymmetric wall systems result in secondary translation in the X direction as well as torsional responses as shall be discussed later. Building-level dynamic responses are described by a simple Multi-Degree-Of-Freedom (MDOF) system as Figure 2 illustrates.

For the ground motion, we utilized the Pacific Earthquake Engineering Research Center (PEER) Ground Motion Database (PEER 2011), and the selected ground motion is scaled up to the intensity at which inelastic responses of the target RCSW building begin to emerge.

## 3  DETAILS OF COUPLED DYNAMIC ANALYSIS

### 3.1  *Microphysical mechanisms harnessed by the coupled dynamic analysis*

The notable strength of the "coupled" dynamic analysis is the inclusion of the microscopic parallel multiscale FEA engine, VEEL (Cho 2013). Through VEEL, nonlinear dynamic behavior of

the 4-story RCSW building is directly derived from microscopic damage phenomena including multi-directional cracks, nonlinear shear over cracked surface, complex confinement effect, steel bar's progressive buckling and yielding, etc. Particularly, the fixed-type multidirectional smeared crack model preserves the physical attribute of real crack and microscopic stresses are defined on each crack surfaces: $\left\langle \sigma_1^c \sigma_2^c \sigma_3^c \ \tau_{12}^c \tau_{23}^c \tau_{13}^c \right\rangle$ Regarding crack-normal stresses, $\sigma_i^c$, we adopted the well-known Thorenfeldt (Thorenfeldt et al. 1987) compression model and the Moelands and Reinhardt (Reinhardt 1984) tension model (see Cho & Porter 2014 for details). To describe nonlinear shear $\tau_{ij}^c$ developing over the cracked concrete, VEEL utilizes a three-dimensional (3D) interlocking mechanism which is based on tribology (i.e., science of frictional motion of contacting surfaces) (Cho 2013). It is noteworthy that the random particle sizes used for the 3D interlocking model is generated based on Gaussian distribution, $N(0.0019 \text{ mm}, 0.00633^2)$, thereby incorporating material heterogeneity.

Departing from the previous experiment-based steel bar models (e.g., Kunnath et al. 2009, Dhakal & Maekawa 2002), information-based steel bar model has been proposed (Cho 2013) to describe "progressive" buckling and other degradation phenomena involving reinforcing bars. Another significant novelty of the VEEL is that it considers the general strength enhancement resulting from reinforcements by the nonlocal information-based confinement model (see details Cho & Hall 2014).

Throughout all the "coupled" dynamic analysis presented herein, all of these advanced microphysical mechanisms are made a play in concert to generate present shear-force resistance of a complex shear wall system.

### 3.2  *Multi-layered parallel grouping for coupled analysis*

The main power of the grouping parallel algorithm arises from the "CPU-grouping" scheme, the structure's sub-domain is handled by a group of CPUs (not communication allows us to perform the coupled building-level dynamic analysis (see Fig. 3). For the present "coupled" nonlinear dynamic analysis, we harnessed 160 CPUs with 4 sub-groups (each per floor), and it took 160 hours to fulfill 400 time steps on CyStorm (HPC cluster of Iowa State University; two 2.2 GHz Intel Opteron quad cores sharing 8 GB memory per node).

### 3.3  *Validations of general applicability*

In terms of general applicability, VEEL has the unique strength requiring a few material properties while the entire geometry and complete

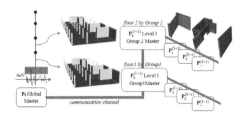

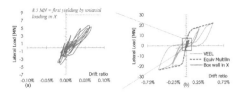

Figure 4. Premature inelastic behavior of the box-shaped wall system in the X direction: (a) Magnified force-horizontal drift responses obtained from building-level dynamic analysis; (b) box-wall's dynamic response overlapped on the quasi-static simulation results.

Figure 3. Parallel grouping algorithms applied to the coupled dynamic analysis of RCSW building. Each group of CPUs handle parallel multiscale FEA of individual wall on the assigned floor.

reinforcement layout are automatically modeled by an in-house finite element preprocessing program. In spite of the small number of input parameters, VEEL has proven to cover a wide range of real-scale RC structures without restriction to complex geometry and reinforcement layout. In all the validation models, concrete and reinforcing steel bars are modeled by 8-node hexahedral solid element and space trusses, respectively. Overall simulation results fit well with the experiment data as well demonstrated in Cho & Porter 2014.

# 4 PHYSICAL RATIONALES BEHIND UNEXPECTED DYNAMIC RESPONSES

## 4.1 *Premature inelastic behavior and stress redistribution within wall*

Although floor-level "force redistribution" over the structural elements on the same story may reasonably take place in the uncoupled analysis, the stress redistribution "within" a wall system is rarely allowed. On the contrary, the coupled analysis naturally allows not only the "force redistribution" over structural elements on a floor, but also the "stress redistribution" within a particular wall system.

This wall-level stress redistribution becomes significant when the wall's dynamic motions occur in considerably different (or unexpected) directions. The misalignment of actual dynamic motions and presumed direction of a major force-resisting mechanism of a wall may be one of the most important sources behind the unexpected nonlinear dynamic behavior of RCSW building.

Particularly for an asymmetric wall system, severe damage on a portion of wall system directly affects the stiffness degradation of the other orthogonal direction. Here, the damage in the Y direction (including cracks, longitudinal bar yielding and localized concrete crushing) appears to give rise to "premature" inelastic behavior of the wall in the X direction. For instance, Figure 4 shows the X-directional lateral force-horizontal drift responses of the box-shaped

wall system. This premature inelasticity development appears to explain the intrinsic difference between "coupled" and "uncoupled" dynamic analysis. In the coupled analysis, the local damage on any portion of a wall in any direction is directly updated through coherent microscopic analysis via multiscale FEA framework. However, in the uncoupled analysis, the X-directional force resistance is predefined by a fixed multilinear model, thereby being independent of Y-directional degradation.

## 4.2 *Aggravated path-dependence of nonlinear responses*

One of the central novelties of the present "coupled" dynamic analysis is that its underlying Parallel Multiscale Finite Element Analysis (P-M-FEA) naturally accommodates unforeseeable equilibrium paths during the dynamic motions. This ability stems from the adopted microphysical mechanisms.

However, this novel feature of the "coupled" multiscale dynamic analysis may engender important computational issue of "sensitivity" to path-dependence. To briefly illustrate this, consider first the internal element force update procedure of the "uncoupled" dynamic analysis. When a macroscopic force-displacement relation $f(.)$ is used, then

$$F_{mc}^{(t)} = f(u^{(t)})$$ (1)

where, $t$ = time step; $F_{mc}^{(t)}$ = internal force vector; $u^{(t)}$ = element displacement vector; $f(.)$ is a macroscopic hysteresis model capable of describing RC element's overall degradation. On the contrary, in the "coupled" multiscale dynamic analysis, greater sensitivity emerges due mainly to the aggravated path-dependence of multiple microphysical mechanisms. In the typical FEA we have

$$F_{mc}^{(t)} = \int_{V}^{\square} B^{T} \left\{ \sigma_{concrete}^{(t)} + \sigma_{steel}^{(t)} \right\} dV$$ (2)

where, $B$ = strain-displacement matrix; $\sigma_{concrete}^{(t)}$ = concrete stress evaluated from current

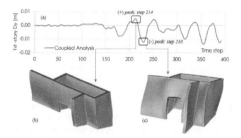

Figure 5. Building-level floor displacement response and the associated deformed shape of box-shaped wall system on garage floor: (a) Y displacement history of the 1st story; (b) Deformation at the first positive peak (time step 214); (c) At the second negative peak (time step 230). All deformations scaled up by a factor of 300.

concrete strain $\varepsilon_{concrete}^{(t)}$; $\sigma_{steel}^{(t)}$ = steel stress. Here, the concrete stress is a complex concomitant of many microscopic mechanisms including multidirectional crack model, 3D interlocking-based shear model, and a general confinement effect model.

## 5 MICROSCOPIC DAMAGES DURING BUILDING-LEVEL DYNAMIC MOTIONS

The central novelty of the "coupled" multiscale dynamic analysis lies in that it can offer unprecedented access to microscopic damage processes, notably "during" the building-level dynamic motions. As shown in Figure 5, the box-shaped wall system undergoes complex deformations during dynamic excitation, and each panel of the box appears to behave in a three-dimensional fashion. Consequently, the resultant resistance stems from the intricate combination of in-plane and out-of-plane deformation, which appears to provide another important rationale behind the intrinsic difference between "uncoupled" and "coupled" dynamic analyses.

## 6 CONCLUSIONS

With the aid of recent technology convergence of multiscale FEA technique and parallel computing algorithms, we compared inelastic dynamic responses of a 4-story RCSW building by fully "coupled" dynamic analysis against those from macroscopic "uncoupled" analysis. At the early phase of an inelastic regime, substantial discrepancy in dynamic responses arises. For the intrinsic difference between the coupled and uncoupled dynamic analyses of RCSW building, the coupled dynamic analysis provides physically

sound rationales in terms of detailed information regarding microscopic damage phenomena of shear wall systems, notably "during" the dynamic motions of the RCSW building.

A physical rationale is that nonlinear damage resulting from one-directional excitation can aggravate the degradation in the other orthogonal direction of general wall system when a wall system involves a substantial level of asymmetric geometry and force-resisting mechanisms. Another physical reason for the difference is the intensified path-dependence of the nonlinear dynamic responses owing to the increased microscopic mechanisms as well as the unexpected bidirectional translational motions of building.

## REFERENCES

Cho, I. & Hall, J.F. 2014. General Confinement Model based on Nonlocal Information. *ASCE Journal of Engineering Mechanics*, 140(6): 04014026.

Cho, I. & Porter, K.A. 2014. Structure-Independent Parallel Platform for Nonlinear Analyses of General Real-Scale RC Structures under Cyclic Loading. *ASCE ASCE Journal of Structural Engineering 140 (SPECIAL ISSUE): Computational Simulation in Structural Engineering*, A4013001.

Cho, I. & Porter, K.A. 2014. Three-Stage Multiscale Nonlinear Dynamic Analysis Platform for Tackling Building Classes. *Earthquake Spectra*; in press.

Cho, I. 2013. Virtual earthquake engineering laboratory for capturing nonlinear shear, localized damage, and progressive buckling of bar. *Earthquake Spectra*, 29(1): 103–126.

Dhakal, R. & Maekawa, K. 2002. Modeling for post-yielding buckling of reinforcement. *ASCE Journal of Structural Engineering*, 128: 1139–1147.

Kunnath, S.K., Heo, Y.A. & Mohle, J.F. 2009. Nonlinear uniaxial material model for reinforcing steel bars. *ASCE Journal of Structural Engineering*, 135(4): 335–343.

Panagiotou, M. & Restrepo, J.I. 2009. Dual-plastic hinge design concept for reducing higher-mode effects on high-rise cantilever wall buildings. *Earthquake Engineering and Structural Dynamics*, 38: 1359–1380.

PEER Ground Motion Database. 2011. Pacific Earthquake Engineering Research (PEER) Center: Berkeley, CA: http://peer.berkeley.edu/peer_ground_motion_database/site.

Reinhardt, H.W. 1984. Fracture mechanics of an elastic softening material like concrete. *Heron*, 29(2): 3–35.

Thorenfeldt, E., Tomaszewicz, A. & Jensen, J.J. 1987. *Mechanical properties of high-strength concrete and applications in design*. Proceedings of the Symp. Utilization of High-Strength Concrete.

Yang, T.Y., Moehle, J.P., Bozorgnia, Y., Zareian, F. & Wallace, J.W. 2012. Performance assessment of tall concrete core-wall building designed using two alternative approaches. *Earthquake Engineering & Structural Dynamics*, 41(11): 1515–1531.

*Advanced Materials, Structures and Mechanical Engineering – Kaloop (Ed.)*
© 2016 Taylor & Francis Group, London, ISBN 978-1-138-02793-0

# Influence of river level fluctuation on the stability of high floodplains bank slope

G.L. Tao, B.Z. Tan, J.Y. Fan, J. Ruan & Z.S. Yao
*College of Harbor, Coastal and Offshore Engineering, Hohai University, Nanjing, China*
*Nanjing Changjiang Waterway Engineering Bureau, Nanjing, China*

ABSTRACT: In this article, high floodplains bank slope in Jingjiang Zhongzhouzi reaches of the Yangtze River is chosen as the research background. According to the features of dual soil structures and the water level variation in Jingjiang reaches, unsaturated-saturated seepage analyses were carried out by using the commercial program SEEP/W to study the change law of the slope seepage field during the river level fluctuation. At the base of the seepage results and shear strength theory for unsaturated soil, the limit equilibrium analysis program (Morgenstern-Price method) was used to evaluate the change in the safety factor of different modes of river level fluctuation. The results of the analysis show that the slope safety factors increase at the rising stage. At the falling stage, the safety factors decrease first and then increase. As the velocity of the descent becomes bigger, the minimum safety factor becomes smaller and the most dangerous river level becomes lower under the same rising conditions. Besides, as the velocity of rising becomes bigger, the minimum safety factor becomes bigger under the same falling condition of the river level. The slope stability at the rising stage will indirectly affect the slope stability at the falling stage.

## 1 INTRODUCTION

Jingjiang reaches of the Yangtze River begin from Zhicheng of Hubei Province and end in Chenglingji, and its total length is 347.2 km. The bank slope of Jingjiang reaches is of complex geological conditions and the evolution of its riverbed is intense. What's more, the seasonal fluctuation of the water level of the Yangtze River makes the seepage field and the stress field of the bank slope change acutely, which will also develop large deformation or even slope failure. In the past, huge landslides had occurred in Xuetangzhou, Guanyin shoal, Qigongling and other places of Jingjiang reaches (Xie et al. 2009).

Factors affecting the slope stability are very complex, including the role of the groundwater that cannot be ignored. Jones investigated the landslides that occurred near the Roosevelt Lake from 1941 to 1953. It was found that 49% of them occurred during the first impounding period of the reservoir, and 30% of them occurred as a result of drawdown of the reservoir (Nakamura 1990). Liao et al. studied the slope stability influenced by the drawdown speed of the water level. Their analysis results showed that the safety factor decline rates were significantly increased as the drawdown speed increased (Liao et al. 2005). Liao et al. also studied the relationship of the seepage field and slope stability with different rising rates of the water level. Their analysis results

showed that the safety factor increased as the water level rose (Liao et al. 2008). Jia and Reto-Schnellmann studied the seepage field and slope stability under the condition of water level fluctuation by developing a physical slope model (Jia et al. 2009; RetoSchnellmann et al. 2010). Jian et al. studied the mechanism and failure process of Qianjiangping landslide in the Three Gorges Reservoir, and found that the water level fluctuation had a significant effect on the slope stability (Jian et al. 2014).

The high floodplains bank slope in Jingjiang Zhongzhouzi reaches of the Yangtze River consists of two kinds of soil. The upper part is the silty clay layer and the lower part is fine sand layer. In this paper, according to the features of dual soil structures and the water level variation in Jingjiang reaches of the Yangtze River, unsaturated-saturated seepage analyses were carried out by using the commercial program SEEP/W to study the change law of the slope seepage field during the river level fluctuation. The obtained pore-water pressure fields were then imported into the limit equilibrium analysis program (Morgenstern-Price method) to evaluate the change in the safety factor under the fluctuations of the river level. The research results have a great significance for the prediction of the slope stability of Jingjiang Reaches, and also provide a reference for the safety during the construction period of waterway regulation projects.

## 2 UNSATURATED-SATURATED SEEPAGE THEORY AND UNSATURATED SHEAR STRENGTH THEORY

Seepage analyses were carried out to simulate the seepage field of the bank slope based on the unsaturated-saturated seepage theory (Fredlund & Rahardjoh 1993). The governing equation is given in Formula (1) as follows:

$$k_1 \frac{\partial}{\partial x}\left(k_x \frac{\partial H}{\partial x}\right) + \frac{\partial}{\partial y}\left(k_y \frac{\partial H}{\partial y}\right) + Q = m_w \gamma_w \frac{\partial H}{\partial t} \quad (1)$$

where $H$ is the total head; $k_x$ is the coefficient of permeability in the horizontal direction; and $k_y$ is the coefficient of permeability in the vertical direction. In unsaturated soil, coefficient of permeability is the function of matric suction, where $Q$ is the boundary flow of application; $t$ is the time; $\gamma_w$ is the unit weight of water; and $m_w$ is the slope of the soil-water characteristic curve.

The change in the river level will result in the change in the pore-water pressure field of the bank slope. And the change in the shear strength with the pore-water pressure can be described by using Equation (2) (Fredlund & Rahardjoh 1993) as follows:

$$\tau_f = c' + (\sigma_n - u_a)\tan\varphi' + (u_a - u_w)\tan\varphi^b \quad (2)$$

where $c'$ and $\varphi'$ are effective stress intensity indices. The pore-air pressure ($u_a$) in the slope is assumed to be zero, and the pore-water pressure ($u_w$) can be obtained from the result of the simulation of the seepage field. The influence of pore-water pressure on the shear strength can be quantified by $(u_a - u_w)\tan\varphi^b$. According to the method of S.K. Vanapalli, unsaturated shear strength can be determined based on the SWCC and saturated stress intensity index (Vanapalli et al. 1996). Thus, this method is adopted in this paper.

## 3 CALCULATION MODEL, MODEL PARAMETERS AND CALCULATION SCHEME

### 3.1 Calculation model

To determine the changes in pore-water pressure in the bank slope under different conditions of river level, unsaturated-saturated seepage analyses were carried out using the software SEEP/W. The obtained pore-water pressure fields were then imported into the limit equilibrium analysis program (Morgenstern-Price method; M-P method) to evaluate the change in the safety factor during the fluctuations of river level in different cases (Morgenstern & Price 1965).

In this calculation model, its length is 100 m and height is 54.5 m. The elevation of its top surface and river bottom is 34.5 m and 13.0 m (yellow sea base level, the same below). This bank slope has three soil layers: the first layer is silty clay of 3.5 m thick, the second layer is incompact fine sand layer of 17 m to 23 m thick and the third layer is medium dense fine sand, as shown in Figure 1. Figure 2 shows the finite element mesh of the slope. The boundary of the seepage model is defined as follows: (1) both sides of the model are specified as zero-flux boundaries above the ground water table and total head boundary below the ground water table. The total head boundary of the left side of the model is equal to the groundwater level value of 32 m, and the right side is equal to the changing river level; (2) the bottom of the model is specified as the zero-flux boundary.

### 3.2 Model parameters

The Soil-Water Characteristic Curves (SWCC) of three kinds of soil are obtained from the database of software Geo-studio based on the types of soil and the saturation volumetric water content. According to the study by Van Genuchten, water permeability functions are determined by the permeability coefficient of saturated soil and the SWCC (Genuchten 1980), which can also be obtained from the database of software Geo-studio.

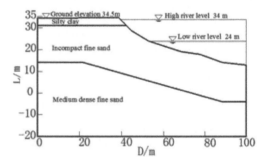

Figure 1.   A typical slope model.

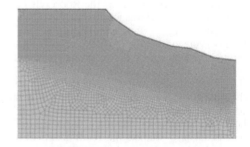

Figure 2.   Finite element mesh of the slope model.

Table 1. Soil parameters.

| Types of soil | Natural unit weight (kN/m³) | Saturated unit weight (kN/m³) | Cohesion (kpa) |
|---|---|---|---|
| Silty clay | 18.6 | 18.79 | 22 |
| Incompact fine sand | 20.2 | 20.44 | 0 |
| Medium dense fine sand | 20.4 | 20.41 | 0 |

| Types of soil | Internal friction angle (°) | Permeability coefficient of saturated soil (m/s) |
|---|---|---|
| Silty clay | 13.8 | $5 \times 10^{-8}$ |
| Incompact fine sand | 29.8 | $5 \times 10^{-6}$ |
| Medium dense fine sand | 32 | $10^{-6}$ |

Table 2. The mode of river level fluctuation of each case.

| | River level: 24 m rise to 34 m | | River level: 34 m descent to 24 m | | River level stay at 24 m |
|---|---|---|---|---|---|
| Case | Velocity of rising (m/d) | Duration (day) | Velocity of descent (m/d) | Duration (day) | Duration (day) |
| 1 | 0.1 | 100 | 0.1 | 100 | 165 |
| 2 | 0.2 | 50 | 0.2 | 50 | 265 |
| 3 | 0.1 | 100 | 0.2 | 50 | 215 |
| 4 | 0.1 | 100 | 0.3 | 33 | 232 |

Table 1 lists the parameters of three kinds of soil used in this model.

### 3.3 Calculation scheme

According to the survey station of water level of the Yangtze River in Jianli, Table 2 presents 4 cases of different combinations of velocity.

The simulation of the seepage field during the water level fluctuation is a transient analysis, which needs the initial seepage field. In this paper, to make a steady seepage analysis as the initial seepage field, the river level is specified as 24 m and the groundwater level as 32 m.

## 4 RESULTS AND ANALYSES

### 4.1 Results from seepage analyses

The pore-water pressure field of the bank slope will change simultaneously with the fluctuation of the river level. What's more, there are different changing features in different modes of river level fluctuation. Figure 3 (a), (c) shows the pore-water pressure field of each case when the river level reaches the top. The phreatic line is relatively smooth in Case 1 and steep in Case 2 when the river level reaches the top.

The rising velocity is higher in Case 2 and the permeability coefficient of silty clay is small, which is the reason why the phreatic line lags behind the river level fluctuation. Figure 3 (b), (d), (e), (f) shows the pore-water pressure field of each case when the slope safety factor is minimum. For Case 1, Case 3 and Case 4, the phreatic lines show that the greater the rate of decline, the greater the slope. The larger hydraulic grade will increase the sliding force, which may induce the slope failure. When the water level of the river rises, the maximum value of pore-water pressure increases and the negative pore water pressure zone (unsaturated zone) reduces. When the river level drops, the maximum value of the pore-water pressure is reduced and negative pore water pressure area and matric suction are increased. There is a break of contour (−50 kpa) on the dividing line between the two soil layers. This is because there are differences between silty clay and fine sand in the SWCC and permeability coefficient.

### 4.2 Results from slope stability analyses

On the basis of the existing seepage field, the slope safety factor of the entire process of the river level fluctuation can be calculated based

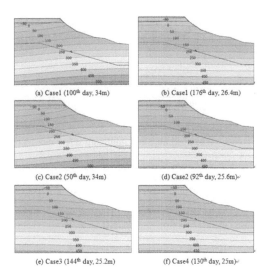

(a) Case1 (100th day, 34m)  (b) Case1 (176th day, 26.4m)

(c) Case2 (50th day, 34m)  (d) Case2 (92th day, 25.6m)

(e) Case3 (144th day, 25.2m)  (f) Case4 (130th day, 25m)

Figure 3. The contour map of four cases at different times (kpa).

on the M-P method. Figure 4 shows that the slope safety factors will increase with the rising river level. At the falling stage, the safety factors decrease first and then increase, which means there exists the most dangerous river level. Table 3 presents the maximum and minimum safety factors of four cases. As the velocity of the descent becomes bigger, the minimum safety factor becomes smaller and the most dangerous river level becomes lower under the same rising conditions.

The difference in the minimum safety factor between Case 1 and Case 2 is small. This is because both the velocity of the descent and rising of Case 2 are bigger compared with Case 1, which means the slope of Case 2 is more stable than that of Case 1 before the drawdown of the river level. What's more, because Case 2 has a bigger velocity of rising, even though the velocity of the descent of Case 2 and Case 3 is the same, the minimum safety factor of Case 2 is bigger. Therefore, the slope stability at the rising stage will indirectly affect the slope stability at the falling stage. The rise or drawdown of the river level cannot be considered separately when calculating their influence on the slope stability.

At the rising stage, the soil shear strength decreases because of the decrease in the unsaturated area and matric suction. Meanwhile, the reversed seepage of the slope is generated, which

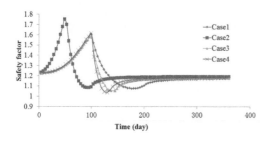

Figure 4. Change in safety factor of four cases.

Table 3. The maximum and minimum safety factors of four cases.

| | The maximum safety factor (rising stage) | | | The minimum safety factor (falling stage) | | |
|---|---|---|---|---|---|---|
| Case | Time (d) | River level (m) | Safety factor | Time (d) | River level (m) | Safety factor |
| 1 | 100 | 34 | 1.609 | 176 | 26.4 | 1.076 |
| 2 | 50 | 34 | 1.754 | 92 | 25.6 | 1.083 |
| 3 | 100 | 34 | 1.609 | 144 | 25.2 | 1.054 |
| 4 | 100 | 34 | 1.609 | 130 | 25 | 1.039 |

is in favor of the slope stability. On the whole, the value of increasing resistant shear force is bigger than that of decreasing driving shear force. Therefore, the slope safety factor increases as the river level rises.

At the falling stage, the seepage towards to the external of the slope is generated, which is bad for the slope stability. Meanwhile, the soil shear strength increases because the unsaturated area increases and matric suction recovers to a certain extent. At the early and mid-period of the falling stage, the slope safety factor decreases because the driving shear force decreases more quickly than the increase in the resistant shear force. But at the later period, the slope safety factor increases because the contribution of the matric suction gradually increases.

## 5 CONCLUSIONS

In this paper, unsaturated-saturated seepage analyses were carried out to study the change law of the slope seepage field during the river level fluctuation. At the base of the seepage results and shear strength theory for unsaturated soil, the limit equilibrium analyses were performed to evaluate the change in the safety factor of different modes of the river level fluctuation.

1. Based on the analysis results, it was found that the change in the phreatic line lags behind the river fluctuation at the silty clay layer when the velocity of rising is relatively high.
2. At the rising stage, the slope safety factor increases. In contrast, at the falling stage, the safety factors first decrease and then increase, which means there exists the most dangerous river level. As the velocity of the descent becomes bigger, the minimum safety factor becomes smaller and the most dangerous river level becomes lower under the same rising conditions.
3. As the velocity of rising becomes bigger, the minimum safety factor becomes bigger under the same falling conditions. The slope stability at the rising stage will indirectly affect the slope stability at the falling stage.

REFERENCES

Fredlund, D.G. & Rahardjoh, H. 1993. *Soil Mechanics for Unsaturated Soils.* (John Wiley & Sons, Inc. America).
Genuchten, M.Th.V. 1980. A Closed-form Equation for Predicting the Hydraulic Conductivity of Unsaturated Soils, *Soil Science Society of America Journal*, 44: 892–898.

Jia G.W. Tony L.T. Zhan, Y.M. Chen, & Fredlund. D.G. 2009. Performance of a large-scale slope model subjected to rising and lowering water levels, *Engineering Geology*, 106: 92–103.

Jian, W.X. Xu Q. Yang, H.F. & Wang, F.W. 2014. Mechanism and failure process of Qianjiangping landslide in the Three Gorges Reservoir, China, *Environmental Earth Sciences*, 72: 2999–3013.

Liao, H.J. Ji, J. & Zeng, J. 2008. Stability analysis of soil slopes considering saturated and unsaturated seepage effect, *Rock and Soil Mechanics*, 29(12): 3229–3234.

Liao, H.J. Sheng, Q. Gao, S.H. & Xu, Z. 2005. Influence of drawdown of reservoir water level on landslide stability, *Chinese Journal of Rock Mechanics and Engineering*, 24: 3454–3458.

Morgenstern, N.R. & Price, V.E. 1965. The analysis of the stability of general slip surfaces, *Geotechnique*. 15: 70–93.

Nakamura, K. 1990. On reservoir landslide, *Bulletin of Soil and Water Conservation*, 10: 53–46.

Schnellmann, R. Busslinger, M. Schneider, H.R. & Rahardjo, H. 2010. Effect of rising water table in an unsaturated slope, *Engineering Geology*, 114: 71–83.

Vanapalli, S.K. Fredlund, D.G. Pufahl, D.E. & Clifton, A.W. 1996. Model for the prediction of shear strength with respect to soil suction, *Canadian Geotechnical Journal*, 33: 379–392.

Xie, L.M. Li, S.X. Yi, F.W. Yin, X.S. & Han, X. 2009. Failure types and influential factors of bank slope along Jingjiang reach of the Yangtze River and control measures, *Yangtze River*, 40: 4–6.

*Advanced Materials, Structures and Mechanical Engineering – Kaloop (Ed.)*
© 2016 Taylor & Francis Group, London, ISBN 978-1-138-02793-0

# Scour protection of the foundations of Offshore Wind Energy Converters

V. Efimova, A. Karelov, A. Nesterov & N. Belyaev
*Peter the Great St. Petersburg Polytechnic University, St. Petersburg, Russia*

ABSTRACT: North and Far-Eastern Russian offshore territories requirements for electric power can be satisfied through the creation of the Offshore Wind Energy Converters (OWEC). The foundation for the offshore settings consists of concrete, steel piles and multi-supporting structures. Protection of a seabed from scouring near the OWEC by using different measures is considered. Conclusions are based on experimental studies.

## 1 INTRODUCTION

A requirement for electrical power for the North and Far-Eastern offshore Russian territories exists, which can be related to both the necessity of the organization of some activity on these territories and the providing of electric power to small settlements. These necessities can be satisfied by the creation of the marine wind energetic settings. Therefore, it is appropriate to design, create and use the OWEC for remote and isolated areas of the country (Mel'nikova 2013). Accommodation of wind farm on sea water area has the following advantages:

- The productivity of wind sets and the quality of the energy supplied to the grid in the region increases;
- Alienation of the landed lands and building of expensive entrance roads is not required in remote locations (e.g. forests, mountains, swamps);
- Mobility, high maintainability and simple recycling at the end of life.

Two varieties of OWEC can be developed: stationary (with support on the seabed) and floating execution.

One of the problems influencing the safe operation of such structures is the danger of scouring.

Experimental and theoretical studies of the scour near the OWEC have not been conducted in Russia until now, and, in general, offshore and port structures used in shallow sea areas have been investigated.

The results of previous experimental studies undertaken in 2013–2014 for the evaluation of the efficiency of some techniques for scour protection near the foundations of offshore structures are used in the present analysis, and the development of recommendations for the most common types of foundations of the OWEC is considered in this research.

The main attention is given to the consideration of the possibility of using new, previously not applied methods of protection.

## 2 THE FOUNDATIONS OF THE SEA-BASED INSTALLATIONS

The foundations of the sea and land-based settings considerably differ in a technical and economic plan from each other. The foundation for the offshore settings consists of concrete, steel piles and multi-supporting structures. The choice of the foundation type is influenced by many factors, such as depth, currents, waves' heights, and ice danger, which are equally very important.

The concrete basket is fixed on a seabed by the weight of foundation (Fig. 1a). The basket is made in a dry dock from steel and concrete, transported by a ship to the place of setting and after immersion is filled with sand and gravel. The advantage of such concrete boxes is a lot of resistance in the case of ice movements. The disadvantages include the high cost at great depths. These foundations have been tested only in shallow water (less than 10 m depth); at great depths, their use is not economically feasible.

The next type is one-pile structures from steel, which are the simplest method for offshore foundations (Fig. 1b). They consist of a steel pipe, which is driven into the seabed. This method is especially expedient for settings at a depth of about 15–20 meters. They can be installed relatively easily and quickly using pile drivers.

Also, the foundation may be in the form of a truss structure (Fig. 1c). Such structures have a lattice structure and are used at great depths (over 20 m). This method is used in the construction of drilling platforms. The tower of wind turbines is associated with a frame structure made of pipes,

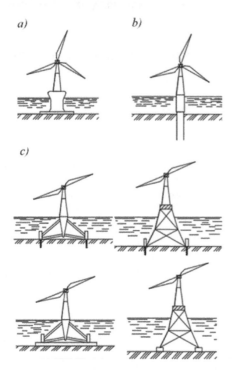

Figure 1. Some types of foundations under the OWEC: a—concrete basket; b—one-pile structure; c—structures of tripod and truss construction.

and distributes the load on the legs or on multiple grids or is fastened to the box or to the piles. For the pile, a less section can be used than at one-pile, which makes the process of pile driving much easier.

There is so-called bucket-foundation that is steel beaker, upside-down. This bucket is installed on the seabed, and then the air is pumped out of it. The created vacuum pulls the beaker into the seabed. The bucket-foundation is used only for homogeneous soils. During installation, it does not require pile driving. Therefore, this method is sparing to the environment. After the end of foundation life, it is very easy to dismantle by supplying a compressed air into a bucket.

Floating foundations are widely used in the practice of building. At depths greater than 50 meters, it is difficult to establish a foundation rigidly mounted at the seabed. Therefore, the idea to make a floating foundation appeared. Considerable experience of exploitation of such foundations is accumulated in the oil industry. But the introduction of this technology in the building of offshore wind farms should take into account the fact that a much larger force will operate at the platform (Mel'nikova & Radchenko 2013).

## 3 SCOUR PROTECTION OF THE OWEC FOUNDATIONS

Scour protection of bottom soils near the OWEC by using different measures is considered in the present work. During the protracted exploitation of this type of buildings, the considerable scour of the bottom is possible directly at fundamental support, which can result in the loss of stability of all construction. The formation of voids due to the washing of soil results in the necessity of expensive technological measures on its strengthening (Alhimenko et al. 2003, Belyaev 2009, Chugunova & Belyaev 2014, Nesterov 2014). A justification of the choice of protective measures, providing faultless work of gravity offshore structures, is a very actual and practically meaningful task (Alhimenko et al. 2003).

The methods of protection are actual and of restoration, which are used after the occurrence of scouring. Furthermore, the known protection methods from seabed scour can be classified into two types (Alhimenko et al. 2003, Belyaev 2009, Chugunova & Belyaev 2014, Nesterov 2014, Simakov et al. 1989): passive and active methods. In addition, all measures of protection can be divided into two groups by the method of propagation:

a. with mainly horizontal distribution (e.g. protection by granular materials, mats, mattresses);
b. with mainly vertical distribution («skirts», hip protection, change of the form of building corps).

All existing methods of protecting the soil from scouring have some disadvantages (Chugunova & Belyaev 2014, Nesterov 2014, Shchemelinin et al. 2014, Technical Report 2012, Vølund 2005).

The proposals for the installation from rubber, geotextile and other flexible elements appeared recently. Advantages of mats are as follows: low dead load compared with stones; produced ashore (high quality); can be easily dismantled.

The new method of protection was developed recently by a company «TWELL». It consists of manufacturing and laying on the seabed surface mat of synthetic, water-resistant, non-woven material, which is a two-layer cloth, quilted along its entire length, which allows to form longitudinal channels, and fills the ballasting material from the dispensing hopper located on the supply vessel (Nesterov 2014, Information on http://twell.ru/marine-technologies/).

## 4 EXPERIMENTAL STUDIES

The experimental studies were undertaken in 2013–2014 to evaluate the efficiency of some techniques for scour protection of the seabed soil near the foundations of offshore structures. They were performed at special-purpose test rigs of model basins in three scales. The detailed description of the experimental setups and equipment has been reported in previous publications (Babchik et al. 2014, Gaydarov et al. 2014, Shchemelinin et al. 2014). A number of local scour protection options were examined for the foundations of drilling platform and barge (Fig. 2): underwater berm around the structure model perimeter and four riprap ledges; deflectors; skirts; synthetic mats with ballast weights proposed by «TWELL» Ltd. The test data indicate that well-known protection systems such as riprap berm provide reliable scour protection. Considering the erection costs are comparatively high in some cases, the use of other methods is preferred (Babchik et al. 2014, Shchemelinin et al. 2014).

The soil scours protection technique suggested by TWELL Ltd using synthetic mats with ballast put in the longitudinal grooves of the mats was investigated in laboratory conditions for the first time. The mats are fitted around the structure foundation perimeter. The ballast is water-hardening mineral-magnesian composition injected as a liquid into the mat grooves immediately during the installation of the mat itself.

Advantages of the new technology are as follows:

– Covering of the underwater surface has a low cost and high durability;
– Filling of the mat with the ballasting material is produced simultaneously with its setting on the seabed;
– Technology can be used to strengthen the seabed in order to protect it from damaging by external influences, in particular from the effects of ship jet propulsion, waves, and currents.

Figure 2. Scour protection techniques in the experimental studies: a—underwater berm with riprap ledge; b—deflectors; c—skirts; d—synthetic mats.

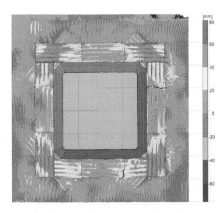

Figure 3. Relief of the bottom with synthetic mats after the experiment.

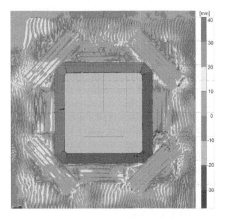

Figure 4. Relief of the bottom after removing of synthetic mats following the experiment: there are areas of local scour under the mats and foundation of the model.

The experiment shows synthetic mats with kentledge made on the technology of «TWELL» Ltd, which may protect the base of marine structures from local scour in principle (Fig. 3). However, the reliability of the device in question is essentially dependent on the type and quality of its laying on the seabed on the perimeter of the hydraulic structure (Fig. 4).

Experiments have shown that the presence of unprotected areas at the structures base can nullify the protective properties of the device under consideration (Technical Report 2012).

5 RECOMMENDATIONS

After analyzing all the known methods of the seabed protection from scouring at the base of the

OWEC and the results of the experimental studies conducted in 2013–2014, we reached the conclusion that the choice of method depends largely on the construction of foundation.

In particular, it is better to use an active method with artificial seaweed for truss bases. A method is based on that seaweed protect subject to scour seabed, reducing the turbulence of the current at the bottom. Artificial seaweed as bunches of fibers fastens to the steel, and concrete blocks are stacked in a specific order on the protected seabed (Simakov et al. 1989). This method is effective and can cause the accumulation of sedimentations and formation of sandy berm even.

For one-pile structures, the most suitable measures of protection are the so-called «skirts» (Kuzina et al. 2014). Such constructions are realized on the base of foundation on its perimeter and strengthen with an internal edges intersecting the area of the base. A skirt and internal ribs must be thin enough to submerge in soil during the setting of foundation, and durable enough to pass to the soil horizontal loadings that act on the base of the OWEC (Vølund 2005). In some cases from the data of report (Report 2012) for the skirts immersion to the calculated depth, an additional load was used.

It is possible to offer a new protection method for the concrete basket-type foundation. The ballasting material for a super heavy kentledge is made from the mobile magnesia-mineral-salt composition, directly on board the boat, using outboard salt water for shutting, and as a weighting agent – iron or lead shot (Figs. 2, c, 3, 4).

## 6 CONCLUSIONS

The results of this work are the guidelines for choosing the options for protective measures against seabed scour for the specific types of OWEC foundations based on the analysis of conducted modeling studies. When selecting a scheme of protection from scouring, an unchanged requirement remains in the need to extend this protection beyond the zone with high bottom velocity. The development of detailed connection of the edge areas with natural seabed is also important to prevent possible undermining of the edges.

The choice of the protection method also depends on the presence and cost of materials, facilities of delivery and installation. Selecting the optimal variant of protection in each case requires conducting experiments on models.

## REFERENCES

Alhimenko, A.I. Belyaev, N.D. & Fomin, Y.N. 2003. *Safety of the marine hydrotechnical buildings*, St. Petersburg, 288.

Babchik, D. Belyaev, N. & Lebedev, V. et al. 2014. *Experimental investigations of local scour caused by currents and regular waves near drilling barge foundations with the cutout in the stern*. Proceedings of 5th International Conference "Coastlab14". Varna, Bulgaria, 114–124.

Beliaev, N.D. 1997. Review of protection methods against propeller erosion. Proceedings of the IV Int. Seminar on Renovation and Improvements to Existing Quay Structures. *Technical University of Gdansk, Poland*, 1: 5–12.

Belyaev, N.D. 2009. *Protecting of foundations of ice-resistant platforms from scour*. The prevention of the failures of buildings Moscow, 228–236.

Chugunova, V.V. & Belyaev, N.D. 2014. *Analysis of the methods of protecting from washings away at marine GTS*, Week of the science of SPbPU: materials of NPK with international participation, SPbPU, 80–83.

Gaydarov, N.A. Zakharov, Y.N. & Ivanov, K.S. et al. 2014. *Numerical and Experimental Studies of Soil Scour Caused by Currents near Foundations of Gravity-Type Platforms*. International Conference on Civil Engineering, Energy and Environment (CEEE2014) Hong Kong. Information on http://twell.ru/marine-technologies/

Kuzina, A.D. Smolenkova, A.V. & Cherniy, A.V. et al. 2014. *Protection of Gravity Offshore Platforms from Scouring with "Skirts"*, Week of the science of SPbGPU: materials of NPK with international participation, SPbPU, 115–117.

Mel'nikova, A.A. 2013. *Marine floating wind power-stations for energy providing of off-shore territories*, Master's degree dissertation, SPbPU, 119.

Nesterov, A.A. 2014. *New proposals for the protection of marine bases of hydraulic structures using a variety of coatings*, Week of the science of SPbPU: materials of NPK with international participation, SPbPU, 103–105.

Radchenko, P.M. 2013. Offshore floating wind farm, *The magazine «Science and transport»*. 1(5): 82–86.

Report, 2012. Decommissioning of offshore Concrete Gravity Based Structures (CGBS) in the OSPAR maritime area/other global regions. *International Association of Oil & Gas Producers*. 484: 50.

Shchemelinin, L.G. Utin, A.V. & Belyaev, N.D. et al. 2014. Experimental studies of means efficiency for protection of the sea bed soil from erosion caused by external factors near offshore structures. *Proceedings of the ISOPE. Busan, Korea*, paper 14TPC-0320, 2: 625–631.

Simakov, G.V. Shkhinek, K.N. & Smelov, V.A. et al. 1989. *Marine Hydraulic Structures on the Continental Shelf*, Sudostroenie, Leningrad, 328.

Technical Information, 1988. *Phenomenon of the local scours of soil of foundation at supports of the drilling settings and measures on it prevention*. Technical Information No. 2, Leningrad, 97.

Technical Report, 2012. *Rhiannon Wind Farm. Stage 2. Preliminary Environmental Information*. Main Technical Report. APPENDIX 7.5, 1: 5.

Vølund, P. 2005. Concrete is the future for offshore foundations. *Wind Engineering*. 29: 531–539.

*Advanced Materials, Structures and Mechanical Engineering – Kaloop (Ed.)*
© 2016 Taylor & Francis Group, London, ISBN 978-1-138-02793-0

# Vector field modeling of seismic soil movement in building footing

Z. Tatiana & S. Pavel
*Astrakhan Institute of Civil Engineering, Astrakhan, Russia*

ABSTRACT: This paper considers various issues of seismic impact presentation on building footing with the help of vector analysis. Seismic impact evaluation of a building is based on a research of vector function behavior that asymptotically defines appropriate soil slewing and translational movements and reflects the dimensional and temporal characteristic of earthquake random field influence. The acceptance of the random processes steadiness hypothesis within the modeling of soil seismic movement allows laying down the input and output spectra at carrier frequencies.

## 1 INTRODUCTION

The main research area of this paper is a solution to the issue of forecasting the industrial buildings reliability, taking into account the changes in the operation of their stress-strain state under the influence of different load combinations (Zolina & Sadchikov 2013, 2015). The solution will give the wrong result if the impact of extreme short-term impacts would be left unattended both at the design stage and during building operation. In the case of impact considering a seismic wave, there is a necessity of creation of a decomposing model of oscillatory processes as the trigonometric series depending on the earthquake intensity level.

The seismic impact can be represented by the force vector and moment vector of force pairs, the components of which are characterized by the module and the direction cosines that are functions of the coordinates and time. Since natural impacts create perturbation fields of the wave character with a finite spreading velocity, within a building, there appears an uneven distribution of these impacts. The propagation of seismic waves in the ground motion is based on uniform structures expanded (dilated) and irregular (rotational) components. The formalization of such a decomposition leads to the consideration of the vectors of translation and rotation of the soil mass. The definition of these vectors and the relations between them depends on the nature of the impact and reduces to the study of the properties of wave fields of ground motion during earthquakes using the vector analysis.

## 2 MODELS AND ALGORITHMS

By analyzing soil movement records during earthquakes, we can distinguish three phases determined with the speed:

Longitudinal P-wave being non-vortex causes high-frequency soils movement with small amplitudes, not resulting in a significant building damage;

Transverse S-wave being vortex causes soil volume expansion and rotation, defining main earthquake power, resulting in significant damages for buildings;

Surface L-wave of vortex character with low-phase spreading velocity but sufficient enough for damages with soil oscillative power.

In accordance with the phases of earthquakes determined with spreading of P-, S- and L-waves within seismic acceleration implementation, there are three segments (Fig. 1), which can be outlined as follows:

1. Initial segment caused the activity of longitudinal seismic waves, the length of which is directly dependent on the distance from the epicenter;
2. The main phase—the most intense segment of the oscillation amplitude, the transition to which expressed on the accelerogram and caused with lateral and surface waves with similar periods of oscillation;
3. The final phase characterized with a gradual irregular change in oscillation amplitude with longer periods in the absence of explicit transition from the middle segment.

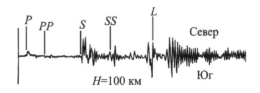

Figure 1. Typical earthquake accelerogram (11 June 1957, Kermadec Islands).

When simulating seismic disturbances for approximation, two types of random processes can be applied (Housner 1952, Barshtein 1960): cut white noise and process with hidden periodicity. In general, the study of soil movement is a nonlinear issue in the elasticity theory. However, at present, the experimental data on soil rotational properties during earthquakes are negligible. Therefore, to describe the vector field of soil seismic movement is advisable to use a linear stochastic model for such a description.

## 3 APPLICATIONS OF THE METHOD

At small rotation angular values, it is assumed that the orientation of the axles in space does not change. This assumption implies a commutative property; at that rotation angles, a vector and its derivatives (vectors of angular velocities and accelerations) are formed. The relations between the parameters of translational and rotational orientation vectors within the modeling of the intensity level of soil movements can be represented as follows:

$$\vec{a}_0(t) = \frac{1}{2} \ \text{rot} \ \vec{X}_0(t), \qquad (1)$$

where $\vec{a}_0(t) = [a_{01}(t), \ a_{02}(t), \ a_{03}(t)]$ is the angular rotation vector; $\vec{X}_0(t) = [X_{01}(t), \ X_{02}(t), \ X_{03}(t)]$ is the movement vector.

Vectors $\vec{a}_0(t)$, $\vec{X}_0(t)$ are defined on an orthogonal basis $(\vec{i}, \vec{j}, \vec{k})$, thus the rotation operator $\vec{X}_0$ at point $A(x_1, x_2, x_3)$ can be expanded as follows:

$$\text{rot} \, \vec{X}_0 = \begin{bmatrix} \vec{i} & \vec{j} & \vec{k} \\ \dfrac{\partial}{\partial x_1} & \dfrac{\partial}{\partial x_2} & \dfrac{\partial}{\partial x_3} \\ X_{01} & X_{02} & X_{03} \end{bmatrix} = \left( \dfrac{\partial X_{03}}{\partial x_2} - \dfrac{\partial X_{02}}{\partial x_3} \right) \vec{i} +$$

$$+ \left( \dfrac{\partial X_{01}}{\partial x_3} - \dfrac{\partial X_{03}}{\partial x_1} \right) \vec{j} + \left( \dfrac{\partial X_{02}}{\partial x_1} - \dfrac{\partial X_{01}}{\partial x_2} \right) \vec{k}$$

$$(2)$$

Therefore, rotation angles vectors can be expressed as follows:

$$a_{01}(t) = \frac{1}{2} \left[ \frac{\partial X_{03}(t)}{\partial x_2} - \frac{\partial X_{02}(t)}{\partial x_3} \right],$$

$$a_{02}(t) = \frac{1}{2} \left[ \frac{\partial X_{01}(t)}{\partial x_3} - \frac{\partial X_{03}(t)}{\partial x_1} \right],$$

$$a_{03}(t) = \frac{1}{2} \left[ \frac{\partial X_{02}(t)}{\partial x_1} - \frac{\partial X_{01}(t)}{\partial x_2} \right]. \qquad (3)$$

In the general vector field of seismic soil, the movement is the one changing in time and space, i.e. the non-stationary random field characterized at each point with the population $\{\vec{X}_0(\vec{r},t), \vec{a}_0(\vec{r},t)\}$. Its modeling requires extensive statistical data of seismic soil movement records. Lack of relevant information directs the research area to a field of asymptotic approximations with representation (Esteva & Rosenblueth 1964, Nikolaenko & Nazarov 1988):

– Either in the form of stationary random process for one-dimensional design and dynamic models,
– Or in the form of a stationary random field without considering the correlation between its components for two- and three-dimensional design and dynamic models.

## 4 IMPLEMENTATION OF THE METHOD

The use of stationary random processes and fields in contrast to the non-stationary ones does not cause significant computational difficulties, and allows the evaluation of the seismic impact on an object with examining of the vector function behavior as follows:

$$\tilde{U}(t) = \left[ \tilde{X}_1(t), \ \tilde{X}_2(t), \ \tilde{X}_3(t), \ \tilde{a}_1(t), \ \tilde{a}_2(t), \ \tilde{a}_3(t) \right], \quad (4)$$

where $\tilde{a}_i(t)$, $\tilde{X}_i(t)$ are vector components, which asymptotically determine, respectively, rotational and translational seismic soil movements. The function reflects the spatio-temporal characteristic of the random earthquake field influence.

According to the seismic process steadiness hypothesis (Pshenichkina et al. 2010), spectral composition change during an earthquake can be neglected. This neglecting can be explained by the fact that high-frequency shocks, which usually conclude an earthquake, have no significant impact on the change in the stress-strain state of the building compared with the strong low-frequency shocks. This hypothesis is well implemented for large earthquakes that are usually of long duration and their intensity varies slightly.

Considering a behavior of function $\tilde{U}(t)$ at the function input to a system as a white noises group, several realization alternatives of seismic impact model on a building can appear:

1. From averaged characteristics of spectral density,
2. Provision of seismic spectra as polyextremum functions,
3. Representation in a view of random processes set, in which spectral densities fill a specified band of carrier frequencies.

Within the choice of a third realization alternative of the seismic impact model on a building, a search of statistical characteristics for generic coordinates should be organized using the method of canonical decomposition proposed by V.S. Pugachev. Values of carrier frequencies of canonical decomposition $\lambda_j (j = 0 .. n)$ should be chosen so that their common band $[0, \Omega]$ will include not less than 90% of spectral density square in an earthquake output $S^{BX}(\lambda)$, i.e.

$$\int_0^\Omega S^{BX}(\lambda)\, d\lambda > 0{,}9 \tag{5}$$

It is typical for random processes of seismic soil movements that as a rule, their expected values are an order less that the appropriate standard ones, i.e. such process can be treated as aligned ones with the expected values.

In-phase $\Psi_{ij}^s(t)$ and quarter-phase $\Psi_{ij}^s(t)$ components of co-spectral densities of the soil seismic movement field can be treated as follows (Zolina & Sadchikov 2013):

$$\begin{cases} \Psi_{ij}^s(t) = \dfrac{1}{\omega_i} \displaystyle\int_0^t e^{-\gamma\tau} \sin(\lambda_j \tau)\, e^{-c(t-\tau)} \sin\omega_i(t-\tau)\, d\tau \\[4mm] \Psi_{ij}^c(t) = \dfrac{1}{\omega_i} \displaystyle\int_0^t e^{-\gamma\tau} \cos(\lambda_j \tau)\, e^{-c(t-\tau)} \sin\omega_i(t-\tau)\, d\tau \end{cases} \tag{6}$$

where $\omega_i$ is the natural oscillation frequencies of building; $c$ is the dissipation factor; and $\gamma = \gamma_1 = \gamma_2 = \gamma_3$ are phase angles of co-spectral densities.

Approximating $S^{BX}(\lambda)$ and taking into account the hidden periodicity of seismic acceleration process, a rated spectral density with parameters $a, m, \alpha$ can be found as follows:

$$S_j^{BX} = \frac{2\alpha}{\pi} \frac{m^2 + \lambda_j^2}{m^4 + 2a\lambda_j^2 + \lambda_j^4}, \tag{7}$$

Considering the curvilinear trapezoid column square $d_j$, bounded with function $S^{BX}(\lambda)$ in separated carrier frequencies bands, we obtain

$$d_j = \int_{\lambda_j - \Delta\lambda}^{\lambda_j} S^{BX}(\lambda)\, d\lambda, \text{ where } \Delta\lambda = \lambda_{j1} - \lambda_{j-1}, \tag{8}$$

Dispersions are defined for generic coordinates $D_i(t)$ with each natural oscillation frequency of building in separate time moments, which is as follows:

$$D_i(t) = D \sum_{j=1}^n d_j \left[ \left(\Psi_{ij}^s(t)\right)^2 + \left(\Psi_{ij}^c(t)\right)^2 \right], \tag{9}$$

where $D$ is the seismic load dispersion.

Output spectral for a time moment in which a maximum of system $(t = t_{max})$ movements is reached, can be calculated in the same manner as follows:

$$S_i^{BbIX}(\lambda_j) = d_j \left[ \left(\Psi_{ij}^s(t)\right)^2 + \left(\Psi_{ij}^c(t)\right)^2 \right]. \tag{10}$$

Within the earthquake modeling of the design intensity dynamic factor value $\beta_i(t)$ of the system, the seismic vulnerability of frame structures can be evaluated as follows:

$$\beta_i(t) = \omega_i^2 \sigma_i(t), \text{ where } \sigma_i(t) = \sqrt{D_i(t)}. \tag{11}$$

The inertial load standard for each point $x_k$ with the weight $m_k$ depending on the $i$ th oscillation mode can be found out with the axis $X_i(x_k)$ as follows:

$$\sigma_{S\,ik}(t) = \beta_i(t)\, X_i(x_k)\, m_k, \tag{12}$$

And appropriate elements of the coefficient matrix are given as follows:

$$J_{ik} = \frac{X_{ik} \displaystyle\sum_{j=1}^n m_j\, X_i(x_j)}{\displaystyle\sum_{j=1}^n m_j\, X_i^2(x_j)}. \tag{13}$$

After obtaining the matrix $J = (J_{ik})$, there is a possibility to evaluate qualitatively a building oscillation period $T_{3k}$ under the action of a seismic load, which is as follows:

$$T_{3k} = 2\pi \sqrt{\frac{\displaystyle\sum_{j=1}^n \sum_{i=1}^N X_{ik}^2 J_{ik}^2 S_i(\lambda_j)}{\displaystyle\sum_{j=1}^n \sum_{i=1}^N X_{ik}^2 J_{ik}^2 S_i(\lambda_j)\lambda_j^2}} \tag{14}$$

Realization of presented decompositions is used within the automation of the appropriate calculations in changes evaluation of the stressed state of industrial buildings under the action of loads system. We developed a self-checking software complex «DINCIB-new» (Zolina & Sadchikov 2012), registered in the Russian State Register of software. Its specification includes calculation organization and conduction of industrial building structures operation equipped with overhead cranes, including at extreme disturbance caused by seismic ground motion.

Choice of the model for a description of wave features of the soil movement field within earthquakes allows us to conduct a quantitative analysis of building performance under random seismic load using classic methods of statistical dynamics and reliability theory. With the processing results of numerous field surveys of engineering workshops, shipyards, metallurgical and mining and processing plants, we analyzed the conclusions reliability of developed techniques that implement the proposed model. Design diagrams of industrial buildings subjected to surveys are presented in the form of three-dimensional systems with the reference points in the details of the intersection of columns and crane equipment brake structures, frames and top longitudinal axis. Comparative analysis of the oscillation characteristics design values of the process that involves building frame under the influence of a seismically active wave with real indications shows deviations in the range of not more than 4%.

For example, in the case of the hull shop of Astrakhan Maritime Shipyard, natural frequency values of the most heavily loaded and moving transverse frames in the level of bridge rail and top were obtained. The results for the first five ones are summarized in Tables 1 and 2.

Conducted probabilistic calculations demonstrate the possibility of using the vector field model of seismic ground motion in the ground developed in this paper to assess the reliability of designed and operated buildings and structures exposed to influences of a given intensity.

## 5 CONCLUSIONS

Considering seismic ground motion as the stationary random process, we structured a probabilistic calculation, including the search of:

- In-phase and quadrature components of the vector field spectral density;
- Input and output spectra;
- Dispersion of generalized coordinates for each of oscillation natural frequencies of the building;
- Waveforms coefficient matrices;
- Oscillation effective period of building frame structures under the influence of seismic loads;
- Failure rates at the designed level of significance;
- Summary dispersion in all forms of oscillations;
- Conditional, external and total seismic risks.

The proposed model allows us to evaluate the seismic vulnerability of structures in the implementation of a given earthquake intensity level. Problems arising with the implementation of an appropriate algorithm due to the complexity of probabilistic calculations are resolved by automated control systems.

Table 1. Values of building oscillation natural frequencies.

| Oscillation frequency number $i$ | 1 | 2 | 3 | 4 | 5 |
|---|---|---|---|---|---|
| Natural frequency $\omega_i$, $c^{-1}$ | | | | | |
| Design | 4,131 | 4,394 | 24,569 | 25,405 | 25,548 |
| Recorded | 4,11 | 4,43 | 24,62 | 25,32 | 25,34 |
| Deviation, % | 0,5 | 0,8 | 0,2 | 0,3 | 0,8 |

Table 2. Motions values of the most loaded transversal frame at each oscillation natural frequency.

| Oscillation natural frequency number $i$ | 1 | 2 | 3 | 4 | 5 |
|---|---|---|---|---|---|
| Motion at the bridge rail level, m | | | | | |
| Design | 0,0511 | 0,0706 | 0,0006 | −0,0008 | 0,0027 |
| Recorded | 0,0502 | 0,0712 | 0,0006 | −0,0008 | 0,0028 |
| Deviation, % | 1,8 | 0,8 | 0 | 0 | 3,6 |
| Motion at the top level, m | | | | | |
| Design | 0,0880 | 0,1194 | 0,0008 | −0,0021 | 0,0026 |
| Recorded | 0,0874 | 0,1180 | 0,0008 | −0,0021 | 0,0027 |
| Deviation, % | 0,7 | 1,2 | 0 | 0 | 3,7 |

## REFERENCES

Barshtein, M.F. 1960. Application of variety method to a building design in relation to seismic impacts, *Constructional mechanics and buildings design*, 2: 6–14.

Esteva, L. & Rosenblueth, E. 1964. Espectros de Temblores a Distancians Moderadas y Grandes. *Boletin Sociedad Mexicana de Ingenieria Sesmica*, 2(1): 1–18.

Housner, G.W. 1952. *Spectral Intensities of Strong—Motion Earthquakes*. Proc. Symp. Earthq. and Blast Effects Structures. C.M. Duke and M. Feign, (eds.) Los Angeles: University of California, 21–36.

Nikolaenko, N.A. & Nazarov, J.P. 1988. *Dynamics and building seismic stability*. Moscow: Stroyizdat, 312.

Pshenichkina, V.A. Belousov, A.S. Kuleshova, A.N. & Churakov, A.A. 2010. *Building reliability as space complex systems at seismic impacts*, Pshenichkina V.A. Volgograd: VolgASU, 180.

Zolina, T.V. & Sadchikov, P.N. 2012. Automated calculation systems of an industrial building for crane and seismic loads. *Industrial and civil building*. 8: 14–16.

Zolina, T.V. & Sadchikov, P.N. 2013. Variety approach to an evaluation of industrial building seismic stability. *Vestnik MGSU*, 11: 42–50.

Zolina, T.V. & Sadchikov, P.N. 2013. Evaluation methodology of remaining operational lifetime of an industrial building equipped with overhead cranes. *Newsletter of VolgGASU. Construction and architecture*, 32(51): 51–57.

Zolina, T.V. & Sadchikov, P.N. 2015. Revisiting the Reliability Assessment of frame constructions of Industrial Building. *Applied Mechanics and Materials*, 752–753: 1218–1223.

*Advanced Materials, Structures and Mechanical Engineering – Kaloop (Ed.)*
© 2016 Taylor & Francis Group, London, ISBN 978-1-138-02793-0

# The effect of ocean flow on reconstruction accuracy in the test based on the NAH

Q.W. He
*College of Power Engineering, Naval University of Engineering, Wuhan, Hubei Province, China*

P. Hu
*No. 92474 Unit of Navy, Sanya, Hainan Province, China*

X.W. Liu & S.C. Ding
*College of Power Engineering, Naval University of Engineering, Wuhan, Hubei Province, China*

ABSTRACT: In this paper, the impact of ocean flow on the hydrophone array is studied in the progress of making the real-boat test by the technology of submarine noise test based on the plane NAH. The effect of ocean currents on the hydrophone array is, respectively, divided into two types, namely random offset and tilt offset, and analyzes the influence of different offsets on the reconstruction accuracy based on the NAH, to provide a reference for selecting the right sea conditions to do the real boat test.

## 1 INTRODUCTION

Since the use of the acoustic holography technology to measure submarine radiated noise, which was proposed first by Williams et al. (1985, 1999) from the Naval Research Laboratory, military powers have carried on acoustic holography technology research and military applications. However, the countries that have truly made this technology practical include only the USA and France, and a handful of countries.

American and French navy have respectively reported their submarine noise measurement systems based on NAH and NAH-processing software. But due to confidentiality, details have been reported. According to published data, the Australian Defence Science and Technology Organisation and the Naval Undersea Technology Center jointly launched the submarine test based on the NAH in the mooring state. Research conducted in Europe and other countries has shown that testing based on the NAH technology can effectively reconstruct the acoustic image of the submarine surface, which contributes to locate the submarine acoustic weaknesses, and then guide the submarine design, maintenance and support. The above work shows that the NAH technology in foreign has been widely studied and applied, and has had successful application cases in the field of acoustic measurements, and thus is a new and matured generation of technology of submarine radiated noise test.

Since the 1990s, many experts and research institutes in China have started a research of submarine NAH testing technology (He et al. 2003, Zhang et al. 2006), but the application of acoustic holography technology has really took from laboratory research to complex environment in the outfield, which is very difficult. This needs to solve many problems, in particular, the effect of the water flow on the posture of acoustic array sued into acoustic holography measurement cannot be ignored. In laboratory, because water is stationary, acoustic array can be fixed according to the design of the shape; however, in the ocean environment, ocean flow must make a tilt of the array or irregular deformation. In response to these problems, this paper studies the effects of different flow velocities of sea water on the accuracy of the NAH test by simulation, to provide a reference for selecting the right sea conditions to do the real boat test (Wang & Chen 2004).

## 2 THE BASIC PRINCIPLES AND STEPS OF SFT-NAH

The SFT-NAH (Spatial Fourier Transform Near-field Acoustic Holography; Chen & Bi 2013) is a technology of sound field transform based on fluctuation acoustic. It uses the measurement results of a variable acoustic in the sound field of a region near the sound source, for example the distribution of complex sound pressure, to reconstruct an acoustic hologram of other regions including the sound source surface and the spatial distribution of a variety of variable acoustical in the near field and the far field. The basic flowchart of the PNAH is shown in Figure 1. In Figure 1, the solid line shows the holographic measuring surface, and the dashed line represents the conversion surface that is the sound source surface or any plane in the near field and the far field. The process of holographic transformation can be divided into four steps as follows:

1. Obtaining the spatial distribution of complex sound pressure on the measurement surface $p(x, y)$ by measuring.
2. Performing dimensional Fourier transform for $p(x, y)$ to obtain complex sound pressure $P(k_x, k_y)$ in the k-space. The transformation formula of planar NAH is as follows:

$$p(x,y,z_s) = F_x^{-1}F_y^{-1}\{F_xF_y[p(x,y,z_h)] \times G_D^{-1}(k_x,k_y,z_h - z_s)\} \quad (1)$$

$$v_z(x,y,z_s) = \frac{1}{\rho_0 c_0 k} F_x^{-1}F_y^{-1}\{F_xF_y[p(x,y,z_h)] \times G_N^{-1}(k_x,k_y,z_h - z_s)\} \quad (2)$$

where $p(k_x, k_y, z_s)$ represents the sound pressure on the conversion surface; $p(k_x, k_y, z_h)$ represents the sound pressure on the measurement surface; $G_D(k_x, k_y, z_h-z_s)$ and $G_N(k_x, k_y, z_h-z_s)$ are, respectively, the results of spatial Fourier transform of Green function according to the boundary conditions of Dirichlet and Neumann; and $v_z(x, y, z_s)$ represents the particle vibration velocity at the z direction.

3. Selecting the appropriate Green function as a transfer function H according to the different boundary conditions, to obtain a complex sound pressure $P(k_x, k_y)$ on the conversion surface.
4. Finally, performing inverse Fourier transform of the above results, to ultimately obtain a complex sound pressure $p(x, y)$ on the conversion surface.

According to transformation formulas, it is obvious to possibly reconstruct and forecast sound intensity and other acoustic variables.

# 3 THE EFFECT OF RANDOM OFFSET OF MEASUREMENT POINT ON THE RECONSTRUCTION SOUND FIELD

In practice, the velocity of flow at different depths is often inconsistent. In extreme cases, even the flow direction is opposite. In this condition, the effect of ocean flow on the measurement array can be viewed as the random effect. The purpose of the simulation is to study the influence of random offset caused by ocean flow on reconstruction results.

## 3.1 *The model and parameters of simulation*

The sound model description is as follows: ship model, shown in Figure 2, has a total length of 70 m, and 10 m shell diameter. Inside the submarine model, there is a cylindrical shell with hats at both ends, and the cylindrical shell and the shell are directly connected by seven ring ribs. Inside the cylindrical shell, there is a clapboard, and mechanical equipment are connected with the clapboard through the damping device. The quality system is used to simulate mechanical equipment. Spring damping structure is used to simulate the pipe, and the support structure is used to connect the mechanical equipment and the clapboard pedestal.

Description of the shell material property is as follows: the elastic modulus is $2.06e11$ N/m$^2$; Poisson's ratio is 0.3; density is 7800 kg/m$^3$; shell thickness is 30 mm; central rib thickness is 30 mm; and the thickness of the upper separator of the internal cylindrical shell is 30 mm. The meshing principle is as follows: the number of grids is 63 and the length of the unit is 200 mm.

Description of the measuring surface is as follows: the measuring surface is a rectangular form of 12 m × 30 m, which is 1 m from the model; the length of the hydrophone array is 12 m; the interval of the hydrophone array element is 0.5 m; the scanning distance is 30 m; and the interval of measurement points is 0.5 m.

Description of the random offset is as follows: add random amount [−0.03 m~0.03 m], [−0.06 m~0.06 m], [−0.12m~0.12m], [−0.18m~0.18m], [−0.24m~0.24m], respectively, to each measurement point on the

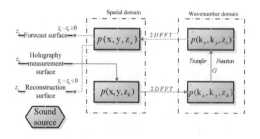

Figure 1. The basic flowchart of the PNAH.

Figure 2. Model diagram.

measurement surface, and use a measurement surface with the random amount as a hologram surface for NAH transforming.

## 3.2 *The analysis of simulation results*

First, different random amounts are added to every measurement point. Then reconstruction sound press of every measurement point is calculated through the NAH and compared with the reference value to study the effect of the holographic surface tilt on reconstruction results. Sound pressure amplitudes on a tangent through the center of the reconstruction surface at different frequency are shown in Figure 3, Figure 4 and Figure 5. Error analysis of the reconstruction results for each frequency point is shown in Figure 6.

By comparison, we can find that when adding random offsets to every measurement point, reconstruction results will have errors. And this error increases with the increasing frequency and offset. When the array offset is less than 0.12 m, the reconstruction error is less than 3dB, which can meet the engineering

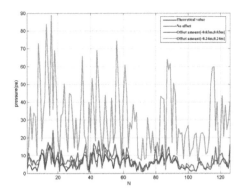

Figure 5.   Reconstruction error when f = 100 Hz.

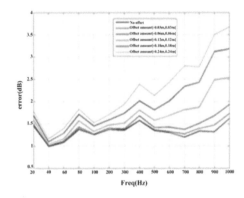

Figure 6.   Reconstruction error.

requirements. But when the absolute value of the offset is more than 0.18 m, the reconstruction error significantly increases at a high frequency. At this time, the upper limit of analysis frequency should be reduced according to the needs.

## 4   THE EFFECT OF THE TILT OF MEASUREMENT ARRAY ON THE RECONSTRUCTION SOUND FIELD

If the speed and direction of the ocean flow is consistent at different depths in the same position, ocean flow will make a holography array tilt to one side. The purpose of the simulation is to study the influence of the tilt of the measurement array caused by the ocean flow on reconstruction results.

### 4.1 *The model and parameters of simulation*

The model and parameters of simulation is the same with 2.1. Description of the array tilt amount is as follows: total tilt amounts of the holography array are $-3$ m, $-2$ m, $-1$ m, $0$ m, $1$ m, $2$ m, and $3$ m, respectively, as shown in Figure 7.

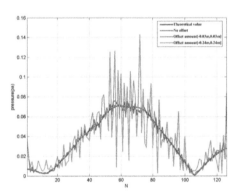

Figure 3.   Reconstruction error when f = 20 Hz.

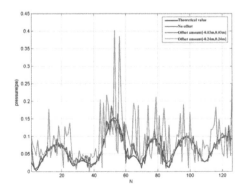

Figure 4.   Reconstruction error when f = 100 Hz.

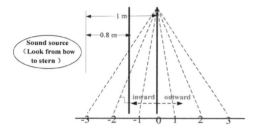

Figure 7.   Diagram of the tilted measuring plane.

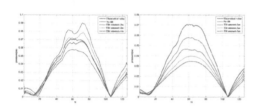

Figure 8.   Comparison of the reconstructed results when frequency is 20 Hz.

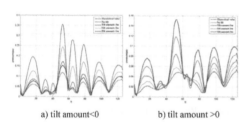

a) tilt amount<0          b) tilt amount >0

Figure 9.   Comparison of the reconstructed results when frequency is 100 Hz.

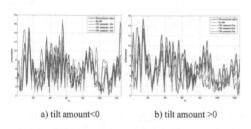

a) tilt amount<0          b) tilt amount >0

Figure 10.   Comparison of the reconstructed results when frequency is 1000 Hz.

### 4.2   *The analysis of simulation results*

When adding the tilt amounts of −3 m, −2 m, −1 m, 0 m, 1 m, 2 m, and 3 m to the holography surface, respectively, the reconstruction sound pressure of the holography surface can be calculated through the NAH and compared with the reference value to study the effect of the holographic surface tilt on reconstruction results. Sound pressure amplitudes

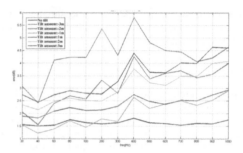

Figure 11.   Reconstruction decibel error during the tilted measuring plane.

on a tangent through the center of the reconstruction surface at a frequency of 20 Hz, 100 Hz, and 1000 Hz are shown in Figures 8–10. Error analysis of the reconstruction results for each frequency point is shown in Figure 11.

Figure 8–11 show that the error of reconstruction sound pressure increases with the increase in the tilt amount of the array. And the error caused by tilting inwardly is more than that caused by tilting outwardly. When the array tilts outwardly less than 2 m or tilts inwardly less than 1 m, the reconstruction error is less than 3dB, which can meet the engineering requirements. On the contrary, the reconstruction error is too large to carry out the test work.

## 5   CONCLUSIONS

For the question that in the progress of making the real boat test based on the NAH, ocean flow has an influence on the hydrophone array, we carried out the related work by simulating. The effect of ocean flow on the hydrophone array is, respectively, divided into two types, namely random offset and tilt offset, to analyze, so that studies have shown that the array offset is the main factor for the error of NAH reconstruction, and that the larger the offset is, the greater the error is. So, in the actual test, we must pay attention to posture of the array, and once the offset is beyond the acceptable range of error, we should suspend testing. This study provides ideas for selecting the right sea conditions to do the real boat test and provide a reference for relevant test standards.

## ACKNOWLEDGMENT

This work is funded by the Zhejiang Provincial Natural Science Foundation of China under Grant No. LQ12E09002. Project (51308497) was supported by the National Natural Science Foundation of China.

## REFERENCES

Chen, X. & Bi, C. 2013. *Near-field acoustical holography technology and its application*. Beijing: Science press.

He, Y. He, Z. & Shang, D. 2003. Full spatial transformation of sound field based on planar II, Experiment of NAH for submerged large area planar acoustic holography transmitting array. *ACTA. ACUSTICA*, (1): 45–51.

Wang, Z. & Chen, J. 2004. *Warship Noise Measuring And Analyzing*. Beijing: National Defence Industry Press.

Williams, E.G. 1999. *Sound radiation and near-field acoustical holography*. San Diego, California: Cadexiica Press.

Williams, E.G. Henry, D.D. & Richard, G.F. 1985. A technique for measurement of structure-borne intensity in plates. *Acoustical Society of America*, 78(06): 2061–2068.

Zhang, L. He, L. & Zhu, S. 2006. Review on the Methods of Identification of Submarine Main Noise. *Noise and Vibration Control*, (4): 7–10.

*Advanced Materials, Structures and Mechanical Engineering – Kaloop (Ed.)*
© 2016 Taylor & Francis Group, London, ISBN 978-1-138-02793-0

# Effect of different modification methods on ground motions

C.W. Jiang
*Hunan University, Changsha, P.R. China*

D.B. Wang
*China Railway Eryuan Engineering Group, Chengdu, P.R. China*

ABSTRACT: Ground motion modification techniques can be used to alter a selected, typically recorded, ground motion and match its acceleration response spectrum to the target response spectrum. Ground motion modification provides a basis for reducing the number of ground motions required for analyzes. The impact of different modification techniques on ground motion characteristics and results of seismic geotechnical analysis is investigated. Fourteen motions were selected and scaled and also modified using both Time Domain (TD) and Frequency Domain (FD) techniques. It is shown that TD modification of ground motion slightly decreased PGV but had little impact on PGD, $I_a$ and $D_{5-95}$. TD modification resulted in PGV and PGD that were larger than those for FD modification, but smaller than those for scaled ground motions. FD modification decreased PGV and PGD but had little impact on $I_a$. Similar $I_a$ were produced by both modification techniques.

## 1 INTRODUCTION

In the present seismic design of bridge structures, the ground motions are generally used to carry out nonlinear time-history response analyzes. As the selected ground motions will affect the nonlinear time-history response of the structure, it is crucial for the seismic design to select or generate ground motions close to the site response spectrum. There are three ways to select the time histories of accelerations of the ground motions at present: 1) direct application of the strong motion earthquake record; 2) synthetic time histories of ground accelerations; 3) standardized time histories of ground accelerations. In general, the chosen should be the priority and preliminary field seismic record of similar geological conditions. However, due to the limited records, synthetic ground motions, and standardized ground motions are more widely used. Of the latter two ground motions, standardized ground motions which are provided by corresponding specifications have not been used in China. As a result, synthetic time-history ground accelerations given by the seismic safety evaluation reports of the sites of bridges have been used. In the seismic safety evaluation reports from the construction site, it is necessary to provide the time-history ground accelerations that fit the design response spectra and meet the requirement of the seismic design.

Two kinds of methods are generally used to fit the target spectra including the scaled ground motion methods and the ground motion modification methods. The scaled ground motion methods linearly amplify or reduce the ground motions along the entire time history based on one of the peaks of the ground motions. Ground motion modification methods which can be divided into frequency domain techniques and time domain techniques adjust both the frequency component and amplitudes of the ground motions to fit the target spectra. As Bommer et al. (2004) found that certain ground motion modification techniques might produce some unreasonable ground motions which lead to a large difference in some ground motion intensity indices between these ground motions and the original ground motions, it is important to study their influence on the time histories of the ground motions.

At present, the seismic design philosophy has developed from strength-based seismic design to performance-based seismic design for which one key point is to select an index which can represent the structural performance. As for the structures, Riddle (2007) and Mollaioli (2011) found that structural shift rate is closely related to the peak acceleration, peak velocity, peak displacement and Arias intensity of the earthquake wave. Zekkos & Saadi et al. (2012) found a large correlation between the cyclic stress ratio in liquefaction analysis and the duration and average period of the earthquake wave. The research mentioned above all emphasize that ground motion intensity indices should keep unchanged as far as possible during ground motion modifications to make sure that sudden changes will not appear in structural earthquake response. A target response spectrum is generated in this paper and fourteen measured ground motions are selected to test the time domain and frequency domain modification methods. The impacts of these two methods on the ground motion intensity indices are studied.

## 2 GROUND MOTION MODIFICATIONS

The ground motion modification methods can be divided into two kinds which are time domain methods and frequency domain methods. Although the procedures of calculation are different, they both include the procedure of adding the relevant frequencies to the original ground motions. Frequency domain methods usually calculate the ratio between the target response spectrum and the ground motion response spectra first and use this ratio as the accuracy of each iteration. The new time-history of the ground motion is obtained through Fourier transform and inverse Fourier transform. Many methods are available for spectrum modifications. As Fourier spectra include contributions of both positive and negative components and can modify phases, the amplitude spectra modification method presented by Hu et al. (1986) is applied to modifying the time histories of the accelerations. When the spectrum modification methods are used to modify the ground motions, defective points or points which are difficult to converge usually occur especially for long periods, which make researchers tend to apply time domain methods to modify ground motions.

Time domain methods calculate the difference between the response spectrum of the ground motion and the target response spectrum to generate wavelets of various amplitudes, frequencies, and durations. These wavelets are iteratively added to the original ground motions to fit the target response spectrum. The time domain method is proposed by Lilhanand & Tseng et al. (1988) and after years of improvement the basic calculation procedure is given by Al-Atik and Abrahamson et al. (2010). The time domain method also includes an iterative process. Assuming that $a(t)$ is the original time-history of the ground motion, the increment of the ground motion used in each step can be calculated as

$$\delta a(t) = \sum_{j=1}^{N} b_j f_j(t) \qquad (1)$$

where, $b_j$ is the amplitude of the adjustment function, $f_j(t)$ is the adjustment function and $N$ is the number of points in the response spectrum that need adjustment.

$b_j$ can be expressed as

$$b = C^{-1} \delta R \qquad (2)$$

where, $\delta R$ is the difference between the selected ground motion and the target spectrum, $C$ is a matrix with a constant determinant. The entries of $C$, $c_{ij}$, denotes the response at the adjustment point for the single-degree-of-freedom system and can be calculated with the following equation.

$$c_{ij} = \int_0^{t_i} f_j(\tau) h_i(t_i - \tau) d\tau \qquad (3)$$

where, $h_i(t)$ is impulse response function of the acceleration and can be written as,

$$h_i(t) = \frac{-\omega_i}{\sqrt{1-\beta_i^2}} \exp(-\omega_i \beta_i t) \dots$$
$$\dots \left[ \left( 2\beta_i^2 - 1 \right) \sin\left( \omega_i t \sqrt{1-\beta_i^2} \right) \dots \right.$$
$$\left. \dots -2\beta_i \sqrt{1-\beta_i^2} \cos\left( \omega_i t \sqrt{1-\beta_i^2} \right) \right] \qquad (4)$$

where, $\omega_i$ is the frequency of the $i$-th point in the target response spectrum, $t$ is the time and $\beta_i$ is the damping of the $i$-th point in the target response spectrum.

In addition, time histories of the velocity and displacement of the ground motion are obtained by integration. Equation (1) indicates that inappropriate adjustment functions may lead to divergence. The study by Al Atik and Abrahamson et al. suggests the following form.

$$f_j(t) = \cos\left[ \omega_i \sqrt{1-\beta_i^2} \left( t - t_j + \Delta t_j \right) \right] \dots$$
$$\dots \exp\left[ -\left( \frac{t - t_j + \Delta t_j}{\gamma_j} \right)^2 \right] \qquad (5)$$

where, $t_j$ is the time when the acceleration reaches its peak, $\Delta t_j$ is time and $\gamma_j$ is relevant to frequencies and is linearly related to $t_j$. The time histories of the velocity and displacement of the ground motion converge this way. After a series of iterations, the response spectrum calculated with the adjusted $a(t)$ will match with the target response spectrum.

## 3 SELECTION OF GROUND MOTIONS

One site in central China, 34 degrees northern latitude and 114 degrees east longitude, is selected for analysis in this research. The earthquake return period is 2475 years (50 years of 2%). Performance based seismic design requires to carry out structural vulnerability analysis and risk analysis and selects measured ground motions, which is different from the practical seismic design in China where synthetic ground motions generated by response spectra are used. Furthermore, in some projects from engineering, at least one-third of the ground motions should be measured on site. To meet the future development, a certain amount of measured ground motions are selected for analyses based on the target response spectrum in this research. The ground motions are selected from the European ground motion database. The ground motions should meet the following requirements: 1) epicentral distance should be larger than 10 km and smaller than 35 km, 2) earthquake magnitude should be larger than M 5 and smaller than M 7,

and 3) the average shear wave velocity should be larger than 360 m/s.

Figure 1 illustrates the response spectra of the fourteen selected ground motions based on the target response spectrum. Table 1 lists the fourteen ground motions used in this paper and through scaling the response spectra of the ground motions match with the target response spectrum. Time domain and frequency domain methods are then respectively applied to modify the fourteen ground motions and their influence on the ground motion characteristics will also be studied. Ground motion characteristics can be represented by many indices and the generally used ones include Peak Ground Acceleration (PGA), Peak Ground Velocity (PGV), Arias Intensity ($I_a$) and strong motion duration $D_{5-95}$. PGA is a parameter that has poor core relationship with structural damage, and thus Peak Ground Velocity (PGV), Peak Ground Displacement (PGD), Arias Intensity ($I_a$) and strong motion duration $D_{5-95}$ is chosen in this paper.

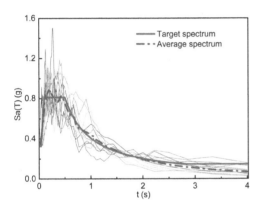

Figure 1. The response spectra of the fourteen ground motions selected.

## 4 DATA ANALYSES AND DISCUSSION

The time-domain and frequency domain methods are applied to modify the scaled ground motions. Peak Ground Velocity (PGV), Peak Ground Displacement (PGD), Arias intensity ($I_a$) and strong motion duration $D_{5-95}$ are then calculated. The corresponding mean values and standard deviations can be obtained.

Table 2 lists the mean values and standard deviations of the ratios between the intensity indices from time domain and frequency domain methods respectively and the original ground motion. This table shows that the mean ratio of PGV

Table 1. The fourteen ground motions selected.

| Earthquake name | Time | Earthquake magnitude | Fault mechanism | Epicentral distance [km] | PGA_X [m/s²] | Scaling factor |
|---|---|---|---|---|---|---|
| Southern Iwate Prefecture | 2008-6-13 | 6.9 | Reverse | 23.08 | 2.9434 | 1.4597 |
| Honshu | 1996-8-10 | 5.7 | Strike-slip | 27.06 | 0.083365 | 3.9578 |
| Eastern Fukushima Pref | 2011-4-11 | 6.6 | Normal | 26.24 | 1.883 | 1.7902 |
| Central Nagano Pref | 2011-6-29 | 5 | Strike-slip | 18.26 | 0.13634 | 2.3322 |
| Parkfield | 2004-9-28 | 6 | Strike-slip | 9.57 | 1.4845 | 3.3918 |
| Parkfield | 2004-9-8 | 6 | Strike-slip | 15.23 | 1.3863 | 1.4181 |
| W Tottori Prefecture | 2000-10-06 | 6.6 | Strike-slip | 19 | 2.6792 | 2.0894 |
| Emilia_Pianura_Padana | 2012-5-29 | 6 | Reverse | 24.98 | 0.79029 | 4.7744 |
| Mid Niigata Pref | 2011-3-11 | 6.2 | Reverse | 23.05 | 1.3399 | 2.7788 |
| Emilia_Pianura_Padana | 2012-5-20 | 6.1 | Reverse | 13.36 | 2.5745 | 1.2352 |
| Erzincan | 1992-3-13 | 6.6 | Strike-slip | 8.97 | 4.8594 | 0.65441 |
| Superstition Hills | 1987-11-24 | 6.6 | Strike-slip | 19.5 | 1.6852 | 1.8871 |
| App. Umbro Marchigiano | 1998-4-03 | 5.1 | Normal | 19.67 | 0.445 | 7.1461 |
| NW Kagoshima Prefecture | 1997-5-13 | 6 | Strike-slip | 26.81 | 1.8956 | 1.6776 |

Table 2. Mean values and standard deviations of the ratios between the intensity indices from the modified ground motions and the original ground motion.

|  | PGV | PGD | $D_{5-95}$ | $I_a$ |
|---|---|---|---|---|
| *TD modification* | | | | |
| Mean values | 0.94 | 0.96 | 0.99 | 0.97 |
| Standard deviations | 0.23 | 0.27 | 0.16 | 0.41 |
| *FD modification* | | | | |
| Mean values | 0.86 | 0.64 | 1.11 | 0.91 |
| Standard deviations | 0.33 | 0.31 | 0.17 | 0.33 |

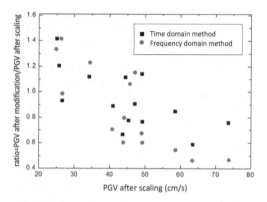

Figure 2. Ratios of PGV from the modification methods and the scaling method.

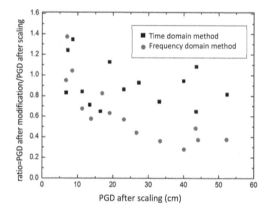

Figure 3. Ratios of PGD from the modification methods and the scaling method.

from the time domain modification method is 0.94 and the mean ratio of PGD is 0.96. Both of them are smaller than 1 but are both larger than those from the frequency domain modification method. The mean ratios of PGV and PGD are respectively 0.86 and 0.64 after the frequency domain modification. The mean ratio of $D_{5-95}$ after the frequency domain modification is 1.11 which is larger than 0.99 from the time domain modification method. The mean ratios of $I_a$ from both methods are respectively 0.97 and 0.91 which are the closest to each other. Further analyses show that the mean ratios of PGD, $I_a$ and $D_{5-95}$ from the time domain modification method are all close to 1 and the error is within 4% while that of the PGV is 6% which of the same level as those from the scaling ground motion method. The mean ratios of PGV and PGD from the frequency domain method, 0.86 and 0.64 respectively, are much smaller than 1, indicating that the values of PGV and PGD from the frequency domain method are much smaller than those from the scaling ground motion method. As for the standard deviations, all the values from the frequency domain modification method are larger than those from the time domain modification method except $I_a$ whose is close from both methods.

In Figure 2, the general trend is that when PGV is relatively small after scaling, PGVs from the two modification methods increase with the decrease of the PGV after scaling. When PGV is relatively large after scaling, PGV after modification will decrease with the increase of PGV after scaling. Meanwhile, the ratios of PGVs from the time domain modification method are mostly larger than those from the frequency domain modification method.

In Figure 3, the general trend is that when PGD after scaling is relatively small, PGDs from the modification methods remarkably increase with the decrease of PGD after scaling. The different trend from that of PGV is that when PGD from the scaling method is relatively large, with its increase PGD after the frequency domain modification remarkably decreases while that after the

time domain modification fluctuates. Table 2 and Figure 3 show that although the results from the two modification methods is relatively close to each other, the frequency domain modification method increases continuous waves of small amplitudes for the whole earthquake period and thus remarkably decreases the values of PGD while the time domain modification method slightly reduces the values of PGD. For most of the ground motions, the values of PGD from the time domain method are larger than those from the frequency domain method especially for the cases when the values of PGD after scaling are relatively large.

With the increase of strong motion duration $D_{5-95}$, the ratios of $D_{5-95}$ approach 1. Figure 4 shows that the ratios of $D_{5-95}$ from the frequency domain modification method are generally larger than those from the time domain modification method as is expected. This is because the frequency domain method adds the time histories of

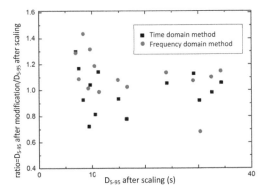

Figure 4. Ratios of $D_{5-95}$ from the modification methods and the scaling method.

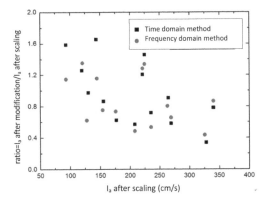

Figure 5. Ratios of $I_a$ from the modification methods and the scaling method.

accelerations to the whole earthquake period while the time domain method uses wavelets. When the values of $D_{5-95}$ after scaling are large enough, the values of $D_{5-95}$ from the two modification methods are almost the same as they are both close to 1.

Figure 5 shows that with the increase of $I_a$ after scaling, values of $I_a$ from the two modification methods both decrease. When the ratios of $I_a$ are relatively large, 0.5 s spectra need to be inflicted to the ground motions so as to match the target response spectrum. As the acceleration spectrum in 0.5 s has a large influence on the results, the amplitudes of accelerations of those ground motions increase and $I_a$ increase correspondingly.

Generally, PGV, PGD and $I_a$ from the time domain modification method are larger than those from the frequency domain modification method but $D_{5-95}$ is quite the opposite. Compared with the scaling method, the influence of the time domain modification method on PGD, $I_a$ and $D_{5-95}$ are minor and the influence of the frequency domain

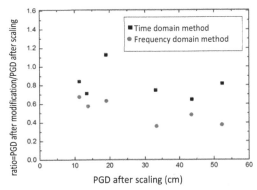

Figure 6. Ratios of PGD from the modification methods and the scaling method.

modification method on $D_{5-95}$ is relatively great. In addition, the latter method reduces PGD and PGV and makes them smaller than the mean values from the scaling method. The two modification methods both reduce $I_a$ slightly. It is worth mentioning that the above conclusions are not tenable when there are not enough ground motions, as is shown in Figure 6.

## 5 CONCLUSIONS

Fourteen ground motions from one real earthquake are used in this paper to evaluate the effect of time domain and frequency domain modification methods on the ground motion intensity indices. The results show that it is feasible to select ground motions with ground motion modification methods. The following specific conclusions can be drawn based on the results in this paper:

1. The time domain method makes no difference to PGD, $I_a$ and $D_{5-95}$ but decreases PGV slightly. Furthermore, PGV and PGD obtained with the time domain method is larger than those from the frequency domain method while smaller than the mean values from the scaling method.
2. The frequency domain method reduces PGV and PGD but has little effect on $I_a$. The values of $D_{5-95}$ from the frequency domain method are larger than those from the time domain method. However, further investigations and study are needed to determine whether the increase of the strong motion duration will greatly affect the structural response.
3. The general trend is that PGV after modification will decrease with the increase of PGV after scaling. With the increase of $D_{5-95}$, the ratios of $D_{5-95}$ (after modification/after scaling) approach 1 and with the increase of $I_a$ after scaling, values

of $I_a$ from the two modification methods both decrease.

## REFERENCES

Al Atik, L. & Abrahamson, N. 2010. An improved method for nonstationary spectral matching, *Earthquake Spectra*, 26(3): 601–617.

Athanasopoulos-Zekkos, A. & Saadi, M. 2012. Ground motion selection for liquefaction evaluation analysis of earthen levees. *Earthquake Spectra*, 28(4): 1331–1352.

Bommer, J.J. & Acevedo, A. 2004. The use of real earthquake accelerograms as input to the dynamic analysis. *Journal of Earthquake Engineering*. 8(1): 43–91.

Chopra, A.K. 2007. *Dynamics of Structures: Theory and Application to Earthquake Engineering, 3rd ed.*, Pearson Prentice Hall, Upper Saddle River, New Jersey.

Hu, Y.X. & He, X. 1986. Phase angle consideration in generating response spectrum-compatible ground motion. *Earthquake Engineering and Engineering Vibration*, 6(2): 37–51.

Lilhanand, K. & Tseng, W.S. 1988. *Development and application of realistic earthquake time histories compatible with multiple-damping design spectra*. Proc. of the 9th World Conference on Earthquake Engineering, Tokyo, Japan.

Mollaioli, F. Bruno, S. Decanini, L. & Saragoni, R. 2011. Correlations between energy and displacement demands performance based seismic engineering. *Pure and Applied Geophysics*, 168(12): 237–259.

Riddell, R. 2007. On ground motion intensity indices. *Earthquake Spectra*, 23(1): 147–173.

*Advanced Materials, Structures and Mechanical Engineering – Kaloop (Ed.)*
© *2016 Taylor & Francis Group, London, ISBN 978-1-138-02793-0*

# The effects of the effective support length of the stern bearing on ship shafting vibration transmission paths

H.F. Li, Q.C. Yang, S.J. Zhu & X.W. Liu
*College of Power Engineering, Naval University of Engineering, Wuhan, Hubei Province, China*

ABSTRACT: The ship propulsion shafting is simplified to mass point unit, elastic supporting elements and beam elements with the distributed parameters and the expression of field transfer matrix of the ship propulsion shafting is deduced by the transfer matrix method based on the modified Timoshenko beam theory. Then, the solution of the bearing force and the displacement response of the propulsion shafting is carried out by introducing the corresponding boundary conditions. Finally, the power flow of each bearing of the propulsion shafting is analyzed numerically from the perspective of energy. The analysis results show that it is feasible and effective to calculate the propulsion shafting bending vibration by the transfer matrix method based on modified Timoshenko beam theory; with the increase of the stern bearing support length, the natural frequencies corresponding increase; and the aft stern bearing support length has the biggest influence on ship shafting vibration transmission paths.

## 1 INTRODUCTION

The stern bearing, an important component of ship propulsion system, is one of the main factors influencing the normal operation of the shafting. When the ship shafting operates, the operating conditions of the stern bearing constantly change due to the un-even dynamic load produced by the propeller, causing a variety of shafting vibration. As to ship propulsion shafting, the changes of the support length of stern bearing have a significant impact on the vibration and the transfer characteristics of shafting (Chen 1987, Zhou 2006). Therefore, it is of great significance to study the influence of stern bearing support length on ship shafting vibration transmission paths in order to ensure the normal operation of the shafting system and protect the safety of navigation of ships.

Previously, Zhou Chunliang and Zhu Junchao (Zhou et al. 2007, Zhu et al. 2012) studied the influence of shafting bearing support length on ship shafting vibration characteristics, but did not study the influence on ship shafting vibration transmission paths. The transfer matrix method (Xiang 1999, Rosignoli 1999) is widely applied to analyze the dynamic characteristics of continuous beams. It is clear and easy to program to solve the structural vibration response and internal forces through the product of the point matrix and the field matrix. A certain type of ship propulsion shafting is simplified to mass point unit, elastic supporting elements and beam elements with the distributed parameters and the expression of field transfer matrix of each beam element which presents the relationship

between the left side state vector of the beam and the right side state vector of the beam is deduced through the differential expression between internal forces and displacements of beam elements based on the modified Timoshenko beam theory. Then, the expression of point transfer matrix of mass point unit and elastic supporting elements is obtained according to the force and displacement conditions on either side of the point. Finally, from the perspective of the power flow of each bearing of the propulsion shafting, the impact of bearing support length on ship propulsion shafting on transmission paths is analyzed.

## 2 CALCULATION MODEL AND PRINCIPLE

### 2.1 The transfer matrix calculation of ship propulsion shafting

Stern bearings are mainly affected by the vertical vibration, so this calculation model only considers excitation in the vertical direction. The support position of the aft stern bearing is always 1/3 away from the end side of the bearing contact (Cao 2008), and the other bearing support position is selected in the center of the bearing contact. It is more practical for computing to simplify the bearing to multi-spring support, because the stern bearing support length is relatively long.

In this paper, the calculation model is shown in Figure 1. The propulsion shafting bearing is composed of the aft stern shaft bearing, the front stern shaft bearing and the thrust bearing.

The shaft is simplified to the seven-span continuous beam, and it is divided into seven units, the corresponding node number is 0, 1, 2, 3, 4, 5, 6. Also the propeller is simplified to mass point unit, and the stern bearings are simplified to multi-spring support. The aft stern bearing is simplified to three vertical springs, the fore stern bearing is simplified to two vertical springs and the thrust bearing is simplified to a vertical spring. The propulsion shafting is mass distribution, whose cross-section is hollow circular, as well as the bending stiffness is EI, the mass per unit length is m, and the cross-sectional area is A.

The relational expression between the left side state vector of the shaft $[y\ \alpha\ M\ Q]_0^T$ and the right side state vector of the shaft $[y\ \alpha\ M\ Q]_7^T$ is established according to the corresponding boundary conditions and the force equilibrium condition of each node when the propeller is excited by a vertical harmonic force $f(t) = f_o \sin wt$.

$$[y\ \alpha\ M\ Q]_7^T = [T] \cdot [y\ \alpha\ M\ Q - f_0]_0^T \tag{1}$$

$[T]$ which is $4 \times 4$ square matrix is the field transfer matrix between the left side state vector of the shaft and the right side state vector of the shaft, as follows

$$T = P_7 T_7 P_6 T_6 P_5 T_5 P_4 T_4 P_3 T_3 P_2 T_2 P_1 T_1 P_0 \tag{2}$$

where, $P_i$ ($i = 0, ..., 7$) is the point transfer matrix of mass point unit and elastic supporting elements, and $T_j$ ($j = 0, ..., 7$) is the field transfer matrix of beam elements.

According to the free bending vibration equation of the modified Timoshenko beam (Chen et al. 2005)

$$EI \frac{\partial^4 y}{\partial x^4} + m \frac{\partial^2 y}{\partial t^2} - \frac{mI}{A} \frac{\partial^4 y}{\partial x^2 \partial t^2} - \frac{mEI}{k'AG} \frac{\partial^4 y}{\partial x^2 \partial t^2} = 0 \tag{3}$$

where, $k'$ is the effective shear coefficient.

The field transfer matrix of the $i$ ($i = 1, ..., 7$) beam element can be obtained by (3).

$$[T_i] = \begin{bmatrix} \dfrac{C_2 + \delta C_3}{C_1} & \dfrac{L_i C_5}{C_1} & -\dfrac{L_i^2 C_3}{EI C_1} & -\dfrac{\delta C_5 + s_1 s_2 C_6}{mL_i \lambda^2 C_1} \\[2ex] \dfrac{C_4 C_6}{s_1 s_2 L_i C_1} & \dfrac{C_7 - \delta C_3}{C_1} & \dfrac{L_i (s_1 s_2 C_5 - \delta C_6)}{EI s_1 s_2 C_1} & \dfrac{C_3 C_4}{mL_i \lambda^2 C_1} \\[2ex] \dfrac{EI C_3 C_4}{L_i^2 C_1} & \dfrac{EI (\delta C_5 + C_8)}{L_i C_1} & \dfrac{C_7 - \delta C_3}{C_1} & -\dfrac{EI C_4 C_5}{mL_i^3 \lambda^2 C_1} \\[2ex] \dfrac{mL_i \lambda^2 (\delta C_6 + C_9)}{s_1 s_2 C_1} & -\dfrac{mL_i^2 \lambda^2 C_3}{C_1} & -\dfrac{mL_i^3 \lambda^2 C_6}{EI s_1 s_2 C_1} & \dfrac{\delta C_3 + C_2}{C_1} \end{bmatrix} \tag{4}$$

where, $\sigma = \dfrac{m\lambda^2 L_i^2}{EA}$, $\delta = \dfrac{m\lambda^2 L_i^2}{k'AG}$, $\beta^4 = \dfrac{m\lambda^2 L_i^4}{EI}$.

$$s_{1,2} = \left\{ \left[ \beta^4 + \frac{1}{4}(\delta + \sigma)^2 \right]^{\frac{1}{2}} \mp \frac{1}{2}(\delta + \sigma) \right\}^{\frac{1}{2}}$$

$C_1 = s_1^2 + s_2^2$, $C_2 = s_1^2 \cos s_2 + s_2^2 chs_1$,

$C_3 = \cos s_2 - chs_1$, $C_4 = (\delta + s_1^2)(\delta - s_2^2)$,

$C_5 = s_1 shs_1 + s_2 \sin s_2$, $C_6 = s_1 \sin s_2 - s_2 shs_1$,

$C_7 = s_1^2 chs_1 + s_2^2 \cos s_2$, $C_8 = s_1^3 shs_1 - s_2^3 \sin s_2$,

$C_9 = s_1^3 \sin s_2 + s_2^3 shs_1$

The expression of point transfer matrix of mass point units and elastic supporting elements is showed according to the force and displacement conditions of the connected node between the $i$ ($i = 1, ..., 7$) beam element and the $i + 1$ ($i = 1, ..., 6$) beam element.

$$P_0 = \begin{bmatrix} 1 & 0 & 0 & 0 \\ 0 & 1 & 0 & 0 \\ 0 & 0 & 1 & 0 \\ m_0 \lambda^2 & 0 & 0 & 1 \end{bmatrix}, \quad P_i = \begin{bmatrix} 1 & 0 & 0 & 0 \\ 0 & 1 & 0 & 0 \\ 0 & 0 & 1 & 0 \\ -k_i & 0 & 0 & 1 \end{bmatrix}$$

The boundary conditions at both sides of the shafting shown in Figure 1 are as follows:

The left side is a free boundary: shear and bending moment is respectively zero,

$$M_0 = Q_0 = 0 \tag{5}$$

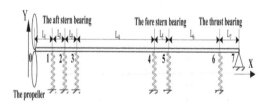

Figure 1. Simplified model of propulsion shafting structure.

The right side is simply supported: displacement and bending moment is zero,

$$y_7 = M_7 = 0 \tag{6}$$

The Formula (5) and (6) into Equation (1) can be obtained

$$\begin{cases} T_{11}y_0 + T_{12}\alpha_0 = T_{14}f_0 \\ T_{31}y_0 + T_{32}\alpha_0 = T_{34}f_0 \end{cases} \tag{7}$$

The left side of the shaft state vector can be obtained by solving Equation (7)

$$\begin{cases} y_0 = \dfrac{T_{14}T_{32} - T_{12}T_{34}}{T_{11}T_{32} - T_{12}T_{31}}f_0 \\ \alpha_0 = \dfrac{T_{14}T_{31} - T_{11}T_{34}}{T_{12}T_{31} - T_{11}T_{31}}f_0 \end{cases} \tag{8}$$

The solution of the force and displacement response of the propulsion shafting bearing is carried out by substituting Equation (8) to Equation (1) and in accordance with the point transfer matrix $P_i$ $(i = 0, \ldots, 7)$ and the field transfer matrix $T_j (j = 0, \ldots, 7)$.

### 2.2 Power flow analysis

The steady power flow is defined as the mean energy of a cycle for a harmonic motion (Hussein & Hunt 2006, Feng et al. 2009)

$$P = \frac{w}{2\pi}\int_0^{2\pi/w} Re(f)Re(v)dt = \frac{1}{2}Re\left[F(w) \cdot V^*(w)\right]$$
$$= \frac{1}{2}Re\left\{F(w) \cdot \left[jwU^*(w)\right]\right\} \tag{9}$$

where, $Re(\ )$ represents the real part, the symbol "*" represents the complex conjugate, $U(w)$ and $V(w)$ are the generalized displacement vector and the generalized velocity vector respectively.

Thus, the transmitted power flow at path $i$ $(i = 1, \ldots, 6)$ in Figure 1 is carried out

$$P_i = \frac{1}{2}Re\left[F_i(w) \cdot V_i^*(w)\right] \tag{10}$$

$F_i(w)$ and $V_i(w)$ are the transmitted force at path $i$ $(i = 1, \ldots, 6)$ and the transmitted velocity response at path $i$ $(i = 1, \ldots, 6)$ respectively.

### 3 NUMERICAL RESULTS AND ANALYSIS

The actual structural parameters of the propulsion shafting shown in Figure 1 are listed in Table 1.

Table 1. Parameters of propulsion shafting structure.

| Parameter | Unit | Value |
|---|---|---|
| Length $L_1$ | m | 0.52 |
| Length $L_2$ | m | 0.45 |
| Length $L_3$ | m | 0.45 |
| Length $L_4$ | m | 7.75 |
| Length $L_5$ | m | 0.6 |
| Length $L_6$ | m | 3.9 |
| Length $L_7$ | m | 0.64 |
| *The cross-sectional radius* | | |
| Outer radius | m | 0.11 |
| Inner radius | m | 0.065 |
| Young's modulus E | Pa | 2.11E+11 |
| Shear modulus G | Pa | 8.11E+10 |
| Density $\rho$ | kg/m³ | 7850 |
| Mass of propeller $m_1$ | kg | 4000 |
| Stiffness of the aft stern bearing | N/m | 6E+09 |
| Stiffness of the fore stern bearing | N/m | 4E+09 |
| Stiffness of the thrust bearing | N/m | 3E+09 |

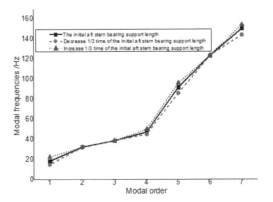

Figure 2. The influence of the aft stern shaft bearing support length on the inherent characteristics of shafting.

In this paper, the transmitted power flow of the propulsion shafting bearing is carried out to analyze the effects of bearing support length to ship propulsion shafting on transmission paths by increasing or decreasing 1/3 time of the initial stern bearing support length, and the other factors remain unchanged.

### 3.1 The influence of the aft stern bearing support length on the vibration transmission path

The influence of the aft stern bearing support length on the inherent characteristics of the propulsion shafting is shown in Figure 2, with the increase of the aft stern bearing support length, the first-order, the fifth-order and the seventh-order natural frequency corresponding increases, the corresponding impact on other order natural frequency is little, largely because increasing the length of the support will have a greater bending stiffness.

The transmitted power flow of the thrust bearing, the fore stern shaft bearing and the aft stern shaft bearing is shown in Figure 3 under different support length. As we can see from the figure, the transmitted power flow amplitude of the aft stern shaft bearing which is the main transmission path is maximum when the propeller is excited by the vertical excitation. This is because the aft stern bearing which is closest to the propeller is influenced by

the uneven load seriously. The increase of the aft stern bearing support length can effectively reduce the power flow of the three paths, especially the impact on the transmission characteristics of the fore stern shaft bearing is largest.

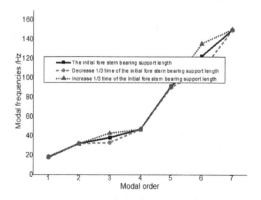

Figure 4.   The influence of the fore stern bearing support length on the inherent characteristics of shafting.

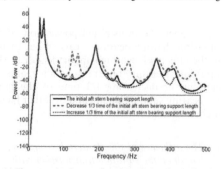

(a) The transmitted power flow through the aft stern bearing

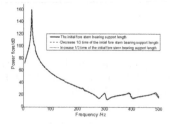

(a) The transmitted power flow through the aft stern bearing

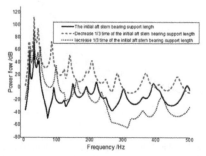

(b) The transmitted power flow through the fore stern bearing

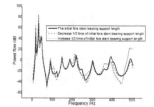

(b) The transmitted power flow through the fore stern bearing

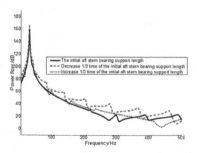

(c) The transmitted power flow through the aft stern bearing

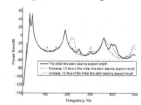

(c) The transmitted power flow through the thrust bearing

Figure 3.   The influence of the aft stern bearing support length on the transmitted power flow through three bearings.

Figure 5.   The influence of the fore stern bearing support length on the transmitted power flow through three bearings.

### 3.2 The influence of the fore stern bearing support length on the vibration transmission path

The influence of the fore stern bearing support length on the inherent characteristics of the propulsion shafting is shown in Figure 4, with the increase of the fore stern bearing support length, the third-order and the sixth-order bending natural frequency corresponding increases, the corresponding impact on other order natural frequency is little.

The transmitted power flow of the thrust bearing, the fore stern bearing and the aft stern bearing is shown in Figure 5 under different fore stern bearing support length. As we can see from the figure, the transmitted power flow amplitude of the aft stern bearing which is the main transmission path is maximum when the propeller is excited by the vertical excitation. The increase of the fore stern bearing support length can effectively reduce the peak of the power flow response curve of the thrust bearing, but the impact on the transmission characteristics of the aft stern bearing and the fore stern bearing is little.

## 4 CONCLUSIONS

This paper studies the influence of the stern bearing support length on the inherent characteristics of the propulsion shafting, and the power flow of each shafting bearing is carried out to analyze the effects of bearing support length to ship propulsion shafting on transmission paths. The analysis results show that:

1. With the increase of the stern bearing support length, the natural frequencies corresponding increase, but the effects of the fore and the aft stern bearing support length on the natural frequencies is different: the aft stern bearing support length mainly affects the first-order, the fifth-order and seventh-order natural frequency of the shafting, and the fore stern bearing support length mainly affects the third-order and the sixth-order natural frequency of the shafting.

2. The aft stern bearing is the main transmission path when the propeller is excited by the vertical excitation; the change of the aft stern bearing support length has greater impact on shafting vibration transmission characteristics, especially affects the transmission characteristics of the fore stern bearing; with the increase of the aft stern bearing support length, the transmitted power flow through the thrust bearing, the fore stern bearing and the aft stern bearing decreases. The change of the fore stern bearing support length has little effect on the shafting vibration transmission characteristics.

## REFERENCES

Cao, Y. 2008. *Study on underwater structure vibration and radiated noise control caused by propeller exciting force.* Harbin: Harbin Engineering University.

Chen, R. Wan, C. & Xue, S. 2005. Modification of motion equation of Timoshenko beam and its effect. *Journal of TongJi University (Natural Science),* (336): 711–715.

Chen, Z. 1987. *Ship propulsion shafting vibration.* Shanghai: Shanghai Jiaotong University Press.

Feng, G.P. Zhang, Z.Y. & Chen, Y. 2009. Research on transmission paths of a coupled beam-cylindrical shell system by power flow analysis. *Journal of Mechanical Science and Technology,* 23: 2138–2148.

Hussein, M.F.M. & Hunt, H.E.M. 2006. A power flow method for evaluating vibration from underground railways. *Journal of Sound and Vibration,* 293: 667–679.

Rosignoli, M. 1999. Reduced-transfer-matrix method for analysis of launched bridges. *ACI Structural Journal,* 96(4): 603–608.

Xiang, Y. 1999. The exact form of transfer matrix for analysis of free vibration of structures. *Journal of Vibration and Shock,* 18(2): 69–74.

Zhou, C. 2006. *Vibration research on ship shafting system.* Harbin: Harbin Engineering University.

Zhou, C. Liu, Z. & Zheng, H. 2007. Bearing support length and spacing to ship shafting system vibration performance. *Ship Engineering,* 29(5): 16–18.

Zhu, J. Zhu, H. & Yan, X. 2012. The effect of the effective contact length of stern tube bearing on shaft vibration. *Lubrication Engineering,* 37(2): 25–28.

*Advanced Materials, Structures and Mechanical Engineering – Kaloop (Ed.)*
© 2016 Taylor & Francis Group, London, ISBN 978-1-138-02793-0

# Development of a water-collecting device for collecting water at a certain depth

Y.H. Chen, Y.P. Xu, J.B. Jiang, Z.T. Ni & X.L. Li
*Institute of Oceanology, Chinese Academy of Sciences (Nantong), Nantong, Jiangsu Province, China*

ABSTRACT: The collection of water is the basic means to carry out water research. In some cases, we need to obtain the water at a certain depth. So, a water-collecting device is invented, which is comprised of a scale rope, hammer, recovery cylinder, and lead fish, and the recovery cylinder is composed of two end covers, a barrel body, a component used to fix scale rope, opening and closing mechanism and a water discharge component. The fixed component includes two fixed blocks and two pressing screws while the opening and closing mechanism comprises an elastic piece, rope buckles near the ends, two other buckles, the stick body, a U block, a stop block, a straight pin, a folding pin and a spring. Before collecting the water, lead fish, recovery cylinder, and the hammer are fixed on the scale rope in turn. When collecting, the recovery cylinder is driven to the needed depth with the scale rope released. Then, the hammer is released by hand and hit to the opening and closing mechanism, which will make the cylinder end cover closed. Thus, the water at this depth is blocked up in the recovery cylinder. Finally, with the help of a shipboard winch, the cylinder is driven to the deck and the collected water is used for analysis. The invention has the advantages of collecting water quickly, simple production and obtaining multilayer water.

## 1 INTRODUCTION

Analysis of microorganisms, capture of suspended particles and identification of chemical constituents are important for humans to understand and carry out the research on rivers, lakes, and oceans. To achieve this purpose, their water samples and the surface water should be easy to obtain, but it is difficult to collect water at a certain depth. The existing water samples are complex, expensive and difficult to operate. So, we decided to develop a water sampler with the advantage of collecting water quickly, simple production and obtaining multilayer water.

## 2 MECHANICAL LAYOUT

To meet the need, a new kind of water-collecting device is invented, which comprises a scale rope, hammer, recovery cylinder, and lead fish. The scale rope is connected to the shipboard winch. The other end of the scale rope is followed by the lead fish, and the hammer and recovery cylinder ride on the scale rope in turn (Fig. 1).

The recovery cylinder is composed of two end covers, a barrel body, a component used to fix scale rope, opening and closing mechanism and a water discharge component. The fixed component includes two fixed blocks and two pressing screws, while the opening and closing mechanism

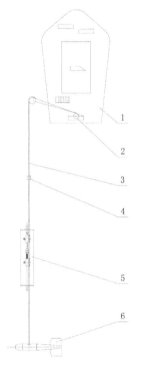

Figure 1. The overall structure of the water-collecting device.
(1-the survey ship, 2-shipboard winch, 3-scale rope, 4-hammer, 5-recovery cylinder, 6-lead fish).

comprises an elastic piece, rope buckles near the ends, two other buckles, the stick body, a U block, a stop block, a straight pin, a folding pin and a spring.

As shown in Figure 2 and Figure 3, opening and closing mechanism includes a stick, a U block, a limiting block, and a spring. The U block is

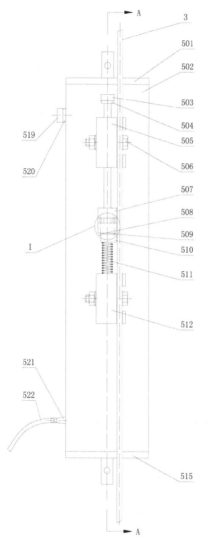

Figure 2.   The structure of the recovery cylinder.
(501-the upper end cover, 502-cylinder, 503-stick, 504-Oring A, 505-the upper fixed block, 506-compression screw, 507-U block, 508-fold pin, 509-straight pin, 510-limit block, 511-spring, 512-fixing block, 513-elastic band, 514-O ring B, 515-the lower cover, 519-air intake valve, 520-Oring C, 521-tap, 522-rubber tube, 523-out bulge, 524-the inner bulge, 525-holeB, 526-grooves, 527-hole A, 528-U hole.).

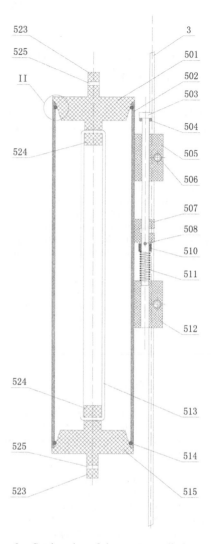

Figure 3.   Section view of the recovery cylinder.

arranged on the cylinder, and the limiting block is movably positioned between the U block and the fixing block. The stick passes through the U block, a limiting block, and the spring, and then it inserts into the fixing piece, which can move up and down. The limiting block and the U block are connected together, and they can move with a stick. At both ends of the spring are the limiting block and a fixing block, respectively.

The outside of the upper cover and the lower cover are connected through the rope buckle, and the rope buckle is connected to the pin. The insides of the upper cover and the lower cover are connected together in the cylinder by the elastic piece (Fig. 4).

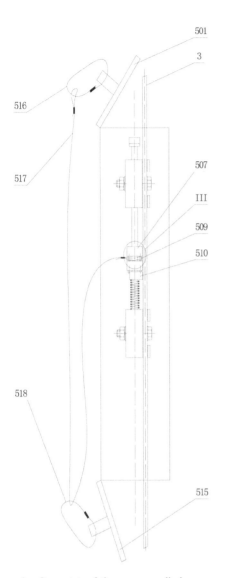

Figure 5. Application of the water-collecting device.

Figure 4. Open state of the recovery cylinder.

## 3 APPLICATION

In 2014, we used a water-collecting device for the observation and investigation in Haiyang's offshore, Shandong province, and we collected a large number of water samples, which was used for analyzing the sediments in water (Fig. 5).

When using the device, the following steps should be taken:

Step 1: Select a long enough scale rope with one end on the shipboard winch, and the hammer is crossed and the lead fish is tied up on the other end of the rope.

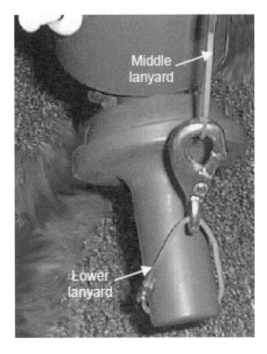

Figure 6. Correct—connection centered on the bottom end cap.

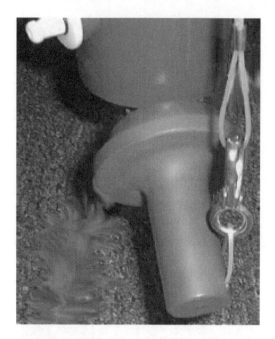

Figure 7. Incorrect—connection on the side of the bottom end cap.

Step 2: Mount the recovery cylinder to the scale rope. Clip the lower lanyard to the middle lanyard. The end cap handle should angle out away from the scale rope, not in towards the middle of the rope. Center the connection on the bottom end cap. The difference between Figure 6 and Figure 7 should be noted.

Step 3: Deploying and operating: hold the water sampler stationary before releasing the hammer to take the samples. If the water sample does not stop before firing recovery cylinder, the water in the cylinder is a mixture of water from many meters below the firing point (assuming that water samples are taken speedily). If moving at 1 m/sec, the cylinder's flushing constant is typically five to eight volumes, with water flushing slowly at the

cylinder inside wall and faster toward the cylinder center. For a 5-liter cylinder, the trapped sample contains a mixture of water from a cylinder in the water column with a diameter equal to the cylinder's inner diameter. Therefore, the standard practice is to stop the recovery cylinder to allow the cylinder to flush freely for several minutes before closing the cylinder.

## 4 SUMMARY

In summary, the water-collecting device proposed in this paper is a simple, reliable water sampling system, which can meet the demand of offshore water sampling.

ACKNOWLEDGMENT

The authors thank the senior engineer SiRen Li for his technical help. This research was supported by the Nantong Science and Technology Development Plan under Grants No. BK2013002 and the Strategic Leading Science and Technology Special Project of China Academy of Sciences (Class A) under Grants No. XDA11040306.

REFERENCES

Doherty, K.W. Frye, D.E. & Liberatore, S.P. et al. 1999. A Moored Profiling Instrument. *Journal of Atmospheric and Oceanic Technology*, 16: 1816–1829.
George, A.F. Greg, R.S. & Simon, J.P. 2004. An Energy-conserving Oceanographic Profiler for Use Under Mobile Ice Cover: ICYCLER. *International Journal of Offshore and Polar Engineering*. 14: 176–183.
Grant, M. & James, O. 2009. A novel method for estimating vertical eddy diffusivities using diurnal signals with application to western Long Island Sound. *Journal of Marine Systems*, 77: 397–408.
Rhinefranka, K. Agamloha, E.B. & Jouannea, A. 2006. Novel ocean energy permanent magnet linear generator buoy. *Renewable Energy*, 31: 1279–1298.

*Advanced Materials, Structures and Mechanical Engineering – Kaloop (Ed.)*
© 2016 Taylor & Francis Group, London, ISBN 978-1-138-02793-0

# An empirical analysis of the influence of international oil price's fluctuations on the exchange rate of RMB

J. Deng & J.Q. He
*College of Economics and Management, Nanjing University of Aeronautics and Astronautics, Nanjing, Jiangsu, China*

ABSTRACT: Considering RMB's internationalization and oil price's impact on the economy, this paper tries to find out the influence of international oil price on RMB exchange rate with cointegration model. The model shows that increasing oil price brings about RMB appreciation pressure. Then ECM model is applied to test the cointegration-model's stability. After that is the economic analysis of the relationship between RMB and oil price, finding energy structure is one of the main reasons. In order to minimize the international oil price's impact on the national economy and the exchange rate of RMB, the paper offers some policy advice for China.

## 1 INTRODUCTION

Petroleum is an important raw material and energy source, as it is applied in various industries in the 21st century and is called "black gold". With the rapid growth of the national economy in China, the demand of petroleum is increasing rapidly while the dependence on foreign oil of China also increases every year. In August 2011, the data published on the website of MIT shows that China's dependence on foreign oil has surpassed the United States' and reached 55.2%. If the trend continues, the impact of international oil price on China will increase. Meanwhile, RMB is facing a severe appreciation pressure, and the influence can't be ignored. Therefore, the study of the impact of the international oil price fluctuations of the exchange rate of RMB is necessary.

Hotelling (1931) studied the oil price's tendency, oil price fluctuates widest and the most frequently among the exhaustible resources. Applying his model, the price is the best when resource price increasing rate equals the discount. Based on his study, many works of literature build up different models to analyze the reasons of oil price's fluctuation and tendency according to the market structure and subjects' behaviors.

Krugman (1981) studied the relationship between oil price and USD exchange rate and find shares in OPEC influence the exchange rate. Based on his study, Golub (1983) built a model to analyze the relationship between oil price and USD exchange rate, different from Krugman, he added the oil demand elasticity. Amano (1998) analyzed the relationship between USD exchange rate and oil price

empirically. And found the oil price is the main cause of exchange rate changes which supported Krugman and Golub's opinions. Lastrapes (1992) tested the real and nominal impact on the USD exchange rate and found real impact such as oil price fluctuation is an important influence on the exchange rate.

Yu Weirong and Hu Haixin (2004) applied ADF to test the oil price fluctuation's influence on RMB exchange rate and found the oil price influence USD, RMB exchange rate is mainly influenced by USD exchange rate and the price level of main trade partners. Wu Lihua and Fu Chun (2007) studied the relation between RMB exchange rate and Oil price. They argued oil price influenced RMB exchange rate and RMB exchange rate has little influence the on the oil price.

Based on previous literature, this paper would apply cointegration regression to find whether there is a stable relationship between RMB exchange rate and international oil price. And base on the empirical statistics, try to find out the reason of the relationship and provide policy advice for China.

## 2 EMPIRICAL ANALYSIS

### 2.1 *Model introduction*

As for the paper studies the influence of international oil price on the exchange rate of RMB, we choose international oil price and RMB exchange rate as variables. Because America is the biggest oil consumer country, and WTI (West Texas Intermediate) oil price is usually used as the benchmark price by investors to measure the oil price fluctuation. For this reason, this paper use WTI spot price (Cushing, OK WTI Spot Price FOB, derived from http://www.eia.gov/) as the international oil price. And the nominal effective exchange rate of RMB is used to represent the exchange rate of RMB (derived from http://www.bis.org/). As for China reformed the price management system of crude and refined oil from 1998 to 2001, the oil price management system has been stable since 2001. Thus, this paper uses monthly data from 2002.01–2012.03 to analyze the influence of international price on RMB exchange rate.

The basic model is as follows:

$$lneer_t = \alpha lop_t + \beta \qquad (1)$$

$Lneer_t$ represents the nominal effective exchange rate of RMB, and $lop_t$ represents the international oil price. In order to avoid heteroscedasticity, variables such as RMB exchange rate and international oil price are logarithmic.

## 2.2 Variables' stationary

This paper adopts the usual unit root test method ADF to test the variables. The results are showed in Table 1.

After ADF unit root test, we find the two variables—exchange rate and oil price are nonstationary time series. Then we test their first difference series and find the results in Table 2.

Table 1. The results of ADF test of variable *lneer* and *lop*.

| Variable | Exogenous (c, t, k) | ADF test statistic | Critical value | Conclusion |
|---|---|---|---|---|
| *lneer* | (c, t, 4) | −2.78754 | −3.44835 | Non-stationary |
| *lop* | (c, t, 2) | −3.28094 | −3.4477 | Non-stationary |

*c, t, k in the test form is constant, linear trend, and lag length. **The critical values are calculated by Eviews, at the confidence level of 5%. ***lneer* is the logarithmic RMB nominal effective exchange rate, *lop* is the WTI spot price.

Table 2. The results of ADF test of first-differenced variable *lneer* and *lop*.

| Variable | Exogenous (c, t, k) | ADF test statistic | Critical value | Conclusion |
|---|---|---|---|---|
| *Dlneer* | (0, 0, 6) | −4.31871 | −1.94364 | Stationary |
| *Dlop* | (0, 0, 5) | −5.22668 | −1.94361 | Stationary |

*c, t, k in the test form is constant, linear trend, and lag length. **The critical values are calculated by Eviews, at the confidence level of 5%. ***dlneer* is the logarithmic RMB nominal effective exchange rate, *dlop* is the WTI spot price.

When in the regression without linear trend and constant, the ADF test statistics of two variables are less than the critical values at the confidence level of 5%. Therefore, the first difference of the original time series are stationary, and the original time series are first-order cointegration, which means RMB nominal effective exchange rate and international oil price have a stable relationship in the long run. The following is the quantitative analysis of the influence of international oil price on RMB exchange rate.

## 2.3 Cointegration regression

Johansen test includes five regression formulations, this paper choose the one that series $y_t$ excludes linear trend and regression function includes the intercept. Besides, the lag length is eight according to AIC (Akaike information criterion) and SC (Schwarz Criterion). As for the lag length of Johansen test in E-views is the lag length of variables' first difference, the lag interval is seven. The results are shown in Table 3.

Table 3 includes unrestricted cointegration rank trace test and maximum eigenvalue test, then we find the statistics of the two tests are larger than their 5% critical value, therefore, we reject the null hypothesis $H_{00}$; and they are both less than the critical values in the second line, so we accept the second alternative hypothesis $H_{10}$. As a result, we conclude there's only one independent cointegration variable, which means variable *dlneer* and *dlop* have a stable relationship in the long run.

After regression, we have Equation (2)

$$lneer_t = 0.262427731 lop_t + 3.51492035 \qquad (2)$$
$$(0.06520) \qquad (0.26364)$$
$$[-4.02485] \qquad [-13.3323]$$

Figures in parentheses are standard errors; figures in square brackets are t-statistics. Usually, when a simple's size is large than 30, we take it as a large sample and replace t-statistic with z-statistic. And the z-test statistic critical value is 1.64 at the confidence level of 5%, which means

Table 3. Johansen test results.

| Test method | Eigenvalue | Statistic | 5% critical value | Equation number |
|---|---|---|---|---|
| Trace | 0.140281 | 21.95294 | 20.26184 | None |
|  | 0.038966 | 4.570678 | 9.164546 | Most 1 |
| Max eigenvalue | 0.140281 | 17.38226 | 15.8921 | None |
|  | 0.038966 | 4.570678 | 9.164546 | Most 1 |

142

the variable lop's parameter is significant and RMB exchange rate correlates with international oil price positively.

However, the fact that variable's parameter is significant doesn't mean it is an independent variable, we should apply Granger causality test to find which is the cause.

### 2.4 Granger causality test

Cointegration regression results show the nominal effective exchange rate of RMB correlates positively with the international oil price in the long run. But it doesn't determine which one is the cause and should be the independent variable; does RMB exchange rate promote international oil price increase or the oil price cause exchange rate changes? Then we should proceed with Granger causality test. Table 4 is the results of Granger causality test.

According to Table 4, we can conclude that the nominal effective exchange rate is the cause of WTI spot oil price changes; and WTI spot oil does not cause the RMB exchange rate to change. Which means lop should be the independent variable in the regression equation.

### 2.5 Error correction model

ECM (Error Correction Model) includes lagged variables, and when the variable series are not stationary, ECM could help to avoid spurious regression. Cointegration equation describes the long-run relationship between two of variability. Like attraction of gravitation, cointegration would draw disequilibrium condition to equilibrium. To test the tendency once it's not in the equilibrium, we apply ECM:

$$\Delta lneer_t = -0.0285\ ECM_{t-1} - 0.0259\ \Delta lop_t$$
$$+ 0.2708\ \Delta lneer_{t-1} \tag{3}$$

$$(-3.531)\ (-2.232)\quad (3.237)$$
$$ECM_{t-1} = lneer_t - 0.2624 lop_t - 3.5149 \tag{4}$$

Figures in parentheses under Equation (3) are t-statistics of the parameters of variables. Because the sample is a large simple, we replace z-statistics with t-statistics. And we can find z-statistic critical

value is 1.64 at the confidence level of 5%, thus parameters are significant. $ECM_{t-1}$ is the error correction term and is the minus, which accords with the theory. Besides, the parameter of $ECM_{t-1}$ is −0.028, which means the correction speed is 0.028—when nominal effective exchange rate is in disequilibrium, the variables lneer and lop tend to the equilibrium, the power is about 0.028 times of its deviation from the equilibrium.

## 3 ECONOMIC ANALYSIS

RMB nominal effective exchange rate and international oil price have a stable relationship in a long run according to the cointegration regression equation. When the oil price is increased by 1%, RMB nominal effective exchange rate will increase by 0.26%. Besides, the oil price is the cause of RMB exchange rate changes according to Granger causality test at the confidence level of 5% which is contrary to some economic opinions. In general opinions, there are two ways for international oil price influencing a country's economy, which is the balance of payment and domestic economy. In the way of the balance of payment, international oil price increasing would enlarge China's expenditure and stimulate the demand for foreign currencies. Therefore, foreign currencies will appreciate and the NEER of RMB will decrease. In the way of domestic economy, oil price increasing will raise the cost of industries such as chemical, transportation, tourism, as petroleum is a widely applied energy source and material. Therefore, the cost push inflation, with RMB devaluate and the NEER decrease. As a result, there should be a negative relationship between the RMB NEER and international oil price, which mean if the international oil price increases, the NEER should decrease, and vice versa.

However, the empirical statistics are contrary to the theory above. According to the empirical analysis and reality, the reasons may be as follows.

In the way of the balance of payment, international oil price's fluctuation does influence our demand for foreign currencies. However, China's exchange reserves are very huge. Until 2009, China's exchange reserves reached 2,399 billion dollars according to the statistics from National Bureau of Statistics of the People's Republic China and People's Bank of China. In the same year, the refined and crude oil import was 106 billion dollars, accounting for 4.43% of the exchange reserves. And China's exchange reserves reached 2,847 billion dollars when the refined and crude oil import was 157 billion dollars, accounting for 5.53%. Therefore, international oil price's fluctuation influences a little on RMB exchange rate with the rapid increase of China's exchange reserve.

Table 4. Granger causality test results.

| Null hypothesis | F-statistic | Critical value | Probability |
| --- | --- | --- | --- |
| lop doesn't cause lneer | 2.73088 | 2.03 | 0.00922 |
| lneer doesn't cause lop | 1.50941 | 2.03 | 0.16368 |

143

From the way of domestic economy, the oil price's increase would push the inflation, however, it influences not only China but also China's main trade partner such as United States, European Union, Japan, South Korea. Thus, foreign countries also inflate. Oil-import dependency and energy consumption structure determine the degree of inflation. Take 2007 as an example, China's oil-import dependency is 50% and the U.S. is 63.9%, and China's main energy consumption is coal other than petroleum in the U.S. As a result, China's inflation pressure by oil price is less than the U.S., thus RMB appreciated compared with the U.S. dollar. Although China's oil-import dependency was 55.2% in 2011 according to MIIT (Ministry of Industry and Information Technology of the People's Republic of China), more than America's 53.5%, the difference is small. And the oil took 18.58%, 17.62% of energy consumption in China in 2009 and 2010, while raw coal took 70.62%, 70.45%; meanwhile, crude oil took 30.63% and 37.19% in America, almost twice of China. To sum up the aspects of oil-import dependency and energy consumption structure, America's inflation pressure caused by oil price increase is larger than China, consequently, oil price increasing makes RMB appreciates compared with USD. Other countries are the same with the analysis.

As the USD is used as the world currency and oil price is marked by USD, USD's value influences the oil price directly, and RMB exchange rate changes influence oil price by the way of USD, so the influence of RMB exchange rate on international oil price is little.

For the reasons mentioned above, international oil price increasing will bring about the pressure of RMB appreciation. As the structure of energy consumption is stable is the short run, the relationship between RMB exchange rate and international oil price will remain.

## 4 CONCLUSIONS

After empirical analysis of the influence of international oil price on RMB exchange rate, this paper concludes that international oil price and NEER have a positive relation. Because China's main consumed energy is coal other than oil, the inflation

pressure caused by oil price increase on China is less than other countries. Thus, NEER is positively related with international oil price.

However, as a huge oil-consumption economy, China influences little on the international oil price and is always a price-taker. However, as international oil price fluctuates in last several years, China should find more ways other than administrative orders to avoid the risk of oil price. China could ensure its oil reserves from three aspects:

1. Using multiple energies, reducing the dependency on oil, promoting the development of solar energy, wind energy, nuclear energy and so on. After the oil crisis in the 21st century, countries around the world are trying to replace oil with other energy resources. China could build wind station in the northwest; use more geothermal energy in the southwest; use more wave energy in the east such as Fujian, Zhejiang.
2. Establishing crude oil forward, market to increase China's influence on the international oil price. Despite the market risk, it could help to make the oil price stable and is a way to avoid the oil price risks. If there's mature regulation, the forward market's advantage is obvious.
3. Increasing oil import resources, making oil reserves stable. China may focus on the countries with large oil resources and production, including seven Asian countries, six African countries, seven mid-east countries and ten west countries, and decide proper import weight of each one.

## REFERENCES

Amano, R.A. & Norden, S. 1998. Oil prices and the rise and fall of the US real exchange rate. *Journal of international money and finance*. 17: 299–316.

Golub, S.S. 1983. Oil prices and exchange rates. *The economic journal*, 93(371): 576–593.

Krugman, P. 1980. *Oil and the dollar*. National bureau of economic research working. 554.

Lastrapes, W.D. 1992. Sources of fluctuations in real and nominal exchange rates. *Review of economics and statistics*, 74(3): 530–539.

Wu, L. & Fu, C. 2007. Research on the relation between oil price and exchange rate. *Fujian finance*. 3: 15–17.

Yu, W. & Hu, H. 2004. Research on the oil price impact on the exchange rate between America and China. *Studies of international finance*. 12: 33–39.

*Advanced Materials, Structures and Mechanical Engineering – Kaloop (Ed.)*
© *2016 Taylor & Francis Group, London, ISBN 978-1-138-02793-0*

# LQR control of seismic-excited tall building with ATMD

Y.M. Kim & K.P. You
*Department of Architecture Engineering, Chonbuk National University, Jeonju, Korea*
*Long-Span Steel Frame System Research Center, Jeonju, Korea*

J.Y. You
*Department of Architecture Engineering, Songwon University, Gwangju, Korea*

D.K. Kim
*Department of Civil and Environmental Engineering, Kunsan National University, Kunsan, Korea*

ABSTRACT:   Active Tuned Mass Damper (ATMD) is used for the control of a tall building under earthquake loads. Active control force acting on ATMD is estimated by Linear Quadratic Regulator (LQR) algorithm. Optimal parameters of tuning frequency and damping ratio of ATMD for minimizing rms response of the damped main system derived by Krenk were used. The performance of ATMD for suppressing rms response of a tall building under earthquake loads is compared with that of a single Tuned Mass Damper (TMD) without control force. It is found out that a building/ATMD system is more effective than that of TMD system in reducing the response of a tall building under earthquake loads. Comparing the response of a building without TMD with a building/TMD system, a building/TMD system did not show the efficiency for suppressing response under earthquake loads.

## 1   INTRODUCTION

Suppressing the seismic-excited responses of a tall building has been much attention to structural engineers. In last three decades, many control devices, passive, active and hybrid, have been developed to control structural vibration due to dynamic loads such as earthquakes and wind loads (Chang 1995). Tuned Mass Damper (TMD) is a classical vibration control device consisting of a mass, a spring and a damper supported by the main vibrating system (Lee et al. 2006). The concept of TMD is that if the natural frequency of TMD is tuned to the natural frequency of the vibrating main system, then the vibration energy is dissipated through the damping in the TMD, so reducing the main system's vibration (Ayoringde 1980). The original idea of TMD is from Frahm in 1909, who invented vibration control device called a vibration absorber using a spring supported mass without damper (Frahm 1909). That was only effective when absorber's natural frequency was tuned to a frequency close to the excitation frequency. This shortcoming was improved by introducing damper in the spring supported mass by Ormondroyd and Den Hartog (Ormondroyd 1928). Later Den Hartog derived optimal frequency tuning and damping ratio for the undamped main system under harmonic load (Den Hartog 1956).

While Den Hartog considered harmonic loading only Warburton and Ayroinde derived optimum parameters of TMD for a undamped main system under harmonic and white noise random excitations (Ayoringde 1980). If the mass ratio is small and the main system's damping ratio is small relative to that of TMD, Krenk derived the optimum parameters for tuning frequency and applied damping ratio of TMD for damped main system (Krenk 2008). A number of TMDs have been installed in tall buildings to suppress wind-induced vibrations of a tall building (Housner et al. 1997)

At that time, it was accepted as a fact that the performance of TMD could be enhanced by incorporating a feedback controller through the use of an actuator as an active control force in the design of TMD, which was called Active Tuned Mass Damper (ATMD) (Chang 1980, 1995). ATMD design for suppressing wind-induced vibration of a tall building using assumed deterministic harmonic wind load was presented by Chang and Soong (1980). That is a first active control study for mitigating wind-induced vibration of a tall building installed in ATMD using Linear Quadratic Regulator (LQR) controller. Then many advanced studies including Linear Quadratic Gaussian (LQG), $H_2$ and $H\infty$ control theories for reducing dynamic responses of a tall building under earthquakes and wind loads have been developed and a number of tall buildings are currently implemented with active control devices systems (Yang 2002).

In this study, LQR control algorithm is used for the control of dynamic responses of a tall building with ATMD subject to El-Centro 1940 N-S component excitation. The optimum parameters of tuning frequency and damping ratio for ATMD derived by Krenk (2008) are used.

## 2 MODELING OF BUILDING-ATMD SYSTEM

A tall building installed in ATMD at the top floor level with an active control force device such as an actuator is shown in Figure 1. The dynamic analysis of a multi-degree-of-freedom model of a tall building can be simplified if the contribution of the higher modes is ignored. In that case a tall building can be represented by its fundamental mode generalized system in the model (Kareem 1995).

The building modeled as an equivalent Single-Degree-of-Freedom (SDOF) system is consist of fundamental modal mass $m_1$, stiffness $k_1$ and damping constant $c_1$. The mass, damping, and stiffness constants of TMD are represented by $m_d$, $c_d$, and $k_d$. $u(t)$ is an active control force.

The dynamic equations of motion of the system can be written as,

$$m_s \ddot{x}_s + c_s \dot{x}_s + k_1 x_1 - c_d \dot{x}_d - k_d x_d = -m_s \ddot{x}_g + u \quad (1)$$

$$m_d \ddot{x}_d + c_d \dot{x}_d + k_d x_d + m_d \ddot{x}_s = -m_d \ddot{x}_g + u \quad (2)$$

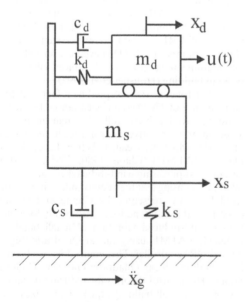

Figure 1. Building-ATMD system.

where, $z(t) = x_d - x_s(t)$ is the displacement of $m_d$ relative to that of $m_s$. $\ddot{x}_g$ = seismic acceleration.

This equation can be written in terms of the state-space form as follows (Dorato 1995, Lewis 2012).

$$\dot{X}(t) = AX(t) + Bu(t) + Hf(t) \quad (3)$$

where, $X(t) = \begin{bmatrix} x_s & z & \dot{x}_s & \dot{z} \end{bmatrix}^T$ denotes the state vector of the system with,

$$A = \begin{bmatrix} 0 & 0 & 1 & 0 \\ 0 & 0 & 0 & 1 \\ -\dfrac{k_s}{m_s} & \dfrac{k_d}{m_s} & -\dfrac{c_s}{m_s} & \dfrac{c_d}{m_s} \\ \dfrac{k_s}{m_s} & -\left(\dfrac{k_d}{m_d} + \dfrac{k_d}{m_s}\right) & \dfrac{c_s}{m_s} & -\left(\dfrac{c_d}{m_d} + \dfrac{c_d}{m_s}\right) \end{bmatrix} \quad (4)$$

is a system dynamic matrix.

$$B = \begin{bmatrix} 0 & 0 & -\dfrac{1}{m_s} & \dfrac{1}{m_s} + \dfrac{1}{m_d} \end{bmatrix}^T \quad (5)$$

is a location vector of $u(t)$.

$$H = \begin{bmatrix} 0 & 0 & -1 & 0 \end{bmatrix}^T \quad (6)$$

is a location vector of $f(t)$, where, $f(t) = \ddot{x}_g$.

## 3 OPTIMUM PARAMETERS OF TMD

Ayroinde & Warburton derived optimum parameters of tuning frequency $f_{opt}$ and applied damping ratio $\xi_{opt}$ of TMD for undamped single-degree-of-freedom main system under white noise random excitations (Ayroinde 1980). Krenk (2008) derive optimum parameters of TMD for single-degree-of-freedom damped main system if the mass ratio is small and the main system's damping ratio is small relative to that value of TMD as.

$$f_{opt} = \frac{1}{1+\mu} \quad (7)$$

$$\xi_{opt} = \frac{\sqrt{\mu}}{2} \quad (8)$$

where, $\mu$ = mass ratio of $m_d/m_s$; $m_s$ = mass of main structure; $m_d$ = mass of TMD.

It is shown that the optimum damping and tuning frequency ratio is only influenced from the mass ratio (Krenk 2008).

## 4 LINEAR QUADRATIC REGULATOR CONTROLLER

Linear Quadratic Regulator (LQR) algorithm is a widely used optimal control theory in structural vibration control problem (Dorato 1995, Lewis 2012).

In LQR control law, all continuous time state-space variables are available and linear dynamic equations of motion can be represented by the state-space formulation as in Equation (3) (Dorato 1995, Lewis 2012).

The object of LQR control law is sought to find out a state-feedback optimal control force $u(t)$ that minimize the deterministic cost functional $J$ maintaining the state close to the zero state. The cost functional $J$ is given by (Dorato 1995, Lewis 2012).

$$J = \int_0^\infty (X(t)^T Q X(t) + u(t)^t R u(t)) dt \qquad (9)$$

where, $Q$ = positive semi-definite state weighting matrix; $R$ = positive definite control weighting matrix.

The term $X(t)^T Q X(t)$ in Equation (9) is a measure of control accuracy and the term $u(t)^T R u(t)$ is a measure of control effort. Minimizing $J$ with keeping the system response and the control effort close to zero needs appropriate choice of the weighting matrices $Q$ and $R$ (Lewis, 2012). If it is desirable that the system response be small, then large values for the elements of $Q$ should be chosen. If it wants the control energy be small, then large values of the elements of $R$ should be chosen (Lewis, 2012).

The state-feedback optimal control force $u(t)$ is derived as (Dorato 1995, Lewis 2012).

$$u(t) = -KX(t) \qquad (10)$$

where, $K = R^{-1} B^T P$

In Equation (10), $K$ is called an optimal controller gain and $P$ is the unique, symmetric, positive semi-definite matrix solution to the Algebraic Riccati Equation (ARE) given by

$$A^T P + PA - PBR^{-1} B^T P + Q = 0 \qquad (11)$$

Then the closed-loop system using the optimal control force $u(t)$ becomes.

$$X = (A - BK)X(t) = AcX(t) \qquad (12)$$

where, $Ac$ = the closed-loop system matrix.

## 5 NUMERICAL EXAMPLE

This numerical example is from "Numerical Examples" in reference (Solari 1993). The tall building's height H = 180 m, width B = 60 m, depth D = 30 m. The modal mass, damping, and stiffness constants are $2.4*10^7$ kg, $1.220832*10^6$ N-s/m, and $6.9001425*10^7$ N/m. Natural frequency $f_1$ = 0.27 Hz, damping ratio $\xi_1$ = 0.015. The optimum parameters of ATMD are considered as the same value of the passive TMD. The optimum parameters of TMD, that is, mass ratio $\mu$ = 0.015, optimum tuning frequency $f_{opt}$ = 1.0, and optimum damping ratio $\xi_{opt}$ = 0.05.

Figure 2 shows the actual accelerogram of El-Centro 1940 N-S component earthquake ground motion. The Peak Ground Acceleration (PGA) of the El-Centro 1940 is 3.421 m/s².

Dynamic response of a tall building without TMD is shown in Figure 3. The rms response value is 0.0595 m. And rms response with TMD is shown in Figure 4. The rms value is 0.0586 m, which shows the reduction effect using TMD is about 2%, so the reduction effect of TMD for suppressing seismic-excitation of a tall building cannot be obtained. The relative displacement of TMD with respect to tall building is in Figure 5, which shows that the relative displacement of TMD is only increased

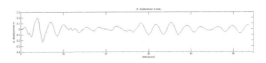

Figure 2. Accelerogram of El-Centro 1940 N-S component earthquake ground motion.

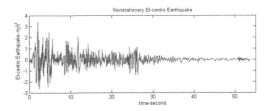

Figure 3. Dynamic responses without TMD (rms = 0.0595 m).

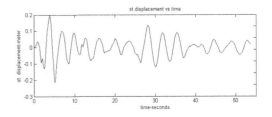

Figure 4. Dynamic responses with TMD (rms = 0.0586 m).

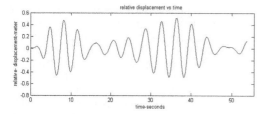

Figure 5. Relative displacement of TMD (rms = 0.2413 m).

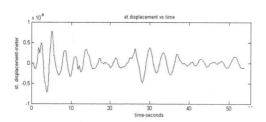

Figure 6. Dynamic responses with ATMD (rms = $1.9671*10^{-9}$ m).

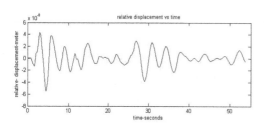

Figure 7. Relative displacement of ATMD (rms = $1.4451*10^{-8}$ m).

4 times over the rms response of tall building. That shows TMD would not be worked well under that condition. Therefore, the TMD effect for reducing the seismic excitation of tall building cannot be obtained (Ricciardelli 2003).

When the weighting matrix $Q$ and $R$ for LQR controller are chosen as

$$Q = \begin{bmatrix} 100 & 0 & 0 & 0 \\ 0 & 1 & 0 & 0 \\ 0 & 0 & 100 & 0 \\ 0 & 0 & 0 & 1 \end{bmatrix}, R = 1.0* 10^{-15}$$

The reduced responses with ATMD using LQR controller is shown in Figure 6. And the relative displacement of ATMD with respect to tall building is shown in Figure 7. The active control force of LQR controller applied to ATMD is shown in Figure 8.

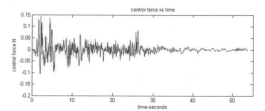

Figure 8. Active control force (N).

As shown above results, the control effect for suppressing the seismic-excitation of tall building can be obtained using ATMD system. And TMD system could not obtain such a control effect for suppressing the seismic excitation of a tall building.

## 6 CONCLUSION

The performance of ATMD with LQR controller and TMD without any controller for suppressing seismic excitation of a tall building. is investigated. ATMD with LQR controller system is superior to that of TMD system for suppressing seismic excitation of a tall building. However, comparing the reduced rms response with TMD system (rms = 0.586 m) with that of without TMD system (rms = 0.0595 m), the reduced ratio of rms response is only about 2%, which shows insignificant effect of TMD system for reducing seismic–excitation of a tall building. Therefore, TMD system is not effective for the control of seismic excitation of a tall building.

## ACKNOWLEDGMENT

This work was supported by the National Research Foundation of Korea Grant funded by the Korean Government (NRF-2014R1A2A1A10049538).

## REFERENCES

Ayoringde, E.O. & Warburton, G.B. 1980. Minimizing structural vibrations with absorbers, *Earthquake Engineering and Structural Dynamics*, 8: 219–236.
Chang, C.C. & Yang, T.Y. 1995. Control of building using active tuned mass dampers. *Journal of Engineering Mechanics*, 121(3): 355–366.
Chang, J.C.H. & Soong, T.T. 1980. Structural control using active tuned mass dampers. *Journal of Engineering Mechanics,* 106: 1091–1098.
Den Hartog, J.P., 1956. *Mechanical Vibration, 4th edn.* McGraw-Hill, New York. (Reprinted by Dover, 1985).
Dorato, P. Abdallah, C. & Cerone, V. 1995. *Linear-Quadratic Control*, Prentice Hall, New Jersey.

Frahm, H. 1909. *Device for damped vibration of bodies.* U.S. PATENT No. 989958. October 30.

Housner, G.W. Bergman, L.A. Caughey, T.K. Chassiakos, A.G. Claus, R.O. Masri, S.F. Skelton, R.E. Soong, T.T. Spencer, B.F. & Yao, J.T.P. 1997. Structural control: past, present, and future, *Journal of Engineering Mechanics ASCE*, 123(9): 897–971.

Kareem, A & Kline, S. 1995. Performance of multiple mass dampers under random loading, *Journal of Structural Engineering ASCE,* 121(2): 348–361.

Krenk, S. & Hogsberg, J. 2008. Tuned mass damped structures under random load, *Probabilistic Engineering Mechanics*, 23: 408–415.

Lee, C.L. Chen, Y.T. Chung, L.L. & Wang, Y.P. 2006. Optimal design theories and applications of tuned mass dampers. *Engineering Structures,* 28: 43–53.

Lewis, F.L. Vrabie D.L. & Syrmos, V.L. 2012. *Optimal Control, 3rd edn.* John Wiley & Sons.

Ormondroyd, J. & Den Hartog, J.P. 1928. The theory of the dynamic vibration absorber, *Transactions of ASME,* 50 (APM-50-7): 9–22.

Ricciardelli, F. Pizzimenti, A.D. & Mattei, M. 2003. Passive and active mass damper control of tall buildings to wind gustiness, *Engineering Structures,* 25: 1199–1209.

Solari, G. Gust Buffeting. II. 1993: Dynamic alongwind respose. *Journal of Structural Engineering, ASCE,* 119(2): 383–398.

Yang, J.N. Lin, S. Kim, J.H. & Agrawal, A.K. 2002. Optimal design of passive energy dissipation systems based on H and H2 performances, *Earthquake Engineering & Structural Dynamics.* 31: 921–936.

*Advanced Materials, Structures and Mechanical Engineering – Kaloop (Ed.)*
© *2016 Taylor & Francis Group, London, ISBN 978-1-138-02793-0*

# Use of nonconventional liquid fuel in heat power industry

A. Lesnykh & K. Tcoy
*Far Eastern Federal University, Russia*

ABSTRACT: The article demonstrates the possibility of using fish processing industry subproducts instead of conventional liquid fuels. The results of fish oil burning research are demonstrated in this article. There is a comparison of burning processes and polluting emissions while burning conventional and alternative liquid fuels. The article formulates conditions for effective use of fish oil in boiler plants.

## 1 INTRODUCTION

### 1.1 *Price increase on liquid hydrocarbons*

The most industrial enterprises have installed boiler plants working on liquid fuel. In the current fuel market situation the price on liquid hydrocarbons increases constantly. The cost of heating oil reaches 390–400 USD/ton. The most food industry enterprises have accumulated waste products or secondary products of their production. As for fish processing enterprises, such kind of product is mainly fish oil. Boilers of these enterprises function on liquid hydrocarbon fuel, which is brought to them from far away. It can be explained by the fact that these enterprises have a remote location. Fuel is shipped to them by water transport only which is the most effective way of liquid fuel supply.

### 1.2 *Fish oil reserves*

Average fish processing enterprise processes about 80 thousand tons/year of fish and accumulates within a season around 6.4 thousand tons of fish oil. Total annual reserves of fish oil reached 105 thousand tons at the Russian Far East fish processing enterprises in the year 2014. The volume of fish oil utilized by medical and food industries is not included.

Based on the above-said, the utilization of fish oil by way of it's burning in the boiler furnaces is very perspective. The experience of American power industry shows the possibility of using fish oil in the stationery diesel engines (Blythe 1996, Steigers 2002).

## 2 RESEARCH OF FISH OIL BURNING PECULIARITIES AT THE LABORATORY BENCH

### 2.1 *The comparison of fuel properties and obtained torches*

Researches of change the conventional liquid fuels to fish oil were conducted in several stages.

The first stage was the comparison of fish oil burning and diesel fuel in the experimental laboratory burner. Fuel parameters are shown in Table 1. Fish oil from reserves of Yuzhno-Morskoy fish processing plant was used as an experimental sample.

During the comparison of fuel parameters it is worth paying attention to a difference in calorific capacity. Net calorific value of fish oil is about 6,000 kJ/kg lower than of diesel fuel. Also, an important fact is that fish oil has bigger amount of oxygen combined with multivitamin compositions (about 3%).

The first attempt of burning fish oil was made on turbo cyclone burner of Kiturami 17r water boiler (Fig. 1). This burner is equipped with mechanical atomizer. The attempt was unsuccessful: the torch failed and died down immediately. The changes a of fuel supply scheme and burner stabilizers did not give any results.

The next experiment was conducted on GierschG20 burner (Fig. 2). This burner is of an ejector type. Fuel is atomized by air from burner's topping compressor. Fuel consumption on this burner depends on primary air pressure. After heating of fish oil up to 70°C we managed to obtain the stable torch.

All experiments were conducted on laboratory bench shown on Figure 3. The bench consists of fire

Table 1. Comparison of fuel parameters and contents.

| Setting | Diesel fuel | Fish oil |
|---|---|---|
| $C^r$ | 86% | 79% |
| $H^r$ | 13% | 14% |
| $O^r$ | 0.9% | 3.9% |
| $N^r$ | 0.002% | 0.0% |
| $S^r$ | 0.2% | 0.0% |
| $A^r$ | 0.01% | 0.0% |
| $W^r$ | 0.1% | 3.0% |
| $Q_i^r$ | 42,727 kJ/kg | 36,025 kJ/kg |

Figure 1. Turbo cyclone burner Kiturami 17r.

Figure 2. Ejection burner Giersch G20.

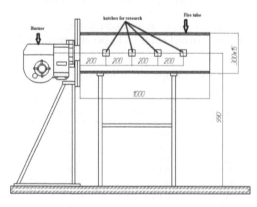

Figure 3. Laboratory bench.

tube on supports and a rack with removable panels, to install different burners. Along its length, the fire tube has hatches to conduct measurements.

For comparison purposes, the same burner was started on diesel fuel. The geometrical comparison of diesel fuel and fish oil torches obtained on the GierschG20 burner with the same fuel consumption was shown on Figure 4. The change of capacity on this burner is made by primary air pressure; that is why in all experiments we are going to adjust on this value.

As it is shown on Figure 4, fish oil torch is almost 2 times shorter than diesel fuel torch, whereas the capacity is the same. When burner functions on fish oil, unsatisfactory atomization is observed, which is proven by the presence of large fuel droplets in all the work range. This fact could be explained by big difference in fish oil and diesel fuel density.

## 2.2 Analysis of fish oil and diesel fuel burning processes

Chart for temperature changes depending on primary air pressure value is shown on Figure 5.

In the process of research, we have taken readings in the representative points of fire tube for fish oil and diesel fuel in all burner's capacity range.

As it can be seen on the chart, the temperature of exhaust gasses during burning diesel fuel is about 40°C higher than temperature of exhaust gases during burning fish oil in all the range of capacity.

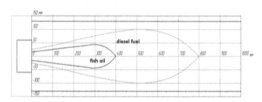

Figure 4. Comparison of fish oil and diesel fuel torches.

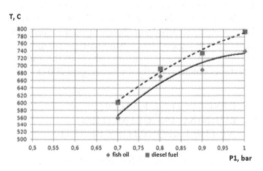

Figure 5. Dependency of combustion products temperature on primary air pressure.

Dependency of NOx emissions on primary air pressure was measured at the tube outlet of experimental bench. After that the obtained figures were reduced to excess air ratio α = 1 and shown on Figure 6.

As it can be seen on the Figure, in all the range of capacity NOx emissions during burning of fish oil is not less than 200 ppm lower than when burning diesel fuel. NOx formation mechanism is very complicated, but one may conclude that emission reduction is explained by lowering of temperature in the active burning zone (Shtym et al. 2015).

Fuel combustion quality is estimated according to CO concentration in combustion products. Comparison of this parameter is shown on Figure 7. As it can be seen on the figure, fish oil burnout is more complete than diesel oil burnout in spite of its worse atomization. And CO emissions are 2–3 times lower.

Based on the above analysis one may conclude that fish oil burning is more complete, and pollutant emissions are less than when diesel fuel burns.

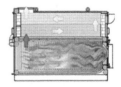

Figure 8. Water heating boiler ТГВ-1.

Table 2. Technical and economic indexes for boiler ТГВ-1 operation.

| Name | Heating oil | Fish oil |
| --- | --- | --- |
| Exhaust gases temperature, °C | 315 | 303 |
| Excess air ratio | 1.54 | 1.43 |
| Efficiency ratio % | 80 | 82 |

## 3 RESEARCH OF FISH OIL BURNING IN THE WATER BOILER

The second stage has become fish oil burning in the water boiler with a capacity of 1 MW (Fig. 7), equipped with fuel heaters. The rotary principle burner has been installed on the boiler, atomization is made with the help of the rotating bowl, to which the fuel flows.

Before experiment, load balance tests were conducted on this boiler with heating oil, based on a method proposed by the authors of (Trembovlya 1991). Comparison of technical and economic indexes of boiler operation is shown in Table 2.

As it can be seen from a comparison of technical and economic indexes, excess air ratio α when working on fish oil is 0.11 less then when working on heating oil. It is explained by the big quantity of fuel-combined oxygen, which reacts with combustible components while heating. That leads to a reduction of air quantity that is necessary for burning. Exhaust gases temperature is 12°C lower. The efficiency ratio for ТГВ-1 boiler is 2% higher when working on fish oil. Whereas specific reference fuel consumption was reduced at 5 kg of reference fuel/Gcal.

## 4 CONCLUSIONS

The conducted research may lead to the following conclusions:

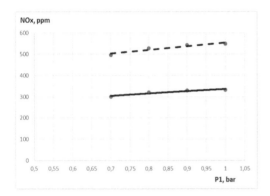

Figure 6. Dependency of NOx concentration in combustion products on primary air pressure.

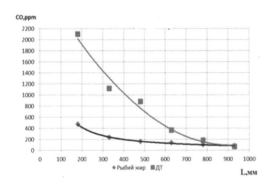

Figure 7. Analysis of Carbon Oxide (CO) burnout along the length of experimental tube.

1. Burning of fish oil is possible in steam and water heating boilers, water-tube as well as shell type boilers, equipped with low pressure burners,

with air atomization or rotary principle ones (Belikov 2003).

2. Fish oil burning compared with heating oil burning helps reduce water wall surface pollution when sustaining boiler inlet temperature not lower than 70°C, or when changing boiler circuit (feeding via convection surface).

3. Normal boiler functioning on fish oil requires fish oil heating not less than up to 75°C.

4. Fish oil burning helps increase boiler efficiency ratio by way of reduction of heat loss with exhaust gases.

5. Fish oil use as a boiler fuel reduces environmental pollution in the area of industrial enterprise, by way of NOx emission reduction. NOx emission reduction begins because of the lowering of temperature in the active combustion zone and a decrease of air oxygen necessary for burning.

As per simplified estimation, a cost of fish oil from the fish processing enterprise will reach 130–170 USD/ton. The cost of transportation for a distance up to 200 km is taking into account. It is reasonable to use fish oil as a fuel in the boiler plants located not far from fish processing enterprises. It will allow to avoid 100% fuel cost increase because of transportation cost.

According to the data in the researches of Blythe and Steigers (Blythe 1996, Steigers 2002) it is noted the significant reduction of NOx emissions to 230 ppm (alfa = 1) with increase of the proportion of fish oil in the fuel up to 50%. An analysis of the efficiency of diesel generators with an increase

in the proportion of fish oil is absent. According to the authors it is not advisable to burn fish oil mixed with fuel oil in furnaces of boilers equipped with burners with steam atomization in industrial environments. Increase of fish oil proportion in the mixture up to 20% leads to slight increase in efficiency of 0.1%. Emissions of nitrogen oxides are reduced to 190 ppm (alfa = 1). But the concentration of CO in the flue gas reaches values of more than 1200 ppm. And the quantity of chemical incomplete combustion increases by 0.3–0.4%.

## REFERENCES

Belikov, S.E. 2003. *Burners for heating and industrial boiler plants. Catalog of heat engineering and heating equipment*. Moscow, LLC Quartika Publishing House.

Blythe, N.X. 1996. Fish oil as an alternative fuel for internal combustion engines. *American Society of Mechanical Engineers. Internal Combustion Engine Division (Publication) ICE*, 26: 85–92.

Shtym, K.A. Lesnykh, A.V. & Tcoy, K.A. 2015. Experience of use of fish oil in boiler plants as an alternative liquid fuel. *Energetik*, 2: 22–23.

Steigers, J.A. 2002. *Demonstrating the use of fish oil as fuel in a large stationary diesel engine*, Advances in Seafood Byproducts 2002 Conference Proceedings. Bechtel, P.J. (ed). Alaska Sea Grant, Fairbanks, AK. 187–200.

Trembovlya, V.I. 1991. *Heat engineering testing of boiler plants. Edition 2*, Revised and enlarged. Moscow, Energoatomizdat, 414.

*Advanced Materials, Structures and Mechanical Engineering – Kaloop (Ed.)*
© *2016 Taylor & Francis Group, London, ISBN 978-1-138-02793-0*

# Economic evaluation of hydroconversion of fatty acid methyl ester into renewable liquid hydrocarbons

A. Kantama, P. Narataruksa & C. Prapainainar
*Department of Chemical Engineering, King Mongkut's University of Technology North Bangkok,*
*Bangkok, Thailand*
*Research and Development Centre for Chemical Engineering Unit Operation and Catalyst Design,*
*King Mongkut's University of Technology North Bangkok, Bangkok, Thailand*

P. Hunpinyo
*Department of Chemical Process Engineering Technology, King Mongkut's University of Technology*
*North Bangkok, Rayong, Thailand*
*Research and Development Centre for Chemical Engineering Unit Operation and Catalyst Design,*
*King Mongkut's University of Technology North Bangkok, Bangkok, Thailand*

ABSTRACT: In this study, economic evaluation of renewable liquid hydrocarbon production, in Thailand, from fatty acid methyl ester, FAME, is studied. The production of liquid hydrocarbons or bio hydrogenated diesel, BHD, is processed via hydro-conversion, in a capacity of 180 tonne/day of FAME by using chemical process simulation software. The main products are alkane hydrocarbons, C1–C18, and are categorized as light gas, naphtha and diesel. Techno-economic evaluation of BHD production is carried out and it is found that total capital investment is 63.460 million USD. With the project life of 15 years, the acceptable BHD price is 1.22 USD/L providing a conventional payback period of 7.2 years and a return on investment of 10%, which normally referred to as a minimum acceptance in chemical process industry. The selling price of BHD must be increased to 1.135 and 1.391 USD/L if the investment attractiveness is changed to obtain the payback period of 10 and 5 years, respectively. Finally, the sensitivity analysis to determine the effect of FAME cost, hydrogen cost, operating cost and purchased equipment cost on BHD price is performed.

## 1 INTRODUCTION

Nowadays, renewable biomass-based resources are considered to be a new challenge in the development of renewable energy due to environmental concerns. The first generation of biofuel or biodiesel has been produced commercially by transesterification reaction between vegetable oil, animal fat or wasted cooking oil with methanol to produce fatty acid methyl ester, FAME. However, FAME which mainly consists of c = c bonds and oxygen atoms exhibit undesirable fuel properties, such as, low cloud point, low calorific value, low oxidation stability and limited blending with liquid diesel from fossil (Zuo et al. 2012). In the case of Thailand, the limitation of blending FAME with diesel for transportation use is 7% by volume (2013). In order to improve properties of biodiesel, catalytic partial and full hydrogenation process of polyunsaturated methyl ester into their monounsaturated compound can be introduced. There have been reports of full hydrogenation of vegetable oils including hydrogenation of linseed,

sunflower, soybean (Bouriazos et al. 2010), and palm kernel oil (Bouriazos et al. 2015). Non-edible oil, such as jatropha methyl ester, is also studied in partial hydrogenation (Fajar et al. 2012) and it is found that oxidation stability and cetane number of the products fuel are improved. However, some properties of partially hydrogenated methyl ester do not meet diesel standard (Bezergianni & Dimitriadis 2013).

From limitations mentioned above, Hydrodeoxygenation (HDO) process have been discussed and studied to convert vegetable oils or methyl ester components to liquid alkane hydrocarbons and this fuel can be called 2nd generation biofuel or Bio Hydrogenated Diesel (BHD). HDO of methyl ester compounds into alkane hydrocarbon have been reported, such as, methyl heptanoate (Ryymin et al. 2010), ethyl stearate (Kubičková et al. 2005, Snåre et al. 2007), methyl stearate (Do et al. 2009), methyl laurate (Chen et al. 2014, Shi et al. 2014) and methyl hexanoate or methyl palmitate (Shi et al. 2012, Zuo et al. 2012). In order to increase the blending ratio of biodiesel or FAME with conventional diesel fuel from petroleum, it can be converted to alkane hydrocarbons via HDO process to removed oxygenated compounds from methyl ester. Nevertheless, there are limited studies on the techno-economic assessment of BHD production from FAME via HDO process.

In this work, BHD production from FAME is studied by using chemical process simulation software, Aspen Plus®, to perform material and energy balance. Then, evaluation of the total capital investment, total production cost and product price are carried out in a case study of Thailand. The attractiveness of the project investment is reported as a return of investment and payback period. Finally, sensitivity analysis is performed to determine the effect of the prices of raw materials, operating cost and purchased equipment cost on BHD selling prices.

## 2 METHODOLOGY

### 2.1 *Conceptual design of the BHD process*

Bio hydrogenated diesel production from FAME is simulated using Aspen Plus software. In the model, FAME is obtained from transesterification of Refined Palm Stearin (RPS) in a capacity of 180 tonne/day. It is converted to BHD under the operating conditions of 495 K, 2 MPa using Ni/SAPO-11 catalysts. The reaction time of hydrodeoxygenation reaction is 360 minutes (Zuo et al. 2012). The reaction pathways of the production are hydrogenation and decarboxylation/decarbonylation as reported by Zuo (Zuo et al. 2012) in a batch operation with 12 hours of cycle time. Figure 1 shows the process flow diagram of

BHD production. FAME is fed to the process and heated in a fired heater (Heater-1) and delivered to hydro-deoxygenation reactor (R-101), and reacted with $H_2$. After the reaction is finished, all the products are cooled and sent to a flash drum (F-101) for gas/liquid separation. Then, the liquid products are sent to a decanter for water removal and liquid product free water is fractionated into three groups: light gas, C5–C14 (Naphtha) and C15–C18 (BHD) with low fraction of FAME in a distillation column. Finally, the bottom products from distillation column are sent to isomerization reactor (R-102) for upgrading cold flow properties by isomerization reaction (Ono et al. 2009). The product distribution from simulation model is verified by comparing result from the simulation model to that of the experimental data reported by Zuo (Zuo et al. 2012).

### 2.2 *Economic analysis of BHD process*

Cost estimation of process production is performed by using the study and preliminary estimate method (Peters et al. 2003) to evaluate the Fixed Capital Investment (FCI), the Working Capital (WC) and the Total Capital Investment (TCI), respectively. Mass and energy balance from simulation model in a capacity of 50 million liters per year of BHD is used to determine the equipment sizes and free on board (f.o.b.) purchased equipment costs (Seider et al. 2003). The cost of a distillation column is estimated by Aspen Economic Analyzer®. Equipment costs are updated to 2014 using Chemical Engineering Cost Index (CE index) (Peters et al. 2003). Tables 1 and 2 show key assumptions and prices of raw material, product and utility for economic estimation.

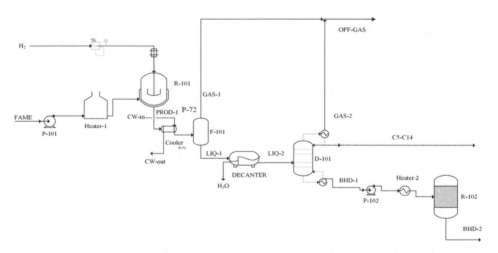

Figure 1. Process flow diagram of BHD production via HDO from FAME in this study.

Table 1. Key assumptions for economic analysis.

| Factors | Value |
| --- | --- |
| Exchange rate | 31.49 THB/USD* |
| Incoming tax rate | 20%** |
| Lifetime of | 15 years |
| Construction period | 2 years |
| Plant capacity | 180 tonne/day of FAME |
| Start-up time | 2 years |
| Inflation rate | 2.2%*** |
| Operating time | 345 days |

*Based on 3/2/15, **https://www.bot.or.th. ***http://www.tradingeconomics.com.

Table 2. The raw material, product and utility price.

| | Value | |
| --- | --- | --- |
| Variables | USD | Unit |
| FAME price* | 0.84 | kg |
| Hydrogen price | 0.48 | $m^3$ |
| Diesel price* | 0.51 | L |
| Gasoline price* | 0.50 | L |
| Electricity | 0.12 | kw |
| Fuel oil | 3.3 | GJ |
| Cooling water | 0.64 | $m^3$ |

*http://www.eppo.go.th/.

# 3 RESULTS AND DISCUSSION

From the process simulation results, a product distribution: alkane hydrocarbons (C1–C18), Dimethyl Ether (DME), Carbon Monoxide (CO), and Carbon Dioxide ($CO_2$) are verified by comparing to the experimental data (Zuo et al. 2012). The root mean square error, mean square error and absolute fraction of variance are 1.276, 0.872 and 0.903, respectively, showing that the product distribution from simulation model can be reliable. The mass fraction of light gas, naphtha and diesel at the output of the hydro-deoxygenation reactor are 3.6 wt%, 2.2 wt%, and 77.7 wt%, respectively. FAME is 0.2 wt% and DME and by-product (CO, $CO_2$ and $H_2O$) are 16.3 wt%. Mass and energy balance of BHD production are used to calculate the total product cost based on the year of 2014.

## 3.1 Product cost estimation and comparison

The cost estimation of a production process, FCI and TCI are 54.09 and 63.46 million USD, respectively. The operating cost is 45,485 USD/day. Unit cost breakdown of BHD in the capacity of 50 million L/year is shown in Figure 2. The cost of BHD from

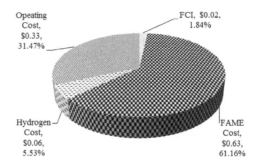

Figure 2. Unit cost breakdown of BHD production cost (USD/L).

the production process is 1.030 USD/L comprising of 0.63, 0.33, 0.06 and 0.02 USD/L of FAME cost, operating cost, $H_2$ cost and FCI respectively.

Cumulative cash flow diagram for different BHD prices from 1.054 to 1.391 USD/L is shown in Figure 3. It can be seen that the minimum selling price giving the payback period at the end of the project life is 1.054 USD/L. If the project requires shorter repayment with the payback period of 5 and 10 years, the selling BHD prices are 1.391 USD/L and 1.135 USD/L, respectively. According to Bao (Bao et al. 2010) and Apostolakou (Apostolakou et al. 2009), the acceptable product price should give at least 10% of ROI, therefore an acceptable price of BHD prices is 1.224 USD/L giving the payback period of 7.2 years.

When comparing BHD cost in this work to the NExBTL renewable diesel from Neste Oil Company (2011), the cost of NExBTL renewable diesel is 220 USD per tonne or 0.17 USD/L without the cost of the feedstock. If the cost of FAME is added to NExBTL renewable diesel, the cost of NExBTL renewable diesel will be approximately 0.88 USD/L which is lower than BHD cost from this work.

If comparing between BHD price from this work (1.224 USD/L) and HDRD prices reported by Miller (Miller & Kumar 2014), HDRD prices are 0.134 USD/L (canola oil) and 0.375 USD/L (camelina oil) lower than BHD price from FAME in this work, respectively.

## 3.2 Sensitivity analysis

The sensitivity analysis of cost factors: FAME cost, $H_2$ cost, operating cost and purchased equipment cost are performed. The change of operating cost is set to ±50%, which may be changed due to the energy lost from the process or the change of utility price. The accuracy of purchased equipment cost could be fluctuated in the range of ±30% according to the study and preliminary estimate method (Peters et al. 2003). The change in $H_2$ cost

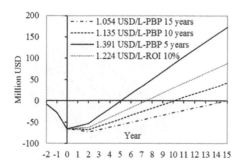

Figure 3. Cumulative cash flow diagram with different BHD prices.

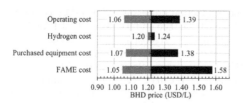

Figure 4. Tornado diagram of BHD price from the sensitivity analysis.

of ±5 USD per cubic meter may be due to the prices from $H_2$ supplier. The FAME cost may be affected by the market price in a range of 0.64–1.27 USD/L. Figure 4 shows the tornado diagram of the sensitivity analysis. It can be seen that the BHD selling price is very sensitive to FAME cost because BHD price decreases to 1.05 USD/L and increases to 1.58 USD/L with FAME cost of 0.64–1.27 USD/L, respectively. The change of operating cost gives BHD price between 1.06–1.39 USD/L and the effect of purchased equipment cost gives the BHD price between 1.07–1.38 USD/L. Hydrogen price has slight effect on the BHD price giving 1.20–1.24 USD/L, when the hydrogen price is changed.

## 4 CONCLUSIONS

FAME in a capacity of 180 tonne/day is used to produce 50 million L/year of BHD in a batch operation. The economic assessment found that FCI and TCI in this work are 54.093 million USD and 63.460 million USD, respectively. The BHD prices provide payback periods at 15, 10 and 5 years are 1.054, 1.135 and 1.391 USD/L, respectively. Additionally, the acceptable BHD prices giving a minimum ROI for a typical chemical industrial investment 10% is 1.224 USD/L. The sensitivity analysis is conducted on BHD price, it is found that BHD price is highly sensitive to FAME

cost, which is the main raw material, following by operating cost and purchased equipment cost and slightly sensitive to hydrogen price.

## ACKNOWLEDGEMENT

Graduate College KMUTNB, TRF (MSD57I0110) and Suksomboon Palm Oil Co., Ltd, Thailand.

## REFERENCES

2011. Neste Oil's capital markets day and short-term outlook. http://www.nesteoil.com: Neste Oil Corporation.

2013. Energy Policy. December 3 2013 ed.: Ministry of Energy.

Apostolakou, A.A. Kookos, I.K. Marazioti, C. & Angelopoulos, K.C. 2009. Techno-economic analysis of a biodiesel production process from vegetable oils. *Fuel Processing Technology*, 90: 1023–1031.

Bao, B. El-Halwagi, M.M. & Elbashir, N.O. 2010. Simulation, integration, and economic analysis of gas-to-liquid processes. *Fuel Processing Technology*, 91: 703–713.

Bezergianni, S. & Dimitriadis, A. 2013. Comparison between different types of renewable diesel. *Renewable and Sustainable Energy Reviews*, 21: 110–116.

Bouriazos, A. Sotiriou, S. Vangelis, C. & Papadogianakis, G. 2010. Catalytic conversions in green aqueous media: Part 4. Selective hydrogenation of polyunsaturated methyl esters of vegetable oils for upgrading biodiesel. *Journal of Organometallic Chemistry*, 695: 327–337.

Bouriazos, A. Vasiliou, C. Tsichla, A. & Papadogianakis, G. 2015. Catalytic conversions in green aqueous media. Part 8: Partial and full hydrogenation of renewable methyl esters of vegetable oils. *Catalysis Today*, 247: 20–32.

Chen, J. Shi, H. Li, L. & Li, K. 2014. Deoxygenation of methyl laurate as a model compound to hydrocarbons on transition metal phosphide catalysts. *Applied Catalysis B: Environmental*, 144: 870–884.

Corporation, N.O. 2011. Neste Oil's capital markets day and short-term outlook. http://www.nesteoil.com.

Do, P.T. Chiappero, M. Lobban, L.L. & Resasco, D.E. 2009. Catalytic Deoxygenation of Methyl-Octanoate and Methyl-Stearate on Pt/Al2O3. *Catalysis Letters*, 130: 9–18.

Fajar, R. Prawoto & Sugiarto, B. 2012. Predicting Fuel Properties of Partially Hydrogenated Jatropha Methyl Esters Used for Biodiesel Formulation to Meet the Fuel Specification of Automobile and Engine Manufacturers. *Natural Sciences*, 46: 629–637.

Kubičková, I. Snåre, M. Eränen, K. Mäki-Arvela, P. & Murzin, D.Y. 2005. Hydrocarbons for diesel fuel via decarboxylation of vegetable oils. *Catalysis Today*, 106: 197–200.

Miller, P. & Kumar, A. 2014. Techno-economic assessment of hydrogenation-derived renewable diesel production from canola and camelina. *Sustainable Energy Technologies and Assessments*, 6: 105–115.

Ono, H. Iki, H. Koyama, A. & Iguchi, Y. *Production of BHD (Bio Hydrofined Diesel) with Improved Cold Flow Properties.* 19th Annual Saudi-Japan Symposium, 2009 Dhahran, Saudi Arabia.

Peters, M.S. Timmerhaus, K.D. & West, R.E. 2003. *Plant design and economic for chemical engineers,* McGraw-Hill.

Ryymin, E.M. Honkela, M.L. Viljava, T.R. & Krause, A.O.I. 2010. Competitive reactions and mechanisms in the simultaneous HDO of phenol and methyl heptanoate over sulphided NiMo/γ-Al2O3. *Applied Catalysis A: General,* 389: 114–121.

Seider, W.D. Seader, J.D. & Lewin, D.R. 2003. *Product and process design principles synthesis, analysis and evaluation,* Wiley.

Shi, H. Chen, J. Yang, Y. & Tian, S. 2014. Catalytic deoxygenation of methyl laurate as a model compound to hydrocarbons on nickel phosphide catalysts: Remarkable support effect. *Fuel Processing Technology,* 118: 161–170.

Shi, N. Liu, Q.Y. Jiang, T. Wang, T.J. Ma, L.L. Zhang, Q. & Zhang, X.H. 2012. Hydrodeoxygenation of vegetable oils to liquid alkane fuels over Ni/HZSM-5 catalysts: Methyl hexadecanoate as the model compound. *Catalysis Communications,* 20: 80–84.

Snåre, M. Kubičková, I. Mäki-Arvela, P. Eränen, K. Wärnå, J. & Murzin, D.Y. 2007. Production of diesel fuel from renewable feeds: Kinetics of ethyl stearate decarboxylation. *Chemical Engineering Journal,* 134: 29–34.

Zuo, H. Liu, Q. Wang, T. Ma, L. Zhang, Q. & Zhang, Q. 2012. Hydrodeoxygenation of Methyl Palmitate over Supported Ni Catalysts for Diesel-like Fuel Production. *Energy & Fuels,* 26: 3747–3755.

*Advanced Materials, Structures and Mechanical Engineering – Kaloop (Ed.)*
© *2016 Taylor & Francis Group, London, ISBN 978-1-138-02793-0*

# DEM investigation of stress field in sand during pile penetration

S. Liu & J. Wang

*Department of Architecture and Civil Engineering, City University of Hong Kong, Kowloon, Hong Kong*

ABSTRACT: A two dimensional discrete element model of driven piles in the crushable sand was developed. A new stress normalization method is adopted to synthesize the data at different driven depth in a deep penetration test. The normalized stresses can be expressed by two-dimensional, axially symmetric functions of the relative position to the pile tip. The normalized vertical stress map has shown qualitative agreement with previous theoretical results in the field, and indicated the vertical stress state that is established in the field is different to what is established in the calibration chamber test.

## 1 INTRODUCTION

Over the past two decades, field and laboratory investigations on instrumented model piles installed in sand have made considerable contributions towards revealing the pile penetration mechanism in the sand. The stresses measured directly at the tip, along the shaft of the pile and in the sand mass in this type of study provide the first-hand information for understanding a range of complex soil behaviors arising from the pile installation.

The full-field stress distribution within the sand mass surrounding the pile could also be obtained by using highly instrumented calibration chamber tests on the driving of model displacement piles (Jardine et al. 2013 a, b). The experimental results of physical modeling tests provide benchmarks for numerical modeling of the pile penetration problems.

In parallel with the experimental investigations, considerable advances have also been achieved on this topic using the discrete element method (Arroyo et al. 2011), whose advantages include the full access to the particle-scale kinetic and kinematic information and the potential of being a virtual testing tool substituting physical model tests (Jiang et al. 2006). In a previous related study (Wang & Zhao 2014), the authors made a detailed discrete-continuum analysis of the pile penetration behavior based on the 2D DEM simulation results. The strain data provided by the model were compared with experimental results from the Particle Image Velocimetry (PIV) measurements of soil deformation in calibration chamber tests by White and Bolton (2004). Noticing the stress state in the field is different to that in the calibration chamber during pile penetration (Pournaghiazar et al. 2012), the current study aims to exploit the stress field developed in deep penetration.

## 2 MODELING APPROACH

### 2.1 Model construction

The DEM model of penetration test is made up of a rectangular container filled with a well compacted, poly-dispersed assembly of round particles and a model pile with a triangular tip (two inclined planes forming 60° with the vertical) pushed gradually into the granular foundation. The friction coefficient of the model pile was set to 0.5. As illustrated in Figure 1a, only the right half of the model is used by taking advantage of the axial symmetry of the problem. All the boundary walls are fixed. The bottom and left wall are set to be frictionless, while the right wall is frictional with a friction coefficient of 0.9 to minimize any relative slip between the particles and the wall. The model has the width 2 W and height H of 480 mm, and a pile diameter B of 16 mm.

The granular foundation consists of two zones: a crushable zone surrounding the pile and an uncrushable zone surrounding the crushable zone (Fig. 1a). The granular material in the uncrushable zone is composed of rigid disks with diameters uniformly varying between 0.6 mm and 1.2 mm. The density of the uncrushable disks is 2650 kg/m³. Each of the agglomerates in the crushable zone is composed of 24~30 parallel-bonded elementary balls with diameters between 0.069~0.278 mm (Fig. 1d). In order to have an identical initial stress state as in the crushable zone, the density of balls is set to be 2200 kg/m³ in the uncrushable zone. More details about the model generation can be found in Wang & Zhao (2014).

### 2.2 Observation and analysis method

As illustrated in Figure 1b, a total number of 1800 particle groups, with a size of $0.5B \times 0.5B$ for each

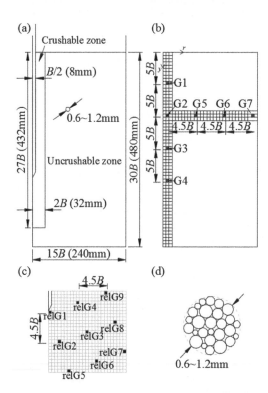

Figure 1. (a) model geometry; (b) layout of the groups at fix positions; (c) layout of the groups at relative positions to the pile tip; (d) a typical agglomerate composed of parallel-bonded disks.

group, are identified before penetration. Every particle group contains about 100 crushable agglomerates or uncrushable disks. The average stress in a group is found using a similar averaging procedure based on the measurement logic in PFC 2D (Itasca Consulting Group Inc. 2008) and the position of each group is represented by the average coordinates of all the particles within the group. It should be pointed out that the stress measurement was made on particle groups identified on the undeformed configuration prior to the pile penetration in this study. This is because the current way of stress measurement allows a direct comparison with the experimentally measured stress data from the laboratory model pile test (Jardine et al. 2013b), in which the stress sensors were embedded within the sand mass before the pile penetration.

Stresses in the sand subject to pile penetration consist of the initial stress and the stress change caused by penetration. A number of different methods (Jardine et al. 2013b, Yang et al. 2014) can be found in the literature to normalize the stresses ($\sigma'$) developed in sand during installation. The initial vertical stress in the field or centrifuge

tests increases with depth, while remains a constant in calibration chamber tests with a constant stress boundary condition. The initial stress near the pile tip is relatively small compared with the stress caused by tip resistance and the impact of penetration decreases rapidly with the increasing distance from the pile tip. If the stresses are normalized for variations in local tip resistance ($q_b$) after eliminating the influence of initial stress, the two-dimensional functions can be expressed as $(\sigma' - \sigma_0')/q_b = \Delta\sigma'/q_b = f(h/R, r/R)$, where $R$ is the pile radius, $r/R$ is the normalized relative radius from the pile axis, and $h/R$ is the normalized relative height from pile tip. A negative value of $h/R$ means below the pile tip.

## 3 SYNTHESIS OF GROUP-BASED STRESS MEASUREMENT

The stress evolutions of particle groups whose initial positions are whether on the same row or the same column (Fig. 1b) is tracked. Figure 2 shows the normalized stresses of four groups of particle positioned at column 2, where $r/R = 1.5$, against the normalized penetration depth ($y_0/B$). Maxima of the normalized horizontal and vertical stress developed, while the sign of the normalized shear

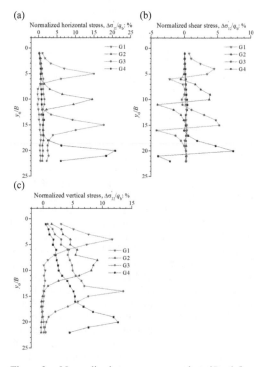

Figure 2. Normalized stresses measured at $r/R = 1.5$.

stress changed, as the pile tip passed the level at the center of each group. The peak magnitudes of the normalized horizontal stress varying between 14.5~20.7%, show a similar level of variations to those of the normalized vertical stress, varying between 9.2~13.6%. The similar pattern of the results indicates that the normalized stresses vary as functions of the normalized relative height from pile tip.

The normalized stresses of four groups of particle positioned at $y/R = 19.5$ (Fig. 1b) are shown in Figure 3. One common feature is the rapid decreasing of maximum stresses with the increasing distance away from the pile tip. This could be described as the normalized stresses are functions of the normalized relative radius from the pile axis. The normalized stresses are in the range of ±2% and little variation is observed in "G6" and "G7", which are positioned at $r/R = 19.5$ and $r/R = 28.5$ respectively.

At various penetration depths, the normalized stresses of nine particle groups whose initial positions are fixed with respect to the pile tip (Fig. 1c) are plotted against $y_0/B$ in Figure 4. At any instance of penetration depth, the positions of any two groups are not on the same row or column so as to obtain a comprehensive and unbiased picture of the normalized stress distribution as function of

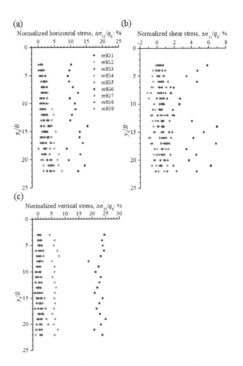

Figure 4. Normalized stresses measured at relative positions to pile tip at various penetration depth.

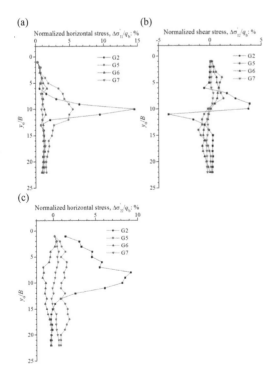

Figure 3. Normalized stresses measured at $y/R = 19.5$.

the relative distance from the particle group to the pile tip. The relationship between the $\Delta\sigma'/q_b$ and $y_0/B$ can be best represented as a unique constant function, which means the normalized stresses can be expressed by two-dimensional, axially symmetric functions of the relative position to the pile tip.

The normalized vertical stress map is generated by calculating the average values of the normalized stresses of all particle groups at various penetration depths and then presenting the average full-field stress distribution around a pile with a generic penetration depth. In Figure 5, the normalized vertical stress vectors above and below −0.25 $h/R$ are magnified by a factor of 10 and 4 respectively, to allow better visualisation of the stress state behind and ahead of the pile tip. The result is very similar to the vertical stress state that is established in the field (Pournaghiazar et al. 2012). The normalized vertical stress has an overall bubble shape below the pile tip. At the relative height of −0.25, the normalized vertical stress reached a stable value of zero when the relative radius is greater than 4. This indicates that the vertical stress spreads into a relatively limited width in the horizontal direction just below the pile tip. With the reducing of relative height, the influence area of the stress caused by penetration is broadening. The normalized vertical

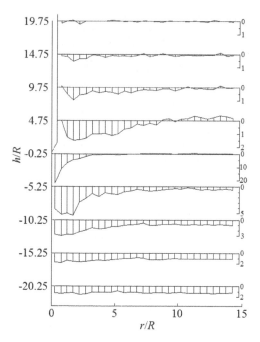

Figure 5. Vertical stress state behind and ahead of the pile tip, shown in %.

stress decreased more rapidly and reached a constant value at $h/R = 14.75$ above the pile tip. The normalized vertical stress map indicates again that the vertical stress state that is established in a calibration chamber test is different to what is established in the field.

## 4 CONCLUDING REMARKS

A 2D DEM simulation has been performed to unravel the unique pile penetration mechanism in the sand which is independent of the pile penetration depth. A new stress normalization method was adopted to synthesize the data at different driven depths. In the group-based stress analysis, the normalized stresses can be expressed by two-dimensional, axially symmetric functions of the relative position to the pile tip.

The normalized vertical stress map is generated by calculating the average values of the normalized stresses of all particle groups at various penetration depths, and has shown qualitative agreement with previous theoretical results in the field, and also indicates that the vertical stress state that is established in the field is different to what is established in the calibration chamber test.

## ACKNOWLEDGMENTS

The study presented in this article was supported by the General Research Fund No. City U 122813 from the Research Grant Council of the Hong Kong SAR, Research Grant No. 51379180 from the National Science Foundation of China and Shenzhen (China) Basic Research Project No. GJHS2012070214054 6194.

## REFERENCES

Arroyo, M. Butlanska, J. Gens, A. Calvetti, F. & Jamiolkowski, M. 2011. Cone penetration tests in a virtual calibration chamber. *Géotechnique*, 61(6): 525–531.

Itasca Consulting Group. 2008. *Guide book: theory and background PFC2D*. Technical report. Itasca Consulting Group, Inc.

Jardine, R.J. Zhu, B.T. Foray, P. & Yang, Z.X. 2013a. Measurement of stresses around closed-ended displacement piles in sand. *Géotechnique*, 63(1): 1–17.

Jardine, R.J. Zhu, B.T. Foray, P. & Yang, Z.X. 2013b. Interpretation of stress measurements made around closed-ended displacement piles in sand. *Géotechnique*, 63(8): 613–627.

Jiang, M.J. Yu, H.S. & Harris, D. 2006. Discrete element modelling of deep penetration in granular soils. *International Journal for Numerical and Analytical Methods in Geomechanics*, 30(4): 335–361.

Pournaghiazar, M. Russell, A.R. & Khalili, N. 2012. Linking cone penetration resistances measured in calibration chambers and the field. *Géotechnique Letters*, 2(2): 29–35.

Wang, J. & Zhao, B. 2014. Discrete-continuum analysis of monotonic pile penetration in crushable sands. *Canadian Geotechnical Journal*, 51(10): 1095–1110.

White, D.J. & Bolton, M.D. 2004. Displacement and strain paths during plane-strain model pile installation in sand. *Géotechnique*, 54(6): 375–397.

Yang, Z.X. Jardine, R.J. Zhu, B.T. Foray, P. & Tsuha, C.H.C. 2010. Sand grain crushing and interface shearing during displacement pile installation in sand. *Géotechnique*, 60(6): 469–482.

*Advanced Materials, Structures and Mechanical Engineering – Kaloop (Ed.)*
© 2016 Taylor & Francis Group, London, ISBN 978-1-138-02793-0

# The water saturation effects on dynamic tensile strength in red and buff sandstones studied with Split Hopkinson Pressure Bar (SHPB)

E.H. Kim & D.B.M. de Oliveira
*Colorado School of Mines, Golden, CO, USA*
*University of Utah, Salt Lake City, UT, USA*

ABSTRACT: Dynamic tensile strength of geomaterials is important for understanding mechanical properties of geomaterials as well as optimizing the design of blasting patterns, oil and gas extractions, demolition, etc. However, there is a little study for quantifying the required energy for the tensile failure of geomaterials under dynamic loading condition. More importantly, as typical geomaterials are supposed to be hydrated in nature, it is essential to precisely estimate how much energy will be needed to break the geostructures in hydrated conditions. Thus, in this study, we analyzed the consumed energy used in deformation of geomaterials using Split Hopkinson Pressure Bar (SHPB), enabling to measure stress and strain responses of geomaterials under dynamic loading condition of high strain rate ($10^2$–$10^4$ m/sec). Red (smaller pore size) and Buff (larger pore size) sandstone samples at the two different saturation levels (dry vs. fully saturated) were tested under the dynamic loading condition. As results, dynamic mechanical strength (maximum tensile stress) was greater in the dry sandstones when compared with the saturated samples. Also, our results showed that Young's modulus obtained from dynamic tensile strength tests was significantly larger in the dry Red sandstone than in the saturated Red sandstone, but insignifciant difference in Young's modulus was observed between the dry and hydrated Buff sandstone samples. This suggests that Young's modulus can be a useful parameter to examine porosity effects of dry and saturated geomaterials on dynamic mechanical properties.

## 1 INTRODUCTION

The Split Hopkinson Pressure Bar (SHPB), for more than a century, has been widely used for studying dynamic mechanical properties of solid materials (Kim & Oliveira, 2015, Changani et al. 2013, Wang et al. 2006, Kubota et al. 2008). Particularly, SHPB can be a useful technique to observe dynamic mechanical behaviors of geomaterials. As geo-materials tend to encounter various dynamic mechanical stresses during excavation processes of mining and tunneling (Kim et al. 2010, 2012a, 2012b), understanding the dynamic tensile strength of geomaterials is important to optimize the excavating processes. Also, comprehensive knowledge of dynamic mechanical properties can provide insights to improve and monitor the safety of geostructures (Schumacher & Kim 2013, 2014).

The mechanical strength of geomaterials under dynamic loading conditions can be assessed with the measurement of dynamic tensile strength (Cho et al. 2003, Zhang & Zhao 2013, Wang et al. 2009 m Kubota et al. 2008). In this paper, to examine water hydration effects on dynamic tensile strength, the dynamic tensile strength of dry and saturated Red (small pore size, 4.7% porosity) and Buff (large pore size, 18.0% porosity) sandstones were measured

with SHPB. As dynamic mechanical responses of geomaterials can vary depending on how stress is applied and what parameters are measured, multiple parameters are needed to precisely analyze dynamic mechanical properties of geomaterials. Thus, we examined maximum stress, maximum strain, maximum strain rate, and Young's modulus of Red and Buff sandstones under dynamic conditions. More importantly, the water hydration effects on these parameters were tested with Red and Buff sandstones having different porosities.

## 2 EXPERIMENT SETUP

### 2.1 *Thin section preparation*

To estimate porosity, thin section analyses of Red and Buff sandstones were performed by TerraTek (Fig. 1). The sandstone samples were impregnated with a low-viscosity fluorescent red-dye epoxy resin under vacuum to highlight porosity. The impregnated samples were surfaced, mounted to standard (24 mm × 46 mm) thin section slides, and ground to a thickness of approximately 30 microns. The thin sections were then stained with a mixture of potassium ferricyanide ($K_3[Fe(CN)_6]$) and Alizarin Red. The prepared sections were examined and

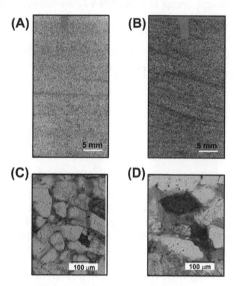

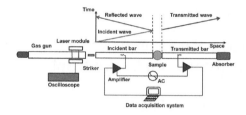

Figure 2. Schematic view of Split Hopkinson Pressure Bar system.

Figure 1. Thin section analysis of Red and Buff sandstones tested for tensile strength. The magenta epoxy is shown between framework grains. (A) Cross-laminated Red sandstone, (B) Cross-laminated Buff sandstone, (C) Magnified Red sandstone, and (D) Magnified Buff sandstone. Scale bars indicate 5 mm (A and B) or 100 μm (C and D).

digitally imaged at various magnifications under plane-polarized and cross-polarized light using a Nikon polarizing binocular microscope equipped with a Spot Insight digital camera, reflected the light source, and UV filters. Void areas stained with pink color were marked as pore space and used to estimate the porosity of the sandstones.

### 2.2 X-Ray Diffraction (XRD)

The powdered Red and Buff sandstone samples were analyzed with a Rigaku Ultima III Advance X-ray diffractometer from 2 to 36 degrees two-theta ($2\theta$) using Cu K-alpha radiation and various slit and filter geometries. The raw data were interpreted with JADE software, which can identify and quantify the mineralogy based on whole pattern fitting and Rietveld refinement methods.

### 2.3 SHPB setup

To assess the dynamic tensile strength of the dry and saturated Red and Buff sandstones, we used SHPB in this study as SHPB test is a relevant technique for examining dynamic mechanical responses of geo-materials. The SHPB is composed of a laser module, gas gun, oscilloscope, projectile (striker), strain gages, amplifiers, two long steel rods (bars), and a data acquisition system (Fig. 2). The striker

is initiated by a gas gun first, which can impact the first rod (incident bar) generating an energy wave. This waveform energy propagates through the incident bar (incident wave) and transfers to the testing sample. At the interface of the sample and the incident bar, part of the wave is reflected (reflected wave). The rest of the wave propagates through the sample and the second rod (transmitted bar) as a transmitted wave. The absorber for safety reasons captures the rest of energy. The wave transformations are recorded in a data acquisition system and assisted with amplifiers, using an AC power supply. The velocity of the striker can be measured with a laser module and oscilloscope. SHPB tests were performed under 25 psi impact pressure. Outliers of the experiments, which were greater by two times or more than average values, were not included in the data analyses.

### 2.4 Sample preparation

The Red (smaller grain size, 4.7% porosity) and Buff (larger grain size, 18.0% porosity) sandstones were prepared with ~0.4 ratio of thickness (~2.1 cm) to diameter (~5.4 cm) using coring, cutting and grinding machines for dynamic tensile strength tests. To prepare saturated Red and Buff sandstone samples, each sandstone sample was soaked into the water for 48 hours inside a vacuum chamber (25 cmHg). The half of fully saturated sandstone samples were placed in a dry oven at 105°C for 48 hours to prepare dry sandstone samples.

### 3 RESULTS AND DISCUSSION

In this paper, we employed Red and Buff sandstones having 4.7% and 18.0% porosity, respectively (Fig. 1) since these two sandstones have relatively homogenous grain and pore size, relevant to dynamic strength tests using SHPB (Lomov et al., 2001). Quartz is the most abundant mineral in both Red and Buff sandstones based on XRD measurements, and the XRD patterns of Red and Buff sandstones are quite similar to each other (Fig. 3).

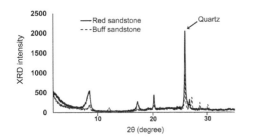

Figure 3. XRD measurement for Red and Buff sandstones used for the dynamic tensile strength test.

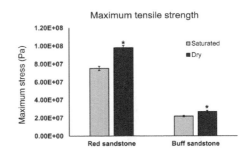

Figure 4. Comparison of the maximum dynamic tensile strength of Red and Buff sandstones between dry and saturated condition. *$P < 0.05$; Student's two-tailed $t$-test ($5 \leq n \leq 9$).

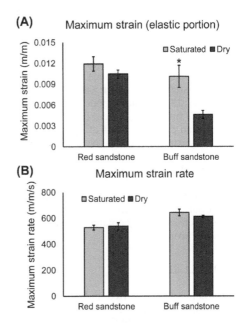

Figure 5. Comparison of maximum strain (elastic portion) and maximum strain rate of Red and Buff sandstones between dry and saturated condition. *$P < 0.05$; Student's two-tailed $t$-test ($5 \leq n \leq 9$).

These results suggest that the porosities of Red and Buff sandstone samples are different, but mineral compositions and contents of Red and Buff sandstones are similar even though their mineral compositions are not identical. Thus, we can suppose that different hydration effect between two sandstone samples on dynamic tensile strength is more likely due to the difference of porosity rather than mineral compositions or contents.

First, the effect of the water saturation (hydration) on dynamic tensile strength was examined in Red sandstone using SHPB. The maximum dynamic tensile stress (Pa) of Red sandstones was ~3.6 fold greater in dry and ~3.4 times larger in saturated conditions when compared with Buff sandstone samples (Fig. 4). Also, the maximum tensile stress was significantly greater in the dry condition than the saturated condition regardless of the porosity of geomaterials. Maximum stress increased ~30.3% in the dry Red sandstone and ~23.2% in the dry Buff sandstone when compared with each saturated sandstone (Fig. 4). These results suggest that smaller pore size sandstones can exhibit greater dynamic tensile strength than larger pore size sandstones, but water saturation effect on maximum dynamic stress does not seriously differ between smaller and larger pore size sandstones.

In Buff sandstone maximum tensile strain (m/m) significantly decreased in the dry state compared with the saturated state, whereas the reduction for maximum strain in Red sandstone was insignificant when dried (Fig. 5A). As more drastic reduction for maximum strain was detected in Buff sandstone than Red sandstone, our result suggest that the hydration effect on maximum strain obtained from dynamic tensile strength tests was enhanced in the larger pore size sample (Buff sandstone) than the smaller pore size sandstone (Red sandstone). This supports the conclusion that water saturation effect on maximum strain is remarkable in more porous geomaterials. However, water saturation does not affect maximum tensile strain rate (m/m/s) in both Red and Buff sandstones (Fig. 5B). This implies that the duration of elastic behavior of Buff sandstones

was shorter in the dry state than in the saturated state, suggesting that maximum strain occurred for a shorter time in the dry Buff sandstone than the saturated Buff sample. This can be explained with ductile behavior of the Buff sandstone as the material shows a relatively longer range of strain (or displacement) with a lower elastic modulus (Wong & Baud 2012, Saksala & Ibrahimbegovic 2014).

Interestingly, in contrast to the maximum stress and maximum strain, Young's modulus (Pa) obtained from dynamic tensile strength tests in Buff sandstone were not significantly affected by

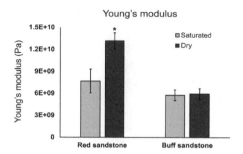

Figure 6. Comparison of Young's modulus of Red and Buff sandstones between dry and saturated condition. $*P < 0.05$; Student's two-tailed $t$-test ($5 \leq n \leq 9$).

water saturation (Fig. 6). On the contrary of Buff sandstone, Red sandstone revealed drastic increase (~77.4%) of Young's modulus in the dry condition compared with the saturated condition (Fig. 6). This result indicates that hydration effect on Young's modulus can be enhanced when the porosity of geomaterials is smaller within the porosity range of 4.7–18.0%. This supports the conclusion that Young's modulus can be a critical parameter to understand porosity effects of hydrated geomaterials on dynamic mechanical behaviors.

In conclusion, dynamic mechanical strength (maximum stress) is greater in the dry sandstones, when compared with the saturated sandstones. More importantly, Young's modulus can be used as a crucial parameter to examine porosity and hydration effects of geomaterials on dynamic mechanical properties.

## 4 CONCLUSIONS

The hydration effects of Red and Buff sandstones having different porosities on dynamic tensile strength were tested with SHPB. First, the maximum dynamic tensile strength of Red and Buff sandstones was greater in the dry state than the saturated state. Second, in contrast to maximum stress, maximum strain rate was not affected by the hydration of sandstone samples regardless of porosity of geomaterials. Third, hydration effects of sandstones on maximum strain were larger and more obvious with an increase of the pore size. However, hydration effects on Young's modulus obtained from dynamic tensile strength tests was very remarkable at the smaller pore size sample (Red sandstone) compared with the larger pore size samples (Buff sandstone). Our findings can provide insights into how hydration state and pore size of geomaterials can influence dynamic mechanical properties of which understanding can contribute greatly to the improvement of safety and cost-effectiveness in geotechnical and industries.

## REFERENCES

Changani, H. Young, A. & Kim, E. *Effect of L/D ratio on dynamic response of Aluminum 7076 and the Natural Motoqua Quartzite Sandstone in Saint George, UT using Split Hopkinson Pressure Bar (SHPB)*. 47th US Rock Mechanics/Geomechanics Symposium, 2013 San Francisco, CA, USA.

Cho, S.H., Ogata, Y. & Kaneko, K. 2003. Strain-rate dependency of the dynamic tensile strength of the rock. International Journal of Rock Mechanics and Mining Sciences, 40: 763–777.

Kim, E. & Oliveira, D.B.M.D. 2015. The Effects of Water Saturation on Dynamic Mechanical Properties in Red and Buff Sandstones having Different Porosities Studied with Split Hopkinson Pressure Bar (SHPB). *Applied Mechanics and Materials*, 752–753: 784–789.

Kim, E. Rostami, J. & Swope, C. 2010. *Full scale linear cutting test to study the rotation of the conical bit*. ARMA 10–181.

Kim, E. Rostami, J. & Swope, C. 2012a. Full scale linear cutting experiment to examine conical bit rotation. *Journal of Mining Science*, 48: 882–895.

Kim, E. Rostami, J. Swope, C. & Colvin, S. 2012b. Study of conical bit rotation using full-scale rotary cutting experiments. *Journal of Mining Science*, 48: 717–731.

Kubota, S. Ogata, Y. Wada, Y. Simangunsong, G. Shimada, H. & Matsui, K. 2008. Estimation of the dynamic tensile strength of sandstone. *International Journal of Rock Mechanics and Mining Sciences*, 45: 397–406.

Lomov, I.N. Hiltl, M. Vorobiev, O.Y. & Glenn, L.A. 2001. Dynamic behavior of berea sandstone for dry and water-saturated conditions. *International Journal of Impact Engineering*, 26: 465–474.

Saksala, T. & Ibrahimbegovic, A. 2014. Anisotropic viscodamage–viscoplastic consistency constitutive model with a parabolic cap for rocks with brittle and ductile behavior. *International Journal of Rock Mechanics and Mining Sciences*, 70: 460–473.

Schumacher, F.P. & Kim, E. 2013. Modeling the pipe umbrella roof support system in a Western US underground coal mine. *International Journal of Rock Mechanics and Mining Sciences*, 60: 114–124.

Schumacher, F.P. & Kim, E. 2014. Evaluation of directional drilling implication of double layered pipe umbrella system for the coal mine roof support with composite material and beam element methods using FLAC3D. *Journal of Mining Science*, 50: 336–349.

Wang, Q. Li, W. & Song, X. 2006. A Method for Testing Dynamic Tensile Strength and Elastic Modulus of Rock Materials Using SHPB. *Pure and Applied Geophysics*, 163: 1091–1100.

Wang, Q.Z. Li, W. & Xie, H.P. 2009. Dynamic split tensile test of Flattened Brazilian Disc of rock with SHPB setup. *Mechanics of Materials*, 41: 252–260.

Wong, T.F. & Baud, P. 2012. The brittle-ductile transition in porous rock: A review. *Journal of Structural Geology*, 44: 25–53.

Zhang, Q.B. & Zhao, J. 2013. A review of dynamic experimental techniques and mechanical behaviour of rock materials. *Rock Mechanics and Rock Engineering*, 1–68.

*Advanced Materials, Structures and Mechanical Engineering – Kaloop (Ed.)*
© *2016 Taylor & Francis Group, London, ISBN 978-1-138-02793-0*

# Economic potential of diesel-like renewable fuel production from the by-product of the palm oil refining process

A. Kantama, C. Prapainainar & P. Narataruksa
*Department of Chemical Engineering, King Mongkut's University of Technology North Bangkok, Bangkok, Thailand*
*Research and Development Centre for Chemical Engineering Unit Operation and Catalyst Design, King Mongkut's University of Technology North Bangkok, Bangkok, Thailand*

P. Hunpinyo
*Department of Chemical Process Engineering Technology, King Mongkut's University of Technology North Bangkok, Rayong, Thailand*
*Research and Development Centre for Chemical Engineering Unit Operation and Catalyst Design, King Mongkut's University of Technology North Bangkok, Bangkok, Thailand*

ABSTRACT: This work examined the economic potential of diesel-like renewable fuel or Bio-Hydrogenated Diesel (BHD) production from Palm Fatty Acid Distillate (PFAD), a by-product from the palm oil refining process. Alternative block flow diagrams for BHD production from different potential raw materials were proposed. Consequently, BHD production from PFAD was studied using a chemical process simulation tool. A capacity of 50 tonnes per day of PFAD feed was used as a case study, as it was the potential feed capacity from an existing refining palm oil plant in Thailand. It was found that when sold at Thailand retail price, PFAD to BHD has the potential of gaining 9.37% higher revenue than PFAD that is directly sold as a feedstock to other industries.

## 1 INTRODUCTION

Bio-Hydrogenated Diesel (BHD), the second generation of diesel-based biofuel technology, has been researched because of its advantages over other alternative fuels, such as its higher cetane number, higher heating value, good cold flow properties, environmental friendliness (Aatola et al. 2008) and perfect blending with diesel fuel from petroleum. These properties make it suitable for being substituted for the first generation of biofuels according to the limitation of the blending ratio between biodiesel and diesel fuel (Bezergianni and Dimitriadis, 2013). BHD consists mainly of paraffinic hydrocarbons and is free of aromatic hydrocarbons, sulfur and oxygenated compounds (Hancsok et al. 2011), and it can be produced from vegetable oils, such as palm oil, sunflower oil, soybean oil, rapeseed oil, and animal fat via catalytic hydro-processing (Mohammad et al. 2013a).

In the hydro-processing process, there are three main reactions for BHD production: hydro-deoxygenation (Equation 1), decarboxylation (Equation 2), and decarbonylation (Equation 3). Side reactions, such as water gas shift and methanation, may also occur (Kiatkittipong et al. 2013a):

$$R - CH_2COOH + 3H_2 \rightarrow R - CH_2 - CH_3 + 2H_2O \tag{1}$$

$$R - CH_2COOH \rightarrow R - CH_2 - CH_3 + CO_2 \tag{2}$$

$$R - CH_2COOH + H_2 \rightarrow R - CH_2 - CH_3 + CO + H_2O \tag{3}$$

Selectivity and reaction pathway in hydro-processing can be affected by operating conditions, $H_2$/oil ratio and catalyst type (Srifa et al. 2014).

As mentioned above, there are several reported feedstock with the potential for BHD production (Mohammad et al. 2013a). In Southeast Asia, palm oil is widely used for the production of edible oil and used as a feedstock for renewable biofuel production because, in this region, palm plantation is on a larger scale than other regions, as well as palm trees have a long life span, are less affected by climate change and are able to be harvested throughout the year (Mekhilef et al. 2011).

In the palm oil refining process via physical refining, Crude Palm Oil (CPO) is fed to the process involving a degumming step to remove gums to obtain Degummed Palm Oil (DPO) and then sent to winterization to remove wax. After that, the color of the oil can be improved by bleaching to obtain Bleached Palm Oil (BPO). BPO is fed to deodorization steps in order to improve odor by steam refining, and refined deodorized palm oil (BDPO) is obtained and Palm Fatty Acid Distillate (PFAD) is removed in this step as a by-product. Finally, BDPO is directed to the fractionation step in order to separate Refined Palm Olein (RPO) and Refined Palm Stearin (RPS) (Kiatkittipong et al. 2013a).

There have been reports studying the potential use of CPO and relevant refining palm oil as feedstock for BHD production (Kiatkittipong et al. 2013a). Palm oil miller effluent (POME) can also be used as a raw material for BHD production (Hasanudin et al. 2012). PFAD is one of the potential feedstock, as it has been reported that it can be used in the hydro-processing process and the yield of diesel-like hydrocarbons of 81% can be obtained on the laboratory scale (Kiatkittipong et al. 2013a).

In this work, BHD production technology from PFAD is preliminarily studied. The motivation of the study is that there is potential PFAD feedstock from existing palm oil refining industries in Thailand. PFAD is currently sold directly to other industries to be used as raw materials. Therefore, this work shows the potential of using PFAD to produce BHD as an alternative renewable energy. Financial comparison between value-added PFAD by BHD and the by-product PFAD from the refining process is reported.

## 2 METHODOLOGY

### 2.1 BHD production technology

There are two options for BHD production technologies: co-processing with an existing distillate hydro-processing unit and a standalone unit, as shown in Figure 1 (Holmgren et al. 2007).

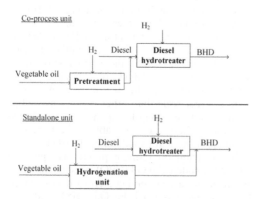

Figure 1. BHD production processes (Mekhilef et al. 2011).

Co-processing has an advantage over the standalone process to some extent, as it has a lower implementation cost, but if vegetable oils contain trace metal contaminants, such as phosphorous, sodium, potassium and calcium, an additional pretreatment unit will be required to remove the contaminants (Holmgren et al. 2007).

In many countries, BHD technology is available from companies such as Naste oil, ConocoPhillips, Petrobras, Syntroleum, and UOP company (Mohammad et al. 2013b). Each company has its own technology, which uses different raw materials; in Europe, the feedstock is usually sunflower or rapeseed oil, while soybean oil is used in the United States. In Southeast Asia, palm oil is the most favorable raw material because the production rate of palm oil is higher than that of other vegetable oils (Holmgren et al. 2007). This work focused on the standalone process.

### 2.2 Block flow diagram for BHD production from relevant refining palm oil

This work focuses on BHD production from palm oil and relevant refining palm oil. Figure 2 shows the potential feedstock of BHD production: CPO, POME, DPO, PFAD, and RPO. Each raw material requires different operating units and operating conditions to achieve a maximum BHD yield.

Palm Oil Mill Effluent (POME) is a solid residue from bunches and the milling process in refining palm oil production. The catalytic hydrocracking of POME to biofuel is conducted in a fixed bed micro-reactor at a temperature of 773 K, Weight Hourly Space Velocity (WHSV) of 10 h$^{-1}$ and the catalyst-to-oil ratio of 0.2. The maximum diesel yield obtained from this process is about 5–10% by using the NMZM catalyst (Hasanudin et al. 2012).

For CPO, there are many reports with a wide range of operating conditions to convert palm oil into biofuel. Sirajudin et al. conducted a catalyst cracking process using the HZSM-5 catalyst in a fixed bed micro-reactor (Sirajudin et al. 2013). It was found that suitable operating conditions that provide the highest yield of gasoline (28.87%), kerosene (16.70%) and diesel (1.20%) are at the temperature of 723 K and the N$_2$ flow rate of $1.67 \times 10^{-6}$ m$^3$/s for 120 minutes of residence time. Diesel yield can be increased by increasing the N$_2$ flow rate, and at the flow rate of $6.67 \times 10^{-6}$ m$^3$/s, the maximum diesel yield (20–25%) is achieved (Sirajudin et al. 2013). Kiatkittipong et al. studied hydro-processing of CPO, with reaction conditions of 673 K and 4 MPa and with the residence time of 3 hours using the Pd/C commercial catalyst, and they obtained the maximum diesel yield (Kiatkittipong et al. 2013b). The final diesel or

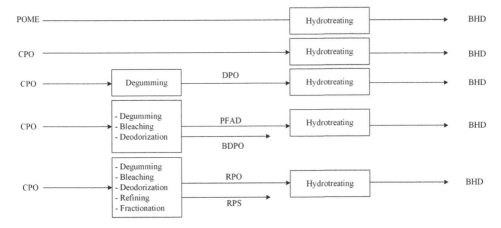

Figure 2. BHD production routes from relevant refining palm oil.

BHD yield under these operating conditions is around 50%, with 90% conversion of CPO. CPO has also been studied in the hydro-deoxygenation process on the pilot scale (Guzman et al. 2010). An industrial hydro-treating catalyst (NiMo/γ-Al$_2$O$_3$) is used to convert CPO into hydrocarbons in the diesel range under the operating conditions of 623 K, 4–9 MPa, and a H$_2$/oil molar ratio of 20:1 with time on stream of 14 days.

DPO, a product from the degumming process in the refining process, is also used as a raw material for BHD production (Kiatkittipong et al. 2013b). The operating conditions for converting DPO into BHD are 673 K, 4 MPa and a residence time of one hour. It is observed that diesel yield from these conditions is higher than that from CPO because DPO is more purified because gums are removed in the degumming step. After hydrogenation reactions, the maximum diesel yield of 70% at 80% conversion from the NiMo/γ-Al$_2$O$_3$ catalyst is achieved.

Palm Fatty Acid Distillate (PFAD), a by-product which is obtained from the deodorization step, has also been studied (Kiatkittipong et al. 2013b). The operating conditions are 648 K, 2 MPa and a residence time of 0.5 hours using the Pd/C commercial catalyst or the NiMo/γ-Al$_2$O$_3$ catalyst. Under these conditions, the water gas shift reaction is promoted to produce CO and causes the deactivation of the catalyst. A diesel yield of 81% and 76% can be observed at 80% conversion by using the Pd/C commercial catalyst and the NiMo/γ-Al$_2$O$_3$ catalyst, respectively.

Refined Palm Olein (RPO), a final product of palm oil refining, is used to produce BHD. In this route, overall pretreatment units and the refining process must be employed prior to the hydroprocessing process. RPO is more suitable for BHD

production than other raw materials because this raw material is free from impurities. The hydroprocessing operates under the conditions of 573 K, 3–5 MPa, and LHSV 1–2 h$^{-1}$, with the NiMoS$_2$/γ-Al$_2$O$_3$ catalyst being recommended (Srifa et al. 2014). The maximum product yield of n-alkane is more than 95.5%.

## 3 PROCESS MODELING

From the literature review and survey from palm oil refining factories, it is found that the attractive BHD production route is from PFAD feedstock because PFAD is a by-product from palm oil refining, which gives a high yield of diesel without affecting the food consumption sector of palm oil. Therefore, BHD production from PFAD is studied using the chemical process simulation tool, Aspen Plus version 7.3. The process flow diagram of BHD production is shown in Figure 3. The suitable operating conditions for hydro-processing of PFAD are 648 K, 2 MPa and residence time of 0.5 hour using the Pd/C commercial catalyst, as has been reported by Kiatkittipong et al. (Kiatkittipong et al. 2013b). For a case study, PFAD fed to the hydro-processing process with hydrogen gas with a capacity of 50 tonnes/day is simulated as a potential feedstock from an existing palm oil refining process in Thailand.

In Figure 3, PFAD (PFAD-1) and H2 (H2-1) are fed to the reactor (R-1). Products from the reactions are first separated in a flash drum (F-1), and water is subsequently removed from the main stream at a decanter (DE-1). Then, it is fed to the separation process at distillation columns: D-101 and D-102. BHD products are then upgraded in an isomerization reactor (R-102). Finally, the product

171

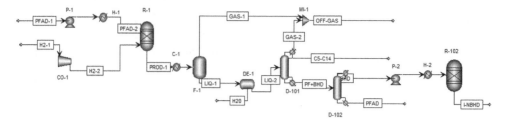

Figure 3.    Process flow diagram of BHD production from PFAD.

Table 1.    Comparison of PFAD and BHD sales.

|  | PFAD supply for other process | BHD products |
|---|---|---|
|  | Tonnes per day | Tonnes per day |
| Feed capacity | 50 | 50 |
|  | Mass yield (%) | Mass yield (%) |
| Product yield | 100 | 18.3 (Gasoline) 50.8 (BHD) |
|  | USD/kg | USD/kg |
| Unit price | 0.53 | 1.03** 0.77*** |
|  | USD/day | USD/day |
| Expected revenue | 26,500 | 28,982 |

*Exchange rate is 33.683 THB/USD (https://www.bot. or.th); **Gasoline retail price; ***Diesel retail price (http://www.eppo.go.th/petro/price/index.html).

BHD is obtained from the I-NBHD stream. The product distribution consists of alkane hydrocarbons C5–C18 and CO, $CO_2$, $H_2O$ as by-products. From these operating conditions at 80% conversion of PFAD, the mass fraction of BHD or alkane hydrocarbons in the diesel range (C15–C18) is 50.8%, and alkane hydrocarbon in the range C5–C14 is 18.3%.

## 4    RESULTS AND DISCUSSION

As can be seen from the process simulation, the main products from this process are alkane hydrocarbons in the diesel range, which can be sold as a diesel fuel. Table 1 presents the comparison of revenues gained from selling PFAD and BHD. If PFAD is sold as a raw material for use in other industries, such as cosmetics or animal feed, the income from this option is 26,500 USD/day or 9.14 million USD per year.

The revenue from the BHD sale is 28,982 USD/ day or 9.99 million USD per year (345 operating days) if the products are sold at Thailand retail price. Additionally, when selling BHD, potential saving from using the fuel gas by-product from the process may be further considered. In this case, it must be noted that the preliminary economic potential is used by comparing the product sale between two alternatives.

## 5    CONCLUSIONS

In this work, block flow diagrams of BHD production from different potential raw materials are proposed for the palm oil refining industry. BHD produced from PFAD feedstock for diesel-like renewable fuel production is studied in order to find the alternative potential market for value-added PFAD. From a BHD process simulation with 80% conversion of 50 tonnes per day of PFAD, it is found that a mass yield of BHD can be achieved at 70% or greater. PFAD to BHD has the potential of gaining a higher revenue than PFAD that is directly sold as a feedstock to other industries. If the product BHD and the by-product gasoline from the BHD production are sold at Thailand retail prices, the total revenue is 9.99 million USD per year.

## ACKNOWLEDGMENT

This work was supported by NRCT (KMUTNB-GOV-55-01) and Suksomboon Palm Oil Co., Ltd, Thailand.

## REFERENCES

Aatola, H. Larmi, M. & Sarjovaara, T. 2008. *Hydrotreated Vegetable Oil (HVO) as a Renewable Diesel Fuel: Trade-off between NOx,* Particulate Emission and Fuel Consumption of a Heavy Duty Engine. Helsinki University of Technology.

Bezergianni, S. & Dimitriadis, A. 2013. Comparison between different types of renewable diesel. *Renewable and Sustainable Energy Reviews*, 21: 110–116.

Guzman, A. Torres, J.E. Prada, L.P. & Nuñez, M.L. 2010. Hydroprocessing of crude palm oil at pilot plant scale. *Catalysis Today*, 156: 38–43.

Hancsok, J. Baladincz, P. Kasza, T. Kovacs, S. Toth, C. & Varga, Z. 2011. Bio Gas Oil Production from Waste Lard. *Biomedicine and Biotechnology*, 2011(2011): 384184.

Hasanudin, Said, M. Faizal, M. Dahlan, M.H. & Wijaya, K. 2012. Hydrocracking of oil residue from palm oil mill effluent to biofuel. *Sustaining Environmental Resources*, 22: 395–400.

Holmgren, J. Gosling, C. Marinangeli, R. Marker, T. Plaines, D. Illinois, Faraci, G. & Perego, C. 2007. *New developments in renewable fuels offer more choices*, Hydrocarbon Processing.

Kiatkittipong, W. Phimsen, S. Kiatkittipong, K. Wongsakulphasatch, S. Laosiripojana, N. & Assabumrungrat, S. 2013a. Diesel-like hydrocarbon production from hydroprocessing of relevant refining palm oil. *Fuel Processing Technology*, 116: 16–26.

Kiatkittipong, W. Phimsen, S. Kiatkittipong, K. Wongsakulphasatch, S. Laosiripojana, N. & Assabumrungrat, S. 2013b. Diesel-like hydrocarbon production from hydroprocessing of relevant refining palm oil. *Fuel Processing Technology*, 116: 16–26.

Mekhilef, S. Siga, S. & Saidur, R. 2011. A review on palm oil biodiesel as a source of renewable fuel. *Renewable and Sustainable Energy Reviews*, 15: 1937–1949.

Mohammad, M. Hari, T.K. Yaakob, Z. Sharma, Y.C. & Sopian, K. 2013a. Overview on the production of paraffin based-biofuels via catalytic hydrodeoxygenation. *Renewable and Sustainable Energy Reviews*, 22: 121–132.

Mohammad, M. Kandaramath Hari, T. Yaakob, Z. Chandra Sharma, Y. & Sopian, K. 2013b. Overview on the production of paraffin based-biofuels via catalytic hydrodeoxygenation. *Renewable and Sustainable Energy Reviews*, 22: 121–132.

Sirajudin, N. Jusoff, K. Yani, S. Ifa, L. & Roesyadi, A. 2013. Biofuel Production from Catalytic Cracking of Palm Oil. *World Applied Sciences*, 26: 67–71.

Srifa, A. Faungnawakij, K., Itthibenchapong, V. Viriya-Empikul, N. Charinpanitkul, T. & Assabumrungrat, S. 2014. Production of bio-hydrogenated diesel by catalytic hydrotreating of palm oil over NiMoS2/γ-Al2O3 catalyst. *Bioresource Technology*, 158: 81–90.

*Advanced Materials, Structures and Mechanical Engineering – Kaloop (Ed.)*
© 2016 Taylor & Francis Group, London, ISBN 978-1-138-02793-0

# Study on acoustic similarity in different fluids

M. Wang
*First Department, Bengbu Naval Petty Officer Academy, Bengbu, China*

Q.D. Zhou, Y.F. Wang, G. Ji, J.B. Xie & Z.Y. Xie
*College of Naval Architecture and Power, Naval University of Engineering, Wuhan, China*

ABSTRACT: Large-scale acoustic model experiments underwater are usually difficult, while air-acoustic experiments in an anechoic room have a better efficiency and are easier to organize. If acoustic similarity in different fluids is discovered, the data gained in air-acoustic experiment can be used to calculate underwater noise. Via dimensional analysis, acoustic similarity in an arbitrary fluid is firstly studied. Then, via FEM/BEM method, using NASTRAN to simulate surface vibration and FORTRAN BEM program to simulate sound radiation, sound pressure of ring stiffened cylinder in water, air, and a virtual fluid is calculated. Theory analysis and numerical calculation both point out that when any 8 parameters in different acoustic systems among non-dimensional material density, non-dimensional material stiffness, Poisson ratio of material, damping coefficient of structure, non-dimensional exciting force, non-dimensional sound velocity in fluid, non-dimensional coordination in sound field and non-dimensional sound pressure are the same, respectively, the last one should have the same value.

## 1 INTRODUCTION

Large scaled model is widely used in the research on vibration and noise depression of submarine and other underwater structures. However, underwater an experiment is not easy to organize. In order to avoid disturbance, the experiment waters should be deep, wide and calm. Man-made ponds can hardly meet these conditions, and large lakes are always far away from cities. In many countries, there may be no such area which satisfies all the requirements at all. Thus, the organization of large scaled model underwater experiment could be very hard and even impossible, while the cost could be very high. Besides, underwater experiment is usually disturbed by bad weather or fishery, etc. Usually, underwater experiment costs a lot, but gets a low SNR (Signal-to-Noise Ratio).

Compared with underwater experiment, air-acoustic experiment in the anechoic room is easier to organize with a much higher SNR. The background noise in anechoic room can be tens of dBs lower than that underwater. While large anechoic room easily satisfies experiment requirements, the experiment cost is not high. If acoustic similarity in different fluids was discovered, experiment data gained in anechoic room can be adopted to calculate underwater noise. Thus, difficult underwater experiment could be replaced by air-acoustic experiment, which is easier, cheaper, and more accurate.

Based on complete similitude scale models, J. Wu (2003) predicted the vibration characteristics of the elastically restrained flat plates subjected to dynamic loads. Pairod Singhatanadgid and Anawat Na Songkhla (2008) organized an experiment to prove the prediction of vibration responses of rectangular thin plates using scaling laws. C.J. Chapman (2000) reported similarity variables for sound radiation in a uniform flow. Via FEM and BEM, M. Yu et al (1993, 1998) studied on vibration and acoustic similarity of elastic stiffened cylindrical shell, and an experiment is organized to prove its vibration similarity (Yu et al. 2002). D. Yang studied on an acoustic similarity of a submerged complex shell (Yang et al. 2005). Up to now, a study on vibration and acoustic similarity in different fluids has not been reported.

In different fluids, structure vibration and acoustic field vary from each other. Thus, seeking similarity laws from governing equations could be very hard. However, acoustic systems in different fluids obey the same basic laws. Thus, dimensional analysis, which only concerns the basic laws, is the best choice in similarity approach.

Via dimensional analysis, acoustic similarity in different fluids is studied, and the similarity laws are given. Numerical simulation on ring-stiffened cylinder proved the theoretical approach.

## 2 THEORETICAL APPROACH

Consider an arbitrary structure stimulated by an arbitrary exciting force in an arbitrary fluid. The fluid-structure coupled vibration can be described as (Everstine et al. 1990):

$$(-\omega^2 M + \omega C + K)X = F + f \qquad (1)$$

where, $\omega$ represents structure vibration frequency, $M$, $C$, $K$ represent structure mass matrix, damping matrix and stiffness matrix, respectively. $F$ is the exciting force applied to the structure and $f$ is the additional force from the fluid.

Velocity potential of the acoustic field can be described as (Zhou & Joseph 2005):

$$\begin{cases} \dfrac{\partial^2 \Phi}{\partial t^2} = c_0^2 \nabla^2 \Phi \\[2mm] \lim\limits_{r\to\infty} |r\Phi| < \infty \\[2mm] \lim\limits_{r\to\infty} r\left( \dfrac{\partial \Phi}{\partial r} + \dfrac{k}{\omega}\dfrac{\partial \Phi}{\partial t} \right) = 0 \\[2mm] \dfrac{\partial \Phi}{\partial n_e}\bigg|_{S_0} = U_n \end{cases} \qquad (2)$$

where, $\Phi$ is the velocity potential at an arbitrary point in acoustic field, $c_0$ is the sound velocity in fluid, $k = \frac{\omega}{c_0}$ is the acoustic wave number, $n_e$ is the inner normal vector of the wet surface $S_0$, $U_n$ is the inner normal component of vibration velocity.

In acoustic field, sound pressure can be calculated by velocity potential:

$$p = i\rho_0 \omega \Phi \qquad (3)$$

Consider the whole acoustic system, from Equation (1), (2) and (3), all the variables that contribute to sound pressure are: material density $\rho_m$, structure geometry scale $l$, material stiffness $E$, material Poisson ratio $\nu$, structure damping coefficient $\mu$, exciting force $F$, frequency $\omega$, fluid density $\rho_0$, sound velocity in fluid $c_0$ and acoustic field coordinate $(r, \theta)$. Thus, the sound pressure of an arbitrary point in acoustic field can be written as:

$$p = \varphi(\rho_m, l, E, \nu, \mu, F, \omega, \rho_0, c_0, r, \theta) \qquad (4)$$

where, $\varphi$ is an undetermined function.

Expand Equation (4):

$$p = \sum_{i=1}^{\infty} k_i \rho_m^a l^b E^c \nu^d \mu^e F^f \omega^g \rho_0^h c_0^l r^m \theta^n \qquad (5)$$

where, $a$ to $n$ are coefficients.

Take force $[F]$, geometry scale $[L]$ and frequency $[T^{-1}]$ as basic dimension, the dimension equation can be written as:

$$[FL^{-2}] = [FT^2 L^{-4}]^a [L]^b [FL^{-2}]^c [1]^d [1]^e [F]^f \\ \cdot [T^{-1}]^g [FT^2 L^{-4}]^h [LT^{-1}]^l [L]^m [1]^n \qquad (6)$$

To ensure dimension of both side of Equation (6) equal, there should be:

$$\begin{cases} 1 = a + c + f + h \\ -2 = -4a + b - 2c - 4h + l + m \\ 0 = 2a - g + 2h - l \end{cases} \qquad (7)$$

Equation (7) can be simplified as:

$$\begin{cases} h = 1 - a - c - f \\ b = 2 - 2c - 4f - l - m \\ g = 2 - 2c - 2f - l \end{cases} \qquad (8)$$

Substitute Equation (8) for Equation (5):

$$\frac{p}{\rho_0 \omega^2 l^2} = \sum_{i=1}^{\infty} \left[ \begin{array}{l} k_i \left( \dfrac{\rho_m}{\rho_0} \right)^a \left( \dfrac{E}{\rho_0 \omega^2 l^2} \right)^c \nu^d \mu^e \\[3mm] \cdot \left( \dfrac{F}{\rho_0 \omega^2 l^4} \right)^f \left( \dfrac{c_0}{\omega l} \right)^l \left( \dfrac{r}{l} \right)^m \theta^n \end{array} \right] \qquad (9)$$

where, the non-dimensional parameters in Equation (9) $\frac{p}{\rho_0 \omega^2 l^2}$, $\frac{\rho_m}{\rho_0}$, $\frac{E}{\rho_0 \omega^2 l^2}$, $V$, $\mu$, $\frac{F}{\rho_0 \omega^2 l^4}$, $\frac{c_0}{\omega l}$ and $\left( \frac{r}{l}, \theta \right)$ can be defined as non-dimensional sound pressure, non-dimensional material density, non-dimensional material stiffness, non-dimensional material Poisson ratio, non-dimensional structure damping coefficient, non-dimensional exciting force, non-dimensional sound velocity in fluid and non-dimensional polar coordinate in acoustic field, respectively.

Among these non-dimensional factors, material Poisson ratio, structure damping coefficient and the angle of polar coordinate are non-dimensional themselves. To ensure acoustic similarity, the value of these three factors cannot be changed, which is widely acknowledged in research on acoustic similarity in the same fluid. Newly-defined non-dimensional factors have the same position with these ones. In acoustic systems in different fluids, when any 8 non-dimensional factors equals, the last one should be the same.

An intuitionistic description is: when the non-dimensional material density, material Poisson ratio, structure damping coefficient, non-dimensional exciting force, non-dimensional sound velocity in fluid and non-dimensional polar coordinate in acoustic field in acoustic systems in different fluids equals, respectively, the non-dimensional sound pressure in different acoustic system equals.

Besides, according to the results in this section, when different acoustic systems are in the same fluid, that is, $\rho_0$ and $c_0$ equals, similarity law can be simplified to the same law that article (Mengsa et al. 1993) described. That is to say, the similarity law concluded in this section is also applicable for the case of the same fluid.

# 3 NUMERICAL EXAMPLES

## 3.1 *Method and model*

Based on FEM/BEM method, the numerical calculation uses NASTRAN to calculate vibration and FORTRAN BEM program to simulate acoustic radiation (Zhou & Joseph 2005). This method is proved by G. Ji etc (Ji et al. 2006, Xie et al. 2009, Wang et al. 2011).

Three cases are calculated whose parameters satisfy the requirements in Part 2, as Table 1.

The parameters of the model in case 1 are as Table 2. The models in case 2 and case 3 are strictly geometry similar with that in case 1.

The FEM model used in simulation and its placement in acoustic field is as Figure 1. Along the gauging circle, a gauging point is laid at every

5 degrees. The element division is the same with article (Zhou & Joseph 2005).

## 3.2 *Numerical results and discussions*

The corresponding frequency in case 2 and case 3 is 34 Hz and 36.25 Hz when the frequency in case 1 is 145 Hz, respectively. At this frequency, the acoustic field directivities in three cases are as Figure 2–Figure 3. For convenience, the reference sound pressure in case 2 and case 3 is set as $1.0 \times 10^{-6} Pa$, the same with that in case 1.

In three cases, there is no difference in acoustic field directivity. At the corresponding angle, the non-dimensional sound pressures $\dfrac{p}{\rho_0 \omega^2 l^2}$ in three cases are almost the same. Thus, at an arbitrary gauging point, its sound pressure level spectrum is representative.

At gauging point No. 1, its sound pressure level spectrums in three cases are as Figure 4.

It is shown in Figure 4 that no difference at the tendencies of sound pressure level spectrum has been detected among the three cases at the same gauging point. While at the corresponding frequency, non-dimensional sound pressures $\dfrac{p}{\rho_0 \omega^2 l^2}$ in three cases are almost the same.

In all, a simulation shows that when the similar condition in Part 2 is agreed, systems in different fluids have the same acoustic laws. That is, acoustic

Table 1. Parameters of different cases.

| Parameters | Case 1 | Case 2 | Case 3 |
|---|---|---|---|
| Fluid | Water | Air | Virtual fluid |
| Fluid density | $1.0 \times 10^3 \, kg \cdot m^{-3}$ | $1.21 \, kg \cdot m^{-3}$ | $0.5 \times 10^3 \, kg \cdot m^{-3}$ |
| Sound velocity in fluid | $1.45 \times 10^3 \, m \cdot s^{-1}$ | $340 \, m \cdot s^{-1}$ | $725 \, m \cdot s^{-1}$ |
| Structure geometry scale (ratio to case 1) | 1 | 1 | 2 |
| Material density | $7.85 \times 10^3 \, kg \cdot m^{-3}$ | $9.31 \, kg \cdot m^{-3}$ | $3.93 \times 10^3 \, kg \cdot m^{-3}$ |
| Material stiffness | $2.06 \times 10^{11} \, Pa$ | $1.37 \times 10^7 \, Pa$ | $2.575 \times 10^{10} \, Pa$ |
| Material poisson ratio | 0.3 | 0.3 | 0.3 |
| Structure damping coefficient | 0.06 | 0.06 | 0.06 |
| Magnitude of exciting force | 4.454 N | 4.454 N | 2.227 N |
| Frequency (ratio to case 1) | 1 | 0.2345 | 0.25 |
| Radium of the gauging circle | 6.096 m | 6.096 m | 12.192 m |

Table 2. Computing parameters of the basic model.

| Item | Value | Item | Value |
|---|---|---|---|
| Length/mm | 1905 | Rib width/mm | 50.8 |
| Radium/mm | 635 | Rib thickness/mm | 17.1.7 |
| Shell thickness/mm | 6.35 | Distance between | 235.125 |
| Bottom thickness/mm | 25.4 | adjacent ribs/mm | |

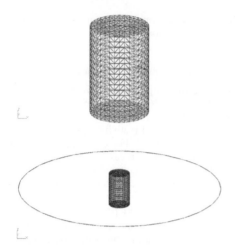

Figure 1.    The FEM model and its placement in acoustic field.

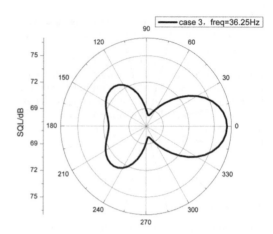

Figure 4.    The directivity pattern of the acoustic field at 36.25 Hz in case 3.

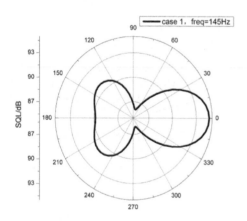

Figure 2.    The directivity pattern of the acoustic field at 145 Hz in case 1.

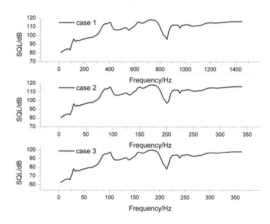

Figure 5.    Sound pressure level spectrum curve of point No.1 in three cases.

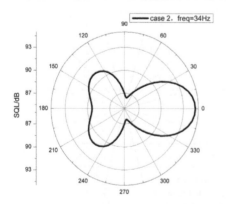

Figure 3.    The directivity pattern of the acoustic field at 34 Hz in case 2.

similarity in different fluids is proved by numerical calculation.

## 4   CONCLUSIONS

Via dimensional analysis, acoustic similarity in different fluids is studied, and its similar conditions and relations are given. Numerical calculation also proved theory approach. The discovery can be described as when the non-dimensional material density, material Poisson ratio, structure damping coefficient, non-dimensional exciting force, non-dimensional sound velocity in fluid and non-dimensional polar coordinate in acoustic field in acoustic systems in different fluids equals, respectively, the non-dimensional sound pressure in different acoustic system equals.

According to the acoustic similarity in different fluids, if experiment data gained in air acoustic experiment are adopted to calculate practical structure noise underwater, the model should be built of a kind of material with a super low density and stiffness. Changing gauging frequency or geometry scale may make stiffness larger, but the super low density cannot be avoided. If the practical structure is made of steel, the most common material, and the model used in air acoustic has the same geometry scale with practical structure, it should be made of a material with $\rho_m = 9.31\,\text{kg}\cdot\text{m}^{-3}$ and $E = 13.70\,\text{MPa}$. In fact, as the density of water is much greater than that of air, whatever material is used in practical structure underwater, the model in air acoustic experiment should be built of a kind of material whose density is comparable to that of gas. Seeking for a suitable material may be much more difficult than organizing an underwater experiment. Thus, adopting data gained in air acoustic experiment to calculate underwater noise is unrealistic until a kind of cheap material which has a super low density and stiffness is discovered. However, if densities of different fluids are comparable, acoustic similarity laws may help.

# REFERENCES

Chapman, C.J. 2000. Similarity variables for sound radiation in a uniform flow. *Journal of Sound and Vibration*, 233(1): 157–164.

Everstine, G.C. & Henderson, F.M. 1990. Coupled finite element/boundary element approach for fluid-structure interaction. *Journal of Acoustical Society of America*, 87(5): 1938–1947.

Ji, Gang, Zhang, W.K. & Zhou, Q.D. 2006. Vibration and Radiation from Underwater Structure Considering the Effect of Static Water Preload. *Shipbulding of China*, 47(3): 37–44.

Singhatanadgid, P. & Songkhla, A.N. 2008. An experimental investigation into the use of scaling laws for predicting vibration responses of rectangular thin plates. *Journal of Sound and Vibration*, 311: 314–327.

Wang, L.C. Zhou, Q.D. & Ji, G. et al. 2011. Approximate Method for Acoustic Radiated Noise Calculation of Sub Cabin Model in Replacing Full-Scale Model. *Chinese Journal of Ship Research*, 5(6): 26–32.

Wu, J.J. 2003. The complete-similitude scale models for predicting the vibration characteristics of the elastically restrained flat plates subjected to dynamic loads. *Journal of Sound and Vibration*, 268: 1041–1053.

Xie, Z.Y. Zhou, Q.D. & Ji, G. 2009. Computation and measurement of double shell vibration mode with fluid load. *Journal of Naval University of Engineering*, 21(2): 97–101.

Yang, D.S. Wang, S.D. Shi. S.G. & Liu, X. 2005. Study of acoustical similitude of a submerged complex shell. *Journal of Harbin Engineering University*, 26(2): 174–178.

Yu, M.S. Shi, X.J. & Chen, K.Q. 1993. Study on Acoustic Similitude of Elastic Stiffened Cylindrical Shells by FEM. *Ship building of China*, 3(2): 65–71.

Yu, M.S. Shi, X.J. & Wu, Y. 1998. Similitude Analysis of Vibration and Acoustic Radiation for Elastic Structures. *Journal of Ship Mechanics*. 2(1): 55–61.

Yu, M.S. Wu, Y.X. & Lv, S.J. 2002. Experimental Research on Vibration Similitude of Elastic Stiffened Cylindrical Shells. *Ship building of China*, 43(2): 50–57.

Zhou, Q. & Joseph, P.F. 2005. A numerical method for the calculation of dynamic response and acoustic radiation from an underwater structure. *Journal of Sound and Vibration*, 283(3/5): 853–873.

*Advanced Materials, Structures and Mechanical Engineering – Kaloop (Ed.)*
© *2016 Taylor & Francis Group, London, ISBN 978-1-138-02793-0*

# Influence of soil temperature conditions on linear construction stability

E.E. Kholoden & S.A. Lobanov
*Far Eastern Federal University, Vladivostok, Russian Federation*

O.M. Morina
*Pacific National University, Khabarovsk, Russian Federation*

ABSTRACT: To identify the reasons affecting the roadway stability, the rate of soil temperature change and its oscillation amplitude at standard depths at Bogorodskoe meteorological station were analyzed. It was determined that during the last almost 50 years (1963–2013) only the soil temperature growth was observed. It was particularly active in upper meter level in January–March and in July–October—up to 2–3°C. In January, at 0.4 m depth, the level of heat supply moves actually into another climatic zone −6°C–3.5°C. At lower levels active growth occurs in July–October, from 2 up to 4.4°C. At 3.2 m depth a temperature range reaches rather high values - 2,5°C. Thus, non-uniform temperature rise and sharp fluctuations of soil temperature profile create the extreme instability of the territory under study. These conditions must be taken into account in preliminary calculations for any type of nature use: during the construction of linear infrastructure and industrial and agricultural facilities.

## 1 INTRODUCTION

At the present stage of economic development of the Far East the active construction of new highways and upgrading of already existing ones occurs, considering high vehicle speeds and road-bed load. In connection with this increase of road constructions and roadbed life cycle gains special importance. The large-scale construction of oil and gas pipelines also occurs.

According to A.I. Yarmolinskiy (1994), the low state of roads repair is due to low account taken to specific natural and climatic conditions, in particular, no measures are being taken to stabilize water supply, growth thermal mode of an existing road-bed. The roadbed in the Far East is now designed for road-climatic zone II, based on the results of numerous studies of water and thermal regime of the European part of Russia. This is due to the fact that such studies were extremely low, particularly on the distribution of soil temperature in the Far East.

Soil temperature can be one of the universal indicators of territory's stability. The relevance of the problem studying is increased by the fact that in this paper these characteristics are analyzed not statistically but dynamically. When studying the natural processes, as well as in anthropogenically modified territories, the practical importance of soil temperature should be taken into account.

Cyclisity in the climatic and landscape systems is generally accepted. Many authors note that modern climate changing has a general trend towards the widespread temperature increase. But a large part of researchers come to quite the opposite conclusion: a cryolithic zone changing its borders, seizing new territories. The classics of compensatory balance theory have the third view on this problem, they claim that changing of conditions in one place is compensated by opposite sign changes at the adjacent territory. The above problems have made it necessary to identify the dynamics of soil temperature. This factor defining the stability of the territory is unjustly neglected.

At the present stage of economic development of the Far East the active construction of new and upgrading of already existing highways occurs, considering high vehicle speeds and roadbed load. In connection with this increase of road constructions and roadbed life cycle gains special importance. The large-scale construction of oil and gas pipelines also occurs.

## 2 DISCUSSED PROBLEMS

The variations (changes) of heat and moisture values from the statistical average, i.e. from the most stable states are estimated as stress or vulnerability of natural ecosystem (Ecological Encyclopedic Dictionary 1999). In the IPCC climate change means any variation over a time period. Changes may be explained by both natural variability and human activity (Kokorin 2005).

Stability is a fundamental property of the eco-system, landscape in dialectical unity with their variability. Along with this, they integrally reflect ecosystems sensitivity to external influences. The notions (concepts) of territory stability and security are widely used as properties, providing the development of planetary diversity of natural and social phenomena (Naprastnikov 2002).

Stability refers to the consistency of the characteristics of the system in time, the ability of landscapes to maintain its parameters values. Criteria for determining the stability of landscapes to technogenic loads are time and speed of their return to a state close to the source (relaxation time) (Armand, 1983). The economic damage caused by the loss of ecosystem stability can be expressed in the costs required to restore its phytomass to the natural state. The processes of over-growth during this period should be estimated by indirect features, phytomass growth, species composition.

## 3 RESULTS AND DISCUSSION

Determination of the territory stability for the linear construction in the Southern part of Khabarovskiy Krai was carried out to identify the vector dynamics of soil temperature for Bogorodskoe meteorological station. The data of meteorological reference books and monthlies were the basic material for the analysis. The data were obtained from the archival materials for 6 standard depths from 1963 till 2013. A differentiated approach for a particular task is the principal methodological point taken in the temperature analysis. The data were analyzed for all months, average annual.

The analysis of 90 diagrams with definition of correlation coefficient showed that significant warming is characteristic of the whole soil layer. The rate of soil temperature growth was selected as one of the main studied indicators. It was found that warming has been particularly active in January–March and in July–September. So, in January, at a depth of 0.4 m the soils pass almost in a different climate zone, i.e. from values (−6°C) to the values −3.5°C (Fig. 1).

In the same months the growth rate of soil temperature in the upper 1-meter layer reaches 3.5°C. At 3.2 m depth the temperature increase is 1–2° C almost for the whole year (Morina 2014) (Figs. 2–7).

Dynamics of soil temperature range at 3.2 m depth is described by a sine curve with a maximum of 3–4°C. In cold season the range does not exceed 2°C (Fig. 8), but it still remains significant for this depth of soil profile.

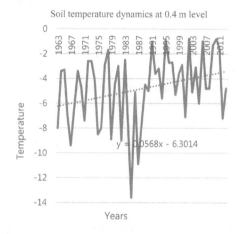

Figure 1. Soil temperature growth at 4 m depth in January, Bogorodskoe meteorological station, 1963–2013, (°C).

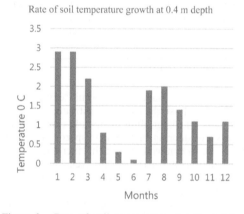

Figure 2. Rate of soil temperature growth at 0.4 m depth, Bogorodskoe meteorological station (°C).

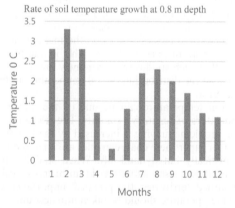

Figure 3. Rate of soil temperature growth at 0.8 m depth, Bogorodskoe meteorological station (°C).

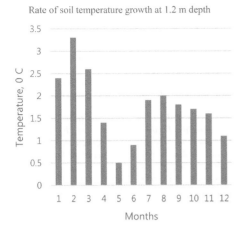

Rate of soil temperature growth at 1.2 m depth

Figure 4. Rate of soil temperature growth at 1.2 m depth, Bogorodskoe meteorological station (°C).

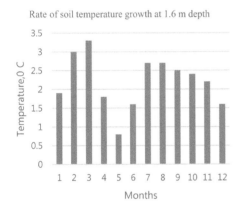

Rate of soil temperature growth at 1.6 m depth

Figure 5. Rate of soil temperature growth at 1.6 m depth, Bogorodskoe meteorological station (°C).

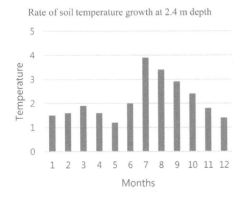

Rate of soil temperature growth at 2.4 m depth

Figure 6. Rate of soil temperature growth at 2.4 m depth, Bogorodskoe meteorological station (°C).

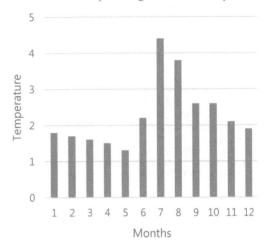

Rate of soil temperature growth at 3.2 m depth

Figure 7. Rate of soil temperature growth at 3.2 m depth, Bogorodskoe meteorological station (°C).

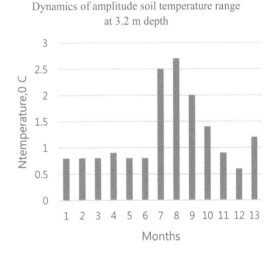

Dynamics of amplitude soil temperature range at 3.2 m depth

Figure 8. Amplitude temperature range at 3.2 m depth (°C).

## 4 CONCLUSIONS

Thus, non-uniform temperature growth both at depths and in certain months, sharp fluctuations of profile soil temperature create extreme instability of the studied territory. Soil temperature fluctuation amplitude can decrease and grow with depth. In some months, even at the depth of 3.2 m, the amplitude of the temperature fluctuations may be higher than in the air that can cause subsidence and soil heaving in a road bed. This condition should be considered in the preliminary design calculations for any kind of environmental management: construction of linear to prolong the projects life.

## REFERENCES

Armand, A.D. 1983. *"Strong" and "weak" systems in geography and ecology. Stability of geosystems.* Moscow: Nauka. 50–61.

Ecological Encyclopedic Dictionary 1999. Moscow: Noosphera, 932.

Kokorin, A.O. 2005. *Climate changing: Review of research information about anthropogenic climate change.* Moscow: RREC, GOF, www-Russia, 20.

Kuraev, S.N. 2006. *Climate change adaptation.* Moscow: RREC, GOF, 16.

Morina, O.M. 2014. *Dynamics of soil and air temperatures in the South of the Amur river region as a reflection of changes of forest site conditions.* Khabarovsk: Pacific National University, 137.

Naprastnikov, V.A. 2002. Criteria of binary assessment of hydrologo-climate stability of geosystems. *Geography and Natural Resiurces*, 3: 22–27.

Novorotzkiy, P.V. 1984. *Heat-moisture exchange in medium-altitude mountain area of BAM zone for different types of weather.* Hydrometeorological survey in the Southern Far East. Vladivostok: FERC AS USSR, 32–42.

Yarmolinskiy, A.I. 1994. *Motor ways of the Far East. Experience of designing and operation.* Transport, 141.

*Advanced Materials, Structures and Mechanical Engineering – Kaloop (Ed.)*
© *2016 Taylor & Francis Group, London, ISBN 978-1-138-02793-0*

# The effects of KH560 addition on brittle/ductile properties of glass fiber/epoxy resin composites

J.S. Sun & J.G. Wang
*Jiangsu Key Laboratory of Advanced Metallic Materials, School of Materials Science and Engineering, Southeast University, Nanjing, P.R. China*

ABSTRACT: Composites containing glass fiber, epoxy resin and 0~15% KH560 (γ-glycidyloxipropyltri methoxysilane) are prepared by a sample hand lay-up method. The effects of KH560 addition on the tensile stress, strain, work of fracture and glass transition temperature ($T_g$) of the samples are investigated. The results show that the addition of 5 wt.% KH560 increases the tensile stress and $T_g$ compared to that of the sample containing no KH560. However, with the increase of KH560 content, the tensile stress and $T_g$ decrease gradually. The strain and work of fracture have the maximum of 21% and 3254 J/m$^2$ under the condition of 15% KH560 content. In other words, the samples have realized a transition from brittle to ductile property by regulating the content of KH560 addition. Scanning Electron Microscope (SEM) images of fracture surfaces explain the possible mechanism of the brittle-ductile transition.

## 1 INTRODUCTION

In recent years, the composites have been widely used in many fields, such as aviation, automotive, shipbuilding, and athletic equipment manufacturing (Chang et al. 2014). Composites are composed primarily of reinforcement and matrix materials. Reinforcement materials are mainly divided into two classes of glass fibers and carbon fibers, and thematrix materials are primarily thermosetting and thermoplastic resins. The reinforcement materials provide support and increase the material strength and hardness, correspondingly, thematrix materials prevent crack propagation. Glass fiber is usually used with thermosetting resin as the substrate responsible for fixing the fibers, sustaining outside stress and passing the forces onto the fibers for loading (Schoßig et al. 2008, SinhaRay & Okamoto 2003).

It is well known that properties of composites are strongly influenced by the type of adhesion between the reinforcement and the matrix. In many cases, failure occurs in the interface region due to a chemical reaction or plasticization when impurities penetrate the interface. Moreover, the properties of composites depend on the ability of the interface to transfer stress from the matrix to the reinforcement. Therefore, if adhesion takes place, the interface must be controlled (Koyanagi et al. 2009, Nishikawa et al. 2008, Ogihara & Koyanagi 2010). Silane coupling agents are the most widely used in varieties of coupling agent. At the same time, it is one of the most important modifiers in glass fiber

composites industry. It is intensively reported as a surface treatment agent on fiber to increase the interlaminar shear strength or as a cross-linking agent to modify the inorganic nanoparticles. (Gao et al. 2008, Khan et al. 2011, Novais et al. 2012, Cui & Kessler 2014). Nonetheless, it is unpractical for industrial manufacture to modify fiber or the inorganic nanoparticles, because it is difficult to soak and dry in the especial fiber fabric.

To overcome the difficulties above mentioned, KH560 coupling agent containing an epoxy group is employed to mix into epoxy matrix directly. This not only improves the operability in the process of manufacturing, but lowers the production cost. Furthermore, we investigate the effects of different KH560 content on tensile strength, fracture energy, and $T_g$ of composites. At the same time, Scanning Electron Microscopy (SEM) analysis is made on fracture surfaces to visualize the damage process: fiber fracture, matrix rupture and interface rupture.

## 2 EXPERIMENTAL

### 2.1 Materials

EPOLAM 2008 EPOXY RESIN (EP2008) and EPOLAM 2008-S HARDENER (EP2008-S) are both commercial products (supplied from AXSON). KH560, supplied from Nanjing Chunshi Chemical Technology co., LTD., the structural formula is shown in Figure 1. Plain glass fiber cloth/ SW100A-90a, supplied from Sinoma Science & Technology Co., Ltd.

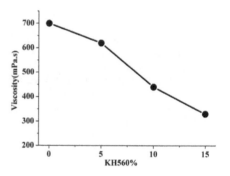

Figure 1. The structural formula of KH560.

Table 1. The weight percentage of detailed ingredients of composites.

| Sample | EP2008 /wt | EP2008-S /wt | KH560 /wt | SW100A-90a /wt |
|---|---|---|---|---|
| GF/E-0 | 100 | 25 | 0 | 55 |
| GF/E-5 | 100 | 25 | 5 | 55 |
| GF/E-10 | 100 | 25 | 10 | 55 |
| GF/E-15 | 100 | 25 | 15 | 55 |

## 2.2 The preparation of samples

At first, KH560 was mixed manually with epoxy resin as per calculated weight ratio (relative to EP2008 0, 5%, 10%, 15%), then the mixture was vigorously stirred until no bubble appeared. At last, EP2008-S was added to the above mixture, repeating the above stirring process. GF/E composites were fabricated by using a sample hand lay-up method, the fiber content was taken control of about 55 wt.%. The entire set-up was then kept 24 h maintained at room temperature, after completion of curing, the laminate was thermally post cured at 60°C for 6 h.

## 2.3 Testing and analysis

The viscosity is measured with the NDJ-l type of rotating viscometer at room temperature. Tensile tests are carried by a CMT5105 (SANS, China) tensile test machine with dumbbell shaped specimens prepared referring to Chinese standard GBT1447-2005, loading rate 5 mm/min. Stress, strain, and work of fracture were calculated based on at least five samples. $T_g$ is tested at a heating rate of 5 °C/min from 25 to 300°C using a thermal mechanical analysis (TMA 402F1, NETZSCH, Germany). Field Emission Scanning Electron Microscopy (FESEM) with the type of XL FEGSFEG-SIRION (FEI Ltd., Netherlands) is applied to investigate the morphological features of the obtained specimens.

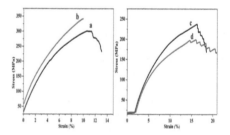

Figure 2. The relationship between the viscosity and the KH560 content.

Figure 3. Stress-strain curves with different composites. (a) GF/E-0 (b) GF/E-5 (c) GF/E-10 (d) GF/E-15.

## 3 RESULTS AND DISCUSSION

KH560 is used as filler, added in epoxy and then coated on glass fibers to form composites. Detailed ingredients of composites are shown in Table 1.

Figure 2 shows the relationship between the viscosity and the content of KH560 for EP-2008/EP2008-S system before curing. With the increase of the KH560 content, the viscosity of matrix is changed from 700 to 330 mPa·S. This indicates that KH560 highly improves the liquidity of EP-2008/EP2008-S system as diluents. It is very advantageous for the operation in the process of hand lay-up.

The tensile stress-strain curves of these samples are given in Figure 3 (a) and (b). Table 2 summarizes the tensile stress, strain, work of fracture and $T_g$ for the different samples. It can be seen that the tensile stress, $T_g$ increased, from 300 to 342 MPa and 149 to

Table 2. The values of stress, strain, work of fracture, and $T_g$.

| Sample | Stress (MPa) | Strain (°C) | Work of fracture (J/m²) | $T_g$ (°C) |
|---|---|---|---|---|
| GF/E-0 | 300 | 13 | 2859 | 149 |
| GF/E-5 | 342 | 10 | 2303 | 207 |
| GF/E-10 | 237 | 18 | 2862 | 134 |
| GF/E-15 | 199 | 21 | 3254 | 127 |

207°C, but the strain and work of fracture decreased from 13 to 10% and 2859 to 2303 J/m² when adding the KH560 of 5%. With the increase of the KH560 content, the reverse has taken place, especially for the content of the KH560 reaching 15%. This mainly because the reactive silanol monomers or oligomers

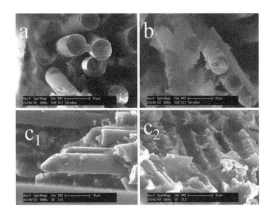

Figure 4. SEM images of fracture surfaces: (a) GF/E-0, (b) GF/E-5, ($c_1$ and $c_2$) GF/E-15.

are physically adsorbed to hydroxyl groups of glass fibers by hydrogen bonds on the fiber surfaces by diffusion, which depends on the molecular size of silanolmonomers/oligomers formed when adding the small amount of KH560. The free silanols also adsorb and react with each other thereby forming a polysiloxane structures linked with a stable Si–O–Si bond and increase the ductility of the matrix when adding the more KH560 much. At the same time, the decrease of $T_g$ is attributed to the increasing free volume with the content of KH560 increasing.

To understand the fracture behavior and the toughening mechanisms in the systems studied here, the fracture surface of GF/E-0, GF/E-5, and GF/E-15 are examined using SEM. From the Figure 4a, it is observed that GF/E-0 exhibits brittle tensile fracture, and a very few resin is adhered on fiber surface after fracture, because no evidence shows that chemical bonding can form between fiber and epoxy matrix in composite made with as-received fiber. Also in modified composite systems containing 5 wt.% of KH560, the fracture morphology is brittle in nature (Fig. 4b). Moreover, we can see that a few resins are adhered on fiber surfaces, and fewer fibers are pulled out after fracture, which demonstrates that good interfacial adhesion is formed by covalentbonding between fiber and matrix through silane coupling agent. However, with increasing KH560 content to 15 wt.%, a brittle to ductile transition in fracture mechanics is observed which is evident from the polymer fibrils pulled (Fig. $4c_1$) out from the matrix with numerous cavities (Fig. $4c_2$) residing among them. Also from Figure $4c_1$, it can be seen that glass fibers are completely debonded from the matrix. This indicates that the interfacial between the fiber surface and matrix weakens with the increase of KH560 content leading to interfacial debonding.

## 4 CONCLUSIONS

The incorporation of KH560 in GF/E is used to regulate stress, strain, work of fracture, and $T_g$ of composites. It is observed that through incorporation of KH560 (GF/E-5) increases the tensile strength as compared to GF/E-0 composites. However, when the KH560 content is up to 10 and 15%, the composites have realized a transition from brittle to ductile property. Morphological observations of fracture surfaces by SEM explain the forming mechanism. It can effectively control the properties and tremendously expand the application range of GF/E composites.

## REFERENCES

Chang, H.L. Chen, C.M. & Lin, K.L. 2014. The effects of chemical resistance for nanocomposite materials via nano-silica addition into glass fiber/epoxy. *Advanced Materials Research*, 853: 28–33.

Cui, H. & Kessler, M.R. 2014. Pultruded glass fiber/bio-based polymer: Interface tailoring with silane coupling agent. *Composites Part A: Applied Science and Manufacturing*, 65: 83–90.

Gao, X. Jensen, R.E. Li, W. Deitzel, J. McKnight, S.H. & Gillespie, J.W. 2008. Effect of fiber surface texture created from silane blends on the strength and energy absorption of the glass fiber/epoxy interphase. *Journal of Composite Materials*, 42(5): 513–534.

Khan, R.A. Parsons, A.J. Jones, I.A. Walker, G.S. & Rudd, C.D. 2011. The effectiveness of 3-aminopropyltriethoxy-silane as a coupling agent for phosphate glass fiber-reinforced poly (caprolactone)-based composites for fracture fixation devices. *Journal of Thermoplastic Composite Materials*, 0892705710391622.

Koyanagi, J. Shah, P.D. Kimura, S. Ha, S.K. & Kawada, H. 2009. Mixed-mode interfacial debonding simulation in single-fiber composite under a transverse load. *Journal of Solid Mechanics and Materials Engineering*, 3(5): 796–806.

Nishikawa, M. Okabe, T. & Takeda, N. 2008. Determination of interface properties from experiments on the fragmentation process in single-fiber composites. *Materials Science and Engineering: A*, 480(1): 549–557.

Novais, V.R. Júnior, S. Cézar, P. Rontani, R.M.P. Correr-Sobrinho, L. & Soares, C.J. 2012. The bond strength between fiber posts and composite resin core: influence of temperature on silane coupling agents. *Brazilian Dental Journal*, 23(1): 08–14.

Ogihara, S. & Koyanagi, J. 2010. Investigation of combined stress state failure criterion for glass fiber/epoxy interface by the cruciform specimen method. *Composites Science and Technology*, 70(1): 143–150.

Schoßig, M. Bierögel, C. Grellmann, W. & Mecklenburg, T. 2008. Mechanical behavior of glass-fiber reinforced thermoplastic materials under high strain rates. *Polymer Testing*, 27(7): 893–900.

SinhaRay, S. & Okamoto, M. 2003. Polymer/layered silicate nanocomposites: a review from preparation to processing. *Progress in Polymer Science*, 28(11): 1539–1641.

*Advanced Materials, Structures and Mechanical Engineering – Kaloop (Ed.)*
© 2016 Taylor & Francis Group, London, ISBN 978-1-138-02793-0

# *In vitro* bioactivity evaluation of Co-Cr-Mo (F-75)/HAP composites

N.M.S. Adzali, S.B. Jamaludin & M.N. Derman
*School of Materials Engineering, University Malaysia Perlis, Perlis, Malaysia*

ABSTRACT: This paper reports on the *in vitro* bioactivity evaluation of Co-Cr-Mo (F-75)/HAP composites fabricated by powder metallurgy. The F-75 alloy is known to be used in the biomedical field because of its excellent biocompatibility when implanted in the human or animal body. Hydroxyapatite (HAP) powders have been used as a filler because HAP is one of the most effective biocompatible materials with similarities to mineral constituents of bones and teeth. HAP powders (chemical formula $Ca_{10}(PO_4)_6$ $(OH)_2$) have been added to the F-75 alloys in the composition of 4 wt.% and 8 wt.%. The mixtures were then milled, cold compacted at 550 MPa, before sintered at 1200°C for 2 hours in a tube furnace. The bioactivity test for the composite was conducted by immersing the composite into Phosphate-Buffered Solution (PBS) for 18 days. SEM, XRD, FTIR and pH analyses were carried out in order to examine the presence of the apatite layer on the surface of the F-75/HAP composites. From the bioactivity test results, all the testing (SEM, XRD, FTIR and pH analyses) proved that the carbonated apatite layer was formed on the surfaces of the composite after 18 days of immersion in the PBS. The bioinert F-75 alloys can be converted into the bioactive F-75 type by adding up to 8 wt.% of HAP. The F-75/HAP bio-composite has an excellent ability to form an apatite layer that can contribute to improved biocompatibility and osteoconductivity of this composite.

## 1 INTRODUCTION

In the biomedical field, the ideal biomaterial for implantation should be the one that is biologically and mechanically compatible with the bone (Li et al. 1995). The bioactivity behavior of the biomaterial needs to be enhanced in order to form a bond between tissues of humans or animals and implant materials. Bioactivity test is an evaluation of apatite formed on the implant material in the Simulated Body Fluid (SBF) solution. According to the British standard (BS ISO 23317:2007), the term SBF means an inorganic solution that has a similar composition to human plasma without organic components.

A variety of materials including metals, ceramics and composites with the appropriate physical properties and biocompatibility are chosen for the fabrication of biomaterials. Cobalt-based alloy has been used as one of the important metallic biomaterials because it exhibits low levels of corrosion, and, to date, this alloy remains a fixture in orthopedic surgery. A Co-Cr-based alloy, which is well known for its high Young's modulus, fatigue strength, wear resistance, good biotolerance and corrosion resistance, is an important metallic biomaterial (Ghazali et al. 2010).

Ceramic, in particular, such as calcium-based biomaterials including hydroxyapatite (HAP), has been developed for implants in orthopedic and dental applications (Hench 1998, Ducheyne & Qiu 1999), due to its excellent biocompatibility, structural and chemical similarity with bone mineral compositions. As reported by Navarro et al. (2008), the application of this ceramic material as bone substitutes started around in the 1970s, and has mainly been used as bone defect fillers.

Several bioactive materials, such as bioglass, fluorapatite and wollastonite have already been studied with metals and alloys (Liu & Ding 2001, Zhang 2008, Oksiuta et al. 2009, Razavi et al. 2010). Therefore, an interest has been focused on hydroxyapatite (HAP) as an addition to the F-75 alloys for the biomedical applications. HAP has been reported as a suitable reinforcing material for the development of metal matrix composite materials such as Ti-HAP, Ti-6 Al-4V-HAP, and AZ91-HAP for implantation to control the corrosion rate (Ning & Zhou 2002, Witte et al. 2007). Interestingly, when implanted in the human body, HAP spontaneously bonds to the living bone via an apatite layer deposited on its surface without forming a fibrous tissue around it (Leonor et al. 2003).

This research reports the fabrication and bioactivity behavior of the F-75 alloy filled with HAP, which is prepared by the powder metallurgy method. This study focuses on the effect of HAP addition on the F-75 alloy related to its microstructure and also to its bioactivity behavior.

## 2 EXPERIMENTAL PROCEDURE

### 2.1 *Materials and preparation of the composite*

This research was conducted using two main raw materials, namely Co-Cr-Mo alloy (ASTM F-75) powder and hydroxyapatite (HAP) powder. The F-75 alloy is in the form of gray solid powder, with an average size of this alloy powder being 10.347 μm. A hydroxyapatite (HAP) powder was used as the filler in the matrix of the F-75 alloy to form a composite of F-75/HAP. HAP is in the form of solid white powder with an average size of 11.048 μm.

The F-75/HAP composites were fabricated by the powder metallurgy method. In the fabrication of the F-75/HAP composite, 0, 4, and 8 wt.% of HAP have been added to the F-75 alloys. The mixtures were milled on a rotation mill for 20 minutes at 154 rpm before cold compacted at 550 MPa using a uniaxial press machine. The samples were then sintered at 1200 °C in a tube furnace for 2 hours.

### 2.2 *Bioactivity test*

The Simulated Body Fluid (SBF) solution was prepared from Phosphate-Buffered Saline (PBS) tablets that were supplied by Sigma Aldrich Company. For 200 ml deionized water, one PBS tablet was diluted to get 200 ml SBF solution. For the bioactivity test, the weight and the dimensions of all the sintered samples were recorded. The surface area of the samples can be calculated from the dimension of the samples and recorded as $S_a$. The volume of SBF for immersing the sample can be determined by using Equation (1) as follows (BS ISO 23317:2007):

$$v_s = S_a/10 \tag{1}$$

where $v_s$ is the volume of the SBF in cubic millimeters and $S_a$ is the apparent surface area of the specimen in square millimeters.

The calculated volume of the SBF (Equation (1)) was poured into a plastic bottle and placed into a water bath. After heating the SBF to 36.5°C in a water bath, the samples were subsequently immersed into the SBF for 18 days without refreshing the soaking medium.

### 2.3 *Characterization*

The morphological observation of the samples before and after the bioactivity test was made by SEM. XRD measurement was done to identify the apatite formation on the sample after immersion, while the FTIR analysis was carried out to reveal the functional group of this sample. The pH value

of the SBF solution with the samples was checked by using a pH meter for up to 18 days, with the periodic interval of 3 days.

## 3 RESULTS AND DISCUSSION

### 3.1 *SEM*

SEM morphologies of the F-75/HAP specimen surfaces before and after immersion in the SBF are shown in Figure 1 (a–c). Figure 1a shows that the sample with no addition of HAP exhibits the corrosion at grain boundaries without the formation of white particles on the corroded surfaces and grain boundaries.

In Figure 1b, when 4 wt.% HAP was added to the samples, the formation of white particles can be clearly seen on the sample surface after immersion for 18 days. A white particle, which appeared in a nodular shape, is precipitated on the surface of the samples after immersion as higher HAP concentration was added to the samples and many pores became invisible due to the coverage of the newly formed deposition. As 8 wt.% of HAP was added,

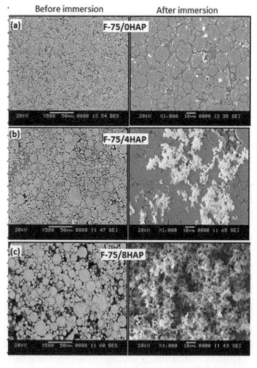

Figure 1. SEM microphotographs of the F-75/HAP samples before and after immersion in the SBF for 18 days with (a) no HAP, (b) 4 wt.% HAP, (c) 8 wt.% HAP.

after immersion for 18 days, the entire surface of the sample (Fig. 1c) was covered by this precipitate.

Figure 2 shows the EDS spectra of the particles/granule on the surfaces of the F-75/HAP composites after 18 days of immersion in the SBF solution. For the sample with no HAP addition (Fig. 2a), the EDS results indicated the samples mainly contained Co, Cr, and Mo. For the samples with the F-75/8HAP composite (Fig. 2b), the EDS results indicate that the precipitated layer (appeared as white particles, as shown in Fig. 1c) observed on the surface of the samples is mainly composed of calcium (Ca), phosphorus (P), Co, Cr, and Mo. These SEM and EDS results implied that the bioactivity behavior of the F-75/HAP composite increased with the increase in HAP content. This result coincides with the XRD, FTIR, and pH analyses, which will be discussed in the next section.

## 3.2 XRD

Figure 3 shows the XRD patterns of the surfaces of the sintered F-75/HAP composite after the immersion in the SBF solution. The diffraction pattern of the F-75/0HAP sample corresponds to that shown in Figure 1a, which is referred to

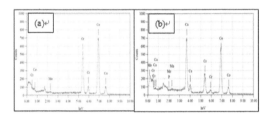

Figure 2. EDS spectra of samples (a) F-75/0HAP and (b) F-75/8 HAP after immersion in the SBF for 18 days.

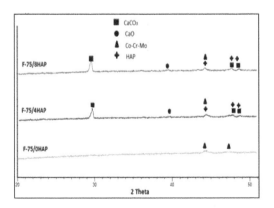

Figure 3. The XRD pattern of the F-75/HAP composite after immersion in the SBF for 18 days.

the sample before immersion in the SBF, which revealed the crystalline peak of Co-Cr-Mo (JCPD 260425). As the addition of HAP increased (4 wt.% of HAP), the intensity of additional peaks also increased, which corresponded to the hydroxyapatite phase (JCPD 9-432), with major angles of 44.5°, 48° and 49.5° and also Co-Cr-Mo peaks. For the samples with the F-75/4HAP and F-75/8HAP composites, the new additional strongest peaks were identified as corresponding to calcium carbonate ($CaCO_3$) with JCPDS card no. 86-0174, which displays the following major diffraction peaks (2Θ [°]): 29.4, 47.8, and 48.8. The formation of this carbonate apatite layer will be discussed in detail in Section 3.3 (Equation (2)). From XRD and FTIR analyses (which will be discussed in the next section), this precipitate layer on the surface of the samples may consist of the carbonate group and the OH group.

## 3.3 FTIR

Figure 4 shows the FTIR analysis of the SBF solution before (day 0) and after the immersion (day 18) of the F-75/8HAP sample.

Two main groups as an active band were detected in the spectra, namely the carbonate group and the OH group. Most of these active bands were located in the range of 3500–1500 cm$^{-1}$. The band in the region from 1650–1600 cm$^{-1}$ was attributed to the $CO_3^{2-}$ group in the carbonated apatite (Elliott 1994). The carbonate group was related to the pure carbonate, which indicates the presence of carbonate ions in the SBF solution (which came from $CO_2$ that reacted with $H_2O$ from water). These findings are related to XRD patterns (Fig. 3) and SEM results (Fig. 1), which revealed the presence of the carbonate apatite layer on the

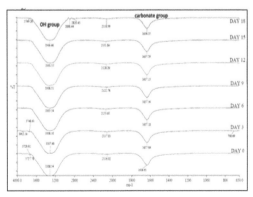

Figure 4. FTIR spectra of the SBF solution before and after sample the immersion at various immersion times for the F-75/8HAP composite.

surface of F-75/HAP composite. This carbonate in the SBF solution will react with $Ca^{2+}$ion released from HAP to form $CaCO_3$, which precipitates on the surface of the composite (Ye et al. 2009), as given in Equation (2):

$$Ca^{2+} + 2HCO_3^- \rightarrow CaCo_3 + H_2O + CO_2 \qquad (2)$$

This $CaCO_3$ detected by XRD will precipitate on the surface of the composite as a carbonate layer. The OH- stretching vibration is observed at the peak in the range of 3000 to 3700 cm$^{-1}$, in a broad band for all the samples. In this research, the formation of the CaO phase (which was detected by XRD, as shown in Fig. 3) would react with $H_2O$ from the SBF, and formed OH-, as given in Equation (3), which would lead to the increase in the pH value of the SBF:

$$2CaO + 2H_2O \rightarrow 2Ca^{2+} + 4OH^- \qquad (3)$$

### 3.4 pH

Figure 5 shows the change in the pH value of the SBF solution after immersion for 18 days. The SBF solution without the samples was used as the reference solution. The data were collected for up to 18 days with a periodic interval of 3 days.

The starting pH value of the reference solution was 7.21. For the SBF solution with the samples, the values of pH increased dramatically from day 0 to day 3. After day 3, all SBF solutions had not much different pH values and maintained at these values until day 18. According to the literature, when the composites Ti/HA (Thirugnanam et al. 2009) and F-75/FA (Fathi et al. 2012) were immersed into the SBF, there was an ionic exchange between the alkaline ions from the sample and the hydrogen ions from the solution (SBF) (Verne' et al. 2009). In this study, the formation of OH- from the SBF solution leads to the increase in the pH value of the SBF (Priya et al. 2010). The formation of OH- is expressed in Equation (3). This increasing value of pH was detected during the first three days for all the samples due to the increase in OH- ion concentrations; the accumulation of OH- can be seen on the surface of the composite. Generally, as more HAP concentration was added to the samples, the values of pH were increased, but were still not much different compared with samples with no HAP. From SEM micrographs (Fig. 1), the samples with no HAP showed no white particles formed on their surface. While samples with HAP addition were covered by white particles and a precipitated layer was detected as the carbonate apatite layer, which also contained CaO (Fig. 3). These covered surfaces will release OH- ions, as given in Equation (3). Samples with no HAP had no CaO, so no OH- could be released from these samples, which will decrease the value of pH compared with the samples with HAP. So, the higher amount of OH- released from the samples with HAP addition resulted from the higher values of pH, as shown in Figure 5.

## 4 CONCLUSIONS

The bioactivity behavior of the F-75/HAP composite was successfully evaluated by immersing the sample in the PBS solution for 18 days. The in vitro bioactivity test for the F-75/HAP composite showed that after immersion, the carbonated apatite layer was formed on the surface of the F-75/HAP composite. The formation of this carbonated apatite layer on the surface of the material was considered as the mark of bioactivity. In conclusion, the bioinert F-75 alloys can be converted to the bioactive F-75 type by adding up to 8 wt.% of HAP.

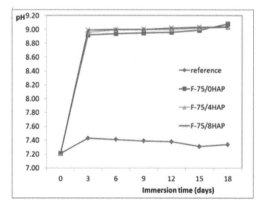

Figure 5. Change in the pH value of the SBF with time after the immersion of the F-75/HAP samples in the SBF for 18 days.

## REFERENCES

British standard (BS ISO 23317:2007) 2009. Implants for surgery—In vitro evaluation for an apatite-forming ability of implant materials.
Ducheyne, P. & Qiu, Q. 1999. Bioactive ceramics: the effect of surface reactivity on bone formation and bone cell function, *Biomaterials*, 20: 2287.
Elliott, J.C. 1994. *Structure and Chemistry of the Apatites and Other Calcium Orthophosphates,* Elsevier.
Fathi, M. Ahmadian, M. & Bahrami, M. 2012. Novel Bioactive Co-based alloy/FA nanocomposite for dental applications. *Dental Research Journal*, 9(2): 173–179.

Ghazali, K. Nurhidayah, A.Z. Dalimin, M.N. Mujahid, A.Z. Shamsul, J.B. & Mahadi, A.J. 2010. The sintering Temperature Effect on the Shrinkage Behavior of Co-Cr Alloy. *American Journal of Applied Sciences,* 7(11): 1443–1448.

Hench, L.L. 1998. Bioceramics. *Journal of American Ceramic Society,* 81: 705–1728.

Leonor, I.B. Ito, A. Onuma, K. Kanzaki, N. & Reis, R.L. 2003. In vitro bioactivity of starch thermoplastic/ hydroxyapatite composite biomaterials: an in situ study using atomic force microscopy. *Biomaterials,* 24: 579–585.

Li, P.J. Bakker, D. & Van Blitterswikj, C.A. 1995. A hydroxyapatite–polymer composite prepared by mimicking bone mineralization. In: L.L. Hench & J. Wilson (eds), *Bioceramics,* 8: 435–439. USA: Pergamon Press.

Liu, X. & Ding, C. 2001. Apatite formed on the surface of plasma sprayed wollastonite coating immersed in simulated body fluid. *Biomaterials,* 22(14): 2007–2012.

Navarro, M. Michiardi, A. Castano, O. & Planell, J.A. 2008. Biomaterials in Orthopaedics. *Journal of the Royal Society Interface,* 5: 1137–1158.

Ning, C.Q. & Zhou, Y. 2002. In Vitro bioactivity of a biocomposite fabricated from HA and Ti powders by powder metallurgy method. *Biomaterials,* 23: 2909–2915.

Oksiuta, Z. Dabrowski, J.R. & Olszyna, A. 2009. Co-Cr-Mo Based Composite Reinforced with Bioactive Glass. *Journal of Materials Processing Technology,* 209(2): 978–985.

Priya, A. Nath, S. Biswas, K. & Basu, B. 2010. In vitro dissolution of calcium phosphate-mullite composite in simulated body fluid. *Journal of Materials Science: Materials in Medicine,* 21: 1817–1828.

Razavi, M. Fathi, M.H. & Meratian, M. 2010. Fabrication and Characterization of Mg-FA nanocomposite for Biomedical Applications. *Materials Characterization,* 61: 1363–1370.

Thirugnanam, A. Sampath Kumar, T.S. & Chakkingal, U. 2009. Bioactivity enhancement of commercial pure titanium by chemical treatments. *Trend Biomaterials Artificial Organs,* 22(3): 202–210.

Verné, E. Bretcanu, O. Balagna, C. Bianchi, C.L. Cannas, M. Gatti, S. & Vitale-Brovarone, C. 2009. Early stage reactivity and in vitro behavior of silica-based bioactive glasses and glass-ceramics. *Journal of Materials Science: Materials in Medicine,* 20: 75–87.

Witte, F. Feyerabend, F. Maier, P. Fischer, J. Stomer, M. & Blawert, C. 2007. Biodegradable magnesium–hydroxyapatite metal matrix composites. *Biomaterials,* 28: 2163–74.

Ye, H. Liu, X.Y. & Hong, H. 2009. Cladding of titanium/ hydroxyapatite composites onto Ti6 Al4V for load-bearing implant applications. *Materials Science and Engineering C,* 29: 2036–2044.

Zhang, D. 2008. In Vitro Characterization of Bioactive Glass (Academic Dissertation, Fac. of Technology, Abo Akademi Uni., Finland, 2008). *Dissertation Abstracts International,* 71.

*Advanced Materials, Structures and Mechanical Engineering – Kaloop (Ed.)*
© *2016 Taylor & Francis Group, London, ISBN 978-1-138-02793-0*

# Formation of nanostructured Ti-Ni-Hf-Cu coatings by high-speed flame spraying of mechanically activated powders

P.O. Rusinov & Zh.M. Blednova

*Kuban State Technological University, Krasnodar, Russian Federation*

ABSTRACT: The object of research are multi-functional surface layers made of one of the most promising materials with shape memory Ti-Ni-Hf-Cu, characterized by elevated temperatures of martensitic transformations and relatively low cost. The subject of the study are functional mechanical properties of the surface layers formed by complex high-energy impacts, including mechanical activation of Ti-Ni-Hf-Cu powder, high-speed gas-flame spraying in protective environment of argon, and subsequent thermal and thermomechanical treatment. We substantiated our choice of elements for multicomponent composition which provides thermal stability and reversible shape memory effect of the formed surface layers. We determined the optimal granulometric structure of the sprayed powders; this provides formation of nanostructures during deposition according to the proven technology with the grain size of 95–140 nm. It is shown that during the subsequent heat treatment the optimal structural-phase state is formed, which provides pseudoelasticity of the surface layers. Experimental studies of functional and mechanical properties of steel 1045 with a surface-modified Ti-Ni-Hf-Cu layer showed an increase of microhardness to $H_\mu = 9{,}5 \div 11{,}8$ GPa, slowdown of damage accumulation under cyclic loading in bending with rotation and increase of fatigue limit in the air by 36.3%. Increase of steel durability by means of surface modification with TiNiHfCu alloy can be explained by the special mechanism of deformation due to a possible "healing" formed during cyclic deformation of defects (which is characteristic for materials with shape memory effect) and by pseudoelastic properties of the surface layer.

## 1 INTRODUCTION

Alloys with Shape Memory Effect (SME) have a set of important characteristics: high strength, unique volume of single and multiple reversible thermo-mechanical memory, a high level of reactive stress, recover and damping, high corrosion resistance and cyclic durability. These characteristics make them irreplaceable despite their high cost, and condition their successful practical use as functional materials of a new generation. Despite all these advantages, the high cost of alloys limits their application in mechanical industry. In these circumstances, it is relevant to study the methods of steel surface modification using materials with SME. Formation of coatings using materials with shape memory effect is currently being implemented by various methods of high-energy impacts (argon arc, laser cladding with local protection, self-propagating high-temperature synthesis, thermal mass transfer and different vacuum plasma methods) (Blednova & Rusinov 2014, Rusinov & Blednova 2015, Rusinov et al. 2015, Blednova et al. 2014). It is known that traditional methods of processing materials with SME lead to the formation in alloys a developed dislocation substructure. Additional possibilities to increase the functional properties of TiNiHfCu

consist in non-traditional combined treatments; including heat treatment and intense plastic deformation, leading to a formation of nanocrystalline or amorphous structure. In this regard, promising results have been published recently (Likhachev 1997–1998, Karabasova 2002, Liakishev & Alymov 2006, Valiev & Alexandrov 2000). In mechanical engineering, the creation of nanocrystalline surface layers with improved performance, functional and mechanical properties will optimize the products, improve their reliability. This will lead to efficient use of resources and improve the strength properties of the products.

Among the alloys with SME, TiNiHf is positioned as a high-temperature alloy with SME and is one of the most promising alloys because of its higher temperatures of martensitic transformations and low cost (Liang et al. 2001, Dalle et al. 2000). Adding copper to TiNiHf alloys with SME improves their thermal stability and gives a reversible SME (Dalle et al. 2000, Meng et al. 2005, Meng et al. 2002). Besides, TiNiHfCu alloy with SME shows excellent glass-forming ability and superplasticity (Meng et al. 2002). This makes it possible to produce very small parts and parts of complex shape in micro-electromechanical systems.

We used a gas flame as a source of thermal energy in this study. The choice of this source of thermal energy can be explained by the possibility to form nanostructures due to extremely short interaction time with the treated surface, i.e. high-velocity heating and cooling of the base metal. Thus, the substrate surface beneath the coating layer to a depth of 1 mm is heated up to 200 °C for approximately 0.01 nanoseconds and then cooled to the initial temperature of the metal for 1.0 ns.

The aim of this work is to study the formation of nanostructured surface layers using materials based on TiNiHfCu by means of high-speed gas-flame spraying of mechanically activated powder and subsequent gradual thermomechanical treatment to provide functional and mechanical properties and create efficient functional materials and components based on them.

Studies were conducted on steel 1045. High-speed gas-flame spraying of PN35T24GF37M4 powder was carried out on a GLC-720 universal machine in argon on cylindrical samples (Ø10 × 50 mm). We used the mechanically activated powder PN35T24-GF37M4 as material for surface modification. The particle size of the PN35T24GF37M4 powder after mechanical activation was 5–20 μm. After mechanical activation, PN35T24GF37M4 powder particles (Fig. 1a, b) are in the shape of deformed discs.

## 2  FEATURES OF STRUCTURE FORMATION OF THE SURFACE-MODIFIED LAYERS OF SHAPE MEMORY TiNiHfCu

The main cause of defects in gas-flame coatings is oxide layers at the interface between the surface layers and the base. Since the high-speed gas-flame sprayed coating is formed by layering of solidified melt drops, i.e. splats, the oxide layers on the boundaries between the solidified melt drops can cause defects and reduce functional properties. Metallographic examination of the TiNiHfCu coating obtained by high-speed gas-flame spraying of mechanically activated powder, showed that the powder particles passing through the flame jet are heated and strike the substrate in the form of solidified deformed discs which have a diameter of 8–25 μm and a thickness of 0.5–1.8 μm (Fig. 2a). Macroanalysis of surface TiNiHfCu alloy layers obtained by the developed technology showed that the structure of the coating is sufficiently dense, it has minimum pores and their size is very small. The interface between the coating and the substrate is without any visible cracks.

At room temperature, the main structural components of the TiNiHfCu surface layer are B2 austenite phase with a cubic lattice, B19' martensite phase with monoclinic lattice, intermetallic $Ti_2Ni$ phases with a cubic lattice, polycrystalline $Hf_2Ni$ phase, Cu, and a small amount of titanium oxide (TiO) and $HfO_2$ comprising less than 3%.

$(Ti, Hf)_2Ni$ phase (Fig. 3) in the TiNiHfCu surface-modified layers with SME significantly improve superelasticity, but prevent the growth of martensite plates. Annealing at 900°C for 1 hour leads to the elimination of $(Ti, Hf)_2Ni$ phase.

Studies of the TiNiHfCu surface layer microstructure by high-resolution scanning electron microscopy showed that the coating is a nano-sized structure with a grain size of 95–140 nm (Fig. 4). In many ways, the formation of such a coating is associated with features of high-speed gas-flame spraying (high-speed collision of particles with the substrate, high-speed cooling, and rapid quenching of the alloy).

The microhardness of the TiNiHfCu layer varies in the range $H_\mu = 9.5 \div 11.8$ GPa. This increase in microhardness is due to the fact that because of high-velocity collisions between the particles and substrate, the high-speed cooling and quenching of the alloy, high strength metastable nanostructure are formed.

(a)

(b)

Figure 1.  The morphology of the PN35T24GF37M4 powder particles after mechanical activation for 0.5 h with a magnification of: (a) ×300; (b) ×30,000.

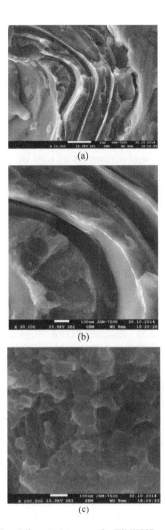

(a)

(b)

(c)

Figure 2. Microstructure of TiNiHfCu coatings obtained by high-speed gas-flame spraying with a magnification of: (a) ×10000; (b) ×30,000; (c) ×100000.

Functional the mechanical properties of surface-modified layers are largely determined by their real structure: the grain size, the presence of impurities, phase composition, structure and level of internal stresses. Therefore, the main challenge in creating new advanced materials is to find out the regularities of the formation of nanostructured surface layers made of materials with SME, their phase-structural states and features of functional and mechanical behavior. Studying the properties of nanocrystalline coatings is a difficult task because of the various factors that affect their characteristics. It is impossible to solve this task without a detailed analysis of the structure and chemical and phase composition. This study discusses some of the problems mentioned above

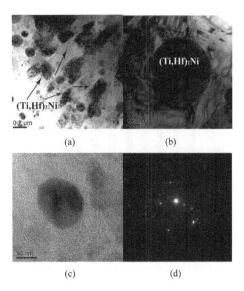

(a)                              (b)

(c)                              (d)

Figure 3. The Microstructure of TiNiHfCu with phase inclusions of (Ti, Hf)₂Ni subjected to high-speed gas-flame spraying with local protection: (a) magnification of ×100000; (b) magnification of ×150,000; (c) tweed contrast shown by electron microscopy; (d) micro-electron diffraction.

and their solutions in relation to the formation of layers using nanostructured materials with SME on the surface of steels.

Studies of the functional and mechanical properties of steel 1045 samples coated with TiNiHfCu was carried out using a bending cycle fatigue test with the rotation of an MIE-6000 drive in accordance with State Standard 19533-74 and 25.502-79. The test results of the samples after coating with TiNiHfCu and without surface modification in air (Fig. 5) show that the surface modification slows down damage accumulation and increases fatigue limit of steel 1045 in air by 36.3%. Pseudoelasticity effect lay at the heart of improving the durability of the material with an SME surface layer based on TiNiHfCu under the cyclic loading.

Improvement in steel alloy durability by surface modification with TiNiHfCu can be explained by the special mechanism of deformation. At relatively low cycle stress, the dominant deformation mechanism in the surface layer is the mechanism where at partial sample unloading deformation significantly reduced. In this part, there may be "healing" of formed defects characteristic for materials with SME. The mechanism of "healing" consists in change of stress field near microconcentrator, which brings to either consistent reorientation of martensite plates due to the stress or to reverse transformation and emergence of martensite in a new place. The structure thus adapts to external influence, preventing the emergence of cracks.

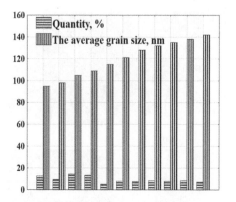

Figure 4. Quantitative distribution of grain size and their percentage in the TiNiHfCu coating.

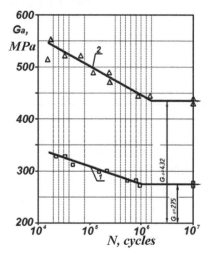

Figure 5. Cycle fatigue curves of steel 1045: uncoated (1), after surface modification with SME alloy TiNiHfCu (2).

## 3 CONCLUSIONS

Complex metallophysical research of surface modified layers (electron microscopy and X-ray analysis) gave new information about nanostructured composition of the surface-modified layer, its mechanical properties, phase composition, which determine the functional properties and allow to find methods of purposeful formation for different operating conditions. The studies found that the complex method of forming the surface-modified layers using materials with SME, which includes high-speed gas-flame spraying of mechanically activated powders based on TiNiHfCu followed by thermal and thermomechanical treatment, made it possible to form surface layers with nanostructured

state having high levels of functional, mechanical, and performance properties. It was established experimentally that after high-speed gas-flame spraying of mechanically activated TiNiHfCu-based powder with SME, the cyclic durability in high cycle fatigue increased by ~ 30–40%.

## ACKNOWLEDGEMENT

This work was performed as part of State Task № 9.555.2014/K, with the financial support of the Ministry of Education and Science of the Russian Federation and President Grant number MK-5017.2014.8.

## REFERENCES

Blednova, Z.M. & Rusinov, P.O. 2014. Mechanical and Tribological Properties of the Composition "Steel—nanostructured Surface Layer of a Material with Shape Memory Effect Based TiNiCu". *Applied Mechanics and Materials*. 592–594: 1325–1330.

Blednova, Z.M. Rusinov, P.O. & Stepanenko, M.A. 2014. Influence of Superficial Modification of Steels by Materials with Effect of Memory of the Form on Wear-fatigue Characteristics at Frictional-cyclic Loading. *Advanced Materials Research*. 915–916: 509–514.

Dalle, F. Pasko, A. Vermaut, P. Kolomytsev, V. Ochin, P. & Portier, R. 2000. Melt-spun ribbons of Ti-Hf-Ni-Re shape memory alloys: first investigations, *Scripta Materialia*. 43(4): 331–335.

Karabasova, J.S. 2002. *New materials*. Moscow: MISA. 736.

Liakishev, N.P. & Alymov, M.I. 2006. Nanomaterials for constructional purposes. *Russian nanotechnology*. 1–2: 71–81.

Liang, X.L. Cheng, Y. Shen, H.M. Zhang, Z.F. & Wang, Y.M. 2001. Thermal cycling stability and two-way shape memory effect of Ni–Cu–Ti–Hf alloys, *Solid State Communications*. 119(6): 381–385.

Likhachev, V.A. 1997–1998. *Materials with shape memory effect: Reference Edition*, St. Petersburg: Publishing NIIH St. Petersburg State University. 1–4.

Meng, X.L. Cai, W. Lau, K.T. Zhao, L.C. & Zhou, L.M. 2005. Phase transformation and microstructure of quaternary TiNiHfCu high temperature shape memory alloys, *Intermetallics*. 13(2): 197–201.

Meng, X.L. Tong, Y.X. Lau, K.T. Cai, W. Zhou, L.M. & Zhao, L.C. 2002. Effect of Cu addition on phase transformation of Ti–Ni–Hf high-temperature shape memory alloys, *Materials Letters*. 57(2): 452–456.

Rusinov, P.O. & Blednova. Z.M. 2015. Technological Features of Obtaining of Nanostructured Coatings on TiNi Base by Magnetron Sputtering. *Advanced Materials Research*. 1064: 160–164.

Rusinov, P.O. Blednova, Z.M. & Chaevsky, M.I. 2015. Options for Forming of Nanostructured Surface Coatings. *Advanced Materials Research*. 1064: 154–159.

Valiev, P.3. & Alexandrov, I.V. 2000. *Nanostructured materials obtained by severe plastic deformation*. Moscow: Logos. 272.

*Advanced Materials, Structures and Mechanical Engineering – Kaloop (Ed.)*
© *2016 Taylor & Francis Group, London, ISBN 978-1-138-02793-0*

# Synthesis of pure zeolite NaP from kaolin and enhancement of crystallization rate by sodium fluoride

J.T. Li, X. Zeng, R.Y. Chen, X.B. Yang & X.T. Luo
*Department of Materials Science and Engineering, College of Materials, Xiamen University, Xiamen, P.R. China*

ABSTRACT: Zeolite NaP, a Gismondine (GIS) framework structure, was synthesized using kaolin minerals as the only silica and alumina sources under a hydrothermal condition of 180°C for 48 hours. The samples were characterized by Fourier Transform Infrared (FTIR) spectroscopy, Powder X-Ray Diffraction (PXRD), Thermogravimetry Analysis (TGA), Differential Thermal Analysis (DTA) and Field Emission Scanning Electron Microscopy (FESEM). FESEM images of synthesized zeolite NaP showed an octahedron structure with clear diamond edges, and the Si/Al ratio of zeolite was found to be 1.2, as indicated by the elemental analysis with Energy-Dispersive X-ray spectroscopy (EDX). In addition, it was further confirmed that sodium fluoride acted as a structure-directing agent and mineralizer, enhancing the crystallization rate of zeolite during synthesis.

## 1 INTRODUCTION

Zeolites are crystalline aluminosilicate with the fundamental building block of tetrahedral $SiO_4$ and $AlO_4$, with sharing of oxygen atoms giving rise to a three-dimensional network. They are widely used in the industry for catalysis, adsorption, ion-exchange and chemical separation because of their molecular sieve properties.

Zeolite P has a Gismondine (GIS) framework with intersecting channels of 0.31 nm × 0.44 nm in [100]c and 0.26 nm × 0.49 nm in [010] (Albert et al., 1998). It is useful for the removal of toxic and radioactive waste species (e.g. Mg, Ca, Cs, Sr, Ba, Pb and U) from wastewater (Huo et al. 2012, Moirou et al. 2000), or the formation of environmental-friendly detergents (Novembre et al. 2011). Commonly, zeolite P is synthesized by using colloidal silica or sodium silicate as silica sources and aluminum hydroxide or aluminum isopropoxide as alumina sources (Huo et al. 2012, Huang et al. 2010). Bohra S. reported the synthesis of cashew nut-like zeolite NaP powders from agro-waste (Bohra et al. 2013).

Kaolin minerals are abundant and inexpensive, and their main component is kaolinite, which has a layered 1:1 dioctahedral structure with the chemical formula $Al_2Si_2O_5(OH)_4$. It has been widely used in the synthesis of zeolites such as SAPOs (Wang et al. 2010), X (Kovo 2012), A (Chandrasekhar and Pramada, 2008), Y (Kovo et al. 2009), mordenite (Mignoni et al. 2008) and ZSM-5 (Pan et al. 2014, Kovo et al. 2009). Atta reported the preparation of analcime from local kaolin and rice husk ash,

accompanied with zeolite P as impurities (Atta et al. 2012).

To the best of our knowledge, we report in this work for the first time the synthesis of the pure zeolite NaP crystal from kaolin. The addition of NaF as a structure-directing agent can greatly enhance the crystallization of zeolite. The present method proposed for the synthesis of zeolite as one-pot mixing of metakaolin and another chemical precursor is easy and cost-effective.

## 2 EXPERIMENTAL PROCEDURE

### 2.1 Materials

Kaolin was supplied by LongYan Kaolin Clay Co. Ltd. The chemical composition (wt%) used was as follows: $SiO_2$ (49.67%), $Al_2O_3$ (35.00%), $K_2O$ (2.49%), $Fe_2O_3$ (0.52%), $TiO_2$ (0.04%), MgO (0.35%), $Na_2O$ (0.35%), CaO (0.19%) and ignition loss (11.50%). All of the other chemical reagents were commercial sources and of analytical grade without further purification.

### 2.2 Synthesis of zeolite NaP

Metakaolin was obtained by the calcination of kaolin minerals at 800°C for 3 hours. The as-calcined metakaolin powders were mixed with NaOH, NaF and water in appropriate ratios, and NaOH or NaF was added or not according to different experiments. Mixtures were stirred at room temperature for 24 hours to a get a homogeneous gel and then transferred to Teflon-lined stainless steel

Table 1. Three experiments and their synthetic conditions.

| Sample | Kaolin | NaOH | NaF | H₂O |
|--------|--------|------|-----|-----|
| A | 1.000 g | 0.107 g | Null | 20 mL |
| B | 1.007 g | Null | 0.506 g | 20 mL |
| C | 1.009 g | 0.106 g | 0.501 g | 20 mL |

| Sample | pH | Temp. (°C) | Duration |
|--------|------|-----------|----------|
| A | 12.63 | 180 | 48 h |
| B | 9.67 | 180 | 48 h |
| C | 12.44 | 180 | 48 h |

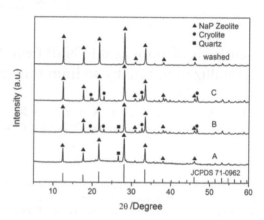

Figure 1. XRD patterns of samples A, B, C and washed sample C.

autoclaves (50 mL capacity). The sealed autoclaves were placed in an air oven maintained at 180°C for 48 hours for the crystallization of zeolite samples. As-prepared samples were filtered, washed with deionized water, and dried at 105°C overnight. Three experiments and their synthetic conditions are summarized in Table 1. The molar ratios of $Al_2O_3$: $SiO_2$: $H_2O$ were equal to 0.42:1:133.87.

### 2.3 Characterization

The powder X-ray diffraction patterns of the samples were carried out on a Bruker-AxsD8 Advance X-ray diffractometer with Cu-Kα radiation (λ = 0.15406 nm), operating at 40 kV and 40 mA. All the diffraction data were collected in the range of 5–60 with the continuous scanning mode. The characteristic vibration bands of the samples were determined by a Thermo Scientific Nicolet iS10 FTIR spectrometer by using the KBr pellet method at a resolution of 4 cm-1 and a scan of 100. The thermal behavior of the samples was studied by Differential Thermal Analysis (DTA) and Thermogravimetry (TG) (Netzsch STA 449 F3 Jupiter®, Germany) from 30°C to 800°C in nitrogen atmosphere at the heating rate of 10°C/min. The morphology of the prepared samples was examined by FESEM (Hitachi SU-70, Japan), operating with an accelerating voltage of 10 kV. The Si/Al ratios of the samples were determined by Energy-Dispersive X-ray spectroscopy (EDX) analysis.

### 3 RESULTS AND DISCUSSION

Figure 1 shows the XRD patterns of samples A, B and C, which were prepared under different conditions. The diffraction pattern of NaP1 zeolite (Baerlocher and Meier, 1972) (JCPDS no. 71-0962, Space group I-4, a = 10.043, b = 10.043) was included for comparison. We can observe that all these samples indicated the characteristic peaks of NaP zeolite, which proved that we had synthesized

NaP zeolite from kaolin minerals successfully. For sample A, the XRD pattern reveals that the powder sample mainly consists of NaP zeolite and a little amount of quartz (2θ = 26.7°), indicating the incomplete crystallization of zeolite. Then, we added NaF instead of NaOH in the experiment to obtain sample B. We can observe from the X-ray pattern of sample B that the zeolite powder still contained some impure components and appeared to be quartz and little amount of cryolite (JCPDS no. 25-0772), which indicated that the crystallinity of NaP zeolite is enhanced compared with sample A. We obtained sample C by mixing both NaOH and NaF with metakaolin. From the XRD pattern, it can be seen that NaP zeolite was synthesized with impure component of cryolite. Moreover, the peaks of quartz (2θ = 26.7°) were not detected in the XRD pattern and the crystallinity of NaP zeolite was the best among the three samples. Interestingly, cryolite was slightly soluble in water and could be removed by washing with a large amount of water, thus we obtained a pure NaP zeolite crystal (Fig. 1).

In order to determine the crystallinity of synthesized NaP zeolite, the washed pure NaP zeolite sample was used as the reference of crystallinity comparison. Herein, the degree of crystallinity of NaP zeolite was defined by utilizing the main X-ray diffraction peak (2θ = 27.7–28.4°). Figure 2 shows the relative crystallinity of as-synthesized zeolite powders, which can be calculated by using the following equation:

$$Crystallinity\ (\%) = \frac{Peak\ area\ of\ product}{Peak\ area\ of\ referent\ sample}$$

(1)

It can be seen from Figure 2 that the crystallinity of the NaP product synthesized without

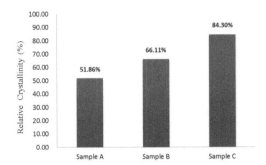

Figure 2. The relative crystallinity of samples A, B and C.

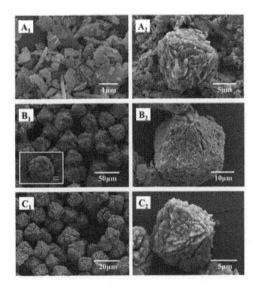

Figure 3. SEM images of samples A, B, and C.

the addition of sodium fluoride was relatively low. When it was added with sodium fluoride, the NaP zeolite with a higher relative crystallinity was obtained. Sodium fluoride acted as a structure-directing agent and mineralizer, which could help to broaden the synthesis methods of zeolite. Compared with samples B and C, the initial large increase in zeolite crystallinity might be related to the dissolution of metakaolin due to the addition of sodium hydroxide.

The morphology of the three samples was studied by using the FESEM technique, as shown in Figure 3. Sample $A_1$ shows the flake and stick morphology, which indicates the existence of incompletely crystallized metakaolin, and there are octahedrons such as particles of zeolite NaP with a diameter about 10 μm (Fig. $3A_2$). Sample B shows a great amount of zeolite particles with uniformly octahedral structures, which indicates the synthesis of a pure zeolite crystal. There still exist impure crystals beside the NaP zeolite (Fig. $3B_1$), which were detected to be cryolite by the elemental analysis with Energy-Dispersive X-ray spectroscopy (EDX) analysis, and the atomic number percentages were F 62.25%, Na 28.31% and Al 9.44% that accord with the cryolite molecular formula $Na_3AlF_6$. Crystal details of NaP zeolite can be observed from a magnifying photo of sample B (Fig. $3B_2$), which shows that NaP zeolite crystallized in little particles and then aggregated to the octahedron structure. There still exist a few uncrystallized impurities among the zeolite crystal particles, revealing the incomplete crystallization of metakaolin. Sample C is fully crystallized and is the purest among the three samples. We can observe a diamond structure with clear crystal edges aggregated to pseudo-octahedron particles, which can be observed by magnifying the figure (Fig. $3C_2$). As mentioned previously, the crystallinity of zeolite NaP increased from sample A to C; this result is in agreement with XRD discussion. The Si/Al ratio of zeolite NaP was found to be 1.2 as detected by

EDX, which was in the range of 1.1–2.5 of typical zeolite NaP; this also conforms to the Si/Al ratio of raw materials, which means that metakaolin is completely converted to zeolite.

In our research, it was suggested that metakaolin could easily convert to zeolite NaP under hydrothermal conditions in alkaline media. Fluoride ions have been considered to act as a structure-directing agent and mineralizer in zeolite synthesis (Egeblad et al. 2007). In the experiment, the addition of sodium fluoride enhanced the crystallization of zeolite NaP. The alkaline environment can also improve the crystallinity of zeolite, and very pure NaP zeolite was also obtained in a contrast experiment of C using ammonium hydroxide adjusting pH value to 12.36 instead of NaOH.

Fourier Transform Infrared Spectroscopy (FTIR) is used to study the chemical environments of Al and Si atoms in the samples. Infrared spectroscopy of sample C, shown in Figure 4, indicates the characteristics absorption bands of NaP zeolite (Zholobenko et al. 1998) at around 420 to 1000 cm$^{-1}$. The absorption bands at around 430–512 cm$^{-1}$ are the bending vibration of internal Al(Si)O$_4$ tetrahedron of NaP zeolite. The bands at around 650–750 cm$^{-1}$ are assigned to the symmetric stretching vibration of Si-O and Al-O bonds, while the peak at around 1000 cm$^{-1}$ is attributed to the asymmetric stretching vibration of Si-O and Al-O bonds. Therefore, the FTIR study confirmed the formation of the pure NaP zeolite crystal.

The TG and DTA curves of sample C are shown in Figure 5. The strongest endothermic peak in the DTA at about 152°C indicated the removal of

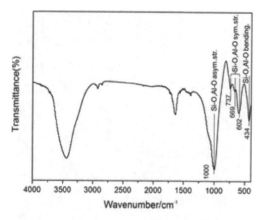

Figure 4. Fourier transform infrared spectra of pure NaP zeolite.

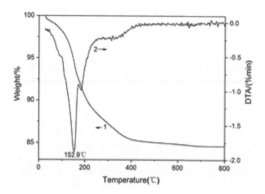

Figure 5. DTA and TG curves of pure NaP zeolite.

absorbed water, which corresponded to a sharp loss of mass by about 15.4% up to 400°C. The total mass loss of sample C is 15.4%, which corresponds to the theoretical value of the NaP zeolite formula $Na_6Si_{10}Al_6O_{32} \cdot 12H_2O$ (the mass loss of water is equal to 16.5%) (Baerlocher & Meier 1972). Practically, there is no obvious mass loss after 400°C, indicating the stability of NaP zeolite up to 800°C.

## 4 CONCLUSIONS

In summary, pure NaP zeolite is synthesized via the hydrothermal method at 180°C for 48 hours. NaP zeolite particles crystallized as the octahedron structure with clear diamond edges. Sodium fluoride acts as a structure-directing agent and mineralizer, and can greatly enhance the crystallinity of zeolite from about 51.8% to 84.3%. Clay mineral kaolin can be directly converted to zeolite, which develops a cheap method in the zeolite industry.

## REFERENCES

Albert, B.R. Cheetham, A.K. Stuart, J.A. & Adams, C.J. 1998. Investigations on P zeolites: synthesis, characterisation, and structure of highly crystalline low-silica NaP. *Microporous and Mesoporous Materials*, 21: 133–142.

Atta, A.Y. Jibril, B.Y. Aderemi, B.O. & Adefila, S.S. 2012. Preparation of analcime from local kaolin and rice husk ash. *Applied Clay Science*, 61: 8–13.

Baerlocher, C. & Meier, W.M. 1972.The crystal structure of synthetic zeolite Na-P 1, an isotype of gismondine. *Journal for Crystallography*. 135(5–6): 339–354.

Bohra, S., Kundu, D. & Naskar, M.K. 2013. Synthesis of cashew nut-like zeolite NaP powders using an agrowaste material as silica source. *Materials Letters*, 106: 182–185.

Chandrasekhar, S. & Pramada, P.N. 2008. Micro-wave assisted synthesis of zeolite A from metakaolin. *Microporous and Mesoporous Materials*, 108: 152–161.

Egeblad, K. Kustova, M. Klitgaard, S.K. Zhu, K.K. & Christensen, C.H. 2007. Mesoporous zeolite and zeotype single crystals synthesized in fluoride media. *Microporous and Mesoporous Materials*, 101: 214–223.

Huang, Y. Dong, D.H. Yao, J.F. He, L. Ho, J. Kong, C.H. Hill, A.J. & Wang, H.T. 2010. In Situ Crystallization of Macroporous Monoliths with Hollow NaP Zeolite Structure. *Chemistry of Materials*, 22: 5271–5278.

Huo, Z.P. Xu, X.Y. Lu, Z. Song, J.Q. He, M.Y. Li, Z.F. Wang, Q. & Yan, L.J. 2012. Synthesis of zeolite NaP with controllable morphologies. *Microporous and Mesoporous Materials*, 158: 137–140.

Kovo, A.S. 2012. Effect of Temperature on the Synthesis of Zeolite X from Ahoko Nigerian Kaolin Using Novel Metakaolinization Technique. *Chemical Engineering Communications*, 199: 786–797.

Kovo, A.S. Hernandez, O. & Holmes, S.M. 2009. Synthesis and characterization of zeolite Y and ZSM-5 from Nigerian Ahoko Kaolin using a novel, lower temperature, metakaolinization technique. *Journal of Materials Chemistry*, 19: 6207–6212.

Mignoni, M.L. Petkowicz, D.I. Machado, N.R.C.F. & Pergher, S.B.C. 2008. Synthesis of mordenite using kaolin as Si and Al source. *Applied Clay Science*, 41: 99–104.

Moirou, A. Vaxevanidou, A. Christidis, G.E. & Paspaliaris, I. 2000. Ion exchange of zeolite Na-P-c with Pb2+, Zn2+, and Ni2+ ions. *Clays and Clay Minerals*, 48: 563–571.

Novembre, D. Di Sabatino, B. Gimeno, D. & Pace, C. 2011. Synthesis and characterization of Na-X, Na-A and Na-P zeolites and hydroxysodalite from metakaolinite. *Clay Minerals*, 46: 339–354.

Pan, F. Lu, X. Wang, Y. Chen, S. Wang, T. & Yan, Y. 2014. Organic template-free synthesis of ZSM-5 zeolite from coal-series kaolinite. *Materials Letters*, 115: 5–8.

Wang, T. Lu, X. & Yan, Y. 2010. Synthesis, characterization and crystallization mechanism of SAPOs from natural kaolinite. *Microporous and Mesoporous Materials*, 136: 138–147.

Zholobenko, V.L. Dwyer, J. Zhang, R.P. Chapple, A.P. Rhodes, N.P. & Stuart, J.A. 1998. Structural transitions in zeolite P—An in situ FTIR study. *Journal of the Chemical Society-Faraday Transactions*, 94: 1779–1781.

*Advanced Materials, Structures and Mechanical Engineering – Kaloop (Ed.)*
*© 2016 Taylor & Francis Group, London, ISBN 978-1-138-02793-0*

# Formation of copper oxide nanostructures by solution-phase method for antibacterial application

C.M. Pelicano, J.C. Felizco & M.D. Balela
*Department of Mining, Metallurgical and Materials Engineering, University of the Philippines, Diliman, Quezon City, Philippines*

ABSTRACT: Copper oxide ($Cu_2O$ and $CuO$) nanostructures were successfully grown on Cu foils at room temperature in aqueous alkaline solution for 12–72 h. The as-grown CuO nanostructures were characterized using X-Ray Diffraction (XRD) and Scanning Electron Microscope (SEM) analysis. The XRD analysis confirmed the formation of highly crystalline copper oxide nanostructures. Copper oxide nano-needles and nanoparticles were both formed on the surface of the substrate after 12 h treatment. When the reaction time was increased up to 72 h, nano-needles assembled into nano-flowers and nano-particles became porous. All of the samples containing copper oxide nanostructures were proven to be effective growth inhibitor of S. *Aureus*.

## 1 INTRODUCTION

Oxidation of copper (Cu) produces two forms of oxides—cuprous oxide ($Cu_2O$) and cupric oxide ($CuO$). $Cu_2O$ is a p-type semiconductor with a direct band gap of 2.17 eV. Being a material with good sensing, catalytic, electric and surface properties, it has found its importance in various fields such as in CO oxidation, photo-catalysis, photochemical evolution of $H_2$ from water, photocurrent generation, antimicrobial agents and solar cells (Gopalakrishnan et al. 2012, Dodoo-Arhina et al. 2010). CuO, on the other hand, is also a p-type semiconductor, but with a narrow indirect bandgap of 1.2 eV (Siddiqui et al. 2014, Zhang et al. 2013). Its superior physical, electrical and optical properties, thermal and chemical stability and low cost production render it usable for a wide variety of applications such as chemical and biosensors, nano-fluids, photodetectors, energetic materials, super-capacitors, superconductors, magnetoresistance materials, solar cells, photo-catalysts and antimicrobial agents (Siddiqui et al. 2014, Zhang et al. 2013, Ekthammathat et al. 2013). The properties of both copper oxides are highly dependent on its structure, particularly on its morphology, size and aspect ratio, thus driving the need for controlling its structure to fit a particular purpose (Siddiqui et al. 2014). Because of this, nanostructured copper oxide materials have currently been gaining interest due to their highly reactive surfaces, which dramatically improves its properties as compared with the bulk material.

There have been numerous efforts in developing synthesis methods of copper oxide nanostructures with varying morphologies—hydrothermal, sono-chemical, solvo-thermal, thermal oxidation, microwave irradiation, solution-phase synthesis and sol-gel method (Zhang et al. 2013, Ekthammathat et al. 2013). Among these, solution-phase synthesis at room temperature is considered to be one of the most promising because of its high efficiency, low cost and potential high quality upscale production (Ekthammathat et al. 2013).

Over the years, it has been established that certain transition metal nanostructures exhibit superior antimicrobial activities such as silver, copper, gold, titanium and zinc (Dizaj et al. 2014). Cu nanostructures, in particular, have been receiving much attention because of its low cost and high abundance. However, due to the inherent low oxidation resistance of Cu, the applicability of Cu nanostructures as antimicrobial agents has been limited. Hence, copper oxide nanostructures have also been investigated, but less commonly explored. In one study, $Cu_2O$ nanoparticles were proven to be effective growth-inhibitors of *Escherichia coli* (Gopalakrishnan et al. 2012). $Cu_2O$ micro/nanocrystals were also reported to be effective against *Bacillus subtilis* (Theja et al. 2014). CuO nanorods were verified to inhibit the growth of Bacillus cereus, Staphylococcus aureus, Escherichia coli and Klebsiella pneumonia (Azam et al. 2012). Flower-shaped CuO nanostructures showed effectiveness as antibacterial agents against Bacillus subtilis, Bacillus thuringiensis, Salmonella paratyphi, Salmonella paratyphi-a, Salmonella paratyphi-b, Escherichia coli and Pseudomonas aeruginosa, as well as antifungal agents against Aspergillus niger, Rhizopusoryzae, Aspergillus flavus, Cladosporium carrionii, Mucor, Penicillium notatum and Alternaria alternate (Mageshwari & Sathyamoorthy 2013). Although the exact mechanism of antimicrobial activity of copper oxide nanostructures is not yet established, its effectiveness against a large number of bacterial species drives a need for developing an efficient, low cost and easily scalable method for synthesizing copper nanostructures for antimicrobial applications.

In this study, flower-like copper oxide nanostructures ($Cu_2O/CuO$) were successfully grown on Cu foils by solution-phase synthesis in alkaline medium at room temperature. The effects of growth time on their morphology and crystallinity were investigated by Scanning Electron Microscopy (SEM) and X-Ray Diffraction (XRD). The antimicrobial activity of the grown CuO nanostructures was also tested against Staphylococcus Aureus.

## 2 EXPERIMENTAL

### 2.1 *Synthesis of $Cu_2O$ and CuO nanostructures*

High purity Cu foils (99.99%, $1 \times 1$ cm) were manually ground using silicon carbide sheets and sonicated in 4 M HCl solution for 15 min to remove surface impurities and native oxide layers. Then, 200 ml deionized water was placed in a beaker, and the solution was adjusted to pH 10 using reagent grade ammonia. The samples were then immersed in this alkaline solution for 12, 24, 48 and 72 hours at room temperature to allow the formation of the copper oxide thin film. Afterwards, the Cu foils were washed using deionized water and annealed in air for 3 hours at 100°C using a hot plate.

### 2.2 *Characterization techniques*

The surface morphologies of the copper oxide nanostructures were examined using scanning electron microscope (JEOL 5300) operating at 25 kV. The composition and crystal structure of the as-grown copper oxide nanostructures were determined by X-ray diffraction (Shimadzu XRD-7000) operated at 40 kV and 30 mA with Cu Kα radiation in the range of 30–80°.

### 2.3 *Antibacterial testing*

The antimicrobial property of the copper oxide nanostructures was tested against Staphylococcus Aureus, a type of Gram-positive bacteria, using the inhibition zone method. The suspending medium used was 0.1% peptone water. The microbial suspension was inoculated in pre-poured Nutrient Agar (NA) plates with a thickness of 3 mm by swabbing the surface of the agar for five times. Four Cu foil samples (12-hr, 48-hr, 72-hr and untreated), with the side containing CuO nanostructures facing down, were then placed directly over the inoculated agar surface. The agar plates were then incubated at 35°C and observed after 24 h. The diameter of the clearing in the agar was measured to determine the inhibition zone.

## 3 RESULTS AND DISCUSSION

### 3.1 *Structural analysis*

The XRD patterns of the resulting CuO nanostructures formed on Cu foil for different lengths of time are shown in Figure 1. Major peaks at diffraction angles (2θ) 34.76°, 35.70° and 37.92° correspond to the (110), (002) and (111) planes, respectively of the monoclinic structure of CuO (JCPDS No. 48–1548). Also, the peaks at 2θ values 43.23° and 50.28° can be attributed to the (111) and (200) planes of face-centered cubic structured Cu substrate (JCPDS No. 04-0836). The intensity of CuO peaks relative to metallic Cu peaks increased with longer reaction time, which denotes an increase in the concentration and crystallinity of CuO on the surface of the foil. Smaller peaks of $Cu_2O$ are also observed at 2θ 44.14° and 64.52°, which are reflections of (200) and (220) planes, respectively.

### 3.2 *Morphological analysis*

The surface morphologies of $Cu_2O/CuO$ nanostructures formed on Cu foil by a solution-phase method at room temperature are shown in Figure 2 (a–d). After 12 h of immersion in alkaline medium, particles with minute needle-like structures covered the Cu substrate as shown in Figure 2a.

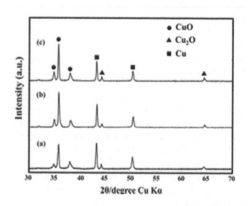

Figure 1. XRD patterns of $Cu_2O/CuO$ nanostructures formed on Cu foils by solution-phase method in alkaline medium for (a) 24 h, (b) 48 h and (c) 72 h.

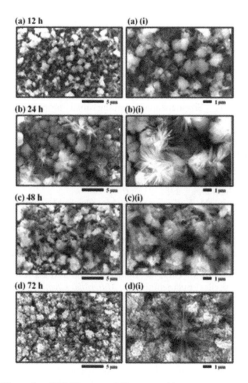

**(a) 12 h**  **(a) (i)**

**(b) 24 h**  **(b)(i)**

**(c) 48 h**  **(c)(i)**

**(d) 72 h**  **(d)(i)**

Figure 2. SEM images of Copper oxide nanostructures formed on Cu foils by a solution-phase method in alkaline medium for (a) 12 h, (b) 24 h, (c) 48 h and (d) 72 h. Images labelled with (a–d) (i) are taken at higher magnification.

The nano-needles have a mean diameter of 50 nm and length of 0.813 μm, while the particles have an average diameter of 0.892 μm. A longer treatment time of 24 hours resulted in the formation of flower-like structures and more uniform particles (Fig. 2b). The average diameter of copper oxide particles increased to about 3 μm. These flower-like structures possibly formed from the accumulation of the initial needle-like structures. Larger micro-flowers were obtained and the porosity of the particles increased upon extended immersion times of 48 h and 72 has demonstrated in Figure 2c and Figure 2d, respectively.

In this study, the formation of CuO nanostructures at room temperature on Cu foil can be divided into two possible major growth mechanisms. First, at the initial stage of synthesis, the Cu substrate discharges $Cu^{2+}$ ions continuously in the solution. These ions would then form the complex $[Cu(NH_3)_4]^{2-}$ immediately upon addition of $NH_3$. At the same time, a fraction of water is deprotonated by ammonia producing ammonium and hydroxyl ions. Finally, the complex reacts with $OH^-$ ions to form $Cu(OH)_2$ nanocrytals.

$$Cu \rightarrow Cu^{2+} + 2e^- \tag{1}$$

$$Cu^{2+} + 4NH_3 \rightarrow [Cu(NH_3)_4]^{2+} \tag{2}$$

$$NH_3 + H_2O \rightarrow NH_4^+ + OH^- \tag{3}$$

$$[Cu(NH_3)_4]^{2+} + OH^- \rightarrow Cu(OH)_2 \tag{4}$$

Another possible mechanism is the direct reaction of $Cu^{2+}$ ions with $OH^-$ ions to produce $Cu(OH)_2$ nanocrystal precipitates.

$$Cu^{2+} + OH^- \rightarrow Cu(OH)_2 \tag{5}$$

$$Cu(OH)_2 \rightarrow^\Delta CuO \tag{6}$$

Furthermore, $Cu(OH)_2$ nanocrystals transformed into CuO through the heat treatment process at 100°C for 3 h in the air (Akhavan et al. 2011). These CuO nanocrystals served as the nuclei for the anisotropic growth of nanoneedles and nanoflowers. On the other hand, a minute concentration of $Cu_2O$ possibly formed from the thermal oxidation process of reacted Cu foils.

### 3.3 Antimicrobial test

After the 24-hr incubation time, the inhibition zones of 12, 24, and 72 h were measured to be 33 mm, 35 mm and 35 mm, respectively. The untreated Cu foil did not produce any clearing zone, which means that it is not an effective inhibitor. It can be observed that with increasing reaction time, the inhibition zone area also increases, as shown in Figure 3. The exact mechanism for the

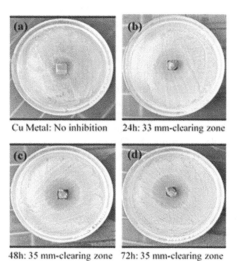

Cu Metal: No inhibition    24h: 33 mm-clearing zone

48h: 35 mm-clearing zone    72h: 35 mm-clearing zone

Figure 3. Antimicrobial activity of copper oxide nanostructures: (a) untreated sample, (b) 24-hr reaction time, (c) 48-hr reaction time and (d) 72-hr reaction time.

205

antimicrobial activity of copper oxide nanostructures has not yet been established.

However, Azam et al. and Ekthammathat et al. both suggested disruption of the vital enzymes of the bacteria through the release of Reactive Oxygen Species (ROS) (Ekthammathat et al. 2013, Azam et al. 2012). First, free $Cu^{2+}$ in copper oxide nanostructures release ROS such as $\bullet O^{2-}$, $\bullet HO_2$, $\bullet OH$ and $H_2O_2$. These ROS penetrate through the outer cell wall of the bacteria, and then further generate free radicals. These radicals cross through the inner cell membrane, and then disintegrate the vital enzymes of the bacteria, leading to cell death (Ekthammathat et al. 2013, Azam et al. 2012).

It was proven in the XRD analysis that the amount of copper oxide increases with increasing reaction time. Also, increasing the reaction time produces porous and more orderly nanostructures, which means that surface area is also increased. Thus, with increasing area of copper oxide in the sample, more $Cu^{2+}$ ions are present in the system. This leads to the availability of more ROS that attacked the pores of the bacterial cell wall. The cell wall of Staphylococcus Aureus, being a gram-positive bacteria, is composed of a thick layer of highly porous peptidoglycan, making it highly susceptible to ROS penetration (Ekthammathat et al. 2013).

## 4 CONCLUSIONS

Copper oxide nanostructures were successfully grown on Cu foils at room temperature under an aqueous alkaline medium for 12–72 h. XRD analysis proved the existence of highly crystalline $Cu_2O/CuO$. The intensity of CuO increased with longer reaction time, which denotes an increase in the concentration and crystallinity of CuO on the surface of the foil. SEM Analysis proved that after 12 h, both nano-needles and nanoparticles are formed on the surface. As growth time proceeds, nano-needles assemble into nano-flowers, while the nanoparticles became porous. All of the samples containing CuO nanostructures were also proven to be effective inhibitors of Staphylococcus Aureus.

## REFERENCES

Akhavan, O. Azimirad, R. Safad, S. & Hasanie, E. 2011. CuO/Cu(OH)₂ hierarchical nanostructures as bactericidal photocatalysts, *Journal of Materials Chemistry*, 21: 9634.

Azam, A. Ahmed, A.S. Oves, M. Khan, M.S. & Memic, A. 2012. Size-dependent antimicrobial properties of CuO nanoparticles against Gram-positive and—negative bacterial strains. *International Journal of Nanomedicine*, 7: 3527–3535.

Christy, A. Nehru, L.C. & Umadevi, M. 2012. A novel combustion method to prepare CuO nanorods and its antimicrobial and photocatalytic activities. *Powder Technology*, 235: 783–786.

Dizaj, S.M. Lotfipour, F. Barzegar-Jalali, M. Zarrintan, M.H. & Adibkia, K. 2014. Antimicrobial activity of metals and metal oxide nanoparticles. *Materials Science and Engineering C*, 44: 278–284.

Dodoo-Arhina, D. Leoni, M. Scardi, P. Garnier, E. & Mittigac, A. 2010. Synthesis, characterisation and stability of $Cu_2O$ nanoparticles produced via reverse micelles microemulsion. *Materials Chemistry and Physics*, 122: 602–608.

Ekthammathat, N. Thongtem, T. & Thongtem, S. 2013. Antimicrobial activities of CuO films deposited on Cu foils by solution chemistry. *Applied Surface Science*, 277: 211–217.

Gopalakrishnan, K. Ramesh, C. Ragunathan, V. & Thamilselvan, M. 2012. Antibacterial activity of $Cu_2O$ nanoparticles on E. coli synthesized from Tridaxprocumbens leaf extract and surface coating with Polyaniline. *Digest Journal of Nanomaterials and Biostructures*, 7: 833–839.

Mageshwari, K. & Sathyamoorthy, R. 2013. Flower-shaped CuO Nanostructures: Synthesis, Characterization and Antimicrobial Activity. *Journal of Materials Science Technology,* 29(10): 909–914.

Siddiqui, H. Qureshi, M.S. & Haque, F.Z. 2014. Structural and Optical Properties of CuO Nanocubes Prepared Through Simple Hydrothermal Route. *International Journal of Scientific & Engineering Research*, 3: 173–177.

Theja, G. Lowrence, R. Ravi, V. Nagarajana, S. & Savarimuthu, P. 2014. Synthesis of $Cu_2O$ micro/nanocrystals with tunable morphologies using coordinating ligands as structure controlling agents and antimicrobial studies. *Cryst EngComm*, 16: 9866–9872.

Zhang, Q. Zhang, K. Xu, D. Yang, G. Huang, H. Nie, F. Liu, C. & Yang, S. 2013. CuO nanostructures: Synthesis, characterization, growth mechanisms, fundamental properties, and applications. *Progress in Materials Science*, 60: 208–337.

*Advanced Materials, Structures and Mechanical Engineering – Kaloop (Ed.)*
© 2016 Taylor & Francis Group, London, ISBN 978-1-138-02793-0

# A method of load extrapolation based on the MCMC and POT

S. You, Y.Q. Wu, Y. Li & Y.S. Zhang
*School of Mechanical Science and Engineering, Jilin University, Changchun, China*

ABSTRACT: The fatigue life of components can be obtained by performing fatigue tests. The load spectrum of a full life cycle is an essential parameter. This paper concentrated on the issue of extrapolating a measured load time history to a longer time period. Through combining the Markov Chain Monte Carlo (MCMC) with the Peak Over the Threshold (POT), a new method was presented to account for the load extrapolation of large cycles and the regeneration of small cycles. The measured load at the axle shaft of a wheel loader was analyzed and extrapolated to a longer time period reasonably. The extrapolated load could be the input of a fatigue test or an FEM fatigue analysis.

## 1 INTRODUCTION

Engineering vehicles, such as excavators and wheel loaders, are widely used in the fields of transportation, mining, and construction. Due to the harsh working environments and bumpy roads, the components of engineering vehicles are often subjected to complicated random loads during their full working life. The fatigue failure is one of the major failure modes of these structures (Wang et al. 2011, Rychlik et al. 1996, Wang et al. 2012). In order to conduct proper fatigue design, it is important to determine the load spectrum that considers real service loads. In general, fatigue life assessments are usually based on the extrapolation of the supposed worst load conditions. Extrapolation of a measured load to a design life has been solved by several authors (Johannesson 2001, Bruni et al. 2003, Moriarty 2012). The simplest method is to repeat the measured load block several times until fatigue failure occurs. This method is disadvantageous because only the measured load cycles will appear in the fatigue test. Johannesson (2006) proposed an efficient extrapolation method to repeat the measured load block, and modified the largest maxima and lowest minima in each block to extrapolate more large cycles than the observed ones. In this method, only the maxima above a high load level and the minima below a low load level are randomly regenerated by the Generalized Pareto Distribution (GPD) based on the POT, but the loads between the thresholds are not processed. Compared with the actual load time history, the repeated load blocks have the lower randomness of the load. Meanwhile, under a given threshold, the exceedances in each load block are same. Because the available loads above a threshold are insufficient, it will lead to a bad fit for the GPD.

This paper focused on the extrapolation of a measured load at the axle shaft of a wheel loader. It would be shown in this case about how a simple extrapolation based on the MCMC and the POT could lead to a better extrapolation in the fatigue life prediction. The load analysis was performed with the WAFO software (Brodtkorb et al. 2000).

## 2 MCMC SIMULATION

In this paper, the experimental measurement of service loadings of a wheel loader was carried out under the condition of spading primary soil. The measured load time history at the axle shaft was illustrated as an example.

### 2.1 *Markov chain model*

If the discrete set $\{X_t\}$ is a random process with finite numbers of elements with state space $E = \{1, 2, \ldots n\}$, $X_t = i$ indicates that the random process is under the state $i$ on time $t$. For each $i, j, t$, it satisfies the Markov condition as follows:

$$P\{X_{t+1} = j \mid X_t = i, X_{t-1} = i_{t-1}, \ldots X_1 = i_1, X_0 = i_0\}$$
$$= P\{X_{t+1} = j \mid X_t = i\} = p_{ijt} \qquad (1)$$

The random process $\{X_t\}$ is called as a Markov chain (Grimmett et al. 1992), where $p_{ijt}$ is the transition probability from the state $i$ at time $t$ to state $j$ at time $t + 1$. If the transition probability $p_{ijt}$ does not depend on time $t$, the Markov chain can be called as a homogeneous Markov chain. Its transition probability is represented as

$$p_{ij} = P\{X_{t+1} = j \mid X_t = i\}, (t = 0, 1, 2 \ldots) \qquad (2)$$

where $0 \leq p_{ij} \leq 1\, i, j \in E; \sum_{j \in E} p_{ij} = 1$.

The load time history $\{X_t\}$ for fatigue analysis can be viewed as a series of turning points, which are constituted by local maxima and minima. These local extremes can be accurately modeled by a Markov chain (Johannesson 1999) as follows:

$$\{X_t\} = \{X_{t_1}, X_{t_2}, X_{t_3}, \ldots\} = \{m_1, M_1, m_2, M_2, \ldots\} \quad (3)$$

where $m_1$, $m_2$ ... represent local minima, $M_1$, $M_2$ ... represent local maxima.

The small load cycles that would not cause fatigue damage should be filtered during the extraction process of turning points. A rule of thumb is to use a threshold range equal to 5% of the global range of the load signal, so the threshold of the rain flow filter $h$ was set as 75 $N.m$, and then the cycles whose amplitude below $h$ were filtered. The measured load possessed 51001 data points while decreased to 490 after the extraction of the turning points and rain flow filter. In practice, in order to establish the Markov model, the load time history is often discretized based on fixed levels. This paper selected 32 levels to discretize the load time history. The cycles of the Markov chain, counted as the Min-Max and the Max-Min transitions, are collected in the Markov matrices. The Markov matrices consist of the frequencies of the Max-Min and the Min-Max cycles:

$$F = [f_{ij}]_{i,j=1}^n, f_{ij} = \Im\{m_k = u_i, M_k = u_j\} \quad (4)$$

$$F' = [f'_{ji}]_{i,j=1}^n, f'_{ji} = \Im\{M_k = u_j, m_{k+1} = u_i\} \quad (5)$$

where $\Im$ represents frequencies; matrix $F$ refers to the transfer cycle from local minimum to local maximum; and matrix $F'$ refers to the transfer cycle from local maximum to local minimum.

### 2.2 Load simulation

It is possible to generate the load time history by simply adding in the local extremes generated from the event probabilities collected in the Markov matrices. With the Metropolis-Hastings sampling method adopted, it is convenient and effective to simulate the load by the Monte Carlo, the main steps are as follows (Grimmett et al. 1992, Johannesson 1999):

i. The Markov matrices $F$, $F'$, are known, calculate the transition probability matrices $P$, $P'$:

$$P(i,j) = p[M_k = u_j \mid m_k = u_i] = \frac{F(i,j)}{\sum_l F(i,l)} \quad (6)$$

$$P'(j,i) = p[m_{k+1} = u_i \mid M_k = u_j] = \frac{F'(j,i)}{\sum_l F'(j,l)} \quad (7)$$

ii. Calculate the cumulative probability matrices $P_{cum}$, $P'_{cum}$, each line of them representing the discrete cumulative distribution function of the next transfer:

$$P_{cum,ij} = \sum_{l=1}^{j} p_{il} \quad (8)$$

$$P'_{cum,ji} = \sum_{l=1}^{i} p_{jl} \quad (9)$$

iii. Calculate the stationary distribution $\pi = (\pi_i)_{i=1}^n$, thus the initial state $m_1 = u_i$ can be obtained:

$$\pi = \pi PP', \sum \pi = 1 \quad (10)$$

iv. Produce a random number in internal (0, 1) by a random generator, then compare with the elements in the row $i$ in $P_{cum}$, if the number is located between the state $j-1$ and state $j$, then $M_1 = u_j$.

v. Produce a random number in internal (0, 1) by a random generator and compare with the elements in the row $j$ in $P'_{cum}$, and then $m_2$ will be simulated; the order of the turning points behind can be simulated in the same manner.

The process of load simulation is shown in Figure 1.

### 2.3 Result analysis

The advantage of a Markov chain is that the rain flow matrix can be computed easily, and in particular, the Markov matrix can be extrapolated, which in turn gives the extrapolated rain flow matrix. Long-term load can be simulated by MCMC, and the scattering characteristic of the simulated load can be decreased through simulating a certain length, thus the simulated load is more approximate to the actual load. First, 1 time simulation was conducted. From Figure 2a, the result was relatively ideal compared with the measured load. However, the results for each new simulation fluctuate severely. This

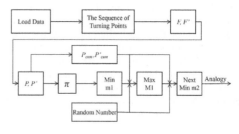

Figure 1. The flowchart of load simulation by MCMC.

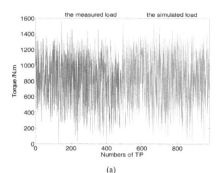

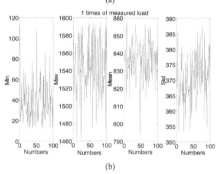

(a)

(b)

Figure 2. Contrasting the measured load with the simulated 1 time; b, Generating 100 new simulations under the 1 time simulation load.

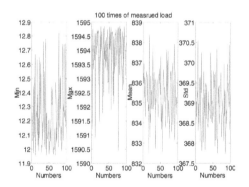

Figure 3. Generating 100 new simulations under the 100 times simulation load.

scatters between different simulations through statistical characteristics (Min, Max, Mean, and Std), as illustrated in Figure 2b. Therefore, it is unreliable that only 1 time simulation is generated.

In order to study how many times the load should be simulated to assure the stability and reliability, $N(10, 20, 50, 100)$ times simulation were also conducted, respectively, and each of them was generated for 100 new simulations. Through analyzing statistical characteristics, under the case of the

Table 1. Statistical characteristics of the measured and 100 times simulation load.

| Load data | Min | Max | Mean | Std |
|---|---|---|---|---|
| The measured load | 11.95 | 1546.89 | 814.17 | 360.62 |
| The simulated load | 12.26 | 1593.82 | 835.45 | 368.85 |

100 times simulation, the change in the results for each new simulation was small and the simulated load was more approximate to the actual load, as shown in Figure 3. The single simulation result can be reliable. The contrasted statistical characteristics of the measured load and 100 times simulation load are summarized in Table 1.

## 3 LOAD EXTRAPOLATION

If the fatigue life analysis and fatigue test need to be performed, it is customary to extrapolate a measured load time history. The simplest method takes the measured load time history as a load block, and then repeats this load block until failure. However, the method failed to obtain the large cycles, which did not appear in the measured load. Especially, large load cycles can be critical for causing most of the fatigue damage. Johannesson improved it by applying the POT extrapolation method based on the extreme value theory to extrapolate the loads both above and below the thresholds.

### 3.1 POT and GPD

POT is a systematic way to analyze the distribution of the exceedances above the high level in order to estimate extreme quantiles outside the range of observed loads. The level should be chosen high enough for the tail to have approximately the standardized form, but not so high that there should remain very few observations above it (Coles 2001). The simplest distribution to fit the exceedances above a level is the GPD with the distribution function:

$$G(x;\sigma,\xi) = \begin{cases} 1-\left(1+\xi\dfrac{x-u}{\sigma}\right)^{-\frac{1}{\xi}}, & \xi \neq 0, x > u \\ 1-exp\left(-\dfrac{x-u}{\sigma}\right), & \xi = 0, x > u \end{cases} \quad (11)$$

where $\xi$ is the shape parameter, which determines the widening degree of the GPD, and the thick tail is paralleled with the increase of $\xi$. When $\xi > 0$, it represents the thick tail distribution and $y \in [0,\infty]$; when $\xi > 0$, it represents the thin tail distribution and $y \in [0, -\sigma / \xi]$; when $\xi = 0$, it represents the exponential distribution.

## 3.2 Extrapolating the measured load

In the process of extrapolation based on the POT, first, $N$ times repetition of the load block was conducted. When selecting a threshold, a default choice is that there are about $\sqrt{N_0}$ exceedances above the threshold, where $N_0$ is the number of cycles in the load block (Johannesson 2006). Under the given threshold, the extreme sample loads were extracted to fit GPD. Because the extrapolated load was $N$ times repetition of the load block, exceedances of each load block were the same after selecting a certain threshold. Thus, only the extreme loads of a load block were available to fit GPD, which will affect the accuracy of the model, as shown in Figure 4a. Furthermore, the maxima above the high load level and the minima below the low load level were randomly regenerated in the extrapolated blocks, but the loads between the thresholds

were not processed. The small cycles were only determined by the measured load, which would lead to less randomness compared with the actual load.

The method used here was to combine the MCMC with the POT for extrapolating the measured load. This paper first used the MCMC to simulate 100 times load, i.e. to generate 100 load blocks, and applied the POT to extrapolate large cycles. In this method, the randomness of the load between the thresholds and the numbers of the exceedances increased after the simulation, and the extreme distribution model of the extreme loads was more accurate, as shown in Figure 4b.

In this case, a 10 times extrapolation was conducted, and the load spectra of the measured load, simulated load, and extrapolated load were compiled, as shown in Figure 5. Compared with the measured

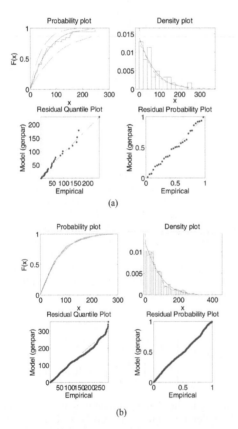

Figure 4. a, Fitting the measured load through GPD under the threshold 1300; b, fitting the simulated load through GPD under the same threshold.

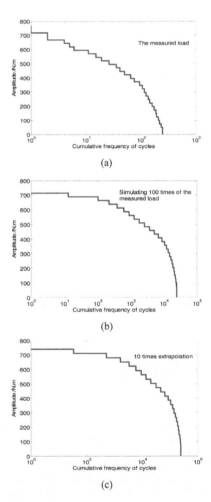

Figure 5. a, Load spectrum of the measured load; b, load spectrum of the simulated load; c, load spectrum of the extrapolated load.

one, the simulated load did not increase the large cycles, but regenerated the small cycles and increased their frequencies. Compared with the simulated one, the extrapolated load increased the amplitudes and quantities of the large cycles, and the frequencies of the small cycles were also increased.

## 4 CONCLUSIONS

If the fatigue life analysis and fatigue test need to be performed, it is essential to efficiently extrapolate the load time history for a longer time, including the generation of small cycles and the extrapolation of large cycles. This paper dealt with the measured load at the axle shaft of a wheel loader by combining the MCMC with POT.

First, the load of 100 times simulation was obtained by MCMC. The loads between the thresholds were regenerated, thus the frequencies of the small cycles increased and the amplitudes of the small cycles were regenerated. Then, a 10 times extrapolation of the simulated load was conducted. The extreme loads both above and below the thresholds were extrapolated, which would result in the generation of the large cycles.

The method proposed for extrapolating the measured load to a longer period of time is proved to work well for the examined load. If the main goal is to generate a longer load block from a measured one, aimed for fatigue tests or fatigue life assessments, an obvious advantage of the method is that a load time history can be obtained directly, but not the amplitude cycles.

## ACKNOWLEDGMENT

This work was supported by the National Natural Science Foundation of China (Grant No. 51375202).

## REFERENCES

Brodtkorb, P.A. Johannesson, P. & Lindgren, G. 2000. *WAFO-a Matlab toolbox for analysis of random waves and loads.* Proceedings of the 10th international offshore and polar engineering conference, 3(4): 343–350.

Bruni, S. Bocciolone, M. & Beretta, S. 2003. Simulation of bridge-heavy road vehicle interaction and assessment of structure durability. *International Journal of Vehicle Design*, 10(1–2): 70–85.

Coles, S. 2001. *An introduction to statistical model-ling of extreme values.* London, Springer Verlag.

Grimmett, G.R. & Stirzaker, D.R. 1992. *Probability and Random Processes.* Oxford University Press, 2nd edition.

Johannesson, P. 2006. Extrapolation of load histories and spectra. *Fatigue and Fracture for Engineering Materials and Structures*, 29: 201–207.

Johannesson, P. & Thomas, J.J. 2001. Extrapolation of rain flow matrices. *Extremes,* 4(3): 241–262.

Johannesson, P. 1999. *Rain flow analysis of switching Markov loads.* Lund University.

Moriarty, P.J. Holley, W.E. & Butterfield, C.P. 2004. Extrapolation of Extreme and Fatigue Loads using Probabilistic Methods, *National Renewable Energy Laboratory.*

Rychlik, I. 1996. Fatigue Stochastic Loads. *Scandinavian Journal of Statistics*, 23(4): 387–404.

Wang, J.X. Hu, J. Wang, N.X. Yao, M.Y. & Wang, Z.Y. 2011. Multi-criteria decision-making method-based approach to determine a proper level for extrapolation of Rain flow matrix. *Proceedings of the Institution of Mechanical Engineers, Part C: Journal of Mechanical Engineering Science.* 226(5): 1148–1161.

Wang, J.X. Wang, N.X. Wang, Z.Y. Zhang, Y.S. & Liu, L. 2012. Determination of the minimum sample size for the transmission load of a wheel loader based on multi-criteria decision-making technology. *Journal of Terramechanics*, 49(3): 147–160.

*Advanced Materials, Structures and Mechanical Engineering – Kaloop (Ed.)*
© 2016 Taylor & Francis Group, London, ISBN 978-1-138-02793-0

# A planar four-bar mechanism demonstrator with variable bar length and connecting rod curve drawing

L.N. Gao, H.Y. Tang & C. Yang
*School of Mechanical Engineering, Chengdu University, Chengdu, Sichuan, China*

ABSTRACT: Aiming at limitation of the teaching device of planar four bar mechanism, a portable telescopic demonstrator of the planar four bar mechanism has been presented in this paper, which is composed of a rack frame, a connecting rod, a fixed length connecting frame rod and a variable length connecting frame rod. By changing the length relationship between the rods, the three basic types of planar four bar mechanism can be achieved. The demonstrator has the functionality of good portability, magnetic adsorption, variable rod length, connecting rod curve drawing and direct demonstration effect.

## 1 INTRODUCTION

Most of the common engineering mechanism is planar mechanism, the planar linkage mechanism is one of the important types and the most common is the four-bar mechanism. The four-bar mechanism is not only applied in a wide range, it is also the foundation of composing the multi-bar linkage. When one rod of the planar four-bar mechanism selected as rack is fixed, it is called the connecting frame rod directly connected to the rack, and that is not connected with the frame is called the connecting rod. If the rod in the four-bar mechanism can be able to do the whole week rotating, it is known as a crank. If the rod in the four-bar linkage can swing back and forth within a certain range of angles, it is called rocker. According to whether the connecting frame rod can do the whole week rotation, the planar four-bar mechanism can be divided into three basic forms, namely crank-rocker mechanism, double-crank mechanism and double-rocker mechanism, which is mainly transformed through the change of the rod length and relationship among the rod length.

Teaching the planar four-bar mechanism is a very important sector in kinematic of machinery. The current teaching equipment of the four-bar mechanism is mainly wooden and fixed rod length structure, or demonstrated through the computer animation soft-ware, that have certain limitations in aspects of suit-able space, the observed effect, a comprehensive functionality.

This paper presents teaching demonstrator that have many functions such as good portability, magnetic adsorption, variable rods length and connecting rod curve drawing. As the lengths of rack, one of the connecting frame rod and the connecting rod are adjustable and marked in scale, the demonstrator

is easy to implement three basic planar four-bar mechanisms in the form of evolution. The back of the rack is equipped with several magnetic blocks, and the demonstrator can be adsorbed on some magnetic objects such as a magnetic blackboard, that makes it very suitable for demonstration and observation of on-the-spot teaching. The chalk cover assembly on the connecting rod enables to draw the connecting rod curve, which further expands the teaching functionality of the planar four-bar linkage. As the use of light-weight material such aluminum or plastic, it is easy to carry.

## 2 GENERAL STRUCTURE

This paper presents a teaching device of planar four-bar mechanism that can demonstrate a variety of planar four bar linkage. It has a convenient conversion and good demonstration effect. As shown in Figure 1, the plane four-bar mechanism demonstrator is composed of rack body, connecting frame rods and connecting rod. There are two connecting frame rods, one is the fixed length connecting frame rod, the other is the variable length connecting frame rod. One end of the fixed length connecting frame rod is sliding and coupled to the rack frame, the other end is hinged with the connecting rod. One end of the variable length connecting frame rod is fixedly connected with the rack frame, the other end is sliding and hinged on the connecting rod. The chalk sleeve assembly is installed on the point of the connecting rod located outside of the hinged point of the variable length connecting frame rod and connecting rod. Several magnetic blocks are stuck to the back of the rack frame. The variable length connecting frame rod is comprised

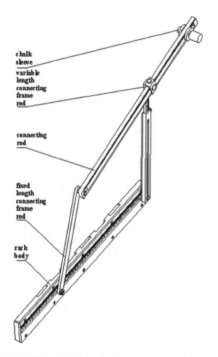

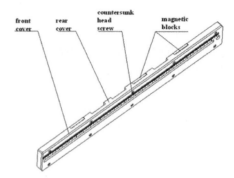

Figure 2. Working principle of rack body.

Figure 1. The general structure of the planar four-rod mechanism demonstrator.

of an inner rod and a rod body, and the inner rod is sliding and mounted on the rod body. The front of the rack frame, the connecting rod and inner rod of the variable length frame connecting rod are marked on the scale. The chalk sleeve assembly is comprised of a hollow stud, an adjusting nut and a fixing nut that are fitted on the connecting rod. When matching the adjusting nut with the hollow stud, a piece of chalk can be arranged at the end of hollow stud, and the fixing nut is screwed on the other end of the hollow stud in which is provided with a spring to make the chalk touch the surface of blackboard to draw the connecting rod curve.

## 3 WORKING PRINCIPLE OF EACH COMPONENT

### 3.1 Rack body

As shown in Figure 2, the rack body assembly consists of a front cover, a rear cover, countersunk head screws and magnetic blocks. The front cover with a rectangle hole and the scale marks is connected to the rear cover with a rectangular groove through ten countersunk head screws. As the magnetic blocks are stuck to the grooves at the back end surface of the back cover, the planar four bar mechanism demonstrator can be adsorbed on the magnetic blackboard for the scene teaching demonstration in classroom.

### 3.2 Fixed length connecting frame rod

As shown in Figure 3, the fixed length connecting frame rod comprises a rod body, a slider, balls, a joint pin, a pin sleeve, a gasket and a connecting nut. After the joint pin is passed through the slider and the front cover of rack body, the pin sleeve and the rod body are placed on the joint pin, and then through the gasket and the connecting nut, the components of the fixed length connecting frame rod are connected together. The joint pin and the slider are placed into the rectangular groove and hole of the rack body before the various components of the rack body are mounted together. In order to reduce the wear, the upper and lower of the slider end are cut with circular grooves and put into the balls, and when the slider moves in the rack body, the frame length changes.

### 3.3 Connecting rod

As shown in Figure 4, the connecting rod is composed of a rod body, a connecting pin, a pin sleeve, a sliding sleeve, a fastening screw and two retaining rings. One end of the rod body is connected to the end of the fixed length connecting frame rod through the pin, the pin sleeve and the two retaining rings. The sliding sleeve slips over the rod body, and when it slides to some position, the corresponding connecting rod length can be obtained by using the fastening screw. The front end of the rod body is provided with the scale marks, so that the length of the connecting rod can be read.

### 3.4 Variable length connecting frame rod

As shown in Figure 5, the variable length connecting frame rod comprises an inner rod, a rod body, a rear cover, countersunk head screws, a screw of limiting position, a fastening screw, two connecting pins, a pin sleeve, a retaining ring and two shaft retainers. The inner rod is arranged in the rectangular groove of the rod body and the rod body is connected with the rear cover through countersunk head screws. The inner rod is provided with the scale marks, the

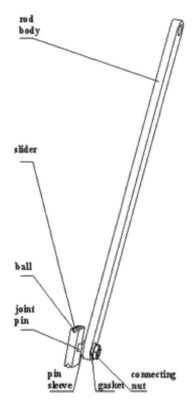

Figure 3. Working principle of fixed length connecting frame rod.

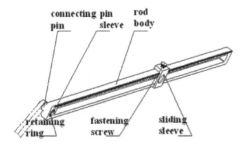

Figure 4. Working principle of connecting rod.

rod body has the "runway" shaped observation hole. When the inner rod slides to a certain position, it can be fixed by using the fastening screw and the length can be read. The screw of limiting position in one end of the inner rod can prevent the inner rod leaving from the rod body. The upper end of the inner rod is connected to the sliding sleeve of the connecting rod through the connecting rod pin, the pin sleeve and the shaft retainer. The lower end of the rod body of the variable length connecting frame rod is hinged on the rack body through another connecting pin and shaft retainer.

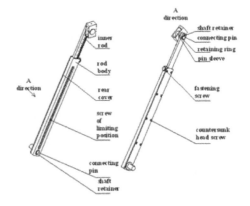

Figure 5. Working principle of variable length connecting frame rod.

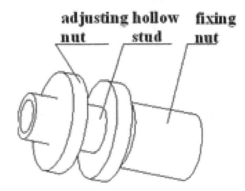

Figure 6. Working principle of chalk sleeve.

### 3.5 Chalk sleeve

As shown in Figure 6, the chalk sleeve mounted on some point of the connecting rod that is located outside of the hinged point of the variable length connecting frame rod and connecting rod, is composed of a hollow stud, an adjusting nut and a fixing nut. The hollow stud is passed through the connecting rod and fixed on the connecting rod by the adjusting nut and the fixing nut. A piece of chalk can be put into one end of the hollow double head stud, and the spring is arranged inside fixing nut that is screwed into another end of the double head stud. The chalk can be put onto the magnetic blackboard to draw a connecting rod curve of four bar mechanism under the current bar lengths relationship.

## 4 MECHANISM DEMONSTRATION

### 4.1 Design demonstration

When meeting the conditions that the combined length of the shortest rod and longest rod is less than

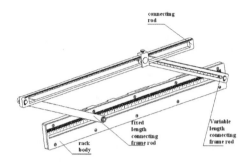

Figure 7. The parallel double crank mechanism when meeting some conditions.

Figure 8. The prototype system of portable planar four-bar mechanism demonstrator.

or equal to the combined length of the remaining two rods and at the same time, one of the adjacent rods connected to the shortest rod is rack, the crank rocker mechanism can be demonstrated. When meeting the conditions that the combined length of the shortest rod and longest rod is less than or equal to the remaining two rods combined length and at the same time, the shortest rod is not connected to the rack, the double rocker mechanism can be demonstrated. As shown in Figure 7, when meeting the conditions that the combined length of the shortest rod and longest rod is equal to the remaining two rods combined length, the lengths of the connecting rod and rack is equal, and the two connecting frame rods are shorter, the parallel double crank mechanism can be demonstrated.

### 4.2 Prototype system

According to the design scheme, a prototype system of portable planar four-bar mechanism demonstrator (as shown in Fig. 8) has been developed in which the non-standard components are made of aluminum

alloy. Under satisfying the conditions of crank rocker mechanism, the connecting rod curve has been drawn on the practical magnetic blackboard.

## 5 CONCLUSION

This paper puts forward a kind of planar four-bar mechanism demonstrator, which comprises a crack body, a connecting rod, a fixed length connecting frame rod and a variable length connecting frame rod. The chalk sleeve is installed on the connecting rod located outside of the hinged point of the variable length connecting frame rod and the connecting rod. A piece of chalk can be put into the chalk sleeve to draw the connecting rod curve. The whole device can be adsorbed on the magnetic blackboard as the magnetic blocks stuck to the back of the rack. The device can demonstrate a variety of planar four-bar mechanism and be used conveniently. It has a good demonstration effect. As the rack structure is unitary, and the two connecting rods located at the same side of the rack, the double crank mechanism cannot be demonstrated in general case, the follow-up work can consider changing the frame structure into the detachable or sliding type, and at the same time, consider the relative position of the connecting frame rods to the rack, in order to improve the double crank mechanism demonstration. In addition, the motor can be considered to add on the active rod to realize automatic operation demonstrator.

## REFERENCES

Chen, J. Ge, W.J. Wang, J.Q. & Zhang, S.F. 2010. Parametric Design and Kinematic Simulation of Plane Four-bar Mechanism. *Manufacture Information Engineering of China,* 39(9): 23–30.

Fang, X.P. & Wen, X.Z. 2012. Research on Simulation Experiment Platform of Plane Four-bar Linkage based on ADAMS. *Journal of Mechanical Transmission,* 36(10): 25–27.

Four-bar linkage—Wikipedia, the free encyclopedia. Information on http://en.wikipedia.org/wiki/Four-bar_linkage.

Four Bar Linkage and Coupler Curve. Information on http://www.mekanizmalar.com/fourbar01.html.

Li, Y.F. & Lu, X.Z. 2013. Design and Kinematic Analysis of Four-bar Mechanism Based on Pro/E. *Agricultural Equipment & Vehicle Engineering,* 51(12): 53–55.

Lin, L.S. 2014. Kinematics Simulation of Four Bar Linkage Mechanism Based on MATLAB/Simulink. *Mechanical Engineer,* (7): 179–180.

Yu, H.Y. Zhao, Y.W. & Xu, D.M. 2015. A path synthesis method of planar hinge four-bar linkage. *Journal of Harbin Institute of Technology,* 47(1): 40–47.

*Advanced Materials, Structures and Mechanical Engineering – Kaloop (Ed.)*
© 2016 Taylor & Francis Group, London, ISBN 978-1-138-02793-0

# Influence of dynamic force on the vibration of the micro-machine tool spindle

M. Guo, B. Li & J. Yang
*Advanced Manufacturing Center of Donghua University, Shanghai, China*

S.Y. Liang
*Georgia Institute of Technology, Atlanta, USA*

ABSTRACT: The micro-scale spindle of the micro-machine plays a pivotal role in determining the performance and reliability. The stiffness of the micro-machine tool spindle is somewhat lower and the stability of the micro-machine is very weak because of the slender features. Thus, the dynamic force would easily excite the tool vibration in a high frequency range, and an understanding of the dynamic behavior is required in order to develop the machine tool and improve the quality of machining parts. This paper developed an FE model considering bearing dynamic stiffness and experimental damping. The influence of dynamic force on the spindle is analyzed based on the finite element model and the simplified dynamic force model. Then, the tool dimension and tool material is studied for vibration reduction. The results show that tool development is the effective way to control the machining vibration. Recommendations for future work on vibration control issues are provided.

## 1 INTRODUCTION

Highly accurate miniaturized components are increasingly in demand for various industries such as aerospace, electronics and automotive. Micro-machining is an ultra-precision material removal process to achieve micro-accuracy and nanometers finish (Liang 2006.). Micro-machine tool is developed for the manufacture of micro-features. And due to the small size of the machine components, the vibration excitation is easier than the traditional machine. The dynamic characteristics of the machine tools and process are critical in the applications of micro-machining technology. Understanding of the dynamic response of the micro-system is critical prior to reducing vibration.

In order to evaluate the dynamics of either macro- or micro-machine tool system, the dynamic response of the tool should be identified. And both the dynamic of the machine tool and the force should be studied in the field. The knowledge of the macro-mechanical machine and force has been well researched and established during last several decades. Guo et al. (2012) studied the dynamic characteristics of the grinding machine spindle system based on experimental modal test techniques and rotating machinery signature techniques. Zhang et al. (2005) developed a nonlinear dynamic model to investigate the dynamic characteristics of the grinding process, and revealed the relationship between grinding force

variations and vibration frequency, thus expanding the opportunities of vibration control of grinding machines. Furthermore, Wang et al. (2010) identified the characteristic twin peaks in the frequency domain with the technique of power spectrum density analysis based on the measurement of cutting force signals, and presented a theoretical and experimental investigation of the influence of tool-tip vibration on surface generation in single-point diamond turning. Besides, Alfares & Elsharkawy (2000) studied the changes in the grinding wheel wear flat area and the work-piece material to analyze the vibration behavior of the grinding spindle by the effects of dynamic changes in the grinding force components.

For micro-mechanical machining, Uhlmann et al. (2013) presented both analytical and experimental studies of the process stability of micro-milling, and carried out experimental modal analyses of the machine tool structure as well as of the end mill. A comprehensive model including the machine tool dynamics as well as the process forces was also developed in the research. Filiz & Ozdoganlar (2010a, b) derived a drilling-dynamics model and experimentally validated it by applying natural frequencies and mode shapes from the model. They presented separately for macro- and micro-drills models and analyzed the effect of geometry diameter, aspect ratio, and helix angle and axial force on the natural frequencies of macro- and micro-drills. However, the micro dynamic issue is still a hotspot research domain, with many unknown issues that need to be explored (Miao et al. 2007).

A great deal of precious work has been accomplished in the field of dynamics of macro- and micro-machine. However, more attention needs to be paid to the influence of dynamic force on the micro-scale spindle and tool. In this work, the finite element model of the micro-machine is first built based on the machine structure and experimentally validated by comparing natural frequencies. The FE model is updated by the damping factor and bearing dynamic stiffness. Then, the dynamic force of the process is studied and simplified, and the dynamic response of the process force is analyzed. With different tool dimensions, Young's modulus in the model, the vibration response amplitude of dynamic force is compared, and the vibration reduction method is studied. Subsequently, the paper concluded the effective methods for vibration reduction and control.

## 2 FINITE ELEMENT MODEL BASED ON EXPERIMENTAL RESULTS AND DYNAMIC STIFFNESS

### 2.1 Original finite element model

To analysis the influence of dynamic force on the micro-machine tool spindle system, a finite element model is developed in this work. The FE model of the spindle system is modeled according to the following procedures:

a. Some small features that would not affect the system modal stiffness and modal mass, for example small holes, shaft shoulders and angle of chamfers, are neglected in this model.
b. The bolt and screw property are specified as the property of the bar element.
c. The clearance fit property in the spindle is modeled as a contact connection with the negative clearance value, and both parts are linked together in the tangential directions.
d. As bearings, the spider element is applied, with one center node representing the inner ring of bearing. The outer ring is connected with it by treating the balls of bearing as springs in the X, Y and Z directions.
e. The spindle system is DOFs restrained to the machine tool.
f. The material of the shaft is 42 CrMo and 45 steel, which are, respectively, used for the spindle box, bearings and other parts.

The FE model of the spindle system is depicted in Figure 1. Because the property of the spindle and the tool is on the micro-scale, the FEM meshing size, especially the tool, should be small to get accurate results. The model, with 91280 nodes and 612690 degrees of freedom, is depicted in Figure 1.

Figure 1. FE model of the micro-machine spindle system.

Figure 2. Experimental setup of impact testing.

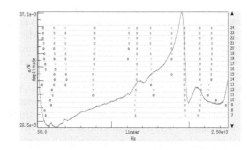

Figure 3. Stabilization diagram of modal analysis.

Moreover, the dynamic characteristics of the spindle system were then studied using Nastran software.

### 2.2 Experimental modal analysis

Machine tool spindle is applied to rotate the tools (drilling, milling and grinding) or workpiece (turning) and transmit the required energy to the cutting zone for metal removal (Abele et al. 2010). The micro-machine tool spindle with a high rotating speed requirement is a complicated vibration system with an almost infinite number of degrees of freedom. After the simplified representation of the machine tool spindle, the impact modal test is applied on the spindle system, as shown in Figure 2. The FRFs obtained in the analysis were studied using the LMS Test Lab software. The PloyMAX, which is a broadband modal identification algorithm developed in recent years (Peeters et al. 2004), is applied to the experimental data to achieve the modal parameters.

In the stabilization diagram of Figure 3, the frequency bandwidth of 50–2500 Hz is applied. And the estimate results of each processing are shown in the stabilization diagram, where the symbols have the following meanings: 'o', new pole; 'f', stable frequency; 'd', stable frequency and damping; 'v', stable frequency and eigenvector; and 's', all criteria stable. Also, the figure shows the sum of the FRFs and the model order in estimation. It is indicated that the modes in the frequency range can be identified based on the stabilization diagrams. And the sum of FRFs also shows the dynamic characteristics of the spindle. Both the EMA and FEM results of the non-rigid modal frequency and damping factors are summarized in Table 1. The modes of EMA and FEM are in acceptable agreement, which in other way indicate that the FE model can be applied to analyze the engineering problem.

## 2.3 Dynamic stiffness of the bearing

Dynamic properties, especially the stiffness of the ball bearing, vary during the operation. So, in the model, the dynamic stiffness is calculated using the implicit differentiation method (Peeters et al. 2004). Applying the bearing parameters in Table 2, the speed-varying stiffness of bearings in the spindle is determined, as given in Table 3.

## 2.4 Dynamic force simplification

In the machining, there are both static and dynamic forces between the spindle and the workpiece. In view of kinematics, the static force leads to the deformation of the tool, while the dynamic force contributes to the vibration of the tool. By filtering the experimental data, the dynamic force can be obtained. And the machining force can be divided into the axial force, the shearing force, and radial force. Due to the slender feature of both the machine and the tool, radial stiffness is much lower than that of the traditional machine. In this paper, the radial force is studied to analyze the radial vibration. It is known that the main frequency of the dynamic force represents the spindle speed (Klocke et al. 2011, Transchel 2012). Due to the high rotating speed, the cutting force magnitudes are always very low (about 1 N) in our test. Therefore, in this paper, to analyze the influence of the dynamic on the spindle system, the force magnitudes obtain a common value of 1 N.

Table 1. The modal results of EMA and FEM.

| | FEM | EMA | | FEM-EMA | |
| Order | Modal frequency [Hz] | Modal frequency [Hz] | Damping factor [%] | Modal shape description | Error [%] |
|---|---|---|---|---|---|
| 1 | 695.6 | 767.3 | 17.32 | X Deflection in the X direction | 9.34 |
| 2 | 1221.6 | 1220.2 | 0.29 | Waving and Bending in the Y direction | 0.11 |
| 3 | 1315.2 | 1327 | 4.66 | Waving and Bending in the X direction | 0.89 |
| 4 | 1445.5 | 1500.2 | 2.57 | Waving in the Z direction | 3.65 |
| 5 | 2162.7 | 1938.2 | 2.54 | Bending in the Y direction | 11.58 |
| 6 | 2203.6 | 2136.6 | 5.58 | Bending in the X direction | 3.14 |
| 7 | 2238.9 | 2289.2 | 9.23 | Deflection and Bending in the Y direction | 2.20 |

Table 2. The parameters of bearing.

| Parameters | Values |
|---|---|
| Density of raceway: $\rho 1$ | 7.8 kg/m³ |
| Density of rolling element: $\rho 2$ | 3.19 kg/m³ |
| Number of rolling elements: Z | 11 |
| Rolling element diameter: D | 3.8 mm |
| Contact angle: a | 15° |
| Pitch diameter: dm | 21.5 mm |
| Outer raceway groove curvature radius coefficient: $f_o$ | 3.619 mm |
| Inner raceway groove curvature radius coefficient: $f_i$ | 3.619 mm |

Table 3. The speed-varying stiffness of bearings.

| Rotating speed (RPM) | Radial stiffness [106 N/m] | Axial stiffness [106 N/m] |
|---|---|---|
| 0 | 11.5 | 5.6 |
| 20000 | 10.98 | 5.34 |
| 40000 | 9.82 | 5.551 |
| 60000 | 8.53 | 3.981 |
| 80000 | 8.65 | 3.534 |
| 100000 | 9.15 | 3.244 |
| 120000 | 9.89 | 3.218 |

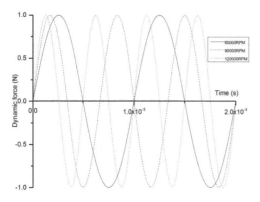

Figure 4. The simplified radial dynamic force model of the different rotating speeds in machining.

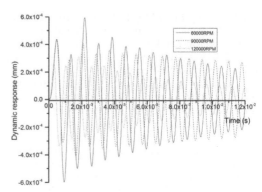

Figure 5. Tool dynamic response of the force.

Table 4. Vibration reduction method comparison.

| Rotating speed (RPM) | Original Value $10^{-4}$ mm | Method 1* Value $10^{-4}$ mm | Ratio [%] | Method 2* Value $10^{-4}$ mm | Ratio [%] | Method 3* Value $10^{-4}$ mm | Ratio [%] |
|---|---|---|---|---|---|---|---|
| 60000 | 6 | 4.4 | 26.7 | 4.9 | 18.3 | 5.6 | 6.7 |
| 90000 | 4.4 | 2.5 | 43.2 | 3.7 | 15.9 | 4.1 | 6.8 |
| 12000 | 3.4 | 2.4 | 29.4 | 2.6 | 23.5 | 3.2 | 5.9 |

*Ratio: Decrease rate of vibration amplitude. 1. Tool shank length reduced by 25%. 2. Tool shank diameter increased by 25%. 3. Young's modulus increased by 25% (applying the tungsten carbide alloy).

And the main frequency characteristic is applied in the force model. The simplified radial force load is shown in Figure 4.

## 3 INFLUENCE OF DYNAMIC FORCE ON THE VIBRATION OF THE SPINDLE SYSTEM

### 3.1 Dynamic response

The original FE model is updated with the experimental damping factor result and the speed-varying stiffness of bearing. And the dynamic force is loaded at the end of the micro-tool. For the machine multi-degree system, the equation of motion can be given by

$$[M]\ddot{x} + [C]\dot{x} + [K]x = F \qquad (1)$$

where [M] is the mass matrix; [C] is the damping matrix; [K] is the stiffness matrix; F is the dynamic force; and x is the displacement, with $\dot{x}$ $\ddot{x}$ being velocity and acceleration.

For excitation of the force in one period, the modal-based transient response of the tool is shown in Figure 5. It indicated that even though the micro-machining force is very small, the dynamic response of the tool seems to be unacceptable to meet the accuracy requirement.

### 3.2 Vibration reduction method

In the mechanical analysis, the deflection of the tool depends on the length, cross-sectional, and the material. To reduce the vibration amplitudes under the dynamic force, the tool is developed by reducing the length, increasing the cross-section and using higher Young's modulus material. Table 4 compares the vibration excited by the dynamic force in different methods; both the vibration amplitude values and the ratio related to the original ones are given.

It is indicated that vibration is greatly cut down related to the original vibration amplitude based on the FEM. All the methods are effective for vibration reduction in this study. And it is obvious that the combination of all the methods would get a better result. However, the tool development still has a limitation, to meet the ultra-high precious requirement; the effect should be made to actively control the vibration between the tool and the workpiece.

## 4 CONCLUSIONS

In this study, the modal analysis of the micro-machine spindle is conducted based on the FEM and EMA. The result of modal parameters obtained an acceptable match, which indicated that the FE modal can be developed to study the dynamic response. Then, the damping factor from the test and the dynamic stiffness model of the bearing is applied in the modal, and the simplified dynamic force in the process is imported as the excitation load. The dynamic response of the tool in the typical speed (60000, 90000, 120000 RPM) is calculated. The results show that even a very low force can cause vibration in the micro-machine.

Furthermore, to reduce the vibration and to improve the machining quality, this paper analyzed three possible ways to develop the tool: reducing the length, increasing the cross-section and using higher Young's modulus material. The results show that development of the micro-tool is an effective way to reduce the vibration caused by the dynamic force.

To gain a better workpiece finish quality, research should also focus on the dynamic force excitation and the vibration response. Further research will concentrate on the active control of tool vibration in the micro-machine tool.

## ACKNOWLEDGMENT

This project was supported by the National 863 High Technology R&D Program (No. 2012AA041309).

## REFERENCES

Abele, E. et al. 2010 Machine tool spindle units. *CIRP Annals—Manufacturing Technology*, 59: 781–802.

Alfares, M. & Elsharkawy, A. 2000. Effect of grinding forces on the vibration of grinding machine spindle system. *International Journal of Machine Tools and Manufacture*, 40: 2003–2030.

Filiz, S. & Ozdoganlar, O.B. 2010. A Model for Bending, Torsional, and Axial Vibrations of Micro-and Macro-Drills Including Actual Drill Geometry—Part I Model Development and Numerical Solution. *Journal of manufacturing science and engineering*, 132: 041017.

Filiz, S. & Ozdoganlar, O.B. 2010. A Model for Bending, Torsional, and Axial Vibrations of Micro-and Macro-Drills Including Actual Drill Geometry—Part II: Model Validation and Application. *Journal of manufacturing science and engineering*, 132: 041018.

Guo, M. et al. 2012. Dynamic performance test and analysis of spindle system of high speed grinding machine tools. *Applied Mechanics and Materials*, 226–228: 720–724.

Klocke, F. et al. 2011. *High frequency force measurement of micro milling operations in hardened steel*. The international conference on advanced manufacturing systems and technology.

Liang, S.Y. 2006. *Mechanical machining and metrology at micro/nano scale*. Proc of SPIE 6280, Third International Symposium on Precision Mechanical Measurements, 628002.

Miao, J.C. et al. 2007. Review of dynamic issues in micro-end-milling. *The International Journal of Advanced Manufacturing Technology*, 31: 897–904.

Peeters, B. et al. 2004. The PolyMAX frequency-domain method: a new standard for modal parameter estimation. *Shock and Vibration*, 11: 395–409.

Sheng, X. et al. 2014. Calculation of ball bearing speed-varying stiffness. *Mechanism and Machine Theory*, 81: 166–180.

Transchel, R. 2012. Effective Dynamometer for Measuring High Dynamic Process Force Signals in Micro Machining Operations. *Procedia CIRP*, 1: 558–562.

Uhlmann, E. et al. 2013. Process Machine Inter-actions in Micro Milling, *Lecture Notes in Production Engineering*: 265–284.

Wang, H. et al. 2010. A theoretical and experimental investigation of the tool-tip vibration and its influence upon surface generation in single-point diamond turning. *International Journal of Machine Tools and Manufacture*, 50: 241–252.

Zhang, N. et al. 2005. The dynamic model of the grinding process, *Journal of Sound and Vibration*, 280: 425–432.

*Advanced Materials, Structures and Mechanical Engineering – Kaloop (Ed.)*
© *2016 Taylor & Francis Group, London, ISBN 978-1-138-02793-0*

# Research of the IEEE802.11a channel coding technology based on Turbo code

Z.Y. Sun, L.Y. Liu & Y. Zhang
*School of Information Engineering, Northeast Dianli University, Jilin, China*

ABSTRACT: With the increase in the transfer rate and communication quality, original convolution coding scheme of IEEE802.11a has become hard to meet the needs of high-speed, high-quality real-time communication. And its decoding complexity is high and power consumption is large. Therefore, this paper proposes the IEEE802.11a channel coding scheme based on Turbo code, and data from the source are encoded first, making it the Turbo code, and then OFDM modulation is made. The simulation analysis proves that the system can get a larger coding gain and has a better performance.

## 1 INTRODUCTION

IEEE802.11a requires physical layer works in the 5 GHZ frequency band. Since the adoption of the Orthogonal Frequency Division Multiplexing (OFDM) modulation technology (Yu et al. 2012, Li & Ke 2013) to transmit data, the highest rate can be up to 54 Mbps. When transmitting data via wireless channels, some sub-channels of the OFDM signal may be destructed due to the fading and white noise interference. So, even if error-free detection has been made on most of the channels, the bit error rate of the whole system is likely to remain high because several sub-channels are affected by deep fading (Wang & Da 2011). To avoid this situation, the channel coding technology can be used to protect the transmission of data. With the increase in the transfer rate and communication quality requirements, the original convolution coding scheme has become hard to meet the needs of high-speed, high-quality real-time communication. Its decoding complexity is high and power consumption is large, so it becomes necessary for the coding system to make further improvement. Thus, an improved IEEE802.11a channel coding scheme is proposed in this paper. In 1993, Berrou proposed a channel encoding method, which is called Turbo, and characterized by using the parallel cascade of encoding, recursive encoding and interleavers, and the iterative decoding. Because of the very good application of the random coding in the Shannon channel encoding's theorem, the decoding performance is almost close to the limit of the Shannon theory.

## 2 MODELS AND DYNAMIC EQUATIONS

A series of main parameters of the OFDM in the system (Shen 2013) is listed in Table 1.

Table 1. Main parameters of the OFDM in the IEEE802.11a standard.

| Data rate | 6, 9, 12, 18, 24, 36, 48, 54Mb/s |
|---|---|
| Modulation | BPSK, QPSK, 16-QAM, 64-QAM |
| Code rate | 1/2, 9/16, 2/3, 3/4 |
| Number of sub-carriers | 52 |
| Number of pilots | 4 |
| OFDM symbol interval | 4 μs |
| Guard interval | 800 μs, 400 μs (optional) |
| Sub-carrier interval | 312.5 kHz |
| Signal bandwidth | 16.66 MHz |
| Channel interval | 20 MHz |

The IEEE802.11a standard chooses the OFDM technology to cope with frequency selective fading and to randomize the burst error caused by the broadband fading channel. The transmission rate choice is determined by link adaptation schemes. In this process, the best code rate and modulation scheme is selected according to the channel conditions (Guo et al. 2010, Amador et al. 2012).

## 3 IMPROVED IEEE802.11A CHANNEL CODING SCHEME

The improved IEEE802.11a channel coding scheme is shown in Figure 1. Data from the source are encoded first, making it the Turbo code, and then OFDM modulation is made.

### 3.1 *The influence of component code on the performance of Turbo code*

Different constraint lengths of the RSC component encoder affect the performance of Turbo code

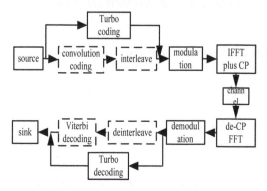

Figure 1. Improved IEEE802.11a channel coding block diagram.

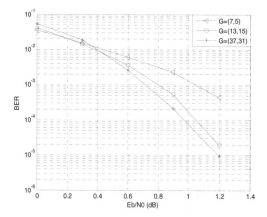

Figure 2. Effects of different constraint lengths on the performance of Turbo decoding.

differently. Figure 2 shows the generator matrix of G = (7,5), G = (13,15), G = (37,31) the Turbo code error rate performance curve, whose coding memory length is 2, 3, 4, respectively, and corresponding component code constraint length is 3, 4, 5, respectively. Simulation parameters are as follows: the interleaving length is 4096 bits, bit rate 1/3, BPSK modulation mode, channel model for AWGN channel and decoding algorithm uses the Max-Log-MAP algorithm.

It can be seen from the figure that increasing the constraint length of the component code will improve the BER performance of Turbo code. When the interleaving length and the code rate are certain, the larger coding constraint length will make Turbo code to show a better performance. But with the increase in constraint length, Turbo decoding computation complexity increases exponentially. As a result, when choosing the component encoder constraint length, we should weigh

the relationship between the decoding performance and the amount of calculation, and select a practically good code.

### 3.2 The influence of different interleaving lengths on the performance of Turbo code

The WLAN is in the bottom of the TCP/IP model, on which are the network layer, transport layer and application layer. The transmission of data involves the processing and encapsulation of the application layer protocol (HTTP, SMTP, FTP, etc.), the transport layer protocol (UDP and TCP) and the network layer protocol (IP). Data that is eventually transmitted to the WLAN Medium Access Control (MAC) layer must add the head of each layer protocol. The head length of the physical layer is used for transmission control (synchronization), which is not considered here. The MAC layer frame length is 12 + 8 (20) + 20 + 28 + effective data = 68 (80) + effective data. Generally, in web browsing and voice service applications, the effective data length is no more than 40 bytes. Because of this, the general data frame length is no more than 128 bytes.

When the frame length is shorter, different interleavers greatly influence the performance of Turbo code. The types of interleaver are random interleaver, block interleaver and spiral interleaver. Random interleaver has the best performance, and also has a largest decoding delay. Block interleaver and spiral interleaver have performance losses, but the complexity is decreased to some extent and also has a smaller decoding delay. But when the decoder bit length is more than 1000 bits, various interleavers have almost a similar influence on Turbo code performance.

With the increase in interleaving length, the time for decoding becomes longer and longer. Therefore, in the specific application, the appropriate interleaving length must be chosen to ensure the transmission quality and time delay requirement.

### 3.3 The influence of different numbers of iteration on the performance of Turbo code

Turbo decoding is iterative. With the increase in the number of iterations, decoding performance will get better, but when the number of iterations is more than six times (at high SNR condition over 3 times), decoding performance changes very little with the increase in the number of iterations, and the decoding delay and power consumption will be increased with the increase in the number of iterations unceasingly. But when the number of iterations increases to a certain value, the improvement in the BER performance of Turbo code tends to be gentle.

### 3.4 The influence of different numbers of iteration on the performance of Turbo code

A lower code rate can improve the BER performance of Turbo code. When the Signal-to-Noise Ratio (SNR) is low, the gap of a different code rate of Turbo BER performance is not big. When the SNR is high, the smaller the code rate is, the relatively better the Turbo code performance is. But the lower code rate means that the Turbo code transmission efficiency is reduced. In the practical application, the corresponding code rate should be selected according to the requirement of the practical business needs.

## 4 PARAMETER SETTING AND PERFORMANCE SIMULATION OF TURBO CODE IN IEEE802.11A

This paper uses the generator matrix of G = (13,15) parallel cascade Turbo encoder. Its coding structure is shown in Figure 3. In the IEEE802.11a coding system, parameter settings of Turbo encoding and decoding scheme are given in Table 2.

Using MATLAB, the performance curve is shown in Figure 4.

The simulation condition is the AWGN channel, the rate of 1/3, decoding length of 1024 bits, using $32 \times 32$ pieces of interleaver, the first component decoder returns to zero and the second component decoder is not zero. When the BER is $10^{-4}$, the required SNR of the Turbo code is only 1.5 dB, and the convolution code has a constraint length of 7, and using Viterbi decoding requires 3.5 dB with coding gain increasing by 2 dB. And when using a random interleaver and the two-component

Table 2. Turbo encoding and decoding scheme parameter settings.

| Parameter | Values |
|---|---|
| RSC encoder | (13,15) |
| Code rate | 1/2, 1/3 |
| Frame length | 576, 1024, 4096 |
| Interleaved mode | Block interleaver |
| Number of iterations | 5 |
| Decoding algorithm | Proposed low-complexity Log-MAP algorithm (Sun & Li 2013) |

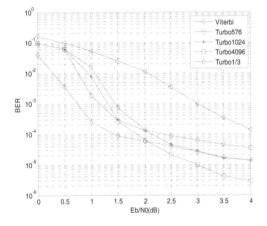

Figure 4. Turbo code performance simulation.

decoder returns to zero, a better performance can be obtained. In addition, half of the Turbo code rate is implemented by the puncturing technique. If not removed, using the 1/3 rate can get the extra coding gain of 0.5 dB without any increase in the operational cost. Getting the 1/3 rate with convolution code, both the encoding and decoding structures need to be changed, and the operation costs increase.

## 5 CONCLUSIONS

This paper first presents the related parameters of OFDM in the IEEE802.11a system, and then presents the IEEE802.11a coding structure diagram adopting Turbo code. The influence of various parameters on the performance of Turbo code is analyzed in detail, and the optimization parameter selection scheme of the IEEE802.11a coding system is obtained. The simulation analysis proves that the system can obtain a larger coding gain and has a better performance.

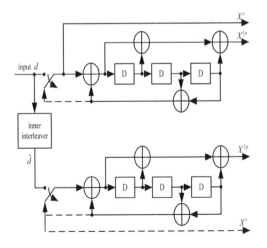

Figure 3. Improved Turbo encoder structure in the IEEE802.11a channel coding system.

# REFERENCES

Amador, E. Knopp, R. Pacalet, R. & Rezard, V. 2012. Dynamic Power Management for the Iterative Decoding of Turbo Codes. *IEEE Transactions on VLSI Systems*, 20(11): 2133–2137.

Guo, J. Li, X. & Zhang, G. 2010. The FPGA Implementation of IEEE802.11a OFDM Baseband Modulation. *Video Engineering,* 34(S1): 90–93.

Li, D. & Ke, F. 2013. Review on LTE System ICI Elimination Technology in the High-speed Mobile Environment Based on OFDM. *Journal of Chongqing University of Posts and Telecommunications (Natural Science),* 25(3): 292–298.

Shen, Z. 2013. *Channel Modeling and Doppler Frequency Estimation in IEEE802.11a System.* Master's Degree Thesis of the Wuhan University of Technology.

Sun, Z. & Li, H. 2013. A Low-complexity LOG-MAP Algorithm. *Video Engineering*, 39(3): 53–56.

Wang, Y. & Da, X. 2011. Adaptive Equalization Algorithm in Turbo-OFDM System. *Computer Engineering*, 37(11): 111–113.

Yu, C. Sung, C. Kuo, C.H. & Yen, M.H. 2012. Design and Implementation of a Low-Power OFDM Receiver for Wireless Communications. *IEEE Transactions on Consumer Electronics*, 58(3): 739–745.

*Advanced Materials, Structures and Mechanical Engineering – Kaloop (Ed.)*
© *2016 Taylor & Francis Group, London, ISBN 978-1-138-02793-0*

# Stability analysis of single-layer hyperbolic shallow lattice shell with different grid reinforcements

Z.Y. Zhang, H.Y. Wei, X. Zhao & X. Qin
*School of Human Settlement and Civil Engineering, Xi'an Jiaotong University, Xi'an, Shaanxi, China*

ABSTRACT: Under the vested steel consumption condition, the effects of different grid reinforcements on the stability of single-layer hyperbolic shallow lattice shell are studied via numerical simulation. In several conditions, such as linear and nonlinear condition as well as initial imperfection, the advisability of each grid reinforcements is illustrated by investigating the stability of hyperbolic shallow lattice shell. The results indicate grid reinforcements have great effects on the stability of single-layer lattice shell and the most reasonable reinforcement is suggested for practical projects.

## 1 INTRODUCTION

Since the end of the 20th century, due to light weight, attractive appearance, direct load transfer path and other benefits, lattice shells have been widely utilized in public buildings. As it is economically sound and easy to mount, lattice shell with quadrilateral grids becomes a common building type, such as the World Trade Center in Dresden German, the Downland Gridshell in England, new train station in Guangzhou China etc (S. Malek et al. 2014). Nonetheless, because of the insufficient shear stiff-ness, lattice shell with quadrilateral grids is prone to instability, which leads to degraded load-bearing capacity, especially under the effect of nonlinearity and initial imperfection.

As for single-layer lattice shell with quadrilateral grids, numerous studies have been made globally to improve degraded load-bearing capacity caused by instability (S.Z. Shen. 1999). Several solutions were proposed, for instance, struts were added to the grid plane including symmetrical struts, herringbone struts etc. to reinforce in-plane shear stiffness. These rigid reinforcements improved the stability of lattice shell with a great amount of steel consumption. In 1989, J. Schlaich (J. Schlaich. 1996, 2004) proposed exerting pre-stressed cables in grids. Compared with the previous methods, this flexible reinforcement improved the stability and reduced steel consumption. Hitherto, there is still a lack of related study Baverl O. 2012). To provide a reference for reasonable and effective grid reinforcement in practical project, study the on the stability of lattice shell with different grid reinforcements is extraordinarily necessary.

In the paper, based on the similar steel consumption, single-layer hyperbolic shallow lattice shell is taken as the research object and the stability under the effects of different grid reinforcements are analyzed via ANSYS. By comparing the load-bearing capacity of lattice shells with different grid reinforcements on the condition that nonlinearity and initial imperfection is considered or not, a comprehensive appraisal is illustrated.

## 2 ANALYSIS MODEL

To study the effects of different grid reinforcements on the stability of single-layer lattice shell, the single-layer hyperboloid lattice shallow shell is adopted as analysis model whose bi-directional span is 50 m and rise is 10 m. All nodes at the perimeter are hinged (four joints at corner are fixed), as Figure 1(a) shows. The lattice shell is composed of Q345 hollow circular steel tube whose standard is $\Phi 140 \times 6$.

Under the vested steel consumption condition, six different grid reinforcements are investigated, shown in Figure 1(b)–(g). The six grid reinforcements contain two different types: a). Flexible reinforcement, pre-stressed cables are exerted in all quadrilateral grids, as Figure 1(b) shows; b). Rigid reinforcement, single diagonal struts are added. According to different layouts of single diagonal struts, several categories of lattice shell are proposed: diagonal strut-braced lattice shell, a central symmetrical tetragon is formed by single diagonal struts, shown in Figure 1(c); symmetrical diagonal strut-braced lattice shell, the layout of struts is symmetrical, as Figure 1(d) shows; herringbone strut-braced lattice shell, composed of positive and negative herringbone arrangement, as Figure 1(e) shows; lamella lattice shell, grids in ordinary lat-

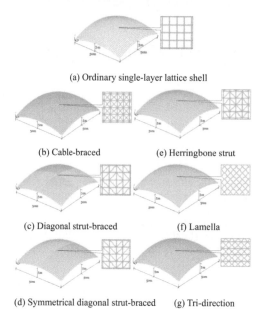

(a) Ordinary single-layer lattice shell

(b) Cable-braced  (e) Herringbone strut

(c) Diagonal strut-braced  (f) Lamella

(d) Symmetrical diagonal strut-braced  (g) Tri-direction

Figure 1. Configurations of seven different 3D lattice shell and corresponding 2D plate form.

Table 1. Steel consumption of different grid reinforcements and diameter of strut/cable.

|  | Steel consumption (kg) | Diameter of strut or cable (mm) |
|---|---|---|
| Ordinary | 71115 | / |
| Cable-braced | 83348 | 20 |
| Diagonal strut-braced | 119110 | $\Phi 140 \times 6$ |
| Symmetrical diagonal strut-braced | 119950 | $\Phi 140 \times 6$ |
| Herringbone strut-braced | 120100 | $\Phi 140 \times 6$ |
| Lamella | 101810 | $\Phi 140 \times 6$ |
| Tri-direction | 135370 | $\Phi 140 \times 6$ |

tice shell are rotated 45 degrees, as Figure 1(f) shows and tri-direction lattice shell, as shown in Figure 1 (g). More details of different grid reinforcements are shown in Table 1.

The FE models are shown in Figure 1. BEAM188 and LINK180 element are used to simulate steel components and pre-stressed cable respectively. The joints are assumed to be rigid. The uniform load is added to full span. For cable-braced lattice shell, pre-stress in the cable is 150 MPa. The constitutive model of steel and cable are perfectly elasto-plastic, with the yield strength of 345 MPa, 1500 MPa respectively. Load factor = practical

load-bearing capacity/(dead load + live load). The standard value of dead load is 0.512 kN/m²; and live load is dominated by snow load whose standard value is 0.5 kN/m².

## 3 COMPARISON OF LINEAR STABILITY

To fully understand the effects of various grid reinforcements on the stability of lattice shell, the linear stability, namely theoretical bifurcation load, is investigated initially (Cai J.G., Gu L.M., et al. 2013). The stability of each lattice shell in different eigenvalue buckling modes in ideal elastic condition is shown in Figure 2.

Figure 2 (a) shows in ideal linear condition, stability of ordinary lattice shell and lamella lattice shell are low, and its theoretical bifurcation load in the first eigenvalue buckling mode are 2.0409 and 2.6317 respectively. However, the theoretical bifurcation load of other reinforced lattice shells in the first eigenvalue buckling mode are higher than that of these two forms. Compared to the ordinary

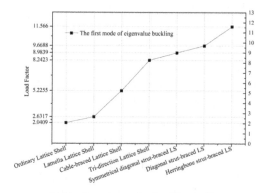

(a) The first eigenvalue buckling mode

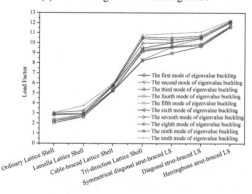

(b) The first ten eigenvalue buckling modes

Figure 2. Comparison of linear stability of each lattice shell.

shell, theoretical bifurcation load of diagonal strut-braced and herringbone strut-braced lattice shell increase to 373%, 466%, reaching 9.6688 and 11.566 respectively. Since the stability load-bearing capacity varies greatly duo to different grid reinforcements, it is concluded that stability of lattice shell can be improved by effective grid reinforcements theoretically. Figure 2(b) shows the critical load of each lattice shell in a higher level of buckling modes which shows the variation law of higher level of buckling modes resembles that of the fundamental mode.

## 4 COMPARISON OF STABILITY UNDER DIFFERENT CONDITIONS

Nonlinearity and imperfections in the actual structure make the stability of lattice shells decrease greatly (S.Z. Shen. 1999). Thus, the actual situation will be reflected more accurately if nonlinearity and initial imperfection are considered at the same time. Figure 3 presents the stability of lattice shell with different reinforcements under the effect of nonlinearity and initial imperfection. And it is illustrated that the nonlinear stability of each lattice shell decreased remarkably with different shape of initial imperfection.

For lattice shell with different grid reinforcements, the shape of the imperfection which leaded to the most unfavorable condition is 2nd (3rd), 5th, 5th (6th), 4th, 6th, 1st and 6th, respectively. For lattice shell whose displacement shape of the loaded structure obtained from geometrical nonlinear buckling analysis was used as a pattern of imperfection, the corresponding stability was not the lowest. As for lattice shell with different grid reinforcements, the stability of each structure in linear stage and nonlinear stage as well as under the influence of initial imperfection is compared, as Figure 4 shows. Technical Specification for Space Frame Structures (JGJ7-2010) suggested that Span/300 is taken as the scale of imperfection and initial imperfection considered to be consistent to the most unfavorable mode. Thus, the stability of different lattice shells in each condition is obtained.

Figure 4 shows that different grid reinforcements affect stability greatly. Furthermore, it is illustrated in Figure 5 that different grid reinforcements have a fairly remarkable effect the on the stability of lattice shell. In accordance with the favorable effects on the stability, a descending order is given as follows: herringbone strut-braced lattice shell, diagonal strut-braced lattice shell, symmetrical diagonal strut-braced lattice shell, tri-direction lattice shell, cable-braced lattice shell, lamella lattice shell and ordinary lattice shell. Compared to ordinary lattice

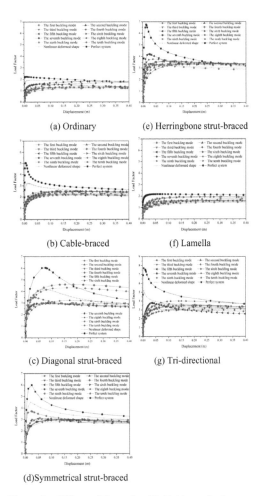

(a) Ordinary    (e) Herringbone strut-braced

(b) Cable-braced    (f) Lamella

(c) Diagonal strut-braced    (g) Tri-directional

(d) Symmetrical strut-braced

Figure 3. Effects of the scale of initial imperfections on the stability of different lattice shell.

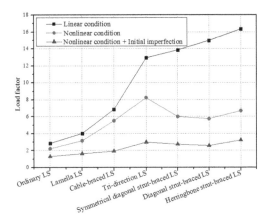

Figure 4. Stability load-bearing capacity of different lattice shells in each condition.

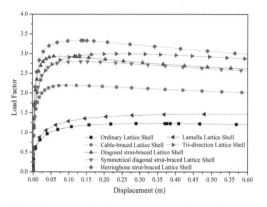

Figure 5. Load-displacement curves for each lattice shell.

Table 2. Increasing rate of reinforcements the on the stability of ordinary lattice shell.

|  | Linear | Nonlinear | Nonlinear + Imperfection |
|---|---|---|---|
| Ordinary | 100.00% | 100.00% | 100.00% |
| Cable-braced | 256.00% | 246.50% | 174.60% |
| Diagonal strut-braced | 473.70% | 296.00% | 234.10% |
| Symmetrical diagonal strut-braced | 440.10% | 294.00% | 222.80% |
| Herringbone strut-braced | 566.70% | 331.10% | 266.50% |
| Lamella | 128.90% | 106.30% | 116.80% |
| Tri-direction | 403.80% | 377.70% | 238.30% |

shell, the increasing rate of stability of reinforced lattice shells is shown in Table 2. As Table 2 indicates, herringbone strut-braced is most favorable grid reinforcement to improve the stability of lattice shell.

## 5 CONCLUSIONS

In the prerequisite that steel consumption is similar, the effects of different grid reinforcements on the stability of lattice shells are studied. The results show that grid reinforcements have a great effect on the stability of lattice shell. The stability of lattice shell can be improved effectively by adding herring-bone struts and pre-stressed cable. Under the circumstance that consumption of steel is close, the stability of lattice shell with herringbone strut-braced is the strikingly higher than others in linear condition, non-linear condition and under the influence of initial imperfection. Moreover, albeit the stability of cable-braced lattice shell is not the highest, yet the corresponding steel consumption is far less than other reinforcements. In various situations, the stability of cable-braced lattice shell is improved greatly than that of ordinary lattice shell, which could be recommended as the optimal reinforcement.

## ACKNOWLEDGEMENT

This work was financially supported by the National Natural Science Foundation of China (51478387).

## REFERENCES

Baverl O. & Francois J. et al. 2012. Gridshells in Composite Materials: Construction of a 300 $m^2$ Forum for the Soliday's Festival in Paris. *Structural Engineering International*, 22:408–414.
Cai J.G. & Gu L.M. et al. 2013. Nonlinear stability analysis of hybrid grid shells. *International Journal of Structural Stability and Dynamics*, 13:1–16
Malek, S. & Wierzbicki, T. 2014. Buckling of spherical cap grid-shells: A numerical and analytical study revisiting the concept of the equivalent continuum. *Engineering Structure*, 75:288–298.
Schlaich, J. 1996. Glass-covered Grid-Shells. *Structural Engineering International*, 6:19–27.
Schlaich, J. 2004. Conceptual design of light structures. *Journal of the International Association for Shell and Spatial Structure*, 45:157–168.
Shen, S.Z. & Chen, X. 1999. *Stability of grid shell*. Science Publishing House, Beijing (In Chinese).
The industry standard of the people's Republic of China. 2010. *Technical specification for space frame structures*. Beijing: China Building Industry Press (In Chinese).

*Advanced Materials, Structures and Mechanical Engineering – Kaloop (Ed.)*
© 2016 Taylor & Francis Group, London, ISBN 978-1-138-02793-0

# Extending stress linearization procedures for direct method design in pressure vessel design

H.J. Li, X. Huang & H. Yang
*School of Mechanical Engineering and Automation, Zhejiang Sci-Tech University, Xiasha Higher Education Zone, Hangzhou, China*

ABSTRACT: The elastic Design by Analysis (DBA) method in pressure vessel design utilizes the stress classification procedure to identify and subsequently limit values of primary and primary plus secondary stress. Stress linearization is used to define constant and linear through thickness stress distributions from continuum stresses that can be used in place of membrane and membrane plus bending stress in DBA. However there have been a few problems when performing stress classification procedure such as how to distinguish primary stress and secondary stress. This paper proposes a new direct method which can avoid the need for stress classification. The assumed through thickness distribution is the general limit state stress distribution in a beam under combined membrane and bending load.

## 1 INTRODUCTION

The elastic Design by Analysis (DBA) procedures specified in the ASME Boiler & Pressure Vessel Code Sections III and Section VIII Division 2(ASME2010), PD5500 Unfired Fusion Welded Pressure Vessels (BSI2006) and EN 13335-3:2002 (BSI 2002) provide a methodology for design against ductile failure based on elastic stress analysis and design criteria developed from the principles of limit and shakedown analysis. In design based on elastic shell analysis, the through wall stress distribution is linear and is treated as the superposition of two stress distributions: membrane stress, which is constant through thickness and bending stress, which varies linearly through thickness and has zero value at the mid-surface. Stress concentration effects due to the presence of stress raisers are accounted for by applying stress concentration factors to the calculated shell stress. In design based on 2D or 3D stress analysis, the effect of stress concentrations is included in the calculated total stress distribution, which is treated as the superposition of three stress distributions: membrane, bending and peak stress.

Two ductile failure modes are considered in the design procedure: gross plastic deformation under static load and incremental plastic collapse, or ratchetting, under repeated or cyclic load. The design criteria associated with these failure modes are based on concepts from limit and shakedown analysis respectively. In design, the criteria are applied implicitly through a procedure called stress classification, in which three classes of stress associated with a distinct failure mechanism are defined: primary, secondary and peak stress.

Primary stress is associated with gross plastic deformation or ductile failure under static load. Secondary stress, when taken with primary stress, is associated with incremental plastic collapse under repeated load. Peak stress is the part of the stress field associated with a local stress concentration effect and when taken with primary plus secondary stress associated with fatigue failure of the component under repeated load. Here consideration is limited to the design procedures for primary and secondary stress.

These DBA procedures for primary and secondary stress are applied in two stages. In the first stage, the vessel subject to the specified mechanical design loads is analyzed. Thermal loads are not included. In this stage the designer is required to ensure that the primary stress limits specified in the procedure are met. In the second stage the vessel subject to the mechanical and thermal operating loads is analyzed. In this stage, the designer is required to ensure that the primary plus secondary stress limits are met. Both stages of the design utilize the stress classification procedure to identify and subsequently limit values of primary and primary plus secondary stress.

The stress classification procedure is amenable to analysis methods that evaluate membrane and bending stresses, such as shell discontinuity analysis or Finite Element Analysis (FEA) using shell elements. The procedure is more difficult to apply to 2D or 3D elastic analysis methods, notably solid FEA, that evaluate the stress throughout the domain in terms of continuously varying stress at a point. There is a fundamental incompatibility between this form of calculated stress and the membrane plus bending plus peak distribution implicit in the design criteria. In practice, this incompatibility is addressed by applying a procedure called stress linearization to the solid FEA stress results.

## 2 STRESS LINEARIZATION

The DBA stress classification procedure requires definition of membrane and membrane plus bending stress distributions through the vessel thickness. These distributions are constant and linear with respect to through thickness position. In 2D or 3D FEA, the stress distribution is not usually linear through thickness. In the stress linearization procedure, linear through thickness stress distributions that result in the same net internal forces and moments as the elastic stress distribution are evaluated by applying a linearization procedure. The calculated constant and linear through thickness distributions are subsequently treated as membrane and membrane plus bending stress distributions in the DBA procedure.

The EN13445 linearization procedure is specified in C.4.4 Decomposition of Stresses. The through-thickness stress distribution is defined by the state of stress at each point along a supported line segment through the vessel wall. The supported line segment, length h, is defined as the shortest straight line segment joining two sides of the vessel wall. Two typical supported line segments and their associated local co-ordinate systems are shown in Figure 1.

The membrane stress $\sigma_{ij,m}$ is defined as the part of stress, constant along the supporting line segment, which is equal to the average value of the elementary stresses $\sigma_{ij}$ along this supporting line segment:

$$(\sigma_{ij})_m = \frac{1}{h} \int_{-\frac{h}{2}}^{+\frac{h}{2}} \sigma_{ij} dX_3 \tag{1}$$

The bending stress $\sigma_{ij,b}$ is defined as the part of stress varying linearly across the thickness of the wall:

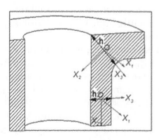

Figure 1. EN13445 supported line segment coordinate system.

$$(\sigma_{ij})_b = \frac{12}{h^3} \int_{-\frac{h}{2}}^{+\frac{h}{2}} \sigma_{ij} X_3 dX_3 \tag{2}$$

As only maximal values of $\sigma_{ij,b}$ at equal and of opposite sign on each side of the wall, i.e. at both ends of the supporting line segment, shall be considered. In this case:

$$(\sigma_{ij})_b = \frac{6}{h^2} \int_{-\frac{h}{2}}^{+\frac{h}{2}} \sigma_{ij} dX_3 \tag{3}$$

Consider a normal stress $\sigma_{(z)}$ acting on a rectangular surface of length h and width $w$. $\sigma_{(z)}$ is constant with respect to $w$ and varies continuously with respect to z over the interval $z = -h/2$ to $z = +h/2$, as illustrated in Figure 2a. A linear stress distribution, $\sigma^L$, acting over the same surface is constant with respect to $w$ and varies linearly over the interval $z = -h/2$ to $z = +h/2$ according to the equation:

$$\sigma^L = a + bz \tag{4}$$

where, $a$ and $b$ are constants. The linear stress distribution is illustrated in Figure 2b. General values of linear stress are defined as $\sigma_m$ at $z = 0$ and at $z = +h/2$.

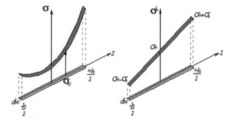

(a) general elastic stress $\sigma_{(z)}$     (b) linearized stress $\sigma^L$

Figure 2. Through-thickness stress distribution.

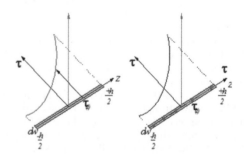

Figure 3. The linearization of in plane shear stresses.

Substituting these general values of stresses into Equation (4) gives an expression for the linear stress distribution in terms of $\sigma_m$ and $\sigma_b$:

$$\sigma^L = \sigma_m + \frac{2\sigma_b}{h}z \tag{5}$$

The two in-plane shear stresses, illustrated in Figure 3, can be linearized in the same manner.

## 3 DIRECT ROUTE

The stress linearization procedure uses an equilibrium analysis and assumed linear stress distribution to determine membrane and bending stresses for stress categorization. In the primary or gross plastic deformation check, mechanical design loads are applied and the membrane and bending stress distributions evaluated. The designer is then required to classify the calculated membrane and bending stress distributions as general primary membrane stress, local primary membrane stress, primary membrane plus bending stress and primary membrane plus secondary bending stress in accordance with Code definitions. Different stress limits are then applied to the general primary membrane stress, Sm, local primary membrane stress, 1.5 Sm, and primary membrane plus bending stresses, 1.5 Sm, and the allowable design load calculated. Sm usually has a value of $2/3\sigma_y$ or less, therefore all primary stress combinations are yield limited.

The stress categorisation procedure for primary stress may therefore be considered to be a form of lower bound limit load analysis. The lower bound limit load theorem (Lubliner 1990) states:

If, for a given load, there exists a statically admissible stress field in which the stress nowhere exceeds yield then that load is a lower bound on the limit load of the structure.

As the primary stress categorization is an equilibrium stress distribution and the limits on primary stress yield-limit the maximum allowable primary stress, the primary stress field satisfies the lower bound limit load theorem and the associated load can be viewed as a lower bound on the limit load of the vessel. This interpretation is valid only if the stress categorization procedure is applied correctly and all primary stresses are identified in the procedure. If a primary stress is incorrectly specified as secondary stress it may exceed the yield-limited (as the primary plus secondary stress limit is 3 Sm or $2\sigma_y$.

In the stress categorization procedure for secondary or incremental plastic collapse assessment, all operating loads are applied (mechanical and thermal) and the membrane and bending stress distributions evaluated. A 3 Sm stress limit is then applied to the membrane plus bending stress and the allowable

operating load calculated. This analysis can be interpreted as satisfying Melan's lower bound (elastic) shakedown theorem, which states: For a given cyclic load set the structure will exhibit shakedown if a constant residual stress field can be found such that the yield condition is not violated for any combination of cyclic elastic and residual stresses.

When a load cycle great enough to cause plastic deformation is applied, stress redistribution will occur when yield is exceeded. Provided the material in the vessel does not experience reverse yielding on unloading, the vessel will shakedown to elastic action. The elastic stress range between yield on loading and yield on unloading is $2\sigma_y$, which equates to 3 Sm. Therefore if the maximum elastic stress range over the loading cycle is less than or equal to 3 Sm. Shakedown will occur.

The main contribution of linearization in the shakedown assessment is to "crop" peak stress from the stress distribution used in the shakedown assessment. Peak stress should not be included.

In principle, the linearized stress distribution could be used directly in a limit analysis or direct route design of the vessel. This would not yield good designs in practice as the solution is generally a poor lower bound due to the assumed form of stress distribution. The stress classification procedure identifies this implicitly through the different values of allowable stress specified for membrane and bending stress. However, it is possible to obtain an improved limit load solution by assuming an alternative form of through thickness stress distribution.

## 4 DIRECT ROUTE PRIMARY STRESS CHECK

The assumed through thickness distribution can be any distribution specified in terms of two independent parameters, to be determined by considering the force and moment equilibrium conditions. The distribution may be continuous or discontinuous (step) in form. Here the stress distribution shown in Figure 4, defined in terms of parameters $\sigma_a$ and $d$, is considered. This form of stress distribution is the general limit state stress distribution in a beam under combined membrane and bending load, the limit state distribution adopted in the construction of the ASME Design region diagram.

Applying force equilibrium:

$$\frac{N_x}{b} = \int_{-\frac{h}{2}}^{\frac{h}{2}} \sigma_x dz = -\int_{-\frac{h}{2}}^{d} \sigma_a dz + \int_{d}^{\frac{h}{2}} \sigma_a dz = -2\sigma_a d \tag{6}$$

$$\frac{M_y}{b} = -\int_{-\frac{h}{2}}^{d} \sigma_a z dz + \int_{d}^{\frac{h}{2}} \sigma_a z dz = \sigma_a\left(\frac{h^2}{4} - d^2\right) \tag{7}$$

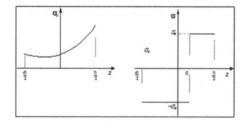

Figure 4. Assumed through thickness stress distribution.

Substituting $\sigma_a$ from (6) into (7) and rearranging gives:

$$\frac{Nd^2}{2} - Md - \frac{Nh^2}{8} = 0 \qquad (8)$$

The roots of the quadratic Equation (8) are,

$$d = \frac{M \pm \sqrt{M^2 + N^2 h^2 / 4}}{N} \qquad (9)$$

## 5  LIMIT ANALYSIS PROCEDURE

A limit analysis line is defined through the vessel in the same manner as the stress classification line defined above.

a. The through wall elastic stress distribution is evaluated in terms of stress components in the line's xyz co-ordinate system.
b. The cut section normal forces and moments are evaluated by numerically integrating Equations (6) and (7).
c. Equation (6), (7) and (9) are applied to $\sigma = \sigma_x, \sigma_y, \sigma_z, \tau_{xy}$ to evaluate an assumed stress $\sigma_a$ for each of these stress components.
d. As the shear stresses $\tau_{xz}$ and $\tau_{yz}$ to moment actions, an equivalent constant stress through the thickness is evaluated in the same way as membrane stress is evaluated in stress linearization.
e. Values of principal stress along the line are calculated based on the assumed stresses $\sigma_a$. The von Mises stress along the line is evaluated from the principal stresses.
f. The maximum value of von Mises stress for the applied load set P, $\sigma_e^P$ is identified.
g. The ratio of $\sigma_e^P$ to the yield stress $\sigma_y$ is a limit load multiplier $\lambda$.

h. The proportional loading limit load $P_L$ is:

$$P_L = \lambda \qquad (10)$$

i. The design load $P_d$ is:

$$P_d = \frac{2}{3} P_L \qquad (11)$$

## 6  SUMMARY

This paper provides a thorough understanding of stress classification procedure and stress linearization in the elastic Design by Analysis (DBA) method in pressure vessel design. It is concluded that the stress categorization procedure for primary stress may therefore be considered to be a form of lower bound limit load analysis. Based on the lower bound limit load theorem, a new Direct Method Design by extending the stress linearization procedure is pro-posed, which does not distinguish between "membrane" and "bending" stress when applying the stress limit, and also could avoid the difficulties in classifying calculated stresses into primary and secondary stresses.

## ACKNOWLEDGEMENT

This research was supported by National Natural Science Foundation of China under Grant No. 51445012 and Zhejiang Provincial Natural Science Foundation of China under Grant No. LQ13E050018.

## REFERENCES

ASME, 2006. *ASME Boiler and Pressure Vessel Code Section III*, The American Society of Mechanical Engineers, New York.
ASME, 2010. *ASME Boiler and Pressure Vessel Code Section VIII Division 2*, The American Society of Mechanical Engineers, New York.
BSI, EN13445–3:2002, *Unfired Pressure Vessels Part 3 Design*, British Standards Institution, London.
BSI, PD5500, 2006. *Unfired fusion welded pressure vessels*. British Standards Institution, London.
Lubliner, J. 1990. *Plasticity Theory*, New York: Macmillan Publishing Company.

*Advanced Materials, Structures and Mechanical Engineering – Kaloop (Ed.)*
© *2016 Taylor & Francis Group, London, ISBN 978-1-138-02793-0*

# Statistical description of fracture pattern from single particle crushing test

B. Zhao & J. Wang
*Department of Architecture and Civil Engineering, City University of Hong Kong, Hong Kong, China*

H. Zhao
*China State Construction Engineering (HK) Ltd., Hong Kong, China*

ABSTRACT:   The spatial distribution of fractures inside sand particles are highly influenced by loading condition, external morphologies and internal microstructures, i.e., initial cracks and voids. To understand this relationship, it is important to obtain quantitative measurements of the spatial distribution of fractures. In this study, the two-point correlation function was utilized to measure fractures. This function was implemented in three different ways, i.e., sample method, exponential method, and point-by-point method. These methods were applied to analysis the fracture of natural Leighton Buzzard Sand (LBS) particle obtained by an in-situ X-ray CT test on single particle crushing. It is found that the fractures are mainly parallel to the loading direction. While the sampling method generates almost the same results as those from the point-by-point method, the exponential method could only reflect the pattern of the main direction.

## 1 INTRODUCTION

Particle breakage is of great importance in determining the constitutive behavior of granular materials, especially at elevated stress level (e.g. Coop & Lee, 1993; McDowell & Bolton, 1998). Single particle crushing tests have been carried out to understand the statistical distributions of particle strength (e.g. Nakata et al., 1999), fracture pattern, morphology evolution and breakage energy dissipation (e.g., Zhao et al., 2015). These observations enlarged our understanding on particle breakage at micro-scale, which helps interpret particle breakage at macro-scale and establish more realistic simulation models.

The fracture observed from single particle crushing tests usually has a complex spatial distribution, as shown in Figure 1. Zhao et al. (2015) showed that the fracture patterns from single particle crushing tests could be influenced by loading condition, particle external morphologies and internal microstructures (e.g., cleavage, impurities, voids). To quantify their relationship, it is important to firstly measure distribution of microstructures and resulted in fractures.

The in-situ X-ray Computed Tomography (CT) has been used to investigate the microstructures and fracture patterns during the loading process without damaging the specimen or disturbing the fracture patterns (Zhao et al. 2015). The obtained X-ray images were processed and quantitatively analysed to obtain the morphology information of the fragments. However, the spatial distributions of fractures are very complex and difficult to be quantified. Microstructures have been statistically described through different kinds of correlation functions. Torquato (2002) shows a review of these methods and their application for different materials. These methods have been used to describe the distribution of microstructures (i.e., voids of concrete) and interpret their mechanical properties (Chung & Han 2010, 2013).

In this study, the statistical correlation method is utilized to describe the distribution of fractures. A two-point correlation function has been developed and applied to the fractures. This method has been implemented through three different ways, i.e., exponential method, sample method, and point-by-point method. The preliminary result shows the potential of this method to provide a

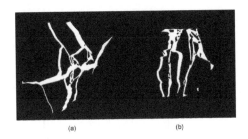

Figure 1.   2D slices of fractures caused by single particle crushing test on LBS sand: (a) slice on x-y plane; (b) slice on x-z plane.

quantitative description of the microstructures and fractures.

## 2 TWO-POINT CORRELATION FUNCTION

The two-point correlation function $P_{ij}(r, \theta, \varphi)$ represents the probability that when a straight line of length $r$ and angular orientation $(\theta, \varphi)$ is randomly placed in the 3D microstructure, the first end is in phase $i$ and the second end is in phase $j$. $i$ or $j$ in $P_{ij}(r, \theta, \varphi)$ could be either void (v) or solids (s). There are some limits for the two-point correlation function:

$$\lim_{r \to 0} P_{ii}(r, \theta, \phi) = f_i \qquad (1)$$

$$\lim_{r \to 0} P_{ij}(r, \theta, \phi) = 0 (i \neq j) \qquad (2)$$

$$\lim_{r \to \infty} P_{ii}(r, \theta, \phi) = f_i^2 \qquad (3)$$

$$\lim_{r \to 0} P_{ij}(r, \theta, \phi) = f_i f_j \qquad (4)$$

where, $f_i$ is the volume fraction of phase $i$.

### 2.1 Sampling method

The properties of heterogeneous microstructures can be computed from numerical data through the statistical sampling method. This approach calculates the two-point correlation value for a given $r$ by randomly dropping a straight line with a length $r$ and specified direction into the material many times, and looking into which phase each end drops in. The binary image, in which 1 represents the fracture and 0 represent the other area, was loaded into Matlab. The voxel was used as a grid for uniformly generating the sampling lines. In this study, three orthogonal directions, i.e., x, y, z, are investigated. The sampling lines are randomly generated in these directions with the length varied from 1 to 100 voxels. After the values at the two ends of all the randomly generated line are checked, the probability of $P_{ii}$ is calculated as the probability that two ends both are equal to one.

### 2.2 Exponential method

The two-point correlation function has been analytically represented based on their fundamental geometric constrains. Gokhale et al. (2005) recommended an easy form of the two-point correlation function for anisotropic materials:

$$P_{ij}(r, \theta, \phi) = f_i f_j \left\{ 1 - \exp\left[ -\left( \frac{P_L(\theta, \phi)_{i,j}}{2 f_i f_j} \right) r \right] \right\} (i \neq j) \qquad (5)$$

where, $P_L(\theta, \varphi)$ is the number of intersections between a test line and the i-j phase interface per unit test line length. $\theta$ is the angle between a test line and z axis, and $\varphi$ is the angle between projection of a test line on the x-y plane and x axis, which are used to define the direction of the two-point probability. Once we obtained $P_L(\theta, \varphi)$ from sampling method, the two-point correlation function could be described analytically.

### 2.3 Point-by-point method for validation

In order to validate the two methods mentioned above, the point-by-point method was performed to go through all possible conditions that generating a testing line with given length $r$ at given direction. This method provides the precise two-point correlation values.

## 3 APPLICATION TO FRACTURES

### 3.1 Fractures of LBS particles

Figure 1 shows the fractures retrieved from a crushed LBS particle. The fractures were generated after the first peak force during compressing the particle. The X-ray CT image has $600 \times 800 \times 344$ voxels with a resolution of 3.89 μm. A series of image processing methods were implemented to reduce noise, segment different phases, and separate attached particles. This has been described in Zhao et al. (2015). In order to retrieve the fracture from the irregular shaped particle, the α-shape was evaluated from the set of the points composing the fragments (Herbert et al. 1983). The α-shape associated with a set of points is a generalization of the concept of the convex hull, i.e., every convex hull is an α-shape but not every α-shape is a convex hull. We use this technique to separate the voids that inside the particle and the voids outside the particle. The fracture was defined as the voids inside the α-shape of all fragments. In the resulted image, 1 represents for fracture, and 0 represents for fragments or voids outsize particle. Figure 2(a) shows

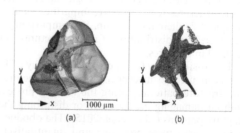

Figure 2. Visualisation of (a) 3D labelled fractured particle; (b) fractures retrieved from this particle.

the 3D view of the particle with the fragments identified by different colors, while its retrieved fractures are shown in Figure 2(b).

Figure 3 compares two-point correlation values obtained from sampling method with different number of sampling lines for a given $r$ at a given direction, namely 10,000 lines, and 5,000,000 lines. It is clear that the two-point correlation values on z direction is greater than that for that on x and y directions. This means that more fractures were along z direction, which is the loading direction. Although fluctuations happened with less sampling lines, the patterns of the two-point correlation functions are consistent.

Figure 4 compares two-point correlation results obtained from three different methods. It can be seen that the results of sampling method and point-by-point method are identical for the values in all the directions. The two-point correlation functions obtained from the exponential method are quite different from the other two methods. However, they all have a consistent pattern reflecting the main fracture distribution direction. It is thought that, although the exponential method developed from geometric constraints could be helpful to analytically represent the microstructural geometry, the sampling method and the point-by-point method obtain the more realistic distributions of two-point correlation function.

There is an intersection point between two-point correlation functions at x and y directions with the testing line of about 32 voxels. Before the intersection point, fracture distribution in y direction is larger than that in x axial direction, and this changed after the intersection point. This could be caused by some parallel fractures along y direction, which have a distance of around 32 voxels (Fig. 2(b)).

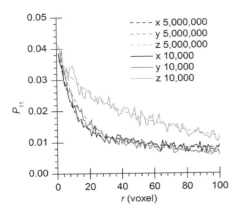

Figure 3. Influence of the number of sample lines on the distributions of two-point correlation function.

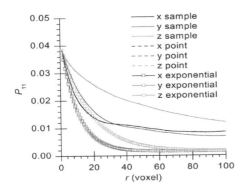

Figure 4. Distributions of a two-point correlation function calculated from three different methods. (sample: sampling method; point: point-by-point method; exponential: exponential method).

## 4 CONCLUSIONS

This paper mainly aims at developing a method to analyze the spatial distribution of fractures in crushed sand particles. Fractures were extracted from the 3D image of a crushed Leighton Buzzard sand particle. The two-point correlation function has been implemented in three different ways, namely sample method, exponential method and point-by-point method as validation. It was shown that two-point correlation values at z axial direction are larger than that at x and y axial directions for a given length of a testing line. It means that the fractures are mainly distributed along z axial direction, which is also the compression direction.

The fewer testing lines of sampling method caused fluctuations, but the trend is consistent with results from much more testing lines. It shows that if the purpose is to obtain the main characteristics of the microstructural geometry, the number of testing lines could be reduced to reduce the computational cost. Results from the exponential method are quite different from those of the other two methods. It shows the limitation of the analytical method developed based on the geometric constrains.

The above results indicate the potential to analyse the spatial distribution of fractures or microstructures by utilizing two-point correlation function. This could be a helpful tool for understanding how the distributions of microstructures affect its material properties and mechanical behaviors, and identifying the key factors that influence its fracture pattern. The future study will show the two-point correlation functions at more directions.

# REFERENCES

Chung, S.Y. & Han, T.S. 2010. Reconstruction of random two-phase polycrystalline solids using low-order probability functions and evaluation of mechanical behavior. *Computational Materials Science*, 49(4): 705–719.

Chung, S.Y. & Han, T.S. 2013. Correlation between low-order probability distribution functions and percolation of porous concrete. *Magazine of Concrete Research*, 65(7): 448–460.

Coop, M.R. & Lee, I.K. 1993. The behaviour of granular soils at elevated stresses. *In Predictive soil mechanics*, London, Thomas Telford: 186–198.

Edelsbrunner, H. Kirkpatrick, D. & Seidel, R. 1983. On the shape of a set of points in the plane. Information Theory, *IEEE Transactions on*, 29(4): 551–559.

Gokhale, A.M. Tewari, A. & Garmestani, H. 2005. Constraints on microstructural two-point correlation functions. *Scripta materialia*, 53(8): 989–993.

McDowell, G.R. & Bolton, M.D. (1998). On the micromechanics of crushable aggregates. *Géotechnique*, 48(5): 667–679.

Nakata, Y. Hyde, A.F.L. Hyodo, M. & Murata, H. 1999. A probabilistic approach to sand particle crushing in the triaxial test. *Géotechnique*, 49(5): 567–583.

Torquato, S. 2002. Statistical description of microstructures. *Annual Review of Materials Research*, 32(1): 77–111.

Zhao, B.D. Wang, J.F. Coop, M.R. Viggiani, G. & Jiang, M.J. 2015. An investigation of single sand particle fracture using X-Ray micro-tomography. *Géotechnique*, accepted for publication.

*Advanced Materials, Structures and Mechanical Engineering – Kaloop (Ed.)*
© 2016 Taylor & Francis Group, London, ISBN 978-1-138-02793-0

# Channel allocation algorithm in multi-hop cellular network based on potential game

Z.Y. Sun, Y. Zhang & L.Y. Liu
*School of Information Engineering, Northeast Dianli University, Jilin, China*

ABSTRACT: To reduce the interference of multi-hop cellular network system, and ensure the normal communication of primary users, this paper proposes an improved distributed channel allocation algorithm based on game theory. In this algorithm, utility function not only considers the interference between the secondary users, but also the interference caused by the secondary users and the main users. The new potential function is obtained. The simulation results show that the algorithm has better convergence and make the primary users obtain higher signal to interference ratio.

## 1 INTRODUCTION

Game theory is a digital theory and method to study the phenomenon of confrontation or competition, which is an important subject in modern mathematics and operations research. The game theory is mainly used to study the model that has the confrontation situation and can find the Nash equilibrium point for the corresponding game process. The literature (Li & Liu 2010) presented a channel assignment algorithm model based on supermodular game theory and introduced an appropriate price function to evaluate the effects of cognitive users to the primary users; The literature (Niyato & Hossain 2007, Niyato & Hossain 2008) presented a channel assignment algorithm based on costs. In this algorithm, the main users rent spectrum to secondary users according to certain price. The literature proposed an adaptive channel allocation algorithm and the algorithm defines 2 different objective functions according to the selfish users and cooperative users. The literature proposed algorithm without considering the impact on the main users in the network, so the allocation results would make the main user obtain poor signal-to-noise ratio, which would affect the communication quality of primary users. Literature (Nie & Comaniciu 2005) proposed adaptive channel allocation algorithm on the expanding research and proposed an improved distributed channel allocation algorithm based on potential game. The algorithm views minimizating interference level of system as the goal. The utility function takes the interference of the secondary users to the primary users into account.

## 2 SYSTEM MODEL

In multi-hop cellular network system with N secondary user transmitter receiver pairs and M main user transmitter receiver pairs, the primary users and the secondary users are randomly distributed in a d × d the region, each primary user has a channel, and secondary users select channel from the network channel set in a distributed manner. This paper assumes that physical layer of the secondary user physical layer can change the data sending rate and the corresponding SNR threshold value by adjusting the modulation mode and the channel encoding rate. If the channel allocation of results cannot meet the primary user and the secondary user's QoS requirements, secondary users can meet the primary user and the secondary user's QoS requirements by adjusting the parameters of its physical layer. In the self-organizing multi-hop Ad hoc—cellular network, this paper supposes the channel gain of secondary user $i$ transmitter to the user $j$ receiver is indicated with $G_{ij}^s$, channel gain of secondary user $i$ transmitter to the main user $m$ receiver is indicated with $G_{im}^{sp}$, channel gain of the main user $m$ transmitter to the receiver is indicated with $G_{mm}^p$, the channel gain of main user m transmitter to the receiver is indicated with $G_{mi}^{ps}$. The calculating formula of the channel gain is $G_{ij} = \frac{10}{d_{ij}^2}$, $d_{ij}$ is the distance of transmitter to receiver, namely channel gain and square of the distance is inversely, which is consistent with the basic characteristics of the wireless channel. In the receiving end, *SINR* of the secondary users $i$ is:

$$SINR_{si}$$
$$= \frac{P_{si}G_{ii}^s}{\sum_{j=1, j \neq i}^{N} P_{sj}G_{ij}^s h(S_j, S_i) + \sum_{m=1}^{M} P_{pm}G_{mi}^{ps} h(S_m^p, S_i) + \sigma^2}$$

(1)

In the receiving end, *SINR* of the main user *m* is:

$$SINR_{pm} = \frac{P_{pm}G_{mm}^p}{\sum_{i=1}^{N} P_{si}G_{im}^{sp}h\left(S_i, S_m^p\right) + \sigma^2} \tag{2}$$

In the formula: $P_{si}$ is the transmit power of secondary users $i (i \in [1, N])$; $P_{pm}$ is the transmit power of main users $m (m \in [1, M])$; $\sigma^2$ is added Gauss white noise power of the receiving end; As interfering function *h*, the definition is as follows:

$$h\left(S_i, S_j\right) = \begin{cases} 1, & S_i = S_j \\ 0, & S_i \neq S_j \end{cases}. \tag{3}$$

In the formula: secondary user $S_i$ and $S_j$ is a channel selection strategy.

# 3 POTENTIAL GAME ANALYSIS

## 3.1 *Nash equilibrium*

The Nash equilibrium provides a prediction on the game output result, when the player $s = (s_i, s_{-i})$ meet a set of strategies (4), *s* is a Nash equilibrium.

$$U_i(s) \geq U_i\left(s'_i, s_{-i}\right), \forall i \in N, s'_i \in S_i \tag{4}$$

In the set of participants, if the participants cannot change the strategy to improve their income, the whole set of participants strategies corresponding is called the Nash equilibrium. As a special game in the game theory, a potential game always is converge to Nash equilibrium. If a potential function of a game can satisfy (5), this game is a potential game.

$$P\left(s_i, s_{-i}\right) - P\left(s'_i, s_{-i}\right) = U_i\left(s_i, s_{-i}\right) - U\left(s'_i, s_{-i}\right) s'_i \in S_i \tag{5}$$

## 3.2 *Utility function*

Because the utility function of literature (Nie & Comaniciu 2005) only considers the mutual interference between secondary users in the network, then the utility function is:

$$U_i\left(s_i, s_{-i}\right) = -\sum_{j=1, j \neq i}^{N} P_{si}G_{ij}^s h\left(s_i, s_j\right) - \sum_{j=1, j \neq i}^{N} P_{sj}G_{ji}^s h\left(s_j, s_i\right) \tag{6}$$

This paper is consider the main users of the network in the presence of interference level, to minimize the system as the goal, so the design of the utility function must consider the interference to

primary users. The utility function is proposed in this paper for:

$$U_i\left(s_i, s_{-i}\right)$$
$$= -\sum_{j=1, j \neq i}^{N} P_{si}G_{ij}^s h\left(s_i, s_j\right) - \sum_{j=1, j \neq i}^{N} P_{sj}G_{ji}^s h\left(s_j, s_i\right)$$
$$- \sum_{m=1}^{M} P_{pm}G_{mi}^{ps} h\left(s_m^p, s_i\right) - \sum_{m=1}^{M} P_{si}G_{im}^{sp} h\left(s_i, s_m^p\right) \tag{7}$$

not only secondary users $I_{ss'} = \sum_{j=1, j \neq i}^{N} P_{si}G_{ij}^s h\left(S_i, S_j\right)$ and $I_{sp} = \sum_{m=1}^{M} P_{si}G_{im}^{sp} h\left(S_i, S_m^p\right)$ has the mutual interference, but the interference of primary users and other users to secondary users $I_{s's} = \sum_{j=1, j \neq i}^{K} P_{si}G_{ij}^s h\left(S_j, S_i\right)$ and $I_{ps} = \sum_{m=1}^{M} P_{pm}G_{mi}^{ps} h\left(S_m^p, S_i\right)$.

## 3.3 *Potential function*

For the utility function is proposed in this paper, construction of the potential function is:

$$P\left(s_i, s_{-i}\right) = \frac{1}{2}\sum_{i=1}^{N}\left(-\sum_{j=1, j \neq i}^{N} p_{si}G_{ij}^s h\left(s_i, s_j\right) - \sum_{j=1, j \neq i}^{N} p_{sj}G_{ji}^s h\left(s_j, s_i\right)\right)$$
$$+ \sum_{i=1}^{N}\left(-\sum_{m=1}^{M} P_{pm}G_{mi}^{ps} h\left(s_m^p, s_i\right) - \sum_{m=1}^{M} P_{si}G_{im}^{sp} h\left(s_i, s_m^p\right)\right) \tag{8}$$

The following paper certify that this potential function satisfies (4) type.

The utility function is expressed as: $U_i\left(s_i, s_{-i}\right) = U_{1i}\left(s_i, s_{-i}\right) + U_{2i}\left(s_i, s_{-i}\right)$

Inside:

$$U_{1i}\left(s_i, s_{-i}\right) = -\sum_{j=1, j \neq i}^{N} p_{si}G_{ij}^s h\left(s_i, s_j\right) - \sum_{j=1, j \neq i}^{N} p_{sj}G_{ji}^s h\left(s_j, s_i\right)$$

$$U_{2i}\left(s_i, s_{-i}\right) = -\sum_{m=1}^{M} P_{pm}G_{mi}^{ps} h\left(s_m^p, s_i\right) - \sum_{m=1}^{M} P_{si}G_{im}^{sp}\left(s_i, s_m^p\right)$$

show: $P\left(s_i, s_{-i}\right) = P_1\left(s_i, s_{-i}\right) + P_2\left(s_i, s_{-i}\right) \tag{9}$

Inside,

$$P_1\left(s_i, s_{-i}\right) = \frac{1}{2}\sum_{i=1}^{N}\left(-\sum_{j=1, j \neq i}^{N} p_{si}G_{ij}^s h\left(s_i, s_j\right) - \sum_{j=1, j \neq i}^{N} p_{sj}G_{ji}^s h\left(s_j, s_i\right)\right)$$
$$P_2\left(s_i, s_{-i}\right) = \sum_{i=1}^{N}\left(-\sum_{m=1}^{M} P_{pm}G_{mi}^{ps} h\left(s_m^p, s_i\right) - \sum_{m=1}^{M} P_{si}G_{im}^{sp} h\left(s_i, s_m^p\right)\right)$$

$P_1\left(s_i, s_{-i}\right)$, that process has been demonstrated in references (Wu & Wang 2008), show:

$$P_1(s_i, s_{-i}) = -\sum_{j=1, j\neq i}^{N} p_{si}G_{ij}^{s}h(s_i, s_j)$$

$$-\sum_{j=1, j\neq i}^{N} p_{sj}G_{ji}^{s}h(s_j, s_i) + p_1(s_{-i})$$

$$= U_{1i}(s_i, s_{-i}) + p_1(s_{-i})$$

Certify: $P_2(s_i, s_{-i})$.

$$P_2(s_i, s_{-i}) = \sum_{i=1}^{N}\left(-\sum_{m=1}^{M} p_{pm}G_{mi}^{ps}h(s_m^p, s_i) - \sum_{m=1}^{M} p_{si}G_{im}^{sp}h(s_i, s_m^p)\right)$$

$$= -\sum_{m=1}^{M} p_{pm}G_{mi}^{ps}h(s_m^p, s_i) - \sum_{m=1}^{M} p_{si}G_{im}^{sp}h(s_i, s_m^p)$$

$$+ \sum_{k=1, k\neq i}^{N}\left(-\sum_{m=1}^{M} p_{pm}G_{mk}^{ps}h(s_m^p, s_k)\right.$$

$$\left.-\sum_{m=1}^{M} p_{sk}G_{km}^{sp}h(s_k, s_m^p)\right)$$

Make:

$$P_2(s_{-i}) = \sum_{k=1, k\neq i}^{N}\left(-\sum_{m=1}^{M} p_{pm}G_{mk}^{ps}h(s_m^p, s_k)\right.$$

$$\left.-\sum_{m=1}^{M} p_{sk}G_{km}^{sp}h(s_k, s_m^p)\right)$$

So:

$$P_2(s_i, s_{-i}) = -\sum_{m=1}^{M} p_{pm}G_{mi}^{ps}h(s_m^p, s_i) - \sum_{m=1}^{M} p_{si}G_{im}^{sp}h(s_i, s_m^p)$$

$$+ P_2(s_{-i}) = U_{2i}(s_i, s_{-i}) + P_2(s_{-i})$$

$P_1(s_i, s_{-i})$ and $P_2(s_i, s_{-i})$ fed into (9) type:

$$P(s_i, s_{-i}) = U_{1i}(s_i, s_{-i}) + P_1(s_{-i}) + U_{2i}(s_i, s_{-i}) + P_2(s_{-i})$$

$P_2(s_{-i})$ show that the secondary users are not influenced by strategy change. When the user $i$ becomes a potential function from the strategy $s_i$ to $s'_i$:

$$P(s'_i, s_{-i}) = U_{1i}(s'_i, s_{-i}) + P_1(s_{-i}) + U_{2i}(s'_i, s_{-i}) + P_2(s_{-i})$$

So

$$P(s_i, s_{-i}) - P(s'_i, s_{-i})$$

$$= \left(U_{1i}(s_i, s_{-i}) + P_1(s_{-i}) + U_{2i}(s_i, s_{-i}) + P_2(s_{-i})\right)$$

$$- \left(U_{1i}(s'_i, s_{-i}) + P_1(s_{-i}) + U_{2i}(s'_i, s_{-i}) + P_2(s_{-i})\right)$$

$$= \left(U_{1i}(s_i, s_{-i}) + U_{1i}(s_i, s_{-i})\right) - \left(U_{1i}(s_i, s_{-i}) + U_{1i}(s_i, s_{-i})\right)$$

$$= U_i(s_i, s_{-i}) - U_i(s'_i, s_{-i})$$

So this game is a potential game, and the game will converge to a Nash equilibrium.

## 4 ALGORITHM STEPS

### 4.1 Initialize

Channel and transmitting power are distributed into users randomly.

### 4.2 Iteration

1. Secondary users will be gaming in accordance by the order of accessing network.
2. The sender of secondary user $i$ calculate $I_{ss'}$ and $I_{sp}$, and the computed value tells the receiver by the common control channel. The receiver calculates $I_{s's}$ and $I_{ps}$, so the received $I_{ss'}$ and $I_{sp}$ could compute each of channel utility function $U_i(s_p, s_{-i})$.
3. The receiving end of secondary user $i$ selects maximum channel by utility function and tell its transmission by the common control channel, then secondary users' new strategy is for $s'_i = \text{argmax}\,(U_i(s_p, s_{-i}))$.
4. The iterative process is repeated until convergence.

## 5 SIMULATION ANALYSIS

It randomly deployed $N = 40$ secondary user transmitter-receiver pairs and $M = 7$ primary user transmitter-receiver pairs in the $200 \times 200$ region. Maximum distance of transmission and receiver is 50 m between primary users and secondary users. The transmit power of secondary users and primary users are randomly [1,3] $mw$ and [2,5] $mw$, then noise power is $\sigma^2 = 10^{-9}\,mw$.

### 5.1 Algorithm convergence

Figure 1 is the policy change of each user through channel allocation algorithm based on the potential game.

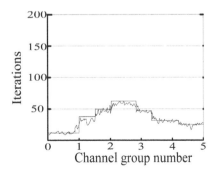

Figure 1. User policy convergence.

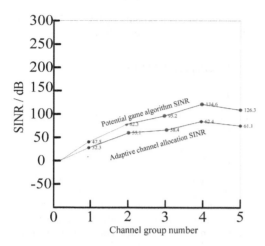

Figure 2. SINR size variation.

interference ratio is higher than the adaptive channel allocation algorithm.

## 6 CONCLUSIONS

This paper presents a new channel assignment algorithm for improving multi-hop cellular network system. In the multi-hop cellular networks, primary users and secondary users exist at the same time, based on this, this paper considers the interference to the primary user of the secondary users in the utility function. The interference situation of the primary users and their own utility maximizing of the secondary users are both taken into consideration. It is proved the game model is the potential game, and simulation results show the convergence property of the proposed algorithm and higher signal to interference ratio.

In Figure 1, the channel group number shows 200 channels, each number represents 40 channels. The coordinate is the number of the implementation of the algorithm. The abscissa is the selectable strategies for each secondary user, namely the available channel.

In Figure 1, when the iterative algorithm is about 60 times, the secondary user strategy does not change, which reaches the equilibrium state. It shows that the algorithm has better convergence.

### 5.2 *Algorithm reliability comparison*

By the reference (Yu et al. 2005), in adaptive channel allocation algorithm the signal to interference ratio is:

$SINR^m(k) = |h^m(k)|^2 \cdot \rho$. Assuming that each subcarrier average signal power of each antenna is P. Defining $\rho = P/\sigma^2$ as the average transmit SNR, and $h^m(k)$ is transmitting antenna to the user M′ receiving antenna channel subcarrier K response. In a different period of time, the bandwidth is 30 $GHz$, carrier number is 1024, and the noise power is $\sigma^2 = 10^{-9} mw$.

In Figure 2, by simulation comparison, potential game channel allocation algorithm for the signal to

## REFERENCES

Li, X.L. & Liu, H.T. 2010. Cognitive radio spectrum allocation algorithm model based on Game Theory. *Journal of Chongqing University of Posts and Telecommunications: Natural Science Edition*, 22(2): 151–155.

Nie, N. & Comaniciu, C. 2005. *Adaptive channel allocation spectrum etiquette for cognitive radio networks*, IEEE DySPAN. Baltimore: IEEE press, 269–279.

Niyato, D. & Hossain, E. 2008. Competitive Spectrum Sharing in Cognitive Radio Networks: A Dynamic Game Approach. *IEEE Transaction on Wireless Communications*, 7(7): 2651–2660.

Niyato, D. & Hossain, E.A. 2007. *Game-Theoretic Approach to Competitive Spectrum Sharing in Cognitive Radio Networks*, IEEE WCNC. Hong Kong: IEEE press, 16–20.

Wu, Y.L. & Wang, B. 2008. Repeated Open Spectrum Sharing Game with Cheat-Proof Strategies. *IEEE Transaction on Wireless Communications*, 8(4): 1922–1933.

Yu, G.D. Zhang, Z.Y. & Ping, L. 2005. The multi antenna OFDM systems with adaptive sub channel allocation and antenna selection. *Journal of circuits and systems*, 10(6): 63–68.

*Advanced Materials, Structures and Mechanical Engineering – Kaloop (Ed.)*
© *2016 Taylor & Francis Group, London, ISBN 978-1-138-02793-0*

# Electroosmotic flow through a microchannel containing salt-free power-law fluids

S.H. Chang
*Department of Mechanical Engineering, Far East University, Taiwan, R.O.C.*

ABSTRACT: The exact analytical solutions for the electroosmotic flow velocity of a salt-free power-law fluid in the planar slit and cylindrical capillary has been obtained by solving nonlinear Poisson-Boltzmann equation and Cauchy momentum equation. The results show that the EOF velocity is larger for flow behavior index $n > 1$ than that for $n < 1$ at smaller scaled surface charge density; however, this trend is reversed at higher scaled surface charge density. Also, the velocity profile is always like a parabolic curve for all value of $n$ at smaller scaled surface charge density. However, such profile substantially deviates from the typical plug-like flow pattern for the shear thinning fluid ($n < 1$) at higher smaller scaled surface charge density, which is different to that observed in a power-law electrolyte fluid.

## 1 INTRODUCTION

A salt-free solution is referred to a special system in which the liquid phase contains only counterions dissociated from the functional groups of the charged surfaces. Such systems occur in lamellar liquid crystals formed by ionic amphiphiles (Engstrom & Wennerstrom 1978), and also when, for example, colloidal particles, clay sheets, surfactant micelles or bilayers whose surfaces contain ionizable groups interact in water (Israelachvili 1991). Even an electrolyte solution with low salt concentration ($<10^{-5}$M) can also be treated as a salt-free solution (vander Heyden et al. 2006). Electrokinetic phenomena such as electrophoresis (Ohshima 2002) and electroosmosis (Chang 2009, 2010, 2012) in a salt-free solution are quite different from those in an electrolyte solution. This difference is caused by the effect of counterion condensation.

Most theoretical studies on the electrokinetic transport phenomena in a salt-free solution are limited to the Newtonian fluids. However, various fluids, which are often analyzed in microfluidic applications, such biofluids, protein chains in solvents, colloids and cell suspension typically display non-Newtonian fluid behavior such as shear-dependent viscosity and viscoelasticity (Chhabra & Richardson 1999, 2008, Tanner 2000). Das & Chakraborty (2006) developed mathematical models for electroosmotic flow of electrolyte power-law fluids in slit microchannels with Debye Huckel linear approximation. Since then, many theoretical (Chakraborty 2007, Berli & Olivares 2008, Zhao et al. 2008; Akgul & Pakdemirli 2008, Afonso et al. 2009, Zhao & Yang 2010, Vasu & De 2010, Misra 2011), numerical (Tang et al. 2009, 2010, Bharti et al. 2009,

Vasu & De 2010) and experimental (Olivares et al. 2009, Berli 2009) works have been reported on the electroknietic flow of electrolyte non-Newtonian fluids in microchannels. All of these studies indicate that the non-Newtonian effect on fluid velocity, friction coefficient, apparent viscosity, and heat transfer cannot be neglected in microscale fluid flow. Since the electrokinetic phenomena in a salt-free solution are different from those in an electrolyte solution, an in-depth understanding of the transport process of the salt-free Newtonian fluids in microchannels is essential.

A theoretical study on the electroosmotic flow of a salt-free power-law fluid in the planar slit and cylindrical capillary has been presented in this paper. The results show that the EOF velocity is larger for flow behavior index $n > 1$ than that for $n < 1$ at smaller scaled surface charge density; however, this trend is reversed at higher scaled surface charge density. Also, the velocity profile is always like a parabolic curve for all value of $n$ at smaller scaled surface charge density. However, such profile substantially deviates from the typical plug-like flow pattern for the shear thinning fluid ($n < 1$) at higher smaller scaled surface charge density, which is different to that observed in a power-law electrolyte fluid.

## 2 ELECTROOSMOTIC FLOW VELOCITY

Consider an EOF in a planar slit ($m = 0$) or a cylindrical capillary ($m = 1$), filled with a salt-free power-law fluid containing only counterions, as shown in Figure 1. Let the surface charge density be $\sigma_s$. Assume a uniform electric field $E$ is applied along the negative axial direction and the gravity

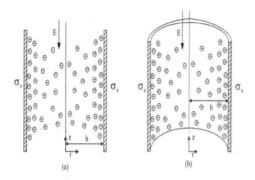

(a)     (b)

Figure 1. Schematic diagrams of the symmetrically charged (a) planar slit and (b) cylindrical capillary, filled with a salt-free power-law fluid containing only counterions.

is neglected. For an axisymmetric, steady, fully developed electroosmotic flow of a power-law fluid, only the axial velocity $u(r)$ exists and the momentum equation becomes

$$\frac{1}{r^m}\frac{d}{dr}\left(-r^m \gamma\left(-\frac{du}{dr}\right)^n\right) + E\varepsilon\varepsilon_0 \frac{1}{r^m}\frac{d}{dr}\left(r^m \frac{d\psi}{dr}\right) = 0$$

(1)

Subject to the boundary condition.

$$\frac{du}{dr}\bigg|_{r=0} = 0, \quad u(r=h) = 0$$

(2)

where, $\gamma$ is the flow consistency index, $\varepsilon$ is the dielectric constant of the medium and $\varepsilon_0$ is the permittivity of the vacuum. Note that $\gamma$ and $n$ are positive. The negative sign is chosen because the velocity decreases with an increase in $r$.

The exact solution to Equation (1) with the boundary conditions (2) can be derived as

$$u(r) = \int_r^h \left(\frac{E\varepsilon\varepsilon_0}{\gamma}\frac{d\psi}{dr}\right)^{\frac{1}{n}} dr$$

(3)

For $m = 0$, the electric potential field has been derived as (Chang 2010)

$$\psi(r) = \frac{k_b T_a}{ze}\ln\sec^2\left(c\frac{r}{h}\right)$$

(4)

from which the EOF velocity is,

$$u(r) = \int_r^h \left(\frac{E\varepsilon\varepsilon_0}{\gamma}\frac{k_b T_a}{ze}\frac{2c}{h}\tan\left(c\frac{r}{h}\right)\right)^{\frac{1}{n}} dr$$

(5)

For $m = 1$ (Chang 2009),

$$\psi(r) = -\frac{2k_b T_a}{ze}\ln\left[1 - \frac{Q}{4+Q}\left(\frac{r}{h}\right)^2\right]$$

(6)

and,

$$u(r) = \int_r^h \left(\frac{E\varepsilon\varepsilon_0}{\gamma}\frac{k_b T_a}{ze}\frac{4Q\,r}{(4+Q)\,h^2 - Q\,r^2}\right)^{\frac{1}{n}} dr$$

(7)

where, $e$ is the elementary charge, $k_b$ is the Boltzmann constants, $T_a$ is the absolute temperature and the constant $c$ can be determined by

$$2c\tan c = \frac{hez\sigma_s}{\varepsilon\varepsilon_0 k_b T_a} = \frac{2h}{\lambda} = Q$$

(8)

Here, the Gouy-Chapman length.

$$\lambda = 2\varepsilon\varepsilon_0 k_b T_a/(ez\sigma_s)$$

(9)

Defines a layer near the charged surface within which most counterions are localized (vander Heyden et al. 2006). Note that $Q$ is positive since $\sigma_s$ and $z$ are in the same sign.

## 3 RESULT AND DISCUSSION

The EOF velocity distributions of power-law fluids containing only monovalent counterions ($z = 1$) in the planar slit and cylindrical capillary are obtained by numerical integration of Equations (5) and (7), respectively. The other parameters used are dielectric constant of medium, $\varepsilon = 78.5$, permittivity of vacuum, $\varepsilon_0 = 8.854 \times 10^{-12}$ CV$^{-1}$m$^{-1}$, absolute temperature of system, $T_a = 300$ K, the applied electric field strength, $E = 30$ kVm$^{-1}$, and flow consistency index, $\gamma = 0.9 \times 10^{-3}$ Pa s$^n$.

Figure 2 shows the velocity distributions of various salt-free power-law fluids in a planar slit at differently scaled surface charge densities $Q$. Obviously, the EOF velocity increases as $Q$ increases. Also, this velocity is larger for $n > 1$ than that for $n < 1$ at smaller scaled surface charge density ($Q < 5$); however, this trend is reversed for $Q > 5$. This phenomenon can be explained by the fact that the EOF velocity for power-law fluids is a nonlinear function of the scaled surface charge density at the power of $1/n$. Also, at smaller $Q$, the velocity profile is like a parabolic curve for all value of $n$. However, such profile substantially deviates from the typical plug-like flow pattern for the shear thinning fluid ($n < 1$) at higher $Q$. Similar trends have been observed for the EOF velocity in a cylindrical capillary as shown in Figure 3.

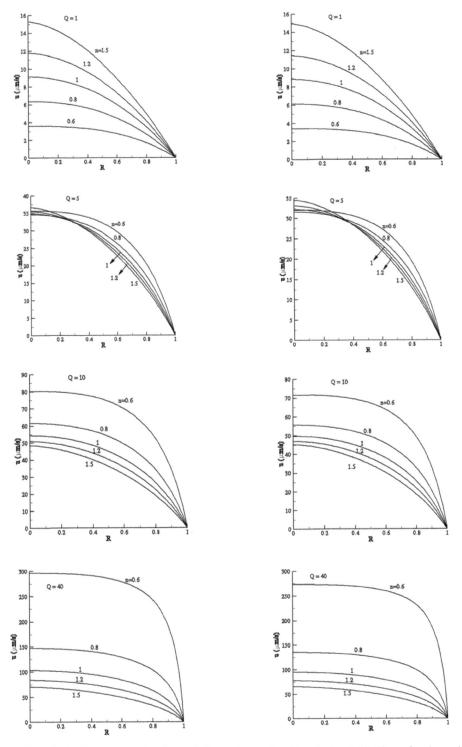

Figure 2. The velocity distributions of various salt-free power-law fluids inside a planar slit at different scaled surface charge densities $Q$.

Figure 3. The velocity distributions of various salt-free power-law fluids inside a cylindrical capillary at different scaled surface charge densities $Q$.

## 4 CONCLUSIONS

A theoretical study on the electroosmotic flow of a salt-free power-law fluid in the planar slit and cylindrical capillary has been presented in this paper. The exact analytical solutions for this EOF velocity are obtained by solving nonlinear Poisson-Boltzmann equation and Cauchy momentum equation. Based on this EOF velocity, the effects of various parameters such as the surface charge density and flow behavior index on EOF velocity are investigated in detail. The results show that the EOF velocity is larger for $n > 1$ than that for $n < 1$ at smaller scaled surface charge density; however, this trend is reversed at higher scaled surface charge density. Also, the velocity profile is always like a parabolic curve for all value of $n$ at smaller scaled surface charge density. However, such profile substantially deviates from the typical plug-like flow pattern for the shear thinning fluid ($n < 1$) at higher smaller scaled surface charge density, which is different to that observed in a power-law electrolyte fluid.

## ACKNOWLEDGEMENT

This work was supported by the National Science Council, ROC under Grant No. NSC 102-2221-E-269-009.

## REFERENCES

Afonso, A.M. & Alves, M.A. & Pinho, F. 2009. Analytical solution of mixed electro-osmotic/pressure driven flows of viscoelastic fluids in microchannels, *Journal of Non-Newtonian Fluid Mechanics*, 159: 50–63.

Akgul, M.B. & Pakdemirli, M. 2008. Analytical and numerical solutions of electro-osmotically driven flow of a third grade fluid between micro-parallel plates, *International Journal of Non-Linear Mechanics*. 43: 985–992.

Berli, C.L.A. 2009. Output pressure and efficiency of electrokinetic pumping of non-Newtonian fluids, *Microfluid Nanofluid*, 8: 197–207.

Berli, C.L.A. & Olivares, M.L. 2008. Electrokinetic flow of non-Newtonian fluids in microchannels, *Journal of Colloid Interface Science*. 320: 582–589.

Bharti, R.P. Harvie, D.J.E. & Davidson, M.R. 2009. Electroviscous effects in steady fully developed flow of a power-law liquid through a cylindrical microchannel, *International Journal of Heat and Fluid Flow*, 30: 804–811.

Chakraborty, S. 2007. Electroosmotically driven capillary transport of typical non-Newtonian biofluids in rectangular microchannels, *Analytical Chimica Acta*, 605: 175–184.

Chang, S.H. 2009. Transient Electroosmotic Flow in Cylindrical Microcapillaries Containing Salt-Free Medium, *Biomicrofludics*, 3: 012802.

Chang, S.H. 2010. Electroosmotic flow in slit microchannel containing salt-free solution, *European Journal of Mechanics—B/Fluids*, 29: 337–341.

Chang, S.H. 2012. Electroosmotic flow in a dissimilarly charged slit microchannel containing salt-free solution, *European Journal of Mechanics—B/Fluids*, 34: 85–90.

Chhabra, R.P. & Richardson, J.F. 1999. *Non-Newtonian Flow*, in Process Industries: Fundamentals and Engineering Applications Oxford: Butterworth-Heinemann.

Chhabra, R.P. & Richardson, J.F. 2008. *Non-Newtonian Flow and Applied Rheology*, Engineering Applications Oxford: Butterworth-Heinemann.

Das, S. & Chakraborty, S. 2006. Analytical solutions for velocity, temperature and concentration distribution in electroosmotic microchannel flows of a non-Newtonian bio-fluid, *Analytical Chimica Acta*, 559: 15–24.

Engstrom, S. & Wennerstrom, H. 1978. Ion condensation on planar surfaces. A solution of the Poisson-Boltzmann equation for two parallel charged plates *The Journal of Physical Chemistry*. 82: 2711–2714.

Israelachvili, J.N. 1991. *Intermolecular and Surface Forces*, San Diego, CA: Academic Press.

Misra, J.C. Shit, G.C. Chandra, S. & Kundu, P.K. 2011. Electro-osmotic flow of a viscoelastic fluid in a channel: Applications to physiological fluid mechanics, *Applied Mathematics and Computation*, 217: 7932–7939.

Ohshima, H. 2002. Electrophoretic mobility of a cylindrical colloidal particle in a salt-free medium, *Journal of Colloid and Interface Science*. 255: 202–207.

Olivares, M.L. Vera-Candioti, L. & Berli, C.L.A. 2009. The EOF of polymer solutions, *Electrophoresis*, 30: 921–929.

Tang, G.H. Li, X.F. He, Y.L. & Tao, W.Q. 2009. Electroosmotic flow of non-Newtonian fluid in microchannels, *Journal of Non-Newtonian Fluid Mechanics*, 157: 133–143.

Tang, G.H. Ye, P.X. & Tao, W.Q. 2010. Electrovicous effect on non-Newtonian fluid flow in microchannels, *Journal of Non-Newtonian Fluid Mechanics*, 165: 435–440.

Tanner, R.I. 2000. *Engineering Rheology*, New York: Oxford University.

vander Heyden, F.H.J. Bonthuis, D.J. Stein, D. Meyer, C. & Dekker, C. 2006. Electrokinetic energy conversion efficiency in nanofluidic channels, *Nano Letters*. 6: 2232–2237.

Vasu, N. & De, S. 2010a. Electroosmotic flow of power-law fluids at high zeta potentials, *Colloids and Surfaces A: Physicochemical and Engineering Aspects*, 368: 44–52.

Vasu, N. & De, S. 2010b. Electroviscous effects in purely pressure driven flow and stationary plane analysis in electroosmotic flow of power-law fluids, *International Journal of Engineering Science*, 48: 1641–1658.

Zhao, C. Zholkovskij, E. Masliyah, J.H. & Yang, C. 2008. Analysis of electroosmotic flow of power-law fluids in a slit microchannel, *Journal of Colloid and Interface Science*. 326: 503–510.

Zhao, C. & Yang, C. 2010. Nonlinear Smoluchowski velocity for electroosmosis of power-law fluids over a surface within arbitrary zeta potentials, *Electrophoresis*, 31: 973–979.

*Advanced Materials, Structures and Mechanical Engineering – Kaloop (Ed.)*
© *2016 Taylor & Francis Group, London, ISBN 978-1-138-02793-0*

# Electrochemical study of Cu nanowire growth in aqueous solution

J.C. Felizco & M.D. Balela
*Department of Mining, Metallurgical and Materials Engineering, University of the Philippines, Diliman, Quezon City, Philippines*

ABSTRACT: The growth of copper (Cu) nanowires formed by electroless deposition in aqueous solution was investigated by in situ mixed potential measurements during synthesis. Higher aspect ratio nanowires were produced at the higher reaction temperature. This can be attributed to enhanced reduction of Cu (II) ions at elevated temperature. On the other hand, increasing the concentration of the structure-directing agent (ethylene diamine) increases the nanowire aspect ratio due to improved capping effect, but does not affect Cu reduction.

## 1 INTRODUCTION

Copper nanowires have currently been gaining interest due to their superior electrical conductivity and high surface area. These are of potential use in chemical and biological sensors, light polarizers, nanoscale electrical interconnects and transparent conducting electrodes (Zhao et al. 2012, Virk 2011, Deng et al. 2013). As the demand for such electronic devices increases, so is the need for finding effective, low cost and easily scalable copper nanowire synthesis process. Over the years, various fabrication methods have been developed, such as hydrothermal method (Zhao et al. 2012), electrodeposition (Deng et al. 2013), chemical vapor deposition (Choi & Park 2004), and electroless deposition (Xu et al. 2014, Chang et al. 2005, Rathmell et al. 2010). Among these, electroless deposition is one of the most promising due to its cost-effectiveness and good scalability (Xu et al. 2014). In this method, metallic deposition is driven by the chemical reduction of metallic ions in an aqueous solution containing a reducing agent (Djokic 2002). Additives such as structure-directing agents and surfactants can influence the reaction rate and final morphology of the products. Wiley et al reported a facile, large scale and low temperature method of producing high aspect ratio Cu nanowires (Wiley & Rathmell 2013). Several succeeding studies discuss the nanowire growth mechanism (Rathmell et al. 2010), quantitative kinetics analysis of growth (Ye et al. 2014), coating with other metals to prevent oxidation (Rathmell et al. 2012), and applicability in transparent conducting electrodes (Rathmell et al. 2010, Ye et al. 2014, Rathmell & Wiley 2011) and electrocatalysts (Du et al. 2015). However, this synthesis process has not yet been investigated from an electrochemical perspective. This kind of

study would allow in depth understanding of the influence of solution conditions on the stability of various Cu species in the solution (Yagi 2011).

Copper deposition by hydrazine is facilitated by the following simultaneous reduction-oxidation reactions:

$$Cu^{2+}_{(aq)} + 2e^- \rightarrow Cu_{(s)} \qquad \text{(reduction)}$$

$$N_2H_{4(aq)} + 4OH^- \rightarrow N_{2(g)} + 4H_2O + 4e^- \quad \text{(oxidation)}$$

$$2Cu^{2+}_{(aq)} + N_2H_{4(aq)} + 4OH^- \rightarrow 2Cu + N_2 + 4H_2O$$
$$\text{(over-all)}$$

In this system, several partial reactions occur simultaneously. Some of these are anodic in nature ($N_2H_4$ oxidation to $N_2$ gas) while some are cathodic (Cu deposition and $H_2$ generation) (Yagi 2011). The potential at the point where the sum of all the currents due to anodic reactions (Ia) equals the sum of all the cathodic currents (Ic) present in the system is called the mixed potential (Yagi 2011). By comparing the mixed potential and the calculated oxidation-reduction potential of Cu species in the solution, the stable Cu species can be determined.

In this work, the growth of copper nanowires in aqueous solution is electrochemically investigated by in situ mixed potential measurements. These measurements are then compared with the calculated oxidation-reduction potential of Cu. The effect of temperature and EDA concentration on the morphology and reduction rate were also examined.

## 2 EXPERIMENTAL

### 2.1 *Synthesis of copper nanowires*

This method was adopted from Wiley et al. (2013). In a 1000 ml round bottom flask, 200 mL of

15 M sodium hydroxide [NaOH, RCI Labscan] and 10 mL solution of 0.1 M copper nitrate hemi (pentahydrate) [$Cu(NO_3)_2 \cdot 5H_2O$, Sigma Aldrich] were mixed. Varying amounts (0, 72, 108 and 180 mM) of ethylenediamine [EDA, Sigma Aldrich] and 0.25 mL hydrazine [$N_2H_4$, 35 wt%, Sigma Aldrich] were added to the solution while stirring at 200 rpm. The solution was then heated to varying reaction temperatures (60, 70 and 80 °C) for 5 min. Then, 25 mL of 0.46 mM poly (vinylpyrrolidone) aqueous solution [PVP, 10,000 MW, Sigma Aldrich] was gently added on top. The flask was then placed in the ice bath for 1 h. The nanowires were collected by centrifugation and washed with 3 wt.% $N_2H_4$ solution. After several washings, the Cu NWs were then stored in 25 mL storage solution containing 1 wt% PVP and 3 wt% hydrazine aqueous solution.

## 2.2 *Electrochemical set-up*

During the 5 min heating time, in-situ mixed potential measurement was done in a potentiostat/galvanostat [Autolab PGSTAT128N] using a two-electrode cell set up with Pt sheet (0.5 cm × 20 cm × 20 cm) as the working electrode and Ag/AgCl electrode (Metrohm) as the reference electrode.

## 2.3 *Characterization techniques*

Morphological analysis of the Cu products was performed using Scanning Electron Microscopy (SEM, JEOL 5300). Structural analysis was done using X-Ray Diffraction (XRD, Shiamdzu XRD-7000).

## 3 RESULTS AND DISCUSSION

### 3.1 *Formation of Cu nanowires*

The SEM images of the Cu nanowires formed by electroless deposition at 60°C at different reaction times are shown in Figure 1. Upon addition of $N_2H_4$ at room temperature, the solution immediately turned from deep blue to milky white. At about 50°C, spherical seeds started to form (Fig. 1a). These seeds acted as nuclei for nanowire formation. When the temperature reached 60°C, several thick and short nanowires have already grown from the spherical seeds. These nanowires have diameters of about 140 nm and lengths of up to 4 μm. The diameter of the nanowires interestingly decreases to about 89 nm as the reaction reached 2–5 min at 60°C. Additionally, the length of the nanowires considerably increases to about 10 μm. Immersion of the solution in the ice bath to delay the reaction resulted to the continued reduction on the diameter of the nanowires up to a final diameter of about 63 nm. The corresponding XRD patterns of the Cu products at varying reaction times as shown are shown in Figure 2. Figure 2a shows the diffraction pattern of the spherical seeds,

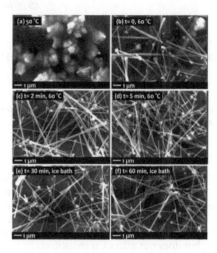

Figure 1. Cu nanowire growth observed at varying stages during the reaction, (a) before 60°C, (b) at 60°C, (c) 2.5 min in 60°C, (d) 5 min in 60°C, (e) 30 min in the ice bath and (f) 1 hour in the ice bath.

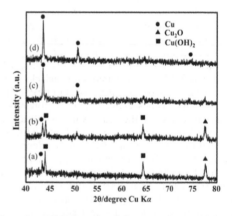

Figure 2. XRD Pattern of Cu nanowires at (a) 0 min at 60°C, (b) 5 min at 60°C, (c) 30 min in the ice bath, (d) 60 min in the ice bath.

which show peak values at 44.1°, 64.5° and 77.6°. The peaks at 44.1° and 64.5° can be attributed to Cu $(OH)_2$ while the peak at 77.6° is due to $Cu_2O$. A small peak of Cu at about 43.5° suggests that Cu (II) reduction has already initiated. After 5 min at 60°C (Fig. 2b), there is an obvious decrease in the intensity of Cu $(OH)_2$ peak at about 44.1° relative to the (111) Cu peak at 43.5°. Additionally, a small peak at about 50.6° attributed to (200) Cu is also indexed as seen in Figure 2b. This indicates that more Cu (II) has been reduced as the reaction proceeds. It is possible that the nanowires formed at the early stages of reaction is a mixture of metallic Cu and Cu $(OH)_2$. Cu $(OH)_2$ is possibly precipitated on the nanowires due to the very high concentration of hydroxyl ions in the solution.

As the reduction reaction proceeds, the Cu (OH)$_2$ deposited on the nanowires are possibly reduced. This explains the sharp decrease in the diameter from 140 to 89 nm as seen in Figure 1b–d. After 30–60 min in the ice bath, sharp Cu peaks are present in the XRD patterns in Figure 2c–d. At this stage, most of the Cu (II) ions and Cu (OH)$_2$ in the solution have been reduced to metallic Cu. This agrees well with the SEM image in Figure 1f where the smallest Cu nanowires of 63 nm are obtained.

## 3.2 *Effect of reaction temperature on Cu deposition and nanowire growth*

Temperature-dependence of the nanowire growth was studied at different reaction temperatures (60, 70 and 80°C) as shown in Figure 3. With decreasing reaction temperature, the nanowire length was observed to increase from 12.9 to 18.5 μm. On the other hand, a sharp drop in diameter from 129 to 63 nm was also observed. At higher temperature, Cu (II) reduction rate is enhanced. Thus, more Cu atoms are deposited at a time, leading to the formation of thick and short nanowires.

Figure 4 shows the mixed potential of the solution at different temperatures. Horizontal lines indicate the calculated oxidation-reduction potential of Cu/Cu (II) at increasing temperature, assuming that Cu$^{2+}$ aqueous ions are in equilibrium with Cu$_2$O. As temperature increases, the mixed potential decreases, and they become more negative than the oxidation-reduction potential of Cu/Cu (II). This implies that the driving force for Cu (II) reduction increases with increasing temperature. It is possible that N$_2$H$_4$ oxidation is enhanced at high temperatures. Consequently, more Cu (II) ions are reduced at a time, leading to the growth of thick and short nanowires. Decreasing the reaction temperature slows down the total reaction rate, which gives time for the anisotropic growth of Cu into nanowires.

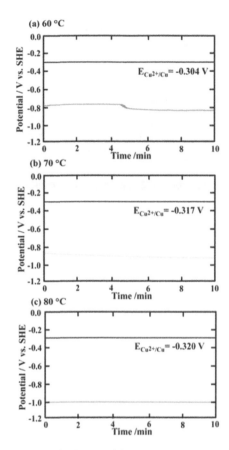

Figure 4. Mixed potential measurements at varying reaction temperatures. (a) 60°C, (b) 70°C, (c) 80°C.

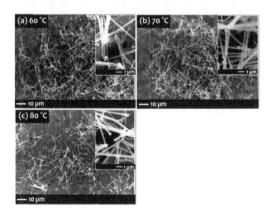

Figure 3. SEM Images of Cu nanowires at varying reaction temperatures. (a) 60°C, (b) 70°C, (c) 80°C.

## 3.3 *Effect of EDA concentration on Cu deposition and nanowire growth*

Figure 5 shows the SEM images of Cu products formed at various EDA concentrations. When no EDA was added, large spherical particles of about 670 nm in diameter were formed. An addition of EDA led to the growth of Cu into nanowires as shown in Figure 5b–d. When the EDA concentration was 180 mM, larger aspect ratio Cu nanowires are produced. EDA capped the sides of the Cu nuclei, which shields them from Cu deposition. Growth of Cu nanowires along its longitude is then favored. These resulted in thinner Cu nanowires. The mixed potential of the solution at varying EDA concentrations is shown in Figure 6. Since EDA has no effect on the temperature and pH of the total solution, the oxidation-reduction potential of Cu/Cu (II) was constant for all EDA concentrations. The mixed potential are below E$_{Cu(II)/Cu}$ = −0.320 V in all EDA concentrations sampled, which suggests that Cu (II) reduction occurs readily in all samples. In addition, there is no significant variation in the

Figure 5. SEM images of Cu nanowires at varying EDA concentrations. (a) 0 mM, (b) 72 mm, (c) 108 mM, and (d) 180 mM.

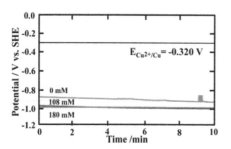

Figure 6. Mixed potential measurements at EDA concentrations of 0 mM, 108 mM and 180 mM.

mixed potential values as the EDA concentration is increased from 0–180 mM. It is possible that EDA only acts as a structure-directing agent and does not influence the rate of Cu reduction in the solution.

## 4 CONCLUSIONS

Electrochemical investigation of Cu nanowire in aqueous solution by electroless deposition was performed using in situ mixed potential measurements. The morphology of the nanowires was proven to be affected by experimental conditions such as reaction temperature and reagent concentration. High reaction temperature enhances the Cu (II) reduction, which leads to thick and short nanowires. Increasing the amount of EDA, on the other hand, greatly enhances the aspect ratio of nanowires, but does not directly affect the Cu (II) reduction rate.

## ACKNOWLEDGEMENT

The authors would like to acknowledge the support of the Engineering Research and Development for Technology (ERDT) Program of the Department of Science and Technology (DOST) in this paper.

## REFERENCES

Chang, Y. Lye, M. & Zeng, H. 2005. Large-Scale Synthesis of High-Quality Ultralong Copper Nanowires, *Langmuir*, 21: 3746–3748.

Choi, H. & Park, S. 2004. Seedless Growth of Free-Standing Copper Nanowires by Chemical Vapor Deposition, *Journal of American Chemical Society*, 126: 6248–6249.

Deng, Y. Wang, N. Ling, H. & Li, M. 2013. *Synthesis of Copper Nanowires on the Substrates in Aqueous Solution*, IEEE, 148–151.

Djokic, S. 2002. *Electroless Deposition of Metals and Alloys,* vol. 35. New York, USA: Kluwer Academic/Plenum Publishers.

Du, J. Chen, Z. Ye, S. Wiley, B.J. & Meyer, T.J. 2015. Copper as a robust and transparent electrocatalyst for water oxidation, *Angewandte Chemie International Edition in English*, 54: 2073–2078.

Rathmell, A.R. & Wiley, B.J. 2011. The synthesis and coating of long, thin copper nanowires to make flexible, transparent conducting films on plastic substrates, *Advanced Materials*, 23: 4798–4803.

Rathmell, A.R. Bergin, S.M. Hua, Y.L. Li, Z.Y. & Wiley, B.J. 2010. The growth mechanism of copper nanowires and their properties in flexible, transparent conducting films, *Advanced Materials*, 22: 3558–3563.

Rathmell, A.R. Nguyen, M. Chi, M. & Wiley, B.J. 2012. Synthesis of oxidation-resistant cupronickel nanowires for transparent conducting nanowire networks, *Nano Letters*, 12: 3193–3199.

Virk, H.S. 2011. *Fabrication and Characterization of Copper Nanowires,* Nanowires Implementations and Applications, Ed., ed: InTech.

Wiley, B.J. & Rathmell, A.R. 2013. *Compositions and Methods of Growing Copper Nanowires*, USA Patent.

Xu, C. Wang, Y. Chen, H. Zhou, R. & Liu, Y. 2014. Large-scale synthesis of ultralong copper nanowires via a facile ethylenediamine-mediated process, *Journal of Materials Science: Materials in Electronics*, 25: 2344–2347.

Yagi, S. 2011. *Potential-pH Diagrams for Oxidation-State Control of Nanoparticles Synthesized via Chemical Reduction,* in Thermodynamics—Physical Chemistry of Aqueous Systems, D.J.C.M. PirajÃ¡n, Ed., ed: InTech.

Ye, S. Chen, Z. Ha, Y.C. & Wiley, B.J. 2014. Real-time visualization of diffusion-controlled nanowire growth in solution, *Nano Letters*, 14: 4671–4676.

Ye, S. Rathmell, A.R. Stewart, I.E. Ha, Y.C. Wilson, A.R. & Chen, Z. et al. 2014. A rapid synthesis of high aspect ratio copper nanowires for high-performance transparent conducting films, *Chemical Communications,* 50: 2562–2564.

Ye, S. Rathmell, A.R. Ha, Y.C. Wilson, A.R. & Wiley, B.J. 2014. The role of cuprous oxide seeds in the one-pot and seeded syntheses of copper nanowires, *Small*, 10: 1771–1778.

Zhao, Y. Zhang, Y. Li, Y. & Yan, Z. 2012. Soft synthesis of single-crystal coppernanowires of various scales, *New Journal of Chemistry,* 36: 130–138.

*Advanced Materials, Structures and Mechanical Engineering – Kaloop (Ed.)*
*© 2016 Taylor & Francis Group, London, ISBN 978-1-138-02793-0*

# Effects of synthesis techniques and initial reagents on compositions and particle morphology of hydroxyapatite

O. Jongprateep & C. Nueangjumnong
*Department of Materials Engineering, Faculty of Engineering, Kasetsart University, Bangkok, Thailand*

ABSTRACT: Hydroxyapatite ($Ca_{10}(PO_4)_6(OH)_2$, HAp), a material commonly used in bone grafting can be synthesized using simple and cost-effective ceramic processing methods, such as through solid-state reaction or through solution combustion techniques. This study was aimed at synthesizing hydroxyapatite using those techniques, as well as at examining the effects of initial reagents (Ca/P ratios) on chemical compositions and morphology of hydroxyapatite particles. Experimental results indicated that powders synthesized by either of the methods contained hydroxyapatite and calcium oxide phases. However, an additional tricalcium phosphate phase ($Ca_3(PO_4)_2$, TCP) was observed in powder synthesized through solution combustion technique. The results have also shown that varying Ca/P ratios (1.7 and 2.3) did not significantly alter the synthesized powder's chemical compositions. Microstructural examination revealed that all synthesized powders were comprised of spherical particles of sizes ranging from 1.38 to 1.79 micrometers. Although neither synthesis technique nor Ca/P ratio affected particle size, the synthesis technique did affect morphology. While the de-agglomerated spherical particles were evident in powders synthesized by the solution combustion technique, inter-particle necking was clearly visible in powders synthesized by the solid-state reaction technique.

## 1 INTRODUCTION

Bone replacement is required when one suffers from bone loss as a result of disease or severe fractures. Materials for bone replacement require that they should not only possess similar characteristics as human bones but also be biocompatible. Thus, hydroxyapatite ($Ca_{10}(PO_4)_6(OH)_2$, HAp), owing to its biocompatibility and nontoxicity properties (Webler et al. 2014) has become one of the most used material for artificial bone production. Hydroxyapatite in powder form, can be used as a coating material (Surmeneva et al. 2015) and as raw material for fabricating three dimensional bone structures (Ramay & Zhang 2003).

Hydroxyapatite can be extracted from animal bones and sea corals or synthesized chemically. Synthesis of hydroxyapatite can be achieved through a number of chemical techniques, such as solid-state method (Guo et al. 2013), solution combustion method (Ghosh et al. 2011), sol-gel method (Tredwin et al. 2013), or high-energy ball milling method (Kheradmandfard & Fathi 2013). Solid-state reaction (Khongnakhon & Jongprateep 2013) and solution combustion are both rapid and cost effective techniques of producing ceramic powders (Zhao et al. 2014). Appropriate synthesis techniques generally provide ceramics powders with the desired chemical composition, particle sizes and morphology, as well as particle size distribution. In addition to the synthesis techniques, initial reagents, specific types and concentrations, can significantly affect chemical compositions and particle morphology of the powders (Jongprateep et al. 2015).

This study, therefore, is aimed at synthesizing hydroxyapatite by solid-state reaction and solution combustion techniques, as well as at examining the effects of initial reagents (Ca/P ratios) on chemical compositions and morphology of hydroxyapatite particles.

This study, therefore, is aimed at synthesizing hydroxyapatite by solid-state reaction and solution combustion techniques, as well as at examining the effects of initial reagents (Ca/P ratios) on chemical compositions and morphology of hydroxyapatite particles.

## 2 MATERIALS AND METHODS

### 2.1 Powder preparation

Solid-state reaction and solution combustion techniques were employed in the synthesis of hydroxyapatite ($Ca_{10}(PO_4)_6(OH)_2$, HAp) powders. To synthesize the powders by the solid-state reaction, calcium nitrate tetrahydrate powder ($Ca(NO_3)_2 \cdot 4H_2O$, Daejung, 97%) and ammonium phosphate dibasic powder ($(NH_4)_2HPO_4$, Daejung, 98.5%) were mixed and ground in a mortar at a Ca:P mole ratio of 1.7 and 2.3. The powder mixtures were subjected to calcination at 800°C for 3 hours at a heating rate of 10°C/min.

To synthesize the powders by the solution combustion technique, aqueous solutions with stoichiometric Ca:P ratios of 1.7 and 2.3 were prepared. Glycine ($NH_2CH_2COOH$, Daejung, 99%), acting as combusting fuel in the powder synthesis process, was added to the prepared solution mixtures. To ensure homogeneity, the prepared solutions were stirred using a magnetic stirrer at a speed of 100 rounds per minute. Combustion was initiated by heating the solution to temperature approximately 400°C. Upon combustion completion, powder products were collected and calcined at 800°C for 3 hours at a heating rate of 10°C/min.

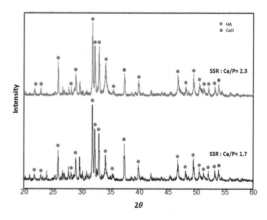

Figure 1. XRD patterns of powders synthesized by the solid-state reaction technique, prepared at Ca/P ratios = 1.7 and 2.3.

### 2.2 *Powder characterization*

Compositions of the calcined powders were examined using an x-ray diffractometer (Bruker, D8 Advance), over angles ranging from 20° to 60° in 2θ, at a step up gradient of 0.01° and a scanning rate of 1.3 °/min. A scanning electron microscope (FEI, Quanta450) was employed in the morphological examination of the particles. Particle sizes were determined using Image J Software.

### 3 RESULTS AND DISCUSSION

#### 3.1 *Chemical compositions*

Chemical compositions of the powders synthesized by the solid-state reaction and solution combustion techniques were examined by x-ray diffraction. The results indicated that all powders contained hydroxyapatite ($Ca_5 (PO_4)_3OH$, JCPDS 01-084-1998) and calcium oxide (CaO, JCPDS 01-070-4068) phases, as shown in Figures 1 and 2. However, it was noted that powders synthesized by the solution combustion technique also contained beta-tricalcium phosphate (Beta-TCP; $\beta$-$Ca_3(PO_4)_2$, JCPDS 01-070-2065) in addition to hydroxyapatite and calcium oxide. Beta-tricalcium phosphate is a phase commonly observed in the synthesis process of hydroxyapatite (Ou et al. 2013). Owing to its biocompatibility and osteoconduction properties (Zhao et al. 2014), it can be used in the production of artificial bone.

The results of the effects of Ca/P ratios on the powder's chemical compositions revealed that no significant change in composition could be noted in powders synthesized by the solid-state reaction but slight decrease in peak intensity of beta-tricalcium phosphate was evident in powders sythesized

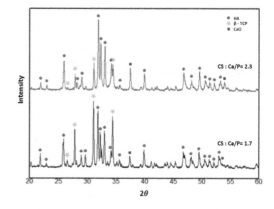

Figure 2. XRD patterns of powders synthesized by the solution combustion technique, prepared at Ca/P ratios = 1.7 and 2.3.

by solution combustion technique particularly when as Ca/P ratio was increased from 1.7 to 2.3. The decrease in beta-tricalcium phosphate peak intensity suggests that higher Ca seems to favor hydroxyapatite synthesis. A similar observation, which suggested that inhibition of beta-TCP nucleation could be achieved in abundant-calcium system (high Ca/P ratio), was reported by Sadat-Shojai et al. 2013.

#### 3.2 *Powder morphology*

Spherical particles were generally observed in all powders, as shown in Figures 3 and 4. Variations in synthesis techniques tended to affect particle morphology. While inter-particle necking was clearly visible in powders synthesized by the solid-state

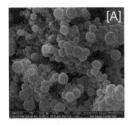

Figure 3. SEM micrographs of the hydroxyapatite powders synthesized by the solid-state reaction technique, [A] Ca/P = 1.7 and [B] Ca/P = 2.3.

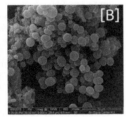

Figure 4. SEM micrographs of the hydroxyapatite powders synthesized by the solution combustion technique, [A] Ca/P = 1.7 and [B] Ca/P = 2.3.

reaction technique, de-agglomerated spherical particles were observed in powders synthesized by the solution combustion technique. No significant effects of Ca/P ratios on particle morphology were observed in this study.

### 3.3 *Particle size and distribution*

There was no clear indication that particle refinement or coarsening was caused by the variation in either synthesis techniques or Ca/P ratios. Table 1 shows that particle size of the powders synthesized by solid-state reaction indicated that average particle sizes of powders prepared at Ca/P of 1.7 and 2.3 were 1.63 and 1.46 micrometers, respectively. For powders synthesized by the solution combustion technique, average particle sizes were 1.38 and 1.79 micrometers for Ca/P of 1.7 and 2.3, respectively.

All powders synthesized by the solid-state reaction technique and the powders synthesized by the solution combustion technique at Ca/P = 1.7 had similar particle size distribution, with size ranging from 0.25 to 3 μm. For powder synthesized by the solution combustion technique at Ca/P = 2.3, a slightly wider particle size distribution, with size ranging between 0.5 and 4 μm, was observed. Typical histograms representing particle sizes of the powders are shown in Figures 5–8.

Table 1. Average particle size of hydroxyapatite powders.

| Synthesis techniques | Average particle size (μm) | |
| --- | --- | --- |
| | Ca/P = 1.7 | Ca/P = 2.3 |
| SSR | 1.63 ± 0.45 | 1.46 ± 0.43 |
| CS | 1.38 ± 0.39 | 1.79 ± 0.76 |

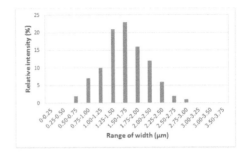

Figure 5. Particle size distribution of the hydroxyapatite powders synthesized by the solid-state reaction technique at Ca/P = 1.7.

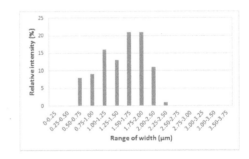

Figure 6. Particle size distribution of the hydroxyapatite powders synthesized by the solid-state reaction technique at Ca/P = 2.3.

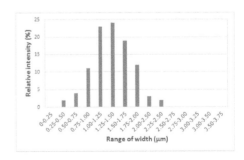

Figure 7. Particle size distribution of the hydroxyapatite powders synthesized by the solution combustion technique at Ca/P = 1.7.

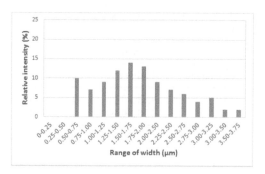

Figure 8. Particle size distribution of the hydroxyapatite powders synthesized by the solution combustion technique at Ca/P = 2.3.

Since the difference in particle size and size distribution of all powders was insignificant, it can be deduced that particle size and size distribution neither depended on synthesis techniques nor Ca/P ratios.

## 4 CONCLUSIONS

Hydroxyapatite powders have been successfully synthesized by solid-state reaction and by solution combustion techniques. The effects of synthesis techniques as well as Ca/P ratios on chemical compositions and particle morphology were also examined in this study. Experimental results revealed that synthesis techniques affected chemical compositions and particle morphology but not particle size. While the two phases of hydroxyapatite and calcium oxide were observed in powders synthesized by either of the techniques, an additional beta-tricalcium phosphate phase, was noted in powder synthesized by the solution combustion technique. Scanning electron micrographs revealed that inter-particle necking was clearly visible in powders synthesized by the solid-state reaction technique, while powders synthesized by the combustion technique appeared as de-agglomerated spherical particles. Experimental results also indicated that Ca/P ratios have no significant effects on chemical compositions, particle size nor particle morphology.

## ACKNOWLEDGEMENT

Equipment supports from the Department of Materials Engineering, Faculty of Engineering, Kasetsart University, and the Department of Science Service, Ministry of Science and Technology, Thailand are gratefully acknowledged.

## REFERENCES

Ghosh, S.K. et al. 2011. Synthesis of nano-sized hydroxyapatite powders through solution combustion route under different reaction conditions. *Materials Science and Engineering B*, 176(1): 14–21.

Guo, X. et al. 2013. Effect of calcining temperature on particle size of hydroxyapatite synthesized by solid-state reaction at room temperature. *Advanced Powder Technology*, 24(6): 1034–1038.

Jongprateep, O. et al. in press. Nanoparticulate titanium diox-ide synthesized by sol-gel and solution combustion techniques. *Ceramics International* 41(S1): S169–S173.

Kheradmandfard, M. & Fathi M.H. 2013. Fabrication and characterization of nanocrystalline Mg-substituted fluorapatite by high energy ball milling. *Ceramics International*, 39(2): 1651–1658.

Khongnakhon, T. & Jongprateep, O. 2013. Ba0.9A0.1 TiO3 (A = Al and Mg) powders synthesized by solid state reaction technique and their dielectric properties. *Advanced Materials Research*, 747: 603–606.

Ou, S.F. et al. 2013. Phase transformation on hydroxyapatite decomposition. *Ceramics International*, 39(4): 3809–3816.

Ramay, H.R. & Zhang, M. 2003. Preparation of porous hydroxyapatite scaffolds by combination of the gel-casting and polymer sponge methods. *Biomaterials*, 24(19): 3293–3302.

Sadat-Shojai, M. et al. 2013. Synthesis methods for nanosized hydroxyapatite with diverse structures. *Acta Biomaterialia*, 9: 7591–7621.

Surmeneva, M.A. et al. 2015. Ultrathin film coating of Hydroxyapatite (HA) on a magnesium–calcium alloy using RF magnetron sputtering for bioimplant applications. *Materials Letters*, 152(0): 280–282.

Tredwin, C.J. et al. 2013. Hydroxyapatite, fluor-hydroxyapatite and fluorapatite produced via the sol–gel method. Optimisation, characterisation and rheology. *Dental Materials*, 29(2): 166–173.

Zhao, J. et al. 2014. Solution combustion method for synthesis of nanostructured hydroxyapatite, fluorapatite and chlorapatite. *Applied Surface Science*, 314(0): 1026–1033.

Zhao, J. et al. 2014. Rietveld refinement of hydroxyapatite, tricalcium phosphate and biphasic materials prepared by solution combustion method. *Ceramics International*, 40(2): 3379–3388.

*Advanced Materials, Structures and Mechanical Engineering – Kaloop (Ed.)*
© 2016 Taylor & Francis Group, London, ISBN 978-1-138-02793-0

# Synthesis and characterization of Fe-Sn nanoparticles via the galvanic replacement method

E.R. Magdaluyo Jr., A.L. Bachiller & P.J. Nisay
*Department of Mining, Metallurgical and Materials Engineering, University of the Philippines, Diliman, Quezon City, Philippines*

ABSTRACT: Tin-iron (Sn-Fe) nanoparticles were synthesized using iron nanoparticles as the template under the hydrothermal environment. Surface morphology of the obtained nanoparticles revealed a spherical crystal habit, with decreasing size as the reaction temperature and time increased. X-ray diffraction of the nanoparticles produced at 80°C and 14-hour reaction time revealed a more crystalline phase with an average diameter of 178 nm. The cyclic voltammetry plot given in these reaction parameters showed a variation in the oxidation peak, with the occurrence of the second peak at around −0.2 V associated with the oxidation of tin into tin oxide. The majority of the component is tin, as revealed by the elemental analyses, and the formation of the Fe-Sn intermetallic alloy nanoparticles can be attributed to the galvanic replacement of tin as it diffuses into the surface of the iron nanoparticle template.

## 1 INTRODUCTION

Nanoparticles are widely used in different industries due to their smaller size, various shapes and increased surface area, and exhibit dissimilar properties than their parent materials. Due to these unique properties, numerous applications can be adapted and tuned for specific use in different fields, but are not limited to biomedicine, catalysis, energy storage and energy generating devices (Reiss et al. 2005, Warsi et al. 2010, Ragupathi et al. 2015, Shahid et al. 2013).

Tin (Sn) has been widely used in lithium (Li) ion batteries due to its high charge capacity at approximately 992 mAhg$^{-1}$. However, synthesizing nanostructured Sn particles has only been done using limited methods such as vapor deposition, chemical reduction of tin salts and thermal decomposition of inorganic or organic precursors (Cao et al. 2014, Glaspell et al. 2006). Other techniques include alloying of Sn with other metals such as Co, Ni and Cu, which will result in the formation of an intermetallic layer. This intermetallic layer serves as a buffer that alleviates stress during lithium insertion/deletion in the charge/discharge cycle (Bing et al. 2014). Moreover, the intermetallic layer prevents the agglomeration of nanoparticles.

Sn has no ability to form an alloy with Li, and forms an inactive matrix with Sn atoms. However, an alloy of Sn and iron (Fe) gives the best electrochemical activity and superior cycling performance (Alexandrescu et al. 2012). Since Fe readily forms oxides when exposed to air, it is natural to have it covered by oxide films that are non-wetted by liquid metals. This thin oxide coating is strongly improved by heating, and, therefore, intermetallic formation is possible by the reaction that likely leads to the replacement of the oxide by the surface of the intermetallic compound (Protsenko et al. 2001).

Thus, this study aims to synthesize Fe-Sn nanoparticles using a simpler method by allowing Sn to coat itself on the Fe nanoparticles. Characterization was done on the surface morphology, phase and composition as well as on the electrochemical properties of the obtained Fe-Sn nanoparticles.

## 2 EXPERIMENTAL PROCEDURE

A sample of 0.2675 g iron chloride hexahydrate ($FeCl_2 \cdot 6H_2O$) was added to 15 mL ethanol. The solution was heated to 60°C at constant stirring. Then, 16.01 mL hydrazine ($N_2H_4$) was added dropwise for 5 minutes. The mixture was then allowed to react for another 25 minutes. The synthesized iron nanoparticles were alternately washed with deionized water and ethanol four times in order to remove any presence of impurities. Subsequently, sonication for 10 minutes and centrifugation for 5 minutes were performed to collect the nanoparticles.

Tin sulfate ($SnSO_4$) was added to the iron nanoparticles, at a 1:1 stoichiometric ratio of Fe and Sn. All the reagents were mixed in a three-necked flask and deionized water was added up to 250 mL. The Fe-Sn nanoparticles were synthesized at three reaction temperatures (60°C, 70°C and 80°C) under constant stirring. Nitrogen gas was purged into the flask for 10 minutes in order to ensure an inert atmosphere. The mixture was allowed to react for a maximum of 14 hours. A 2 mL solution was harvested every 2 hours starting from the 4-hour reaction time. Finally, 3% hydrazine was added to the harvested solution in order to prevent the samples from oxidizing.

The surface morphology of the synthesized nanoparticles was analyzed using a Scanning Electron Microscope (SEM JEOL5300) operating at 20 kV under 15 kX magnification and coupled with Energy-Dispersive X-Ray (EDX) spectroscopy in order to determine the prevalent elemental species in the samples. The Shimadzu X-Ray Diffractometer (XRD-7000) operated at 40 kV and 30 mA with Cu-Kα radiation in the range of 20–70° was employed to determine the phases present.

The cyclic voltammetry plot was also obtained using the PGSTAT128N Potentiostat, with Ag|AgCl as the reference electrode and the platinum as both a working electrode and a counter electrode. *In situ* synthesis of Fe nanoparticles was performed in order to track the chemical reactivity of Fe. For the other samples, the Fe-Sn nanoparticles were immersed in the $Na_2SO_4$ solution. The cyclic voltammogram was performed at four scan rates (200 mVs$^{-1}$, 150 mV s$^{-1}$, 100 mV s$^{-1}$ and 50 mV s$^{-1}$).

# 3 RESULTS AND DISCUSSION

The surface morphology of the Fe-Sn nanoparticles generally has a spherical crystal habit, as shown in Figure 1. Agglomeration is very apparent due to the magnetic properties of the iron nanoparticle template. As the reaction temperature and the time increase, the average diameter decreases. This can be explained by the coating mechanism of tin to iron nanoparticles through the galvanic replacement. The iron present on the surface was replaced with the formation of an intermetallic alloy of tin and iron. A longer reaction time at a high temperature would allow more time for the nucleation of the Fe-Sn alloy to occur, thereby producing smaller particles.

The initial reaction of the iron precursor in ethanol produces an iron hydrate complex, and the addition of hydrazine reduces to form the iron nanoparticles:

$$FeCl_2 \cdot 4H_2O_{(s)} + CH_3CH_2OH_{(l)} \rightarrow [Fe(H_2O)_4]^{2+}{}_{(aq)} \quad (1)$$

$$[Fe(H_2O)_4]_2 +{}_{(aq)} + N_2H_{4(l)} \rightarrow N_{2(g)} + H_2O_{(l)} + Fe_{(s)} \quad (2)$$

The equivalent two half reactions in (2) are represented as follows:

$$4OH^-{}_{(aq)} + N_2H_{4(l)} \rightarrow N_{2(g)} + 4H_2O_{(l)} + 4e^- \quad (3)$$

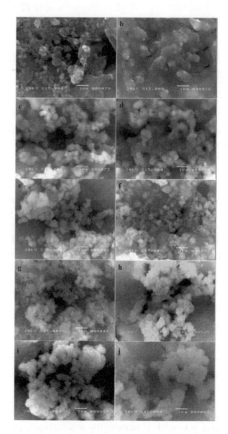

Figure 1. Surface morphology of (a) the iron nanoparticles and (b) the Fe-Sn nanoparticles at 60°C and 6-hour reaction time, (c) Fe-Sn at 60°C and 10-hour reaction time, (d) Fe-Sn at 60°C and 14-hour reaction time, (e) Fe-Sn at 70°C and 6-hour reaction time, (f) Fe-Sn at 70°C and 10-hour reaction time, (g) Fe-Sn at 70°C and 14-hour reaction time, (h) Fe-Sn at 80°C and 6-hour reaction time, (i) Fe-Sn at 80°C and 10-hour reaction time, (j) Fe-Sn at 80°C and 14-hour reaction time.

$$[2e- + [Fe(H_2O)_4]^2+ \rightarrow Fe + 4H_2O]2 \quad (4)$$

The final reduction-oxidation reaction is given as follows:

$$4OH^-{}_{(aq)} + N_2H_{4(l)} + 2[Fe(H_2O)_4]^{2+}{}_{(aq)}$$
$$\downarrow$$
$$2Fe_{(s)} + N_{2(g)} + 12H_2O_{(l)} \quad (5)$$

The synthesized iron nanoparticles were reacted with tin sulfate at different reaction temperatures and times. This would replace the iron with tin at the surface and allows diffusion to form the Fe-Sn alloy:

$$Fe_{(s)} + SnSO_4 \rightarrow FeSO_4 + Sn_{(s)} \quad (6)$$

$$4OH^- + Fe \rightarrow Fe_{(s)} + 2H_2O_{(l)} \qquad (7)$$

Table 1 summarizes the average diameter of the Fe-Sn nanoparticles. The maximum diameter of the Fe-Sn nanoparticles synthesized was 268 nm at 60°C and 6-hour reaction time, and the minimum diameter was 173 nm processed at 70°C and 14-hour reaction time. Moreover, based on the X-ray diffraction of the nanoparticles showing a more crystalline phase, the formation at 80°C and 14-hour reaction time produced an average diameter of 178 nm.

Figure 2 shows the XRD patterns of the Fe-Sn nanoparticles obtained at the 14-hour reaction time. The observed diffraction peaks of the Fe-Sn nanoparticles at 60°C were 25.2°, 44.1° and 64.5° corresponding to (110), (220) and (321) planes, respectively, of the tetragonal crystal structure. Peaks present for the Fe-Sn nanoparticles synthesized at 80°C were 25.78° and 44.08°, corresponding to (110) and (220) planes, respectively. These characteristic peaks are due to the Sn coating of Fe (PDF Card 1071–3772). One characteristic peak was also observed for the Fe-Sn synthesized at 70°C and the diffraction angle of 23.6°.

The diffraction pattern of the Fe nanoparticles structure showed peaks at 32.6° and 36.3°, which

Table 1. Average diameter (nm) of the Fe-Sn nanoparticles.

| Temperature (°C) | Time (hrs) | | |
|---|---|---|---|
| | 6 | 10 | 14 |
| 60 | 268 nm | 245 nm | 229 nm |
| 70 | 207 nm | 192 nm | 173 nm |
| 80 | 185 nm | 183 nm | 178 nm |

Table 2. Average composition of Fe and Sn-Fe nanoparticles based on energy-dispersive X-ray spectroscopy.

| | wt% Fe | wt% Sn | wt% O |
|---|---|---|---|
| Fe NP | 68.70 | – | 31.30 |
| Sn-Fe 60°C, 6 hrs | 14.08 | 85.92 | |
| Sn-Fe 60°C, 10 hrs | 15.82 | 84.18 | |
| Sn-Fe 60°C, 14 hrs | 15.92 | 84.08 | |
| Sn-Fe 70°C, 6 hrs | 23.20 | 76.80 | |
| Sn-Fe 70°C, 10 hrs | 19.70 | 80.30 | |
| Sn-Fe 70°C, 14 hrs | 20.74 | 79.26 | |
| Sn-Fe 80°C, 6 hrs | 3.49 | 96.51 | |
| Sn-Fe 80°C, 10 hrs | 1.98 | 98.02 | |
| Sn-Fe 80°C, 14 hrs | 5.58 | 94.42 | |

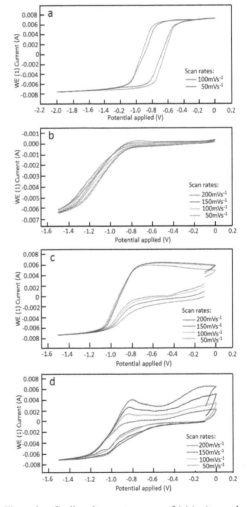

Figure 3. Cyclic voltammetry curve of (a) *in situ* synthesis of Fe nanoparticles (b) Fe-Sn nanoparticles at 60°C (c) Fe-Sn at 70°C and (d) Fe-Sn at 80°C and 14-hour reaction time.

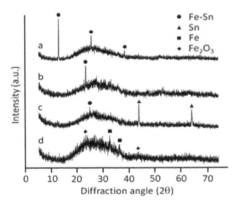

Figure 2. XRD patterns of (a) Fe-Sn nanoparticles at 80°C, (b) Fe-Sn at 70°C, (c) Fe-Sn nanoparticles at 60°C and 14-hour reaction time and (d) Fe nanoparticles.

corresponded to the (100) and (101) planes of the body center cubic structure, respectively. This structure is mainly because of iron (JCPDS No. 87-0721). However, some iron oxide (JCPDS No. 33-0664) was also detected at diffraction angles of 23.3° and 43.5° at (111) and (400) planes, respectively.

The elemental analysis of the Fe-Sn nanoparticles synthesized is given in Table 2. The presence of oxides in the Fe nanoparticle template was confirmed, which was in agreement with the X-ray diffraction analysis. Moreover, the Fe-Sn nanoparticles synthesized at all given temperatures and reaction times showed Sn as the majority composition, which might suggest the occurrence of the coating process of iron nanoparticles.

The electrochemical properties of the Fe-Sn nanoparticles synthesized were investigated using cyclic voltammetry, as shown in Figure 3. The reduction process in the formation of Fe nanoparticles from iron chloride hexahydrate, shown in Figure 3a, started at the initial potential of −2 V to 0 V switching potential, and the cathodic peak potential was around −0.8 V. An identical potential scan was also observed in the oxidation process. A potential window of −1.6 V to 0 V was used to obtain the cyclic voltammetry plot of the Fe-Sn nanoparticles when dissolved in 0.1 M $Na_2SO_4$. No significant change was observed in the values of the cathodic peak potential. However, variation was observed in the oxidation peak in the nanoparticles subjected to the 80°C reaction temperature. The occurrence of the second peak at around −0.2 V might have been due to the oxidation of Sn into SnO.

## 4 CONCLUSIONS

The Fe-Sn nanoparticles were synthesized via the galvanic replacement of Sn in the Fe nanoparticle template. The surface morphology was spherical in nature and the nanoparticles with the highest diameter of 268 nm were synthesized at 60°C and 6-hour reaction time, and those with the minimum diameter of 173 nm were synthesized at 70°C and 14-hour reaction time. X-ray diffraction analyses revealed that the nanoparticles synthesized at 80°C and 14-hour reaction time had more crystalline characteristics with an average diameter of 178 nm. Elemental analyses showed that Sn was the major component in the Fe-Sn nanoparticles synthesized, containing a minimum of 80% composition in all the samples. The results of cyclic voltammetry showed that Sn-coated Fe nanoparticles exhibited best electrochemical properties at 80°C and 14-hour reaction time.

## ACKNOWLEDGMENT

The authors would like to thank and appreciate the Semirara Mining Corporation for the financial support through the Professorial Chair Grant.

## REFERENCES

Alexandrescu, R. et al. Laser processing issues of nano-sized intermetallic Fe-Sn and metallic Sn particles, *Applied Surface Science*, 258: 9421–9426.

Bing, P. et al. 2014. Novel hollow Sn-Cu composite nanoparticles anodes for Li-ion batteries prepared by galvanic replacement reaction, *Journal of Solid State Electrochemistry*, 18: 1137–1145.

Cao, G. et al. 2014. Preparation and electrochemical properties of Fe-Sn (C) nanocomposites as anode for lithium-ion batteries, *Electrochimica Acta*, 129: 93–99.

Glaspell, G. et al. Vapor phase synthesis of metallic and intermetallic nanoparticles and nano-wires: magnetic and catalytic properties, *Pure and Applied Chemistry*, 78: 1667–1689.

Protsenko, P. et al. The role of intermetallics in wetting metallic systems, *Scripta Materialia*, 45: 1439–1445.

Ragupathi, C. et al. 2015. High selective oxidation of benzyl alcohol to benzaldehyde with hydrogen peroxide by cobalt aluminate catalysis: A comparison of conventional and microwave methods, *Ceramics International*, 41: 2069–2080.

Reiss, G. et al. 2005. Magnetic nanoparticles: Applications beyond data storage, *Nature Materials*, 4: 725–726.

Shahid, M. et al. 2013. Structural and electrochemical properties of single crystalline MoV2O8 nanowires for energy storage devices, *Journal of Power Sources*, 230: 277–281.

Warsi, M. et al. 2010. Gd-functionalized Au nanoparticles as targeted contrast agents in MRI: relaxivity enhancement by polyelectrolyte coating, *Chemical Communications*, 46: 451–453.

# Numerical study of micro-reactor inlet geometry for Fischer-Tropsch Synthesis

W. Chaiwang, A. Theampetch, P. Narataruksa & C. Prapainainar
*Department of Chemical Engineering, Faculty of Engineering, King Mongkut's University of Technology North Bangkok, Bangkok, Thailand*
*Research and Development Center for Chemical Engineering Unit Operation and Catalyst Design, King Mongkut's University of Technology North Bangkok, Bangkok, Thailand*

ABSTRACT: In this work, the effect of inlet manifold geometry of a microchannel reactor for Fischer-Tropsch Synthesis (FTS) is studied using a Computational Fluid Dynamics (CFD) model. The model comprises 36 parallel flow channels, with a conical shape of inlet diffuser, and the flow channels are at the micron scale. In order to minimize maldistribution of gaseous reactants ($H_2$ and CO), the geometry, that is, diffuser length and width, of the inlet of the reactor entrance are focused on. The results show that optimal inlet geometry can be obtained with a given degree of flow uniformity. For such design, low deviation of the average velocity in all flow channels, with lower than 0.5% degree of uniformity, is obtained.

## 1 INTRODUCTION

Synthetic fuels produced by Fischer-Tropsch Synthesis (FTS) are considered as high quality fuels. FTS consists of polymerization reactions of hydrocarbons from syngas (Almeida et al. 2013). The micro reactor is a promising technology for FTS due to its benefits, such as its high surface-to-volume ratio enhancing momentum, heat and mass transport (Arzamendi et al. 2010). Flow distribution within a microchannel reactor is one of the most crucial factors affecting reactor performance (Mei et al. 2013), especially for laminar flow in syngas conversion of FTS. The uniformity of flow strongly depends on the structural design of the reactor (Mei et al. 2013, Tonomura et al. 2004). The flow maldistribution can originate from the recirculation zones around the diverging section connecting to the inlet manifold to micro reactor (Agrawal et al. 2012, Tonomura et al. 2004).

The flow maldistribution possibly brings about the unequal residence time of reactants, critically deteriorating reaction efficiency (Agrawal et al. 2012, Tonomura et al. 2004). On the other hand, optimal uniform flow facilitates higher reaction rate and selectivity of the process (Agrawal et al. 2012).

According to many researchers, there have been many attempts to achieve the best fluid distribution in terms of homogeneity in a micro device with channeled plates (E. Vasquez-Alvarez 2010, Tonomura et al. 2004). Alvarez et al. suggested that the inlet region distributing fluid was a challenging part of micro reactor design

(E. Vasquez-Alvarez 2010). Hence, it is important to study the effect of inlet manifold geometric configuration. In this work, the geometry of the reactor entrance, diffuser length (L) and width of the inlet (D), are focused on in order to minimize maldistribution of gaseous reactants.

## 2 CFD MODELLING

The commercial CFD package COMSOL Multiphysics 3.5a is used for simulation. The momentum equations are solved to study the influence of syngas hydrodynamics. The syngas mole ratio ($H_2$/CO) used in this study is 2:1, following to Almeida (Almeida et al. 2013). The simulation is conducted at steady state based on conditions of 508 K and 1 MPa. The corresponding density and viscosity of the syngas are 2.531 kg/m$^3$ and $2.21 \times 10^{-5}$ Pa·s, respectively. The syngas is assumed to be an incompressible gas, and the inlet velocity is determined from Weight Hourly Space Velocity (WHSV) reported by Almeida (Almeida et al. 2013). The resulting velocity in each flow channel is $3.469 \times 10^{-2}$ m/s, which is obtained by $1.7 \times 10^{-8}$ m$^3$/s of volumetric flow rate for a single channel.

Figure 1 illustrates a 2-D simulation model comprising 36 flow channels. The channel width, height, and length are 0.7, 0.7 and 133 mm, respectively. Hence, the hydraulic diameter of a channel is 0.7 mm. The diffuser length is defined as L. In order to decrease wall slip effect of the diffuser, two walls at the edge (E. Vasquez-Alvarez 2010) of the reactor are added to the design.

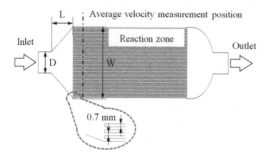

Figure 1.   2-D model of the microreactor.

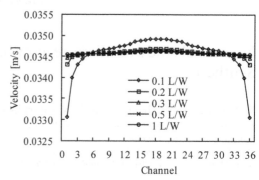

Figure 2. Average velocity profiles in each channel when varying L/W for D/W = 0.3.

The outlet manifold is a hemicircle. With these dimensions, the total volumetric flowrate of fluid flow for 36 channels becomes $6.12 \times 10^{-7}$ m³/s.

In this study, the aspect ratios are defined as L/W and D/W to describe inlet manifold structure in micro reactor. Here, L/W is the ratio of the length of the diffuser to the width of the reaction zone; D/W is the ratio of the width of the inlet to the width of the reaction zone. The average velocity in each single channel is recorded 5 mm from the channel entrance. The degree of flow uniformity (% SD) is used to describe uniformity of flow as shown in Equation 1 (Mei et al. 2013), and then the lower % SD, the lower deviation of average velocity.

$$\sigma_v\% = \frac{\sqrt{\dfrac{1}{N}\sum_{i=1}^{N}\left(v_i - v_m\right)^2}}{v_m} \times 100\% \qquad (1)$$

where, N, $v_i$, and $v_m$ are a number of channels, average velocity in each channel and average velocity of all the channels, respectively.

The design concept of the study is limited such that the maximum length of the diffuser (L) is no longer than the width (W) of the reaction zone, while the width of the inlet (D) is no smaller than 0.2 W.

# 3   RESULTS AND DISCUSSION

## 3.1   Influence of L/W

The diffuser length is one of the most influential parameters, providing distribution area for fluid to the channels. In order to obtain good flow distribution giving uniform average velocity for every single channel, the influence of L/W has to be considered. The influence of L/W on average velocity distribution is shown in Figure 2. For an L/W ratio of 0.1, the middle zone shows the highest

average velocity at the center of the entrance of the channel, while at the edge zone, there is the lowest average velocity due to the limitation of short length of the diffuser. With a short length of diffuser, there is not enough space for fluid flow to distribute before entering the channels. At an L/W greater than 0.2, the results exhibit relatively close average velocity. Thus, it can be concluded that a large L/W ratio gives a long distribution pathway.

## 3.2   Influence of D/W

Liu et al. studied the performance of a micro mixer built from commercial T-junctions and suggested that most micro equipment was difficult to fabricate and construct due to its complex configuration (Liu et al. 2014). In order to provide simplicity and flexibility both for the manufacturer and in operation, the inlet diameter is developed from a commercial 3/8-inch pipe diameter. Then the D/W ratio is varied from 0.2 to 0.5 to study the effect of inlet width on flow distribution. The influence of D/W on average velocity distribution is shown in Figure 3. For a D/W ratio of 0.2, the highest deviation of average velocity is obtained at the middle of the entrance of channels, while the lowest average velocity occurs at the edge of the manifold. As the volumetric flow rate is kept constant while increasing D/W, inlet velocity will decrease with increasing inlet area. This result leads to reduction of maldistribution within the entrance region.

## 3.3   Uniformity of flow distribution

The degree of flow uniformity (% SD) is used as a criterion to consider the uniformity of flow distribution (Mei et al. 2013). A high value of % SD reflects the non-uniform flow. In this work, the % SD is calculated by Equation 1. Figure 4 shows the % SD for different L/W and D/W. When L/W is increased, the diffuser length and area of the

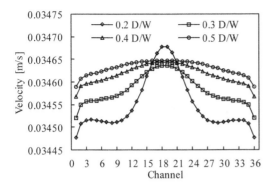

Figure 3. Average velocity profile of microchannel reactor for L/W = 0.5.

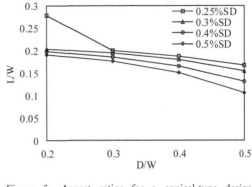

Figure 5. Aspect ratios for a conical-type design entrance of the microreactor at $3.469 \times 10^{-2}$ m/s.

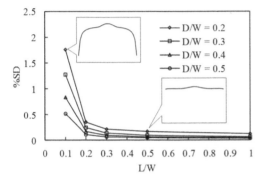

Figure 4. % SD of average velocity at different L/W and D/W with the velocity in a single channel $3.469 \times 10^{-2}$ m/s.

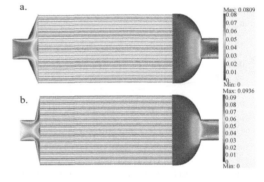

Figure 6. Velocity contours in the microchannel reactor with different dimensions of diffuser: a) L/W of 0.2 and D/W of 0.3, giving % SD lower than 0.5, b) L/W of 0.1 and D/W of 0.3, giving % SD greater than 0.5.

entrance of the manifold are also increased; as a result, decreasing % SD, reflecting more uniform flow, can be seen in the insets. For increasing D/W, the inlet velocity is directly affected by the area of the micro reactor inlet. A D/W ratio of 0.2, inlet velocity is the highest and % SD is higher than that of other D/W ratios. Hence, the higher inlet velocity, the longer the diffuser length required. Furthermore, when considering increasing L/W, the results show a decrease in % SD.

According to Figure 4, in order to design the optimal length of the diffuser and width of the inlet, % SD may be set as a criterion to obtain optimal dimensional parameters. Furthermore, Figure 5 shows different optimal D/W and L/W for designing the entrance of the reactor to obtain uniformity from 0.5% SD to 0.25% SD. The results suggest that lower regions give a high deviation from the average velocity, affecting the non-uniformity in the velocity profile. Consequently, the upper part exhibited very low deviation in velocity and should be the area used to design the entrance of

the reactor. Furthermore, Figure 6 shows velocity contours for suggested design parameter of the diffuser (L/W of 0.2 with D/W of 0.3, giving % SD lower than 0.5%) and velocity contours a design that is not recommended (L/W of 0.1 and D/W of 0.3, giving % SD higher than 0.5%).

### 3.4 Influence of inlet velocity

In this section, the study of increasing average velocity in a channel is considered. To do so, the volumetric flowrate at the inlet is increased. The increased velocity is greater by a magnitude of five times compared to the previous section ($1.734 \times 10^{-1}$ m/s). Velocity is considered as a primary parameter to the design of the geometry of the reactor. In the reaction zone, velocity is related to retention time and leads to reaction activity. Indeed, for FTS, it has been reported that increasing velocity in a channel would affect the conversion of the reaction (Almeida et al. 2013).

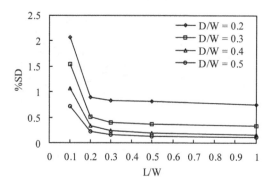

Figure 7. % SD of average velocity at different L/W and D/W with the velocity in a single channel $1.734 \times 10^{-1}$ m/s.

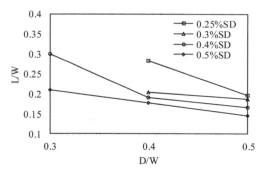

Figure 8. Aspect ratios for a conical-type design entrance of the microreactor at $1.734 \times 10^{-1}$ m/s.

Figure 7 illustrates the influence of velocity in the microchannel on % SD at different D/W ratios. Increasing velocity leads to increasing % SD for all D/W ratios when compared the results in Figure 4.

For 0.5% SD to 0.25% SD in this case, the optimal ratios of L/W and D/W are shown in Figure 8. At $1.734 \times 10^{-1}$ m/s, a small D/W ratio is not recommended, as can be seen from a D/W ratio of 0.2, giving % SD greater than 0.5.

## 4 CONCLUSION

The geometry of the inlet manifold plays an important role in flow distribution and micro reactor performance. In this work, the reactor geometry of the inlet manifold, including diffuser length and inlet width, is studied to obtain optimal dimensions to achieve uniform average velocity. The degree of flow uniformity (% SD) is suggested to be lower than 0.5%. This value is an acceptable deviation giving small effects on hydrodynamic properties of 2-D simulation results. The design guideline can be obtained from the area of the graph where the relation of L/W and D/W with % SD is given, and then the design geometry can be obtained accordingly.

## ACKNOWLEDGEMENT

Authors would like to thanks the Graduate College, KMUTNB, for the financial support.

## REFERENCES

Agrawal, G. Kaisare, N.S. Pushpavanam, S. & Ramanathan, K. 2012. Modeling the effect of flow mal-distribution on the performance of a catalytic converter. *Chemical Engineering Science*, 71: 310–320.

Almeida, L.C. Sanz, O. Merino, D. Arzamendi, G. Gandía, L.M. & Montes, M. 2013. Kinetic analysis and microstructured reactors modeling for the Fischer–Tropsch synthesis over a Co–Re/Al$_2$O$_3$ catalyst. *Catalysis Today*, 215: 103–111.

Arzamendi, G. Diéguez, P.M., Montes, M., Odriozola, J.A., Falabella Sousa-Aguiar, E. & Gandía, L.M. 2010. Computational fluid dynamics study of heat transfer in a microchannel reactor for low-temperature Fischer–Tropsch synthesis. *Chemical Engineering Journal*, 160: 915–922.

Liu, Z. Guo, L. Huang, T. Wen, L. & Chen, J. 2014. Experimental and CFD studies on the intensified micromixing performance of micro-impinging stream reactors built from commercial T-junctions. *Chemical Engineering Science*, 119: 124–133.

Mei, D. Liang, L. Qian, M. & Lou, X. 2013. Modeling and analysis of flow distribution in an A-type microchannel reactor. *International Journal of Hydrogen Energy*, 38: 15488–15499.

Tonomura, O. Tanaka, S. Noda, M. Kano, M. Hasebe, S. & Hashimoto, I. 2004. The CFD-based optimal design of manifold in plate-fin microdevices. *Chemical Engineering Journal*, 101: 397–402.

Vasquez-Alvarez, E. Degasperi, F.T. Morita, L.G. Gongora-Rubio, M.R. & Giudici, R. 2010. Development of a micro-heat exchanger with stacked plates using LTCC technology. *Brazilian Journal of Chemical Engineering*, 27: 483–497.

*Advanced Materials, Structures and Mechanical Engineering – Kaloop (Ed.)*
© *2016 Taylor & Francis Group, London, ISBN 978-1-138-02793-0*

# The efficiency of improving the blade finishing treatment

M. Yu Kulikov, V.E. Inozemtsev, A.A. Bocharov & N.O. Myo
*Moscow State University of Railway Engineering (MIIT), Moscow, Russia*

ABSTRACT: This article considers the technology of the combined finishing treatment of low-melting metals. The method of the machining surface combines the complex action of the cutting tool and the anodic dissolution of metal during the influence of an electric current through the electrolyte solution. The application of this method allows to obtain higher surface quality parameters when machining aluminum alloy and copper unlike the blade processing of the tool. In addition, this method allows to provide the special requirements of some parameters of articles made of ductile material, which are obtained by sintering.

## 1 INTRODUCTION

One of the major world tendencies in the development of engineering industry is technological development and creation of high-tech materials that have the necessary physical, mechanical and chemical properties and, as a rule, having a certain structure. They can be used to solve technological problems related to the analysis of tribological processes and the reasonability of using certain materials with predetermined properties of specific knots, systems, aggregates and modules, promoting the increment of reliability and operating time of individual machine elements, as well as reducing the risks of premature withdrawal knots and aggregates of machines out of service. The use of the high-tech materials can also provide the required quality level of engineering products. In most cases, technologies receipt of these materials allows creating the finished parts without the additional operation of forming, but certain categories of parts require the operation of finishing machining. Therefore, problems associated with the operations of finishing forming are very relevant and require an individual approach for their solution (Kulikov et al. 2013).

Research processes based on blade aluminum alloys have engaged many scientists, including experts from the Clemson University International Center for Automotive Research (Mathew et al. 2010), Technological University of North Texas Discovery Park (Nourredine & Vasim 2012) and others (Anna et al. 2013, Marko et al. 2011). There has been a large assortment of aluminum-based alloys, the amount of which depends on the technological requirements at different levels, which indicates the need to identify best practices and technological approaches to their treatment, reducing the complexity of the processes by reducing the

number of operations and the basic technological time, and by increasing product quality.

## 2 THE METHODS OF IMPROVING THE QUALITY OF MACHINING OF ALUMINIUM ALLOYS

Mechanical processing of materials based on aluminum is accompanied by defects on the surface (Kovensky et al. 2001), resulting from the cutter, as well as sticking of the deleted material on the cutting edge of the tool, which causes reduction in the efficiency of the cutting process and increases the temperature in zone cutting and tool wear. To achieve a high surface quality during the cutting of silumin, the use of the special cutting instrument with adamant-like carbon coating and other wear-resistant coatings is recommended. In this case, the cutting tool must have a large front corner, the small rounding radius of the cutting edge ρ and the small tool apex radius r. In some cases, thin walled sleeves are processed to increase the cooling reasonably to use the lubricant-cooling agent. Also, to reduce roughness, a chemical agent may be added to the lubricant-cooling agent, which will lower the strength of the layer of the removed material. Thus, anodic machining of silumin using the caustic soda solution allows us to obtain the surface roughness Ra of 0.6–0.7, the cutting speed V = 214 m/min, the cutting feed S = 0,05 mm and the cutting depth t = 0,5 mm. A cutting tool uses a replaceable insert of the cup shape, made of cemented carbide. In the capacity of cutting tool, the replaceable carbide like plating form is used.

Operations such as the finishing machining with an axial tool are recommended at low speed. The reaming holes of the aluminum alloy should be accomplished by the reamer with a straight or spiral flute (Beletsku & Kruvov 2005). The reamer with a spiral flute reduces the vibration of the tool and allows the improvement of the surface roughness of the workpiece. Cutting feed for reaming should be chosen depending on the material of the cutting tool, and the feed may be 0.3–1 mm/rev (Kulikov et al. 2013).

There are also non-traditional ways of forming, representing the combining process, which may include mechanical, electrical and chemical actions. The combined method of machining has a wide range of controllable factors, affecting the quality level of the formed surface; therefore, it is convenient and effective to use in the forming process of the intractable and the heterogeneous materials.

There are a wide variety ways of combination machining, which influence the surface with overlapping of an electric field. These include the electrochemical-mechanical combination method of processing (Afonin et al. 2012). This method is a kind of the surface treatment by the anodic process, but the main type of the action is totality action of mechanical force and chemical action. By means of the research confirmed, the electrochemical-mechanical combination machining is effective to achieve the required surface roughness values of a composite metal, produced by casting, such as silumin and other materials.

It was defined by Poduraev (1974) that anode-mechanical machining of aluminum alloys is recommended at a low density of an electric field; therefore, the main effect of its implementation is anodic dissolution mechanisms and mechanical removal of the film by the moving tool. These processes mainly occur on the top of microroughnesses, which are subjected to the most intensive electrochemical action. Only on the tops of microroughnesses occurs the continuous mechanical removal of the film. In the cavity of microroughnesses, a thick layer of the film is formed as a protective part. In the aggregate, this leads to the continuous reducing of roughness, achieving high accuracy and clearness of the surface.

As shown by the results of the studies with the blade, an anode-mechanical machining during cutting silumin formation of oxides on the treated surface substantially degrades the electrochemical processes. This is particularly significant to consider for the anodic dissolution of aluminum alloys, having a high ability to oxidize (Inozemtsev & Kulikov 2011, Bartle 1991).

It is known that electrolyte composition has a large effect on the performance and quality of electrochemical treatment (Atanasyants 1987). The solutions of NaCl and $NaNO_3$ are used as aqueous electrolytes. As the results of an investigation of the processes of the formation (Fig. 1), the most effective is to use 25–30% aqueous solution of

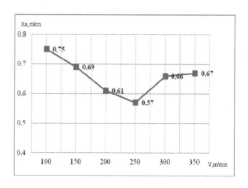

Figure 1. The influence of the concentration of electrolyte NaCl on the surface quality of silumins.

sodium chloride NaCl. Further increasing the concentration of the NaCl solution by more than 30% reduces the effect of the anodic process to achieve the desired degree of surface roughness. According to studies (Bartle 1991, Atanasyants 1987), an increase in the concentration of the electrolyte increases the viscosity of the electrolyte solution, leading to a decrease in the productivity of the process of anodic dissolution.

The changing voltage in the electrical circuit during anode-mechanical machining also affects the roughness of the machined surface. During the experiment, the electrical circuit voltage was varied in the range of 12–24 V. A further increase in voltage would lead to a breakdown in the inter-electrode clearance. From these experimental results, it follows that some increase in voltage in the circuit improves the roughness.

Also, it is established experimentally that in the anode-machining, the speed significantly varies the surface roughness of the formed parts. The large aluminum content in the samples studied (silumin) increases (about 87%) the speed of the process of the anodic dissolution. However, the processing is performed at high speed under such etching conditions that reduce the degree of roughness. The results showed that the minimum surface finish for silumins was achieved at cutting speeds in the range of 200–300 m/min. It was also defined that the roughness of the machined surface reaches Ra < 0,60 microns by using 30% aqueous NaCl solution at a cutting speed of 250 m/min and a voltage of circuit for 24 V.

## 3 THE MODELS FOR DESCRIBING ACTION FACTORS OF ROUGHNESS

The formation of surface layer silumins during anode-mechanical machining consists of two consecutive stages: the removing of shavings by the machining and the following anode dissolution of the metal surface, which appeared from under the cutting tool. Previously, a mathematical model of roughness of the machined surface was developed (Suslov 2000), taking into account the nature of the process, as follows:

$$Rz = h1 + h2 + h3 + h4 - h5 \qquad (1)$$

where Rz is the average height of the roughness profile;

h1 is the component of the roughness profile due to the geometry and kinematic of the movement of the working parts of the tool;
h2 is the component of the roughness profile due to fluctuations in the tool relative to the surface being treated;
h3 is the component of the roughness profile due to plastic deformation in the contact zone between the tool and the workpiece;
h4 is the component of the roughness profile due to the roughness of the working surfaces of the tool; and
h5 is the amount of change in the roughness profile due to anodic dissolution during the anode-mechanical machining.

As a result of statistical processing of experimental data, this model of the roughness of the machined surface for the anode-mechanical machining can be represented as follows:

$$Ra = \frac{\omega^{0,039}}{e^{0,106} \cdot V^{0,06} \cdot U^{0,082}} \qquad (2)$$

where Ra is the roughness (microns); V is the cutting speed (m/min); ω is the electrolyte concentration (%); and U is the voltage of electrical circuit (V).

Also, the optimization of formation's conditions of machined surface's roughness silumins for deploying and method of anode-mechanical machining of threading were provided. The principle of action of the anode-mechanical machining for threading and deploying is not significantly different compared with the anode-mechanical machining to whetting. The main feature of these processes is the formation of surface roughness in a different interaction of cutting tool with the machining surface. Thus, when deploying and threading, the cutting edge of the cutting portion of the tool allows us to get the basic geometric parameters, and the subsequent process of anodic dissolution forms geometric parameters to final surface finishing.

As a result, the mathematical model of reducing the surface roughness for anode-mechanical machining during deployment can be represented as follows:

$$Ra = \frac{e^{0,1545} \cdot \omega^{0,039}}{V^{0,03} \cdot U^{0,35}} \qquad (3)$$

As a result, the corresponding calculations and mathematical models of reducing the surface roughness of anode-mechanical machining for thread take the following form after the transformations:

$$Ra = \frac{e^{8,39} \cdot \omega^{0,66}}{V^{0,72} \cdot U^{3,86}} \qquad (4)$$

## 4 THE RESULTS OF RESEARCHING AND DISCUSSION

The optimal cutting conditions during deploying and threading were achieved by the following values of factors: V—20 m/min, U—24 V and the use of 30% of sodium chloride NaCl solution with the addition 2% of sodium nitrate (roughness for deploying is Ra 0.51; roughness for thread is Ra 0.21).

When finishing of aluminum alloys, the cutting speed of 400–500 m/min is recommended. As studies have shown, for the anode-mechanical machining, the optimal cutting speed is 250 m/min. Tests conducted on the strength of the tool indicate that a decrease in the cutting speed during finishing mechanical machining aluminum alloys reduces intensiveness tool's wear by about 1.5–3 times. It allows to get a much lower surface roughness during the machining (reduced by more than two times). Therefore, the advantage of finishing anode-mechanical machining of silumins is lower surface roughness compared with conventional machining (Ra 0.6 against Ra 1.3). A much lower (1.5–3 times) intensiveness of tool wear can be achieved by reducing the cutting speed. During its operation, the traditional machining reaches the minimal roughness Ra 1.25–1.3. Therefore, to achieve a roughness of Ra 0.6 in this case, it is necessary to introduce additional process steps of finish machining, which will significantly increase the complexity of the process of machining silumins.

The formation's research of the surface quality of parts made from aluminum and aluminum alloys during the milling process also confirms the usefulness of the anode-machining. Given this process of forming, the regularity of shaping can be described by Formula (1), but in this case, it is necessary to take into account the peculiarity of the interaction of the cutting tool with the surface to be treated. So, we should take into account the presence of some of the cutting edges, which may contribute to increasing roughness at the macro level. Research on the increase in the

265

efficiency of milling technology found that there is no close relationship between the roughness and operational parameters of processing. Also, a predominant influence on the roughness parameters is exhibited by the parameters of equipment, the allowable spindle's torque, power, and toughness. Besides, according to the equipment's requirements for milling aluminum alloys, the surface roughness should be in the range of Ra 0.6–0.8 micron.

So, the currently known advantage of using the anode-mechanical machining (Stepanov 2003) can be used to obtain higher quality of parts to provide a considerable increase in the performance and durability of the cutting tool. However, the anode-mechanical milling has not been widespread yet. It is also known (Stepanov 2003) that for finishing operations, anode-mechanical machining is the most effective scheme of cutting, which contains a sequence of mechanical action, and then the electrochemical component.

## 5 CONCLUSIONS

The basis of this method of treatment is the combination of the interaction of electrical contact between the cutting tool and the workpiece (mechanical disruption, or form's change of metal surface produced simultaneously with the heating or melting of the surfaces by an electric current) and the occurrence of the galvanic process (in this case, the anodic dissolution of the metal from the treated surface). The moving tool not only brings a current and removes the softened metal, but also due to vibration, it contributes to the emergence of a lot of intermittent contacts, that is necessary for formation of the arc discharge (Valyetov & Murashko 2006). The electric-processing may be performed in air or in a liquid medium. The processing performance increases almost linearly with the increase in the voltage and power supply.

REFERENCES

Afonin, A.N. Gaponenko, E.B. Erenkov, O.U. et al. 2012. *Progressive mechanical engineering technology.* Moscow. Spectrum. 191.

Anna, C.A. Adriane, L.M. & Fabio, de O.C. 2013. *Micro milling cutting forces on a machining aluminum alloy.* International Conference on Micromanufacturing, university of Victoria.

Atanasyants, A.G. 1987. *The electrochemical production of parts of nuclear reactors.* Energoatomizdat.

Bartle, D. 1991. *The Technology of chemical and electrochemical surface treatment.* Mechanical Engineering.

Beletsku, B.M. & Kruvov, G.A. 2005. *Aluminum alloys (compound, properties, technology, using) Reference book.* Kiev, Kominteh. 180–181.

Inozemtsev, V.E. & Kulikov, M.Yu. 2011. *Research the effects of finishing mechanical processing condition of metal-ceramic sintered material on the formed surface quality. Interuniversity collection of scientific papers "Physic, chemistry and mechanic tribosystem" Ivanovo state university. Tribological center-ivanovo state university.* X. Ivanovo. 88–93.

Kovensky, I.M. Kuskov, B.N. & Prohorov, N.N. 2001. *Structural transformation of metal and alloy by electrolytic action.* Tyumen GNGU. 115.

Kulikov, M.Yu. Inozemtsev, V.E. & Myo, N.O. 2013. *Technological method for the finishing process of fusible alloy Precision Machining VII.* Selected, peer reviewed papers from the 7th International Congress of Precision Machining (ICPM 2013), 2013, Miskolc, Hungary, 224–228.

Kuttolamadom, M. Hamzehlouia, S. & Mears, L. 2010. Effect of Machining Feed on Surface Roughness in Cutting 6061 Aluminum. *SAE International Journal of Materials and Manufacturing*, 3(1): 109–119.

Marko, R. Franc, C. & Uros, Z. 2011. Turning of high quality aluminium alloys with minimum costs. *Technical Gazette 18*, 3: 363–368.

Nourredine, B. & Vasim, S. 2012. Machining using minimum quantity lubrication: A technology for sustainability. *International Journal of Applied Science & Technology*, 2(1): 111.

Poduraev, B.N. 1974. *Cutting hard-to-cut materials.* M.M. Higher school. 469–470, 494.

Stepanov, A.V. 2003. *High-speed milling in modern manufacturing.* CAD/CAM/CAE Observer 4 (13) 2003.

Suslov, A.G. 2000. *The quality of the surface layer of machine parts.* Mechanical Engineering.

Valyetov, V.A. & Murashko, V.B. 2006. *The fundamentals of instrumentation Technology.* St. Petersburg.

*Advanced Materials, Structures and Mechanical Engineering – Kaloop (Ed.)*
© 2016 Taylor & Francis Group, London, ISBN 978-1-138-02793-0

# Study on the hyperbolic sine function creep model

H. Chen & H.D. Jiang
*Department of Thermal Engineering, Tsinghua University, Beijing, China*

H.Y. Chen
*National Research Center of Gas Turbine and IGCC Technology, Beijing, China*

ABSTRACT: In order to solve the creep behavior simulation problem of the structure at high temperature, a hyperbolic sine creep model is proposed, which can describe the three-stage shape of the entire creep curve. First, combined with the advantages of the hyperbolic sine function and the θ projection method, the relationships of the parameters in the model with temperature and stress change are established. Then, the determination process of the parameters in the model is given. Finally, the application ability of the proposed hyperbolic sine creep model is validated using creep test data of a Nickel-based super-alloy at elevated temperature. The results showed that the proposed hyperbolic sine creep model has a good agreement with the experimental data, and can characterize the tertiary stage of the creep behavior.

## 1 INTRODUCTION

High-temperature alloy is usually applied to aircraft engine and heavy duty gas turbine hot end components. These components are often operated at a high-temperature environment. The material creep failure of the hot components under high temperature and harsh environment is one of the key issues in a high-temperature structural strength field. In the design criteria, the creep deformation of the structure in service should not exceed the allowable requirements. In order to meet the design criteria, the creep model in creep deformation analysis should well reflect the creep behavior of materials. Therefore, research on the material creep model is particularly important.

Material creep process is usually divided into three stages: initial creep stage, steady-state creep stage, and accelerated creep stage. Research on considering all the three stages of the creep model has been closely studied. Many pieces of research on the material creep model have been carried out. Evans et al. (1992) studied on the theta projection creep model, and the formula of the creep model was obtained. Evans (2001, 2002) pointed out that the 6θ creep model can better fit the creep curve of a titanium alloy than the 4θ creep model. Based on the theta projection method, the creep behavior of the aluminum alloy has been investigated (Burt & Wilshire 2004, 2005, 2006, Lin et al. 2013), and an aluminum alloy high-temperature creep model has been established. David and Gordan (2013, 2014) analyzed the creep test data of a nickel base alloy, and proposed a modified theta model to simulate

the creep behavior of the nickel base alloy. In the literature (Kim et al. 2008, Harrison et al. 2013, Saijade et al. 2014), the theta projection creep model has been used to evaluate the creep behavior of the nickel-based super-alloy, and it was pointed out that the 4θ creep model in the initial phase of the creep simulation ability is inferior to the 6θ creep model. It should be noted that the theta projection creep model has attracted more and more attention of scholars, because this model can describe the third stage of creep behavior, and the material parameters under different temperature and stress levels can be associated with function. At present, although a lot of research on the creep model for the materials has been carried out, there is no an accepted uniform creep model.

In this paper, a hyperbolic sine creep model is established to describe the creep behavior of materials. The model combines with the advantages of the theta projection model and the hyperbolic sine function, and it can consider the three-phase creep behavior of the material. The proposed model is validated using a high-temperature nickel base alloy creep test data.

## 2 THE PROPOSED HYPERBOLIC SINE CREEP MODEL

Creep model is mainly associated with stress, temperature and time, and a general form of the creep model can be expressed as follows:

$$\varepsilon_c = f(\sigma, t, T) \tag{1}$$

where $\varepsilon_c$ = the creep strain; $\sigma$ = stress; $t$ = time; and $T$ = temperature.

In order to use the hyperbolic sine creep model to simulate the creep behavior of a high-temperature alloy material, a new hyperbolic sine creep model is established. The model is as follows:

$$\varepsilon_c = \beta_1 \left[ \sinh\left(\beta_2 t/t_r\right) \right]^{\beta_3} \tag{2}$$

$$\ln \beta_i = a_i + b_i T + c_i \ln \sigma + d_i T \ln \sigma \ (i = 1, 2, 3) \tag{3}$$

$$\ln t_r = \alpha_0 + \alpha_1/T + \alpha_2 \ln \sigma/T + \alpha_3 \ln^2 \sigma/T \tag{4}$$

where $\beta_1 \sim \beta_3$ = material constant; $t_r$ = creep failure time at a constant temperature and stress.

From the form of the model, it can be found that all parameters change the temperature and stress. If the creep strain under different temperature and stress levels is given, the parameters in the model can be identified easily.

## 3 DETERMINATION PROCESS OF THE MODEL PARAMETER

In the new hyperbolic sine creep model, the parameters of Equations (2)~(4) need to be determined. Determination steps of the parameters are as follows:

1. Get the $\alpha_0 \sim \alpha_3$ value in Equation (4) by fitting the creep failure time under different temperature and stress levels.
2. Calculate $t_r$ using the $\alpha_0 \sim \alpha_3$ value determined in the previous step under different stress levels at a different temperature.
3. Get the $\beta_1 \sim \beta_3$ value by fitting the creep strain time histories and the creep failure time under different temperature and stress levels.
4. Obtain the parameters in Equation (3) by fitting the $\beta_1 \sim \beta_3$ value in the previous step under different temperature and stress level.

In this paper, all parameters are fitted using the nonlinear fitting curves and linear fitting function of the Origin software.

## 4 EXPERIMENTAL VALIDATIONS

The creep test data of a nickel-based super alloy under five different stress levels at 800 °C are used to validate the hyperbolic sine creep model. The order of the stress level is $\sigma_1 < \sigma_2 < \sigma_3 < \sigma_4 < \sigma_5$. Because the only constant temperature test data are adopted, the model does not take into account the temperature correlation in Equations (3) and (4). Equation (3) and Equation (4) can be transformed into:

$$\ln \beta_i = a_i + c_i \ln \sigma \ (i = 1, 2, 3) \tag{5}$$

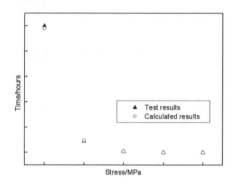

Figure 1. Comparison of failure time between the test results and the calculated results using the fitted parameters.

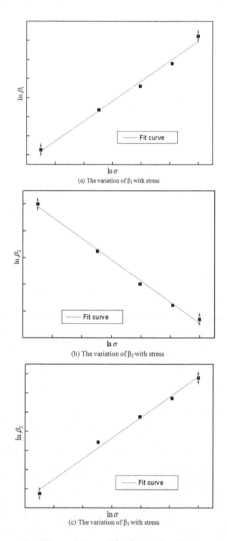

(a) The variation of $\beta_1$ with stress

(b) The variation of $\beta_2$ with stress

(c) The variation of $\beta_3$ with stress

Figure 2. The variation of $\beta_1$, $\beta_2$ and $\beta_3$ with stresses.

268

$$\ln t_r = \alpha_0 + \alpha_2 \ln \sigma + \alpha_3 \ln^2 \sigma \qquad (6)$$

Equation (5) and Equation (6) are used to determine the relevant parameters of the creep model under a constant temperature.

According to the steps, for determining the model parameter, first, the $\alpha_0$, $\alpha_2$, $\alpha_3$ values in Equation (6) should be determined. The creep failure time under the five stress levels needs to be extracted to obtain the $\alpha_0$, $\alpha_2$, $\alpha_3$ values by the Origin software polynomial fitting function for stress and failure time. Comparison of failure time between the test results and the calculated results using the fitted parameters is shown in Figure 1. As can be seen from Figure 1, Equation (6) can well describe the relationship between the stress and failure time.

The $\beta_1\sim\beta_3$ values under different stress levels can be obtained using the nonlinear curve fitting function in the Origin software. The relationships between the $\beta_1\sim\beta_3$ values and stress under the natural logarithmic coordinate are shown in Figure 2. As can be seen from Figure 2, the $\beta_1\sim\beta_3$ values and stress under the natural logarithmic coordinate has a good linear relationship, and $a_i$ and $c_i$ (i = 1, 2, 3) values can be determined from Figure 2.

In order to verify the validity of the model, the creep strain is calculated using the fitted parameters,

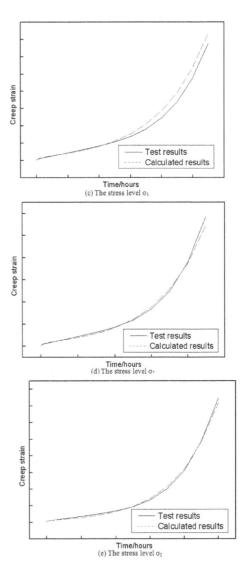

Figure 3. Comparisons between the test results and the calculated results at different stress levels.

and comparisons between the test results and the calculated results are made, as shown in Figure 3. The results show that the hyperbolic sine creep model can well describe the creep behavior of materials under different stress levels, especially the model can well simulate the creep behavior of the third stage.

The model can be used for aircraft engine or gas turbine hot components of creep deformation analysis, and it provides a good theoretical basis. The model will be applied to analyze the creep behavior of the high-temperature components (e.g., high-temperature turbine blades, wheel) in future.

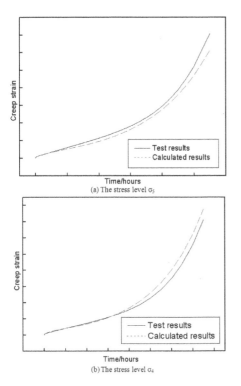

Figure 3. (*Continued*).

# 5 CONCLUSIONS

1. A new hyperbolic sine creep model is proposed and the relationships between the parameters that change with temperature and stress are also established.
2. The results showed that the proposed creep model can well reflect the material creep behavior of the nickel base alloy, and the creep model can be applied for the hot components of creep deformation analysis in future.

# REFERENCES

Burt, H. & Wilshire, B. 2004. Theoretical and practical implications of creep curve shape analysis for 2149 and 2419. *Metallurgical and Materials Transactions A*, 35(6): 1691–1701.

Burt, H. & Wilshire, B. 2005. Theoretical and practical implications of creep curve shape analysis for 8090. *Metallurgical and Materials Transactions A*, 36(5): 1219–1227.

Burt, H. & Wilshire, B. 2006. Theoretical and practical implications of creep curve shape analysis for 7010 and 7075. *Metallurgical and Materials Transactions A*, 37(32): 1005–1015.

David, D. & Gordan, A. 2013. *A modified theta projection creeps model for a nickel-based super-alloy.* Proceedings of ASME turbo expo: turbine technical conference and exposition GT2013-94805.

David, D. & Gordan, A. 2014. *Life fraction hardening applied to a modified theta projection creep model for a nickel-based super-alloy.* Proceedings of ASME turbo expo: turbine technical conference and exposition GT2104-25881.

Evans, M. 2001. The θ projection method and small creep strain interpolations in a commercial titanium alloy. *Journal of Materials Science*, 36(12): 2875–2884.

Evans, M. 2002. The sensitivity of the theta projection technique to the functional form of the theta interpolation/extrapolation function. *Journal of Materials Science*, 37(14): 2871–2884.

Evans, R. Parker, J. & Wilshire, B. 1992. The theta projection concept-a model-based approach to design and life extension of the engineering plant. *International Journal of Pressure Vessels and Piping*, 50: 147–160.

Harrison, W. Whittaker, M. & Williams, S. 2013. Recent advances in creep modeling of the nickel base superalloy, alloy 720Li. *Materials*, (6): 1118–1137.

Kim, W.G., Yin, S.N. & Kim, Y.W. et al. 2008. Creep characterization of a Ni-based Hastelloy-X alloy by using theta projection method. *Engineering Fracture Mechanics*, 75: 4985–4995.

Lin, Y.C. Xia, Y.C. & Chen, M.S. et al. 2013. Modeling the creep behavior of 2024-T3 Al alloy. *Computational Materials Science*. 67: 243–248.

Sajjadi, S.A. Berahmand, M. & Rezaee-Bazzaz, A. 2014. Creep characterization of Ni-based superalloy IN-792 using the 4-θ and 6-θ projection method. *Journal of Engineering Materials and Technology*, 136: 011006-1-6.

*Advanced Materials, Structures and Mechanical Engineering – Kaloop (Ed.)*
© 2016 Taylor & Francis Group, London, ISBN 978-1-138-02793-0

# The calculation and decomposition of embodied energy in Sino-Korean trade: Based on technological heterogeneity

X. Zhang & W.R. Huang
*College of Economics and Management, Nanjing University of Aeronautics and Astronautics, Nanjing, China*

ABSTRACT: Based on the perspective of technological heterogeneity and multi-region input-output model, this paper evaluates the energy embodied in Sino-Korean trade with the two countries' non-competitive input-output tables in 1995, 2000 and 2005. Furthermore, by employing the weighted-average method of Structural Decomposition Analysis model (SDA), this paper analyzes the driving factors of embodied energy in Sino-Korean trade, Results indicate that the embodied energy in China's export to South-Korea and in China's import from South-Korea has been rising since 1995 on the whole, but China is still a net exporter of embodied energy. What's more, both the expansion of export scale and intermediate production technology has promoted the increase of energy embodied in export. Then the improvement of energy utilization efficiency and the structure optimization of export have inhibited the increase of embodied energy. However, it cannot be offset by the increase of embodied energy driven by scale effect.

## 1 INTRODUCTION

The development of social economy cannot go any further without large-scale energy consumption. The 2013 report of BP Statistical Review of World Energy shows that the world's primary energy consumption increased 1.8% compared with that of last year in 2012, which had reached 8.733 billion tons of standard coal. At the same time, China's primary energy consumption was 1.1914 billion that accounted for 21.9 percent of the total primary energy consumption of the world, China has become the world's top energy consumer. However, Chinese environmental deterioration resulted from the massive energy consumption attracted global attention, As a result, China is confronted with the great pressure of energy-saving emission reduction.

With the development of economy globalization, all production processes are distributed into different countries in a complete production circle, which results in a spatial separation of production and consumption. However, latest international studies about measuring energy consumption mostly are based on the production side, if energy consumption of domestic product is measured merely from the perspective of producers, it will ignore the indirect energy consumption during producing these products, which is consumed by importing countries who consumes imported products. As the world's top goods trade country, China is transferring its consumed energy to the outside world through an export trade during the social economic development. Therefore, researching

embodied energy problems of China's foreign trade in the view of consumers can reflect objectively the essence of China's energy consumption situations, which is beneficial to transformation and upgrade of foreign trade growth mode and economic sustainable development (Hong et al. 2007).

Embodied energy is the sum total of the energy necessary for an entire product life-cycle including raw material extraction, transport, manufacture, and so on. In recent years, the embodied energy problems of China's foreign trade have attracted widespread attention from international and domestic academics to research it. However, most literature of related studies are focused on measuring embodied energy of China's overall trade (Chang et al 2010). Input-output analysis is first used in analyzing energy embodied in China's foreign trade, which indicates China is a net exporter of energy during 2002–2006, and the energy embodied in export increased from 240 million tons of standard coal in 2002 to 630 million in 2006, increasing the proportion of China's primary energy consumption from 16% to 25.7% simultaneously (Chen et al. 2008). What's more, input-output method is applied to measure the energy embodied in Chinese import and export products, of which the annual average growth rate is around 20 percent in terms of export value, import value, minimum net value and maximum value (Cui & Wang 2013).

After calculating the energy embodied in China's foreign trade, some scholars also deeply probe into driving factors of its variation. So two-polar forms of Structural Decomposition Analysis model (SDA) is applied to identify five key factors causing the changes of energy embodied in export (Liu et al. 2010). In addition, the SDA model is also utilized (Xie & Jiang 2014) to analyze the impact extent of inducing the variations of energy embodied in exports caused by influencing factors, including energy efficiency effect, input effect, export structure effect and export volume effect.

China and South-Korea are significant trade partners to each other, so the problems of energy embodied in Sino-Korean trade cannot be ignored. As a result, this paper will calculate out the embodied energy consumption in Sino-Korean trade on the basis of input-output tables of Sino-Korean provided by OECD database, then the influencing factors that cause the changes of energy embodied in Sino-Korean trade will be analyzed by weighted average decomposition of SDA model.

## 2 METHODOLOGY AND DATE PROCESS

### 2.1 *The measure methods of embodied energy*

In general, there are two ways to calculate the embodied energy of trade (Yin et al. 2010), one is Single Region Input-Output approach (SRIO), and the other is Multi-Region Input-Output approach (MRIO). The SRIO assumes the domestic level of production technology is identical with the import source country, and the domestic energy intensity can be substituted by other countries. However, since different countries may have different production technology level and energy utilization efficiency, technological homogeneity assumption would affect the accuracy of the embodied energy measurement results. On the contrary, the MRIO based on technological heterogeneity assumption could calculate out one country's energy consumption more accurately, through using energy consumption data, input-output tables and imports and exports from each country.

This paper employs MRIO model based on the technological heterogeneity of Sino-Korean trade, their embodied energy could be estimated in the light of non-competitive input-output tables, the related formulas are given as follows.

1. Direct consumption coefficient ($a_{ij}$): it denotes the product number the $i$th sector consumes directly when the $j$th sector produces one unit product, all the $a_{ij}$ form the direct consumption coefficient matrix $A$. the direct consumption coefficient is computed as:

$$a_{ij} = \frac{X_{ij}}{X_j} \qquad (1)$$

where, $X_{ij}$ represents product number the $i$th sector consumes during the $j$th sector manufacturing process, and $X_j$ is gross output of the $j$th sector.

2. Total consumption coefficient $b_{ij}$: it denotes the sum of the $i$th department's direct and indirect consumption when the $j$th sector produces final products, all the $b_{ij}$ form total consumption coefficient matrix $B$, which can be calculated by $A$.

$$B = (I - A)^{-1} - I \qquad (2)$$

where, $I$ is a unit matrix.

3. Direct Energy Intensity ($EI$): it can be calculated as ratio the $i$th department's total energy consumption $E_i$ to the gross output $X_i$.

$$E_i = \frac{E_i}{X_i} \qquad (3)$$

4. Total energy intensity ($EA$): it is computed by multiplying total energy consumption of producing the one unit production $EA$ with direct energy intensity $EI$.

$$EA = EI \cdot (I - A)^{-1} \qquad (4)$$

5. Embodied energy of trade ($EC$): it is obtained by multiplying total energy intensity of each sector $EA$ with their trade volume $W$.

$$EC = EA \cdot W = EI \cdot (I - A)^{-1} \cdot W \qquad (5)$$

In accordance with MRIO model, the technological difference between China and South-Korea needs considering. When the energy embodied in Sino-Korean export or import is calculated, input-output tables and direct energy intensity are used to calculate total energy intensity, then the results can be obtained by multiplying it with exports or imports.

Superscript symbol "$c$" and "$j$" represent China and South-Korea respectively, and superscript symbol "$e$" and "$m$" denote China's export and import to South-Korea respectively. Then the export embodied energy from China to South-Korea is:

$$EC^e = EI^c \cdot (I - A^c)^{-1} \cdot W^e \qquad (6)$$

6. Net export of embodied energy ($\Delta EC$): it is the difference of volume between the export of embodied energy and the import of embodied energy.

$$\Delta EC = EC^e - EC^m \qquad (7)$$

7. Trading energy terms $\mu$: it denotes the ratio of the embodied energy volume of the unit value

of export and import, which can be written as follows.

$$\mu = \left( EC^e / W^e \right) / \left( EC^m / W^m \right) \quad (8)$$

If $\mu > 1$, it indicates the embodied energy volume of the unit value of China's export to South-Korea is more than that of China's import to South-Korea.

## 2.2 The structural decomposition analysis

When influencing factors causing the changes of embodied energy are analyzed, the main decomposition methods contain Index Decomposition Analysis (IDA) and structural decomposition analysis. In consideration of SDA method is superior to IDA in analyzing direct factor and indirect factor, this paper employs the weighted-average SDA approach to analyze the contribution of various factors to the changes of energy embodied in China's export to South-Korea.

In Equation (6), the formula $(L^c = (I - A^c)^{-1})$ is assumed, the exports of each sector $(W^e)$ can be decomposed into the multiplication of gross exports $(Q^e)$ and export structure $(S^e)$. So the Equation (6) can be transformed to:

$$EC^e = EI^c \cdot L^c \cdot Q^e \cdot S^e \quad (9)$$

where, $S^e$ is the ratio of the exports of each sector $(W^e)$ and gross exports $(Q^e)$.

Then, the changes of embodied energy in China's export to South-Korea can be decomposed into four factors, which contain energy utilization efficiency, intermediate production technology, export scale effect and export structure effect. Subscript numbers 0 and 1 represent separately base period and report period. $\Delta EC^e$ is the changes of the embodied energy in export, it can be computed as:

$$\Delta EC^e = EC_1^e - EC_0^e = EI_1^c L_1^c Q_1^e S_1^e - EI_0^c L_0^c Q_0^e S_0^e \quad (10)$$

If it is decomposed from the base period, the formula becomes:

$$\Delta EC^e = \Delta EI^c L_0^c Q_0^e S_0^e + EI_1^c \Delta L^c Q_0^e S_0^e + EI_1^c L_1^c \Delta Q^e S_0^e + EI_1^c L_1^c Q_1^e \Delta S^e \quad (11)$$

However, if it is decomposed from the report period, the formula becomes:

$$\Delta EC^e = \Delta EI^c L_1^c Q_1^e S_1^e + EI_0^c \Delta L^c Q_1^e S_1^e + EI_0^c L_0^c \Delta Q^e S_1^e + EI_0^c L_0^c Q_0^e \Delta S^e \quad (12)$$

Above all, the Equation (11) can be decomposed into twenty four forms in light of different factors ranking, then all the decomposition forms are calculated by the weighted-average method, the simplified formula of $\Delta EC^e$ is written as:

$$\Delta EC^e = EC^e \left( \Delta EI^c \right) + EC^e \left( \Delta L^c \right) + EC^e \left( \Delta Q^e \right) + EC^e \left( \Delta S^e \right) \quad (13)$$

where, $EC^e \left( \Delta EI^c \right)$ indicates energy utilization efficiency, $EC^e \left( \Delta L^c \right)$ indicates intermediate production technology, $EC^e \left( \Delta Q^e \right)$ indicates export scale effect, $EC^e \left( \Delta S^e \right)$ indicates export structure effect. As a result, the energy utilization efficiency ($EC^e \left( \Delta EI^c \right)$) can be computed as:

$$EC^e(\Delta EI^c) = \frac{1}{4} \Delta EI^c L_0^c Q_0^e S_0^e + \frac{1}{12} \Delta EI^c L_0^c Q_0^e S_1^e$$
$$+ \frac{1}{12} \Delta EI^c L_0^c Q_1^e S_0^e + \frac{1}{12} \Delta EI^c L_0^c Q_1^e S_1^e$$
$$+ \frac{1}{12} \Delta EI^c L_1^c Q_0^e S_0^e + \frac{1}{12} \Delta EI^c L_1^c Q_0^e S_1^e$$
$$+ \frac{1}{12} \Delta EI^c L_1^c Q_1^e S_0^e + \frac{1}{4} \Delta EI^c L_1^c Q_1^e S_1^e$$

$$(14)$$

## 2.3 Data source and processing

The measurement and decomposition of China's trade embodied energy involves the data mainly from three aspects: Sino-Korean input-output tables, sub-department energy consumption data and export-import trade data, which are extracted from OECD database, WIOD database, and OECD database respectively. The classification standard is ISIC Rev. 4. Considering different data source has its own dividing standard, sixteen sectors can be analyzed statistically finally.

## 3 RESULTS AND DISCUSSION

### 3.1 The calculation results of embodied energy in Sino-Korean trade

According to Table 1, the embodied energy in China's export to South-Korea presented growth trend during 1995–2009, increasing from 50.50 million tons to 78.054 million tons with an annual increase of 3.16 percent. The embodied energy in China's export to South-Korea accounting for the proportion of the total energy consumption in China floated from 2.0% to 3.6%. During the increase of he embodied energy in China's export to South-Korea, the embodied energy in China's import from South-Korea increased from 7.77 million tons in 1995 to 31.54 million tons in 2009.

Table 1. Embodied energy in China's export and import to South-Korea during 1995–2009.

| | Embodied energy in Sino-Korean trade | | | Trading energy terms |
|---|---|---|---|---|
| Year | Export | Import | Net export | |
| 1995 | 5050.02 | 776.96 | 4273.06 | 6.58 |
| 1996 | 4811.65 | 891.65 | 3920.00 | 5.08 |
| 1997 | 4480.13 | 1041.29 | 3438.84 | 3.90 |
| 1998 | 3950.16 | 1092.86 | 2857.30 | 3.43 |
| 1999 | 3728.85 | 1169.39 | 2559.46 | 3.29 |
| 2000 | 6171.42 | 1162.00 | 5009.42 | 5.24 |
| 2001 | 6063.80 | 1293.08 | 4770.72 | 4.40 |
| 2002 | 6451.09 | 1717.19 | 4733.90 | 4.10 |
| 2003 | 7777.04 | 2043.43 | 5733.61 | 4.69 |
| 2004 | 8323.27 | 2434.06 | 5889.21 | 4.33 |
| 2005 | 7823.63 | 2863.71 | 4959.92 | 3.21 |
| 2006 | 8272.39 | 3359.87 | 4912.52 | 3.06 |
| 2007 | 8220.50 | 3502.11 | 4718.39 | 3.03 |
| 2008 | 8082.41 | 3388.37 | 4694.04 | 3.04 |
| 2009 | 7805.46 | 3153.65 | 4651.81 | 3.23 |

*Unit: ten thousand tons of standard coal.

Table 2. The structural decomposition of changes in embodied energy in China's export to South-Korea during 1995–2005.

| | Changes of embodied energy*/ contribution rate (%) | | |
|---|---|---|---|
| Type | 1995–2000 | 2000–2005 | 1995–2005 |
| $EC^e (\Delta EI^c)$ | −2640.20/ −235.44 | −3097.95/ −187.50 | −6884.78/ 248.22 |
| $EC^e (\Delta L^c)$ | 631.80/ 56.34 | −162.40/ −9.83 | 1975.30/ 71.22 |
| $EC^e (\Delta Q^e)$ | 2989.40/ 266.58 | 5114.00/ 309.52 | 7947.20/ 286.53 |
| $EC^e (\Delta S^e)$ | 140.40/ 12.52 | −201.44/ −12.19 | −264.12/ −9.52 |
| Total | 1121.40/ 100 | 1652.21/ 100 | 2773.61/ 100 |

*Unit: ten thousand tons of standard coal.

On the whole, during 1995–2009, the embodied energy in China's export to South-Korea was significantly higher than the embodied energy in China's import from South-Korea and China had been a net exporter of embodied energy. In the first sub-period (1995–1999), the embodied energy in China's net export to South-Korea was in a decline stage at an annualized rate of 12.03%. In the second sub-period (1999–2004), the embodied energy in China's net export to South-Korea grew rapidly at an annualized rate of 18.14%. The net export of embodied energy decreased by 4.61% annually during the third period (2004–2009), but it was still higher than the level of 1995.

China's terms of energy trade to South-Korea declined from 6.58 in 1995 to 3.23 in 2009, but it was still much larger than 1. The embodied energy of unit product in export was larger than the embodied energy of unit product in import, which indicated that China's terms of energy trade to South-Korea was unfavorable to China.

### 3.2 The structural decomposition of changes in embodied energy in China's export to South-Korea

Table 2 shows the contribution of energy efficiency, intermediate production technology, export scale and export structure to the embodied energy in China's export to South-Korea. The result shows that export scale is the most important factor influencing the embodied energy in China's export to South-Korea. During 1995–2005, China's total exports to South-Korea expanded rapidly and exports increased from 28.467 billion to 83.986 billion dollars, making embodied energy in export increase by 79.472 million tons, the contribution rate was 286.53 percent. Embodied energy in export was reduced by 68.848 million tons and the contribution rate was −248.22% because of energy efficiency.

The contribution of intermediate production technology to the embodied energy in China's export to South-Korea varies widely. During 1995–2000, the Intermediate production technology increased 6.318 million tons of embodied energy in export, while reduced 1.624 million tons during 2000–2005. From the entire period, the contribution rate of the Intermediate production technology was 71.22%, making the embodied energy in export increased by 19.753 million tons. The input-output relationship among the various departments of China did not significantly improve the status of large embodied energy in China's export to South-Korea. Export structure effect reduced the embodied energy in China's export to Japan. 1995–2005, the embodied energy in China's export to Japan decreased 2.6412 million tons because of the export structure effect.

## 4 CONCLUSIONS

Based on the perspective of technological heterogeneity and the non-competitive input-output tables of China and South-Korea extracted from OECD database, this paper estimates the energy embodied in Sino-Korean trade from 1995 to 2009. At last, this paper analyzes the driving factors of embodied energy in Sino-Korean trade by using the weighted-average method of SDA model. The main conclusions are obtained as follows.

The embodied energy in China's export to South-Korea and import from South-Korea presents growth trend from 1995 to 2009, but the embodied energy in export is always higher than the embodied energy in import, which proves that China has been a net exporter of embodied energy and China undertakes a lot of energy consumption transfer in Sino-Korean trade. In addition, China's terms of energy trade to South-Korea is always greater than one, which indicates that the Sino-Korean trade is very unfavorable to China's energy saving and environmental protection.

Decomposition result shows that the expansion of the export scale is the largest factor impacting the embodied energy in export. China's highly dependent on the South-Korean market leads to the growth of the embodied energy in export. In contrast, the improvement of the utilization efficiency in energy suppresses the embodied energy in export. In recent years, China has obtained remarkable achievements in improving energy efficiency, but compared with South-Korea, China's energy efficiency is still low and it needs further improving. Intermediate production technology promotes the growth of the embodied energy in export, indicating that China's level of production technology is relatively low. The export structure reduces embodied energy, but its influence is the least among them, which illustrates that it still has plenty of room for improvement to optimize the structure of China's export.

## REFERENCES

Chang, Y. Ries, R.J. & Wang, Y. 2010. The embodied energy and environmental emissions of construction projects in China: an economic input–output LCA model. *Energy Policy,* 38(11): 6597–6603.

Chen, Y. Pan, J.H. & Xie, L.H. 2008. Energy embodied in goods of international trade in China: calculation and policy implications. *Economic Research*, (7): 11–25.

Cui, R.M. & Wang, L. 2013. The empirical study on international transfer of energy consumption in China: Based on the four accounting dimensions of embodied energy in import and export trade. *Economic Theory and Business Management,* 2013: 59–68.

Hong, L. Dong, Z.P. & Chunyu, H. et al. 2007. Evaluating the effects of embodied energy in international trade on ecological footprint in China. *Ecological Economics,* 62(1): 136–148.

Liu. H, Xi, Y. & Guo, J. et al. 2010. Energy Embodied in the International Trade of China: An Energy Input-output Analysis. *Energy Policy*. 38(8): 3957–3964.

Xie, J.G. & Jiang, P.S. 2014. Embodied energy in international trade of China: calculation and decomposition. *Economics*, (4): 1365–1392.

Yin, X.P. Huo, D. & Tang, L. 2010. Energy embodied in goods in Sino-Korea Trade: an analysis and policy implications. *World Economy Study*, (7): 32–37.

*Advanced Materials, Structures and Mechanical Engineering – Kaloop (Ed.)*
© *2016 Taylor & Francis Group, London, ISBN 978-1-138-02793-0*

# Design of software for orthogonal packing problems

V.A. Chekanin & A.V. Chekanin
*Moscow State University of Technology "STANKIN", Moscow, Russian Federation*

ABSTRACT: This paper is devoted to the developed applied software intended to solve the optimization orthogonal packing problems. On the basis of the software, the usage of a developed unified class library is laid, allowing the description of any optimization problem of resource allocation. The developed software can be used to solve a variety of practical packing and cutting problems including orthogonal bin packing problem, rectangular packing, one-dimensional cutting stock problem, strip packing problem and many others in industry, economics and engineering.

## 1 INTRODUCTION

A solution of a variety of different practical optimization problems, including resources saving problem, optimization problems in logistics, scheduling and planning, comes down to the orthogonal packing problem. All these problems are combined with a class of cutting and packing problems (C&P). The first survey of C&P problems is given in the paper of Dyckhoff (1990), and, furthermore, this survey was significantly improved (Wascher et al. 2007). Today, the class of C&P problems includes a large set of optimization problems, among which the most popular are orthogonal bin packing problems, cutting stock problems and knapsack problems. Usually, C&P problems take a place in solving problems such as cutting stock problem, trim loss problem, nesting problem, knapsack problem, bin packing problem, strip packing problem, pallet and vehicle loading, assembly line balancing problem, partitioning problem, capital budgeting problem, computer memory allocation problem as well as many other problems in industry, engineering, computer science and economics (Bortfeldt & Wascher 2013, Lodi et al. 2002, Riff et al. 2009, Chekanin & Chekanin 2014b).

In this paper, we consider orthogonal packing problems as one of the most popular problems in practice (Wascher et al. 2007, Chekanin & Chekanin 2014c). The orthogonal packing problem is an actual problem that deals with the optimal placing of a given set of small orthogonal items named by objects into a given finite set of large orthogonal items named by containers.

Let us consider that the statement of any *D*-dimensional orthogonal bin packing problem is given by a set of *N* orthogonal containers (*D*-dimensional parallelepipeds) with dimensions as follows:

$$\left\{W_j^1, W_j^2, \ldots, W_j^D\right\}, j \in \{1, \ldots, N\}.$$

Also, given a set of *n* orthogonal objects (*D*-dimensional parallelepipeds) with dimensions as follows:

$$\left\{w_i^1, w_i^2, \ldots, w_i^D\right\}, i \in \{1, \ldots, n\}.$$

The goal is to find a placement of all objects into a minimum number of given containers under the following conditions of correct placement:

1. all edges of objects and containers are parallel;
2. all packed objects do not overlap with each other, i.e.

$$\forall j \in \{1, \ldots, N\}, \forall d \in \{1, \ldots, D\}, \forall i, k \in \{1, \ldots, n\},$$
$$i \neq k \left(x_{ij}^d \geq x_{kj}^d + w_k^d\right) \vee \left(x_{kj}^d \geq x_{ij}^d + w_i^d\right);$$

3. all packed objects are within the bounds of the containers, i.e.

$$\forall j \in \{1, \ldots, N\}, \forall d \in \{1, \ldots, D\},$$
$$\forall i \in \{1, \ldots, n\} \left(x_{ij}^d \geq 0\right) \wedge \left(x_{ij}^d + w_i^d \leq W_j^d\right).$$

To solve any one-, two- and three-dimensional orthogonal packing problems, we propose a developed applied software that is verified on standard orthogonal packing test instances.

## 2 SOFTWARE FOR ORTHOGONAL PACKING PROBLEMS

All the C&P problems are non-deterministic polynomial-time hard (NP-hard) in the strong sense, and they cannot be solved in polynomial time depending on their size (Garey & Johnson 1979). To solve the C&P problems, optimization metaheuristic algorithms are usually used, the most popular of which are the genetic algorithm (Chekanin & Chekanin 2013a, Kierkosz & Luczak 2014), the anneal simulated algorithm (Loh et al. 2009) and the ant colony algorithm (Fuellerer et al. 2009, Gao et al. 2013). The effectiveness of the application of these metaheuristic algorithms differs among the various C&P problems. Thus, to solve a variety of C&P problems via different optimization metaheuristic algorithms, a unified class library was proposed. This developed class library allows the description of any packing, cutting problem, and assigns an existing optimization algorithm or creates a new combined or modified optimization algorithm to solve the C&P problems.

The developed class library for solving the cutting and packing problems is developed with the high-level object-oriented programming language C++ and is shown in the UML diagram (Fig. 1).

Despite the developed class library, we solve only the orthogonal packing problems by the adjustable parameters, making it applicable for different C&P problems. The parameters are:

- types of packed objects (parallelepipeds, cylinders, spheres);
- dimension of a problem (one-dimensional, two-dimensional, three-dimensional, D-dimensional);
- packing load direction specifying a direction of containers' loading as the priority selection list of the coordinate axes.

The ability to pack all the objects of different types into containers is provided by using a universal node model, which describes the position of any placed object only with one point named by the node (Crainic et al. 2008, Chekanin & Chekanin 2013b). The effective storage of the coordinates of

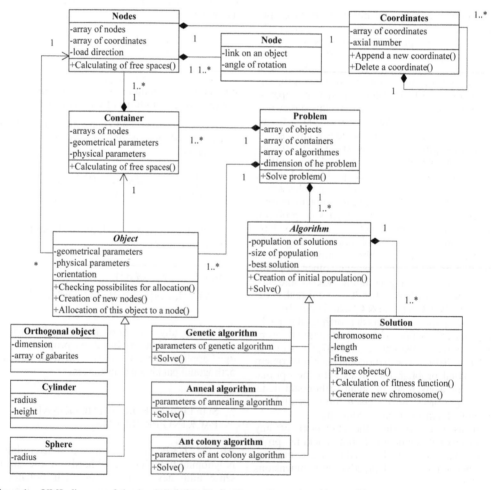

Figure 1.  UML diagram of the developed class library for cutting and packing problems.

278

all nodes is provided by a specially developed new data structure—multilevel-linked data structure (Chekanin & Chekanin 2014a, d), which stores the coordinates of all the nodes as a set of recursively embedded each to other ordered linear-linked lists.

All solutions generated by various metaheuristic optimization algorithms have a unified form and are represented as coded strings (named by chromosomes) containing sequences of objects to be packed into containers. Usage of a unified scheme for coding solutions allows us to use a corresponding unified scheme for decoding of coded solutions of various C&P problems.

The class library for C&P problems is based on the following classes: class of a problem, class of a container and two abstract classes (class of objects, class of algorithms).

The class of a problem is served for managing all operations related to the solving of a C&P problem. It contains information about a problem type, problem parameters, constraints on objects need to be packed, also contains arrays of references to the classes of objects and containers. Additionally, this class contains information about the optimization algorithms chosen to solve the considered C&P problem.

The class of a container contains an array of nodes, which describes the current placement, as well as contains a method for detecting of all free spaces in the container. This method is used in the estimation of the suitability of a found solution of the solved C&P problem.

The abstract class of objects contains information about geometrical (e.g. linear dimensions, tolerances), physical (e.g. weight, fragility) and other parameters (e.g. cost, temperature, urgency of packing, variants of rotation) of objects. In the basic version of the class library from this class, a class of an orthogonal object, a class of a cylinder and a class of a sphere can be inherited. The class of an orthogonal object allows us to set any dimensional orthogonal objects including parallelepipeds, rectangles, and lines.

The abstract class of algorithms contains metaheuristic optimization algorithms, which are used to solve the C&P problems. In the basic version of the class library from this class, a class of the genetic algorithm, a class of the annealing algorithm and a class of the ant colony algorithm can be inherited. This abstract class contains a population of solutions (chromosomes), the size of the population, and the best solution found during optimization.

Using the designed class library, an applied software "Packer" was developed, intending to solve the optimization orthogonal packing problems.

In the work of the developed software "Packer" we demonstrate an example of a three-dimensional orthogonal packing problem. It is necessary to pack 200 parallelepiped objects of 10 types (parameters of the objects are shown in Fig. 2) into 4 parallelepiped containers with the dimensions of $100 \times 100 \times 100$. The best solution of this problem found by a genetic algorithm is shown in Figure 3.

The developed software "Packer" is intended to solve the following orthogonal cutting and packing problems:

– one-Dimensional Bin Packing Problem (1DBPP);
– rectangular non-guillotine cutting problem;
– Strip Packing Problem (SPP);
– two-dimensional bin packing problem (2DBPP);
– three-dimensional bin packing problem (3DBPP).

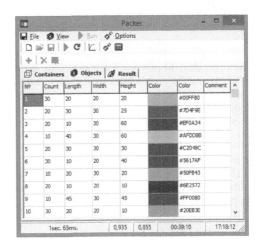

Figure 2. Parameters of a three-dimensional orthogonal packing problem.

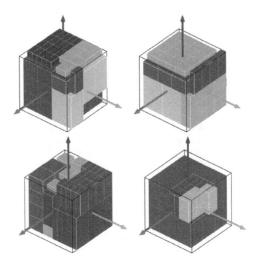

Figure 3. Found solution of the three-dimensional orthogonal packing problem.

The software contains a library of test instances for a variety of C&P problems. This library includes the following standard problem instances:

- SPP instances (classes C1-C6) from the paper of Berkey & Wang (1987);
- SPP instances (classes C7-C10) from the paper of Martello & Vigo (1998);
- 2DBPP instances (Fekete & Schepers 1998);
- 3DBPP instances (Martello et al. 2000).

## 3 CONCLUSIONS

The developed software "Packer" can be used for analyzing the efficiency of the application of various optimization heuristic and evolutionary algorithms to one-, two- and three-dimensional orthogonal cutting and packing problems.

The usage of object-oriented inheritance in the developed class library for C&P problems allows us to expand its possibilities by including into it new types of objects as well as new modified and combined metaheuristic optimization algorithms for solving the C&P problems. This class library can be used in the development of a new applied software intended to effectively solve the problems of resource allocation of any types.

The prospective area of further research in this field includes expanding of a list of cutting and packing problems supported by the developed software.

## ACKNOWLEDGMENTS

This work was financially supported by the Ministry of Education and Science of Russian Federation in the framework of the state task in the field of scientific activity of MSTU "STANKIN" (No. 2014/105).

## REFERENCES

Berkey, J.O. & Wang, P.Y. 1987. Two-dimensional finite bin-packing algorithms. *Journal of the Operational Research Society*, 38(5): 423–429.

Bortfeldt, A. & Wascher, G. 2013. Constraints in container loading—A state-of-the-art review. *European Journal of Operational Research*, 229(1): 1–20.

Chekanin, A.V. & Chekanin, V.A. 2013a. Efficient algorithms for orthogonal packing problems. *Computational Mathematics and Mathematical Physics*, 53(10): 1457–1465.

Chekanin, A.V. & Chekanin, V.A. 2013b. Improved packing representation model for the orthogonal packing problem. *Applied Mechanics and Materials*, 390: 591–595.

Chekanin, A.V. & Chekanin, V.A. 2014a. Effective data structure for the multidimensional orthogonal bin packing problems. *Advanced Materials Research*, 962–965: 2868–2871.

Chekanin, V.A. & Chekanin, A.V. 2014b. Development of the multimethod genetic algorithm for the strip packing problem. *Applied Mechanics and Materials*, 598: 377–381.

Chekanin, V.A. & Chekanin, A.V. 2014c. Improved data structure for the orthogonal packing problem. *Advanced Materials Research*, 945–949: 3143–3146.

Chekanin, V.A. & Chekanin, A.V. 2014d. Multilevel linked data structure for the multidimensional orthogonal packing problem. *Applied Mechanics and Materials*, 598: 387–391.

Crainic, T.G. Perboli, G. & Tadei, R. 2008. Extreme point-based heuristics for three-dimensional bin packing. *INFORMS Journal on Computing*, 20(3): 368–384.

Dyckhoff, H. 1990. A typology of cutting and packing problems. *European Journal of Operational Research*, 44: 145–159.

Fekete, S.P. & Schepers, J. 1998. New classes of lower bounds for bin packing problems. *Integer Programming and Computational Optimization (Lecture Notes in Computer Science)*, 1412: 257–270.

Fuellerer, G. Doerner, K.F. Hardl, R.F. & Iori, M. 2009. Ant colony optimization for the two-dimensional loading vehicle routing problem. *Computers & Operations Research*, 36(3): 655–673.

Gao, Y.Q. Guan, H.B. Qi, Z.W. Hou, Y. & Liu, L. 2013. A multi-objective ant colony system algorithm for virtual machine placement in cloud computing. *Journal of Computer and System Sciences*, 79(8): 1230–1242.

Garey, M. & Johnson, D. 1979. Computers Intractability: A Guide to the theory of NP-completeness, San Francisco: W.H. Freeman.

Kierkosz, I. & Luczak, M. 2014. A hybrid evolutionary algorithm for the two-dimensional packing problem. *Central European Journal of Operations Research*, 22(4): 729–753.

Lodi, A. Martello, S. & Monaci, M. 2002. Two-dimensional packing problems: a survey. *European Journal of Operational Research*, 141(2): 241–252.

Loh, K.H. Golden, B. & Wasil, E. 2009. A Weight Annealing Algorithm for Solving Two-dimensional Bin Packing Problems. *Operations Research and Cyber-Infrastructure (Operations Research/Computer Science Interfaces)*, 47: 121–146.

Martello, S. & Vigo, D. 1998. Exact solution of the two-dimensional finite bin packing problem. *Management Science*, 44(3): 388–399.

Martello, S. Pisinger, D. & Vigo, D. 2000. The three-dimensional bin packing problem. *Operations Research*, 48(2): 256–267.

Riff, M.C. Bonnaire, X. & Neveu, B. 2009. A revision of recent approaches for two-dimensional strip-packing problems. *Engineering Applications of Artificial Intelligence*, 22(4–5): 823–827.

Wascher, G. Haubner, H. & Schumann, H. 2007. An improved typology of cutting and packing problems. *European Journal of Operational Research*, 183(3): 1109–1130.

*Advanced Materials, Structures and Mechanical Engineering – Kaloop (Ed.)*
© *2016 Taylor & Francis Group, London, ISBN 978-1-138-02793-0*

# Flow-induced noise analysis of pipes with variable cross-sections

W. Zhao
*China Ship Development and Design Center, Wuhan, China*

Y.O. Zhang
*School of Naval Architecture and Ocean Engineering, Huazhong University of Science and Technology, Wuhan, China*

Z.L. Guo
*Zhejiang International Maritime College, Zhoushan, China*

ABSTRACT:   Flow-induced noise in pipes has gradually been emphasized by the increase of noise level demand in engineering. The present work applies a Large Eddy Simulation (LES) and Lighthill's acoustic analogy theory hybrid method to simulate the flow-induced noise produced by water flow in pipes with variable cross-sections. Acoustic experiment data show that the computational acoustic method is feasible and correct, thus providing an effective way to compute the hydrodynamic flow-induced noise. In the computation, the turbulence field in pipes is simulated by LES to obtain the velocity distribution. Afterward, ACTRAN codes are applied to compute the flow-induced noise by solving the finite element forms of Lighthill's acoustic analogy equation with the sound source computed with the fluid speed. With this hybrid method, the work analyzes the noise characteristics of pipes with the original and optimized variable cross-sections, and a low noise type of the variable cross-section is proposed.

## 1 INTRODUCTION

Numerical computation and experimental research of flow-induced noise in pipes have gradually been emphasized by the increase of noise level demand in engineering. However, experimental research requires a long cycle, a strict environment, and considerable expense. The development of numerical computation has, therefore, led to its use as an important method for flow-induced noise research. In a pipe system, the main flow-induced noise sources refer to valves, reducers, and pipe branches. Few studies were focused on these components. With hypotheses that include the compact acoustic source and the free far field, the numerical algorithm based on Curle's equation (Curle 1955) or Ffowcs-Williams and Hawkings (FW-H) equation (Ffowcs Williams and Hawkings 1969, Zhang et al. 2014) was unsuitable for computing the acoustic field in pipes.

Therefore, this work applied finite element method to solve Lighthill's acoustic analogy equation with variation form, thus deriving flow-induced noise in variable cross-section pipes. Using Large Eddy Simulation (LES) and Lighthill's acoustic analogy equation with finite element variation form enables the dispersion of the whole noise source to analyze the noise source distribution and sound mechanism in a flow field. Oberai et al. (2000) first applied Lighthill's acoustic analogy equation with variation form to compute the wing noise. The results accorded with Curle's theory. Geng and Liu (2010) solved the cavity hydrodynamic noise by LES and Lighthill's acoustic analogy equation with a finite/infinite element variation form. Kaltenbacher et al. (2010) computed the post disturbance field in a half-free sound field using LES and Scale-Adaptive Simulation (SAS). Zhang et al. (2014) and Zhang et al. (2015) applied the LES/ Lighthill's acoustic analogy theory hybrid method to compute and optimize the flow-induced noise of trash rack and elbow pipes. Research showed that the hybrid method was an effective computational acoustic method.

The present work aims to analyze the property of flow-induced noise in pipes with variable cross-sections, and optimize the design of the pipe to reduce the noise. First, a two-dimensional LES was applied to compute the turbulence field of pipes with variable cross-sections. Then, the turbulence field was imported into the acoustic software ACTRAN to derive the time and frequency domain sound sources. Lastly, the transmission of the frequency domain sound source was computed in the whole pipe. The contrast of simulation results and measured data showed consistent tendency. The cross-section of the pipe was optimized after noise source analysis. Results indicated that flow-induced noise was obviously reduced after optimization.

## 2 FINITE ELEMENT FORM OF LIGHTHILL'S ACOUSTIC ANALOGY EQUATIONS

The finite element form of Lighthill's acoustic analogy equations were firstly proposed by Oberai et al. (2000). We solved it based on volume and surface sources in the present paper rather than sound source items (Escobar 2007). The finite element equation of frequency domain is as follows.

$$-\int_\Omega \frac{\omega^2}{\rho_0 c^2} \psi \delta\psi d\Omega - \int_\Omega \frac{1}{\rho_0} \frac{\partial \psi}{\partial x_i} \frac{\partial \delta\psi}{\partial x_i} d\Omega$$
$$= \int_\Omega \frac{i}{\rho_0 \omega} \frac{\partial \delta\psi}{\partial x_i} \frac{\partial T_{ij}}{\partial x_j} d\Omega - \int_\Gamma \frac{1}{\rho_0} F\left(\tilde{\rho}\tilde{v}_i n_i\right) d\Gamma$$

$$(1)$$

where, $\Omega$ is the integral volume; $\Gamma$ is the integral surface; $\delta\psi$ is the potential function; $T_{ij}$ is the Lighthill's stress tensor; $\rho = -i\omega\psi/c^2$. The first and second terms at the right hand side of the equation are volume and surface sources, respectively. The volume source is used as a discrete solution in the acoustic source field. The surface source plays an important role in the computation of fluid-structure interaction, and the computation of rotary machine. In the work, code ACTRAN is used and the only volume source is computed. In the numerical computation, acoustic source distributions of flow-induced noise can be presented by the regional discretization with the finite element method.

## 3 DEFINITION OF PROBLEMS

The work used variable cross-section pipe models in Lv and Ji (2011) and Lv (2010). Figure 1 showed suddenly shrunken pipes with the axis symmetry structure. In the model, the shrunken pipe has a length of 0.6 m and a diameter of 0.1 m. The lengths of entrance and outlet section are 0.5 m and 7 m, respectively; the flow velocities in the two sections are 23 m/s and 30 m/s, respectively. In the work, two different variable cross-section forms (CASE2 and CASE3) were proposed to optimize flow-induced noise characteristics. Figure 1 showed the concrete structure. CASE2 and CASE3 were conducted with taper and arc transitions.

## 4 COMPUTATION MODEL AND VERIFICATION

Figure 1 showed the computation model of a flow field. In the model, velocity and pressure conditions were used at the entrance and exit, respectively. The steady flow field and turbulence were computed based on $k$-$\varepsilon$ and LES models. The steady computational results after convergence were taken as initial values to compute the unsteady flow field. In order to increase the numerical accuracy, the parameters were set as follows. A residual error was less than $10^{-5}$; time step $4 \times 10^{-4}$ s; grid number 300,000; $y$ + less than 3. After unsteady computation tended to be stable, the process of 1250 steps was saved as the sound source computation file. The frequency spectrum resolution was 2 Hz according to the Fourier transform.

Velocity distribution of flow field was taken to obtain sound source information in the time domain, and was transformed into the sound source signal in frequency domain. At last, the whole sound field result was figured out in the frequency domain. Figure 2 showed the acoustic computation model and observation points. The whole computation domain referred to sound source (grey section) and propagation (red section) regions. The acoustic admittance boundary condition was set to simulate the non-reflection propagation of the sound at both ends of the pipe. Meanwhile, a cosine filter function was applied to reduce false voice produced by truncation on the boundary of the sound source region. There were about 300,000 grids in the acoustic computation.

Figure 3 showed the contrast of computation and experiment Sound Pressure Level (SPL) at different observation points of CASE1 with different entrance velocities. Numerical results indicated the SPL of this broadband noise decreased from about 120 dB to 100 dB with the increase of frequency

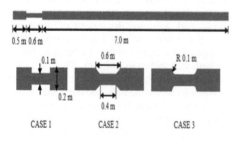

Figure 1. Flow field computation model of the variable cross-section pipe.

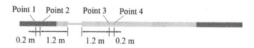

Figure 2. Acoustic computation model of the variable cross-section pipe.

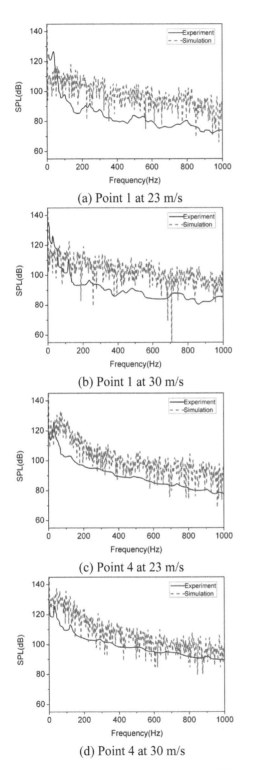

(a) Point 1 at 23 m/s

(b) Point 1 at 30 m/s

(c) Point 4 at 23 m/s

(d) Point 4 at 30 m/s

Figure 3. Computation and experiment results of flow-induced noise in CASE1.

from 100 Hz to 1000 Hz. In addition, SPL of numerical computation was larger than experimental data at two monitoring points of upstream and downstream, but the two SPL frequency spectrum curves had consistent tendency with the experimental data. Therefore, numerical results were able to indicate the acoustic characteristic of variable cross-section pipes. Besides, the experimental SPL was larger than simulation value at the low-frequency stage of upstream observation points. It is partially due to the noise generated by the draught fan.

## 5 RESULT ANALYSIS

Figures 4 and 5 showed that the sound source intensity distribution of CASE1 with different frequencies at the flow velocities of 23 m/s and 30 m/s, respectively. In Figures 4 and 5, the maximum sound source intensity obviously increased with the increase of velocity. The vortex shedding in the entrance recirculation zone formed strong noise sources in sudden contraction section. At the exit, the back-flow caused by jet-flow also produced noise.

Figures 6 and 7 showed sound source intensity distributions of optimized models (CASE2 and CASE3) in the frequency domain. The main noise source was formed by the back-flow caused by jet-flow at the exit of sudden contraction pipe. In sudden contraction pipe, the gradual transition was designed to reduce the vortex pulse formed by the sudden change of section. Compared with the initial model, the optimized model had smaller maximum sound source intensity at the sudden contraction section.

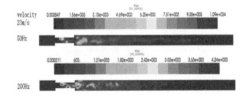

Figure 4. Sound source intensity distribution of CASE1 with different frequencies (23 m/s).

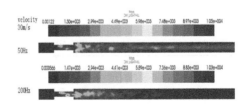

Figure 5. Sound source intensity distribution of CASE1 with different frequencies (30 m/s).

283

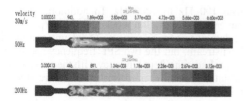

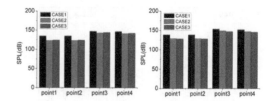

Figure 6.   Sound source intensity distribution of CASE2 with different frequencies (30 m/s).

Figure 9.   Total SPL contrast of three models with different flow velocities.

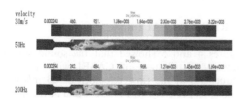

Figure 7.   Sound source intensity distribution of CASE3 with different frequencies (30 m/s).

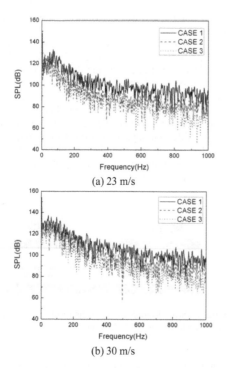

(a) 23 m/s

(b) 30 m/s

Figure 8.   SPL frequency spectrum at monitoring point Point 4 under different working conditions.

Figure 8 showed SPL frequency spectrums of three models at the monitoring point Point 4. In Figure 8, the noises of optimized models were decreased, and CASE3 had the best effect. Noise of the initial model was reduced by 20 dB

comparing with CASE3 at high frequencies. In the downstream, the high frequency section had a better noise reduction effect than the low frequency section. SPL slightly decreased with the increase of frequency at the frequency of over 300 Hz.

Figure 9 showed the total SPL contrast of three models with different flow velocities. After optimization, the SPL at monitoring points were reduced. With a larger decline, the SPL of upstream was reduced by 10 dB. The total SPL difference between CASE2 and CASE3 was less than 3 dB. Therefore, the two models had similar optimization effect according to the acoustic power.

## 6   CONCLUSIONS

The present work applied LES and Lighthill's acoustic analogy theory hybrid method to compute the water flow-induced noise of pipes with variable cross-sections. Meanwhile, the optimized variable cross-section models were proposed and analyzed. The conclusions obtained in this paper are summarized as follows:

1. Flow-induced noise obtained with the two-dimensional computation has a similar tendency with the experimental data. This reliable hybrid method can be applied in engineering analysis.
2. Spectrum analysis of the original model indicates that the main noise source is formed by vortex shedding at the entrance and by jet reflux at the exit.
3. The work proposed two variable cross-section optimization schemes, including tapered and circular transitions. Results show that both optimization models play important roles in noise reduction.

## REFERENCES

Curle, N. 1955. The influence of solid boundaries upon aero-dynamic sound. *Proceedings of the Royal Society of London. Series A,* 231(1187): 505–514.

Escobar, M. 2007. *Finite element simulation of flow-induced noise using Lighthill's acoustic analogy.* University of Erlangen-Nuremberg, Doctor Thesis.

Ffowcs Williams, J.E & Hawkings, D.L. 1969. Sound generation by turbulence and surfaces in arbitration motion. *Philosophic Transactions of the Royal Society A,* 264(1151): 321–342.

Geng, D.H. & Liu Z.X. 2010. Predicting cavity hydro-dynamic noise using a hybrid large eddy simulation-Lighthill's equivalent acoustic source method. *Journal of Harbin Engineering University,* 2: 182–187.

Kaltenbacher, M. Escobar, M. Becker, S. & Ali, I. 2010. Numerical simulation of flow-induced noise using LES/SAS and Lighthill's acoustic analogy. *International Journal for Numerical Methods in Fluids,* 63(9): 1103–1122.

Lv, J.W. & Ji, Z.L. 2011. Numerical prediction and experimental measurement of flow noise in pipes with variable cross-section area. *Noise and Vibration Control,* 1: 166–169.

Lv, J.W. 2010. *Study on prediction and experimental measurement of flow noise in pipes with varying cross-sectional area.* Harbin Engineering University, Master Thesis.

Oberai, A.A., Roknaldin, F. & Hughes, T.J.R. 2000. Computational procedures for determining structural-acoustic response due to hydrodynamic sources. *Computer Methods in Applied Mechanics and Engineering,* 190(3–4): 345–361.

Zhang, C.J. Luo, Y.X. Liang, J.X. Li, L.L. & Li J. 2014. Flow-induced noise prediction for 90 bend tube by LES and FW-H hybrid method. *Scientific Research and Essays,* 9(11): 483–494.

Zhang, Y.O. Zhang, T. Ouyang, H. & Li, T.Y. 2014. Flow-induced noise analysis for 3D trash rack based on LES/Lighthill hybrid method. *Applied Acoustics,* 79: 141–152.

Zhang, T. Zhang, Y.O. & Ouyang, H. 2015. Structural vibration and fluid-borne noise induced by turbulent flow through a 90° piping elbow with/without a guide vane. *International Journal of Pressure Vessels and Piping,* 125: 66–77.

*Advanced Materials, Structures and Mechanical Engineering – Kaloop (Ed.)*
© *2016 Taylor & Francis Group, London, ISBN 978-1-138-02793-0*

# Study and simulation on impact of wind turbine generators on Low Frequency Oscillations in power system

J.R. Xu, F. Tang, Q. Zhang & T. Zhao
*School of Electrical Engineering, Wuhan University, Wuhan, China*

Q.X. Wang
*School of Electrical and Electronic Engineering, Huazhong University of Science and Technology, Wuhan, China*

ABSTRACT: Low Frequency Oscillations (LFOs) occur in large interconnected grid now and then. Nowadays, wind energy is among the fastest-growing renewable energy technologies in the world. With the development of wind energy and power electric control technology, more wind farms with doubly fed induction generator are integrated into a power grid. As the penetration level of wind power increase, the impact of wind turbine generators on low frequency oscillations in a power system should receive enough attention. This paper analyzes the impact of doubly fed induction generator on low frequency oscillations. The simulation of two-area four-generator power system integrated with doubly fed induction generators based on Matlab/Simulink verifies that the system damping decreases after integration of wind power.

## 1 INTRODUCTION

Nowadays, the energy has become more and more important, especially electric energy. However, the development of electric industry has been seriously restricted by the shortage of primary energy and the expansion of traditional power plants may also increase $CO_2$ emission that consequently causes serious environmental problems. Growing environment concerns and attempts to reduce dependency on fossil fuel resources make renewable energy resources becoming the mainstream of the electric power sector. Among the various renewable resources, wind power is assumed to have the most favorable technical and economic viability (Heier 2006) and thus becomes the biggest concern for researchers.

The inter-area low frequency oscillations have always been one of the main threats to security and stability of the power system. The power swing on the transmission lines may undesirably activate the protective relays, separating the system into several pieces with unbalanced generators and loads (Horowitz & Phadke 2006, Li 2008). As a result, the transmission capacity between two areas is limited. When wind energy deployed in small scale, as was done traditionally, the impact of wind turbine generators on low frequency oscillations is minimal. In contrast, the increasing penetration level of wind power promotes the necessity of study on the impact of wind turbine generators on low frequency oscillations in power system.

This paper is focused on the impact of doubly fed induction generators on low frequency oscillations and the simulation of two-area four-generator power system integrated with doubly fed induction generators, which is organized as following. Section 2 introduces low frequency oscillations in a power system and wind power system. In section 3, the impact of doubly fed induction generators on low frequency oscillations is analyzed and the conclusion is verified by the simulation made in section 4. Section 5 presents the concluding remarks.

## 2 LOW FREQUENCY OSCILLATIONS AND WIND POWER SYSTEM

Low Frequency Oscillations (LFOs), having a frequency between 0.1–2.5 Hz, are inherent phenomena of modern interconnected electrical power systems, essentially caused by the continuous exchange of momentum among rotating masses. It is widely recognized that LFOs can be induced by both the occurrence of multiple disturbances and the existence of periodic disturbance sources. LFOs can be classified as local and inter-area mode (Kundur 1994, Pal & Chaudhuri 2006). Local modes are associated with the swing of units at a generating station with respect to the rest of the power system. Oscillations occurred only to the small part of the power system and the typical frequency range is 1–2.5 Hz. Inter-area modes are associated with swinging of many machines in one part of the system against a machine in other parts. It generally occurs in weakly interconnected power systems through long tie lines and its typical frequency range is 0.1–1 Hz.

Large wind farms can consist of hundreds of individual wind turbines which are connected to the electric power transmission network. The mainstream energy conversion system of induction generators based modern wind turbines can be divided into three categories (Tan et al. 2014), which include Squirrel-Cage Induction Generator (SCIG), Doubly Fed Induction Generator (DFIG) and Permanent Magnet Synchronous Generator (PMSG). Because of the differences of their structures, control, network synchronization modes and operating characteristics, their impact on low frequency oscillations varies. As a result, their impact on low frequency oscillations should be discussed respectively.

Among the several wind generation technologies, variable speed wind turbines utilizing doubly fed induction generators are gaining prominence in the power industry. A doubly fed induction machine is a wound-rotor doubly fed electric machine and has several advantages over a conventional induction machine in wind power applications. As the performance is largely determined by the converter and the associated controls, a DFIG is an asynchronous generator. The control of the rotor voltages and currents enables the induction machine to remain synchronized with the grid while the wind turbine speed varies. A variable speed wind turbine utilizes the available wind resource more efficiently than a fixed speed wind turbine, especially during light wind conditions. What's more, the efficiency of the DFIG is very good.

## 3 IMPACT OF DFIG ON LFOS

The variable speed wind turbine generator design consisting of power electronics converter impacts significant effect on the dynamic performance of the DFIG. The introduction of large amounts of wind generation does have the potential to change the electromechanical damping performance of the system, especially when the wind farms with large capacity are in remote part of the power system, which need to access power system through a long transmission line. Then the problems are particularly important that how wind turbines influence LFOs and the system damping and what the law of wind turbines' stable operation is. When wind turbines are integrated into power system, related oscillatory mode shapes are introduced, which are mainly related to the mechanical system and converter control system of the wind turbines.

Since doubly fed induction generators are asynchronous machines, they can primarily affect the damping of electromechanical modes by four mechanisms (Gautama et al. 2009), that is displacing synchronous machines and that have power system stabilizers, impacting major path flows, and DFIG controls interacting with the damping torque on nearby large synchronous generators.

The control system of DFIG has no effect on low frequency oscillations (Lin & Tan 2011) and the current DFIG has no control system established aiming at restraining low frequency oscillations yet. What's more, only through the stator is the DFIG integrated into the power grid. As a result, when LFOs occurs, the change of electrical variables can only be detected through the stator and oscillation current is induced at the rotor side to produce the damping torque. However, because of the small value of the resistance inside the DFIG, the produced damping torque plays a very limited role in restraining LFOs (Tsourakis 2009). Consequently, DFIG has the potential to affect the system damping adversely (Ledesma & Usaola 2005).

## 4 SIMULATION RESULTS

The study is carried out in a two-area four-generator power system integrated with doubly fed induction generators based on Matlab/Simulink. The system is shown as in Figure 1. The parameters of two-area four-generator power system adopted are similar with the demo "power_PSS.mdl" in Matlab and so are the parameters of a wind farm with the demo "power_wind_dfig.mdl" in Matlab.

### 4.1 Case 1

For the two-area four-generator power system without DFIG, load 3 with the capability of 67 MVA is integrated into the system at 4 s. The active power from bus 8 to bus 9 is shown as in Figure 2. Zoom in and the part from 4 s to 4.3 s is shown in Figure 3.

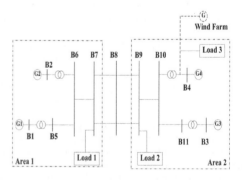

Figure 1. Two-area four-generator power system integrated with DFIG.

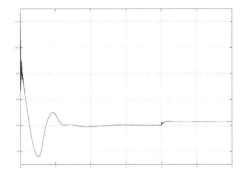

Figure 2. The active power from bus 8 to bus 9.

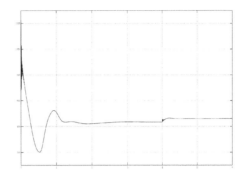

Figure 4. The active power from bus 8 to bus 9.

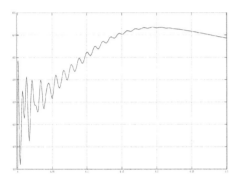

Figure 3. The active power from bus 8 to bus 9 (4 s–4.3 s).

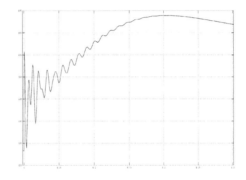

Figure 5. The active power from bus 8 to bus 9 (4 s–4.3 s).

According to Figures 2 and 3, low-frequency oscillations take place when load 3 is integrated into the system at 4 s. The oscillations calm down later because of the PSS.

### 4.2 Case 2

For the system integrated with DFIG whose capacity is 45 MW, load 3 with the capability of 67 MVA is integrated into the system at 4 s. The active power of one transmission line from bus 8 to bus 9 is shown as in Figure 4. Zoom in and the part from 4 s to 4.3 s is shown in Figure 5.

Low-frequency oscillations also take place at 4 s, when load 3 is integrated into the two-area four generator power system. According to Figure 3 and Figure 5, after DFIG integrated into the power system, the oscillation amplitude increases, so the damping of the system decreases.

### 4.3 Case 3

For the system integrated with DFIG whose capacity is 105 MW, load 3 with capability of 67 MVA is integrated into the system at 4 s. The active power

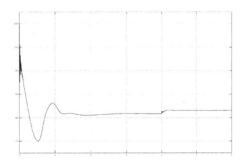

Figure 6. The active power from bus 8 to bus 9.

of one transmission line from bus 8 to bus 9 is shown as in Figure 6. Zoom in and the part from 4 s to 4.3 s is shown in Figure 7.

Low-frequency oscillations also take place at 4 s, when load 3 is integrated into the power system. As we can see, the oscillation amplitude in Figure 7 increases against to Figure 5, so the damping of the system decreases as the penetration level of wind power increase, and the oscillation frequency increases.

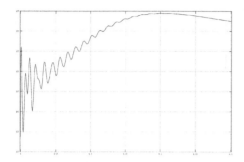

Figure 7. The active power from bus 8 to bus 9 (4 s–4.3 s).

## 5 CONCLUSIONS

In this paper, the impact of doubly fed induction generator and its penetration level on low frequency oscillations is observed for a two-area four-generator system. Analysis is carried out for small disturbance to observe the system damping.

For the system operating conditions considered, the system damping decreases when doubly fed induction generator integrated into the system, and as the penetration level increases, the system damping decreases further.

REFERENCES

Gautama D., Vittal V. & Harbour T. 2009. Impact of Increased Penetration of DFIG-Based Wind Turbine Generators on Transient and Small Signal Stability of Power Systems. *IEEE Transactions on Power Systems,* 24(3): 1426–1434.

Heier, S. 2006. *Grid Integration of Wind Energy Conversion Systems.* Chichester: John Wiley & Sons Ltd.

Horowitz, S.H. & Phadke, A.G. 2006. Third zone revisited. *Power Delivery, IEEE Transaction on,* 21(1): 23–29.

Kunder, P. 1994. *Power System Stability and Control.* N.J. balu & M.G. Lauby (eds). New York: McGraw-hill.

Ledesma, P. & Usaola, J. 2005. Doubly Fed Induction Generator Model for Transient Stability Analysis. *Energy Conversion, IEEE Transaction on,* 20(2): 388–397.

Li, S. 2008. New Method to Quantify the Operation Condition for Zone 3 Impedance Relays during Low-Frequency power Swing. *Journal of Electrical Engineering & Technology,* 3(1): 29–35.

Lin, X. & Tan, H. 2011. Study on the Small Signal Stability of Electric Power System in Respect of Wind Farm Integrated. *Southern Power System Technology,* 5(A01): 55–58.

Pal, B. & Chaudhuri, B. 2006. *Robust Control in Power Systems.* New York: Springer Inc.

Tan, J. Wang, X. & Li, L.Y. 2014. A Survey on Small Signal Stability Analysis of Power Systems with Wind Power Integration. *Power System Protection and Control,* 42(3): 15–23.

Tsouraki, G. Nomikos, B.M. & Vournas, C.D. 2009. Effect of Wind Parks with Doubly Fed Asynchronous Generators on Small-signal Stability. *Electric Power Systems Research,* 79(1): 190–200.

*Advanced Materials, Structures and Mechanical Engineering – Kaloop (Ed.)*
© *2016 Taylor & Francis Group, London, ISBN 978-1-138-02793-0*

# EMD-based event analysis for identifying the influence of the "2008 Financial Crisis" on silver price

L. Chao & S. Yu
*Macau University of Science and Technology, Macau, China*

ABSTRACT: In this paper, the Empirical Mode Decomposition (EMD)-based event study approach is applied to test whether the "2008 Financial Crisis" has influenced the spot silver price or not. First, original spot silver price is decomposed into one residual and several Intrinsic Mode Functions (IMFs). Several IMFs caused by the "2008 Financial Crisis" are summed up to conduct the event study. The empirical results illustrate that the EMD-based event study method provides a feasible solution to estimate the impact of extreme events, and show that the "2008 Financial Crisis" did significantly influence the silver price.

## 1 INTRODUCTION

Large quantities of research have been done on the analysis of silver price by applying models from different areas. Recent research includes Abidin, Banchit, Lou & Niu (2013), Harper, Jin, Sokunle, & Wadhwa (2010) and Kettering (2009). However, silver price consists of many components, which individually represents the influence of different aspects. From this perspective, original data should be processed beforehand to get rid of the disturbance while conducting the analysis to increase the preciseness of the result.

Although silver and gold are always considered as substitutes for each other with respect to risk aversion (Ronald 2009), people begin to doubt the function of gold as the risk-aversion tool and gradually transfer their concentration to silver market for investment, considering that the international gold price has dropped sharply in the first half of year 2013, decreasing from 1911$/ounce in 2011 September to $1,191.21/ounce in 2013 June (Matthew 2013).

Hence, in this paper, analysis of silver price could familiarize them with the components of silver price to make more rational investments in silver market. More importantly, this paper first adopts the EMD-based event study method in researching silver price, which provides a new manner for future studies.

This paper is organized as follows: Section 2 provides the theoretical support for the methods applied and introduces the detailed procedure to conduct the analysis. Section 3 discusses the data, and the economic analysis of the results. Section 4 provides the concluding remarks.

## 2 METHODOLOGY AND DATA

The overall process of the EMD-based event study method of the "2008 Financial Crisis" is illustrated in Figure 1. To begin with, a time series of silver spot price are input into EMD to be decomposed into several IMFs and one residue. Second, further analysis is conducted about which IMFs are chosen to be the basic data for the event study. Next, the "2008 Financial Crisis" is analyzed with appropriate IMFs to test whether it affected the silver price significantly. Lastly, conclusion is made based on previous results. Event study methodology was first proposed by Fama, Fisher, Jensen and Roll in 1969, and became one of the most useful tool to research long-term abnormal returns in capital market. Since the event study is commonly used to investigate the effects of extreme events, the analysis result would be more accurate and convincing if applying the appropriate decomposition technique to decompose the original time series and just analyzing the components that are influenced by those

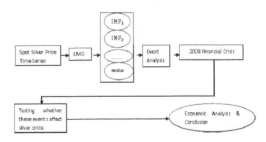

Figure 1. Process of the EMD-based event analysis methodology.

Table 1. Descriptive statistics of the spot daily silver price time series.

| | Number | Mean | Standard deviation | Max | Min |
|---|---|---|---|---|---|
| All data | 5235 | 13.17 | 10.06 | 48.35 | 4.05 |

extreme events. That is why the event study is combined with the EMD technique in this paper.

All the data are gathered from DATASTREAM in this study and daily spot silver price XAG/USD from January 1, 1999 to May 1, 2013 are applied while conducting EMD. High frequency data with a large window size are preferable in order to investigate the data that are widely distributed. What's more, the number of monthly or weekly data is not large to ensure the robustness of the decomposition results and the degree of freedom. Therefore, daily prices of spot silver are applied in this paper.

Table 1 presents the descriptive statistics of the data of XAG/USD. There are totally 5235 samples of daily silver price; the mean is 13.168 and the standard deviation is 10.059, with a maximum of 48.347 and a minimum of 4.050.

While applying the time series to research, it is required to conduct a unit root test to judge whether these series are stationary or not. And as previously mentioned, the EMD technique is used to decompose the non-stationary time series. Therefore, an ADF test is conducted using E-VIEWS on the spot silver price. The result shows there is a unit root in the original time series. Thus, the original series are non-stationary and could be decomposed by EEMD.

## 3 ANALYSIS

A total number of 5235 data points from January 1, 1999 to May 1, 2013 are decomposed using software MATLAB. The decomposed result is shown in Figure 2. As previously mentioned, high-frequency IMFs represent a sudden rise or decent of the original time series, and the sudden rise or drop of silver spot price always follows a drop or rise of price, the total effect of which could be neglected in the long run. This means that the market is a mean-reversion trend in the short term and the effects of these high-frequency IMFs are almost zero for a longer period (Zhang et al. 2009). To judge which specific IMF or IMFs we should choose to analyze the effect of extreme events, we can use a fine-to-coarse reconstruction to inspect the sum of which high-frequency IMFs are not significantly different from zero with the t-test. The detailed procedure is as follows: First, $C_i$ is denoted as the sum of IMFs from 1 to i, for example, $C_3$ means the sum of IMF1 to IMF3. Next, we need to calculate the average of each $C_i$.

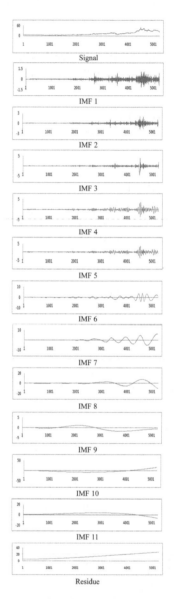

Figure 2. The IMFs and residue for the silver price from January 1, 1999 to May 1, 2013 decomposed by EEMD.

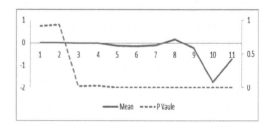

Figure 3. The mean of fine-to-coarse reconstruction with data from January 1, 1999 to May 1, 2013.

Finally, using the t-test, we check from which $C_i$ is significantly different from zero. Then, we add up the left IMFs and treat the result as the component of silver price caused by the "2008 Financial Crisis". Figure 3 shows the result: the means of the reconstruction from IMF1 to IMF2 ($C_1$ and $C_2$) are not significantly different from zero; therefore, IMF1 and IMF2 do not have long-term effects on the price and can be neglected. We should sum from IMF3 to IMF11 to conduct a further test. There are three measures of IMFs that need to be calculated. The first measure is the mean period, which is computed from dividing the total number of points by the number of peaks for each IMF, which means that the length of period can be explained by this IMF (Oladosu 2009).

As can be observed from Table 2, the sum of IMFs from 3 to 11 accounts for more than 55 percent of change in the price, which can also support the result of the above fine-to-coarse reconstruction. The high correlation coefficient between the residue and the original time series proves that the silver price follows the trend of residue, and this trend accounts for a major component of the price. That is to say, although the "2008 Financial Crisis" resulted in the fluctuations of silver price, the average trend is still the main cause to guide silver price.

After having conducted the EMD-based event study method, we can find that the "2008 Financial Crisis" influenced the silver price significantly at 5 percent of significance level.

Also, as for whether the "2008 Financial Crisis" increased the volatility of silver price, a t-test is implemented to test the amplitudes of IMF1 in the event window. If IMF1 s are significantly larger than those in the estimation window, it proves that these events do increase the volatility. The result does support the hypothesis.

Table 2. Measures of IMFs and the residue of spot daily silver price from January 1, 1999 to May 1, 2013.

| | Mean period | Correlation coefficient | Variance as percentage of original silver price |
|---|---|---|---|
| IMF 1 | 3.4 | 0.03 | 0.02% |
| IMF 2 | 9.6 | 0.00 | 0.04% |
| IMF 3 | 15.3 | −0.1 | 0.15% |
| IMF 4 | 65.5 | 0.06 | 0.25% |
| IMF 5 | 115.2 | −0.05 | 1.89% |
| IMF 6 | 226.8 | 0.07 | 1.15% |
| IMF 7 | 431.3 | 0.13 | 3.54% |
| IMF 8 | 1047.1 | 0.31 | 8.64% |
| IMF 9 | 2671.2 | −0.56 | 0.74% |
| IMF 10 | | 0.74 | 34.13% |
| IMF 11 | | −0.6 | 4.95% |
| Residue | | 0.91 | 49.01% |

## 4 CONCLUSIONS

A relatively new method is used in this paper to address the problem of financial time series in terms of nonlinear and non-stationary characteristics. The empirical results of the "2008 Financial Crisis" is also supportive. We obtained the conclusion that the "2008 Financial Crisis" influenced the silver price significantly at 5 percent of significance level. Although the "2008 Financial Crisis" resulted in the fluctuations of silver price, the average trend is still the main cause to guide silver price. However, there also exist some limitations in this approach. To begin with, this method measures the total magnitude of the change caused by the "2008 Financial Crisis"; however, there is no concrete method to measure the estimator like the statistical test or the confidence interval. Thus, the accuracy of the estimates is not guaranteed. Second, the different selections of the event window and the estimation window may also result in different consequences. These and other extensions are left for future studies.

## REFERENCES

Abidin, S. Banchit, A. Lou, R. & Niu, Q. 2013. Information flow and causality between price change and trading volume in silver and platinum futures contracts. *International Journal of Economics, Finance and Management*. 2(2): 241–249.

Han, L. Ding, L. Zheng, G. Yanming, L. & Nianlong, S. 2004. Natural gas load forecasting based on least squares support vector machine. *Journal of Chemical Industry and Engineering (China)*. 5: 026.

Harper, A. Jin, Z. Sokunle, R. Bank, U. & Wadhwa, M. 2010. Price volatility in the silver spot market: An empirical study using Garch applications. *Journal of Finance and Accountancy*. 13: 1.

Ivanov, S. 2013. Analysis of the effects of pre announcement of S&P 500 index changes. *International Journal of Business & Finance Research (IJBFR)*. 7(5): 1–10.

Kettering, R.C. 2009. The effect of international currencies upon gold and silver prices. *Review of Business Research*. 9(2): 138.

Lin, C. Chiu, S. & Lin, T. 2012. Empirical mode decomposition-based least squares support vector regression for foreign exchange rate forecasting. *Econ Model*. 29(6): 2583–2590.

Oladosu, G. 2009. Identifying the oil price–macroeconomy relationship: An empirical mode decomposition analysis of US data. *Energy Policy*. 37(12): 5417–5426.

Park, A. & Chang, C. 2013. Impacts of construction events on the project equity value of the channel tunnel project. *Construction Management & Economics*. 31(3): 223–237.

Zhang, X. Yu, L. Wang, S. & Lai, K.K. 2009. Estimating the impact of extreme events on crude oil price: An EMD-based event analysis method. *Energy Economics*, 31(5): 768–778.

*Advanced Materials, Structures and Mechanical Engineering – Kaloop (Ed.)*
© *2016 Taylor & Francis Group, London, ISBN 978-1-138-02793-0*

# Evaluation of metallurgical parameters of Brazilian lump iron ores with three distinct typologies

J. Januário Mendes
*Materials Engineering (Redemat), Federal University of Ouro Preto, Ouro Preto, Brazil*
*Department of Production Engineering, Federal Institute of Minas Gerais, Minas Gerais, Brazil*

A.A. Cunha
*Materials Engineering (Redemat), Federal University of Ouro Preto, Ouro Preto, Brazil*

F.G. da S. Araújo
*Materials Engineering (Redemat), Federal University of Ouro Preto, Ouro Preto, Brazil*
*Department of Research and Continuing Education (DEPEC) of Gorceix Foundation, Minas Gerais, Brazil*

F.L. von Krüger & R.A. Llobell Sole
*Department of Research and Continuing Education (DEPEC) of Gorceix Foundation, Minas Gerais, Brazil*

ABSTRACT: The evaluation of the behavior lump iron ores during the reduction process in blast furnace reactors is extremely important to determine the system's productivity. To predict this behavior, one has to evaluate lump ores and pellets relative to chemical composition, physical properties, mineralogy, Reduction Degradation Index (RDI) and Reducibility Index (RI). This paper presents the evaluation of the chemical and metallurgical qualities of brazilian lump iron ores, when subjected to the reducing conditions present in the blast furnace. The procedures of the metallurgical tests are standardized by ISO and are performed in automated metallurgical testing ovens that allow real-time monitoring. The results obtained from the metallurgical tests permitted to assess the differences in quality of three typologies of lump ores and how this may influence the productivity of the reactors.

## 1 INTRODUCTION

The Steel Industry is currently facing a new challenge, in which mining companies have difficulty to meet an increasing demand of specifications related to chemical and metallurgical quality of raw materials.

The blast furnace can be fed with three types of metallic charge: natural lump iron ore and two others obtained from fine ores agglomeration processes, e.g., sinter and pellet. The evaluation of the behavior of lump ores, sinters and pellets during the reduction process in blast furnaces is extremely important to determine the system's productivity.

The main factors that influence the descent of burden inside the blast furnace and the gas percolation through it are assessed by the Reducibility Index (RI), the Reduction Degradation Index (RDI) and decrepitation index. Studies in pilot scale are standardized by ISO and try to simulate the conditions in which the phenomena occur and to evaluate the behavior of the raw materials during the reduction.

One characteristic specific to the lump ores is that, when they are loaded into reduction reactors,

they may degrade under heat shock at first contact with the hot gases. This shock, in many ores, can lead to thermal spalling, in a phenomenon known as decrepitation. This originates the generation of fines in the burden, which can reduce its permeability. One advantage of the use of sinters and pellets is that they do not decrepitate.

Evaluation of the decrepitation is made using the assays defined by COISRMJ (Committee for Overseas Iron and Steelmaking of Japan Raw Materials) and ISO 8371. In both tests the sample is subjected to thermal shock by introducing them into a preheated oven at 700 °C. After 30 min the sample is removed for cooling in air and sieved, to measure the decrepitation index, which is the percentage of fines generated below 5 mm (COISRMJ) or 6.3 mm (ISO 8371). Lump ores considered good under thermal shock present decrepitation indexes not exceeding 5% when measured by the COISRMJ test or not exceeding 3% in the ISO 8371 test. However, due to increasing scarcity of lumpy iron ores, there is a widespread effort in the steel industry to enable the use of lump ores with higher decrepitation indexes.

Due to the fact that hematite and magnetite have different crystal structures, the volume expansion during reduction may weaken the ore's structure, which can lead to disintegration in greater or lesser degree, generating a quantity of fines which may hinder the operation. This phenomenon is known as disintegration under reduction in low temperature. In the manufacture of pellets, one can act on raw materials and process parameters in order to mitigate their susceptibility to this disintegration. In the case of lump ores, what can be done is to feed the blast furnace with lump ores of less susceptibility to disintegration under reduction. With this comes the requirement for a quality specification to which the ores must attend, usually known as RDI specification (Reduction Disintegration Index) or LTD (Low-temperature Disintegration) of ore.

The evaluation of the RDI is made primarily using two types of tests, both static, ISO 4696-1 and ISO 4696-2. The ISO 4696-2 test is the mostly used by the steel industry. The RDI index is expressed by the amount of fine material (% <2.8 mm) generated in a tumble test of the material reduced at 550 °C for 30 minutes. The disintegration of lump ores is considered satisfactory when they have less than 20% RDI and for pellets less than 14%. In LTD test, which is dynamic, the reduction is performed at 500 °C simultaneously with the tumble test of the sample. According to Fernandes (2008), and Souza et al. (1998), for lump ores the satisfactory LTD is a percentage of particles greater than 6.3 mm higher than 40 to 45% and for pellets this number should exceed 60%.

The $Fe_2O_3 \rightarrow Fe_3O_4 \rightarrow FeO$ reduction reactions are thermodynamically much more favorable than the final reduction $FeO \rightarrow Fe$. The amount of oxygen per atom of iron to be removed in $FeO \rightarrow Fe$ reaction is much higher than before, and the Fe-O bonds to be broken, are stronger. However, the time available for this reaction before starting the softening of the ore is relatively short. From all this, there is a need to control another important quality of the ore, which is Reducibility Index (RI), which reflects the degree of ease that the ore has to lose (transfer) its oxygen to the reducing gases. Thus, an ore with high reducibility index can be reduced to iron more quickly than another with low reducibility.

The evaluation of Reducibility Index (RI) of ores is made using mostly the ISO 7215 test. In this assay the reduction of the ore sample is done at 900 °C for 3 hours. The final degree of reduction, measured by weight loss, expresses the reducibility index of ore.

Table 1 shows reference values for the metallurgical properties of lump iron ores in both direct reduction and blast furnace, together with physical properties.

This paper presents a study on the metallurgical quality of brazilian lump iron ores of three different typologies. The tests were performed in automated metallurgical testing ovens, simulating the conditions

Table 1. Reference values of metallurgical properties Souza (1998).

| Metallurgical properties | Lump ore direct reduction | Lump ore blast furnace |
|---|---|---|
| Reference values of the metallurgical properties | | |
| Reducibility (% reduction) | >40% | >45% |
| RDI (% <2,80 mm) | <25% | <25% |
| Tumble test (TI: >6,30 mm) | >90% | >75% |
| Abrasion (AI: % <0,50 mm) | <10% | <25% |
| CracklingIndex (<6,30 mm) | <5% | <5% |

inside a blast furnace. The results obtained from the metallurgical testing permitted a comparison of the performances of the different iron ore typologies under reduction in a blast furnace.

2 PROCEDURES

Three samples of lumpy iron ores of different origins, consisting of high grade hematites, were collected for the determination of their microstructures, typologies, physical properties and metallurgical properties.

The samples of three lump iron ores were milled below 0.3 mm and prepared for examination by optical microscopy.

The physical and metallurgical tests were performed according to ISO standards, and performed in automated metallurgical testing ovens that allow real-time monitoring, to simulate the conditions present in the reduction zone of the blast furnace.

The tests of metallurgical quality of a given material seek to generate technical information for reference and comparison, make the quality of materials become uniform in their respective production process, to group the types of materials, database for a correct determination of what material to use in a particular process, to serve as a comparative reference between different test sites and use as neutralizer between customer and supplier.

Samples of three lump iron ores were milled unless 0.3 mm and prepared for examination by light microscopy and electronic.

In the photomicrographs of specularite iron ore (Fig. 1), there are hematite particles in a curious fibrous structure which is divided into lamellas, which indicates an increased susceptibility to abrasion of the material.

3 RESULTS AND DISCUSSION

The samples of lumpy iron ore collected were consisted of three different typologies, related

Specularite 50x

Specularite 50x

Figure 1. Micrographs of specularite iron ore.

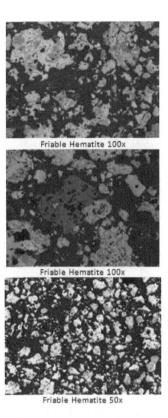
Friable Hematite 100x

Friable Hematite 100x

Friable Hematite 50x

Figure 2. Photomicrographs of the iron ore friable hematite.

to the characteristics of the deposit and to the microstructure of the hematite crystals. Here the hematite types are separated into specularite hematite, friable hematite and rolled hematite.

In the photomicrographs of specularite iron ore (Fig. 1), there are hematite particles in a fibrous structure which is divided into lamellas, which may suggest some susceptibility to abrasion of the material.

The photomicrographs of the friable hematite iron ore (Fig. 2) show particles of martitized magnetite, with martite hematite in white and magnetite in light gray color. In the central photo a large particle of goethite in dark gray color with porosity also appears.

Photomicrographs of the rolled hematite iron ore (Fig. 3) show magnetite particles with different degrees of martitization and a very small portion of quartz appears in difuse gray, without porosity.

Table 2 shows the results of the metallurgical and physical tests of the lump ores.

The results in Table 2 show that the decrepitation index for all types are extremely low, and this behavior indicates that the levels of internal stresses, that could lead to spalling upon heating, are very low. For the tested ores there is no significant difference in the decrepitation index and all of them present values much below the upper limit, of 5%.

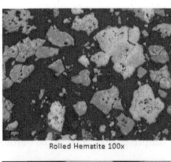
Rolled Hematite 100x

Rolled Hematite 500x

Figure 3. Photomicrographs of iron ore rolled hematite.

Table 2. Metallurgical test of the lump ore.

| Sample | | Decrepitation Index | | | LTD/RI Index | | | RDI | Tumble Test | Abrasion |
|---|---|---|---|---|---|---|---|---|---|---|
| | | <6,35mm | <3,15mm | <0,5mm | >6,35mm | <3,15mm | <0,5mm | % | % | % |
| Friable Hematite | 1 | 0.36% | 0.36% | 0.25% | 15,60% | 69.40% | 57.70% | 51,10% | 62,20% | 23.70% |
| Friable Hematite | 2 | 0,30% | 0.30% | 0.28% | 17,00% | 66.30% | 54.20% | 51,60% | 57,40% | 29.30% |
| Friable Hematite | 3 | 0.38% | 0.38% | 0.28% | | | | | 57,70% | 28.30% |
| | Average | 0.30% | 0.33% | 0.27% | 16,30% | 67.85% | 55.95% | 51.35% | 59.80% | 26.50% |
| | | | | | | | | | | |
| Rolled Hematite | 1 | 0.36% | 0.36% | 0.25% | 44.70% | 46.80% | 38.70% | 32.50% | 78.60% | 15.90% |
| Rolled Hematite | 2 | 0.30% | 0.30% | 0.28% | 46.40% | 44.60% | 36.80% | 30.40% | 82.70% | 13.20% |
| Rolled Hematite | 3 | 0.38% | 0.38% | 0.28% | | | | | 80.60% | 14.70% |
| | Average | 0.33% | 0.33% | 0.27% | 45.55% | 45.70% | 37.75% | 31.45% | 80.65% | 14.55% |
| | | | | | | | | | | |
| Specularite | 1 | 0.05% | 0.05% | 0.05% | 66.80% | 28.30% | 22.60% | 20.80% | 80.90% | 16.40% |
| Specularite | 2 | 1.46% | 0.84% | 0.14% | 66.70% | 29.10% | 22.40% | 21.80% | 81.30% | 16.50% |
| Specularite | 3 | 0.12% | 0.12% | 0.08% | | | | | 81.00% | 16.80% |
| | Average | 0.76% | 0.45% | 0.10% | 66.75% | 28.70% | 22.50% | 21.30% | 81.10% | 16.45% |

LTD tests showed that the ores have a pronounced generation of fines in the fraction <0.5 mm. These fines impair permeability and consequently the performance of the blast furnace. Among these materials, the friable hematite lump iron type does not meet the minimum requirements in terms of LTD results, the rolled hematite type barely comply with the specifications and the specularite iron ore type is the most suitable for the use in a blast furnace, with a high LTD index (>>>45%), but its high percentage of fines below 0.5 mm is a point of concern.

The RDI index values shown in Table 2 demonstrate that friable hematite (51.35%) and rolled hematite (31.45%) have a high RDI in comparision with the desired. Also in this specification only the specularite hematite iron ore type meets the ideal values for blast furnace reduction.

The minimum desired tumble index for granular materials is 75%, therefore the friable hematite ore does not comply with the specification, whilst the rolled hematite and specularite ores are suitable for the use in blast furnaces. It is observed that the three tested lump ores produce large amounts of fines, but both the rolled hematite and the specularite hematite iron ore types could be used in some proportions, with caution to control the negative effect of the presence of fines in the burden.

## 4 CONCLUSIONS

The scarce availability on market of high quality lump iron ores, with high iron content and physical and metallurgical properties suitable for the reduction reactors, lead the industry to adapt to the new materials available. This requires the constant monitoring and assessment of the properties of the metal components, seeking to ensure quality and implement corrective measures. The results of this study showed that the ores used in the form of lump ore have quality characteristics that strongly influence the reduction process. All lump ores presented low decrepitation indexes and tumbling results inside the specifications. However, only the specularite lump ore obtained RDI and LTD indexes in the desired range.

## REFERENCES

ABNT NBR ISO 7215, 2009. Iron ore as raw material for blast furnace—Determination of the reducibility by the final degree of reduction index.

ABNT NBR ISO 4696-1, 2013. Iron ore as raw material for blast furnace—Determination of disintegration rates under reduced at low temperature by static method—Part 1: Reduction with CO, $CO_2$, $H_2$ e $N_2$.

ABNT NBR ISO 4696-2, 2008. Iron ore as raw material for blast furnace—Determination of disintegration rates under reduced at low temperature by static method—Part 2: Reduction with CO e $N^2$.

ABNT NBR ISO 13930, 2014. Iron ore as raw material for blast furnace—Determination of reduction-disintegration indices at low temperature by dynamic method.

ABNT NBR ISO 7992, 2014. Iron ore as raw material for blast furnace—pressure reducing Determination.

Bakker, T. 1999. *Softening in blast furnace burden process: Local melt formation as the tigger for softening of iron-bearing burden material.* Phd Thesis—Delf University of Tecnology, Netherlands.

Cardoso, M.B. 1981. *Decrepitation of iron ore from the Iron Quadrangle.* Masters dissertation—CPGEM/EEUFMG—Belo Horizonte.

Fernandes, E.Z. & Araújo, A.C. 2008. *Characterization Physics, Chemistry, Mineralogical and Metallurgical of iron ore Products*—Doctoral Thesis—School of Engineering UFMG—Belo Horizonte.

Pimenta, H.P & Seshadri, V. 2002. Characterization of structure of iron and its behavius during reduction at low temperatures. *Ironmaking and Steel Making*, 29(3): 169–174.

Souza, C.C. Almeida, R.M. & Pereira, J.F. 1998. *History of physical and metallurgical properties of the blast furnace of MBR.* Internal technical report. Nova Lima, MG.

*Advanced Materials, Structures and Mechanical Engineering – Kaloop (Ed.)*
© 2016 Taylor & Francis Group, London, ISBN 978-1-138-02793-0

# The issue of balancing of eccentric-type vibrators

I.M. Yefremov, D.V. Lobanov & K.N. Figura
*Bratsk State University, Bratsk, Russian Federation*

ABSTRACT:   The problem of balancing of unbalanced rotating mass arises in the technique very often. This problem is especially real in industries related to the use of vibratory machinery. This is due to the fact that the unbalanced oscillating masses may damage the machine, disable it, or even lead to the injury or illness of service personnel. For proper balancing of rotating unbalanced masses, we need to know the value and place of the application of the resultant centrifugal forces. The article presents the methods of calculating these quantities. These methods are based on the fundamental laws and principles of theoretical mechanics. The results were obtained for the regularities connecting the main mass and dimensional characteristics, as well as the rotation frequency of the eccentric vibrator with centrifugal forces. These regularities include the most common schemes of eccentric-type vibrators.

## 1 INTRODUCTION

Currently, the production of multi-component mixtures and powder composites is one of the most important trends in various industries. Technology of preparation of compounds of this type not only takes one of the large segments of the market, but also is a promising basis for the development in the field of technological equipment, in which the special preference is given to the different types of mixing apparatus. Analysis of various papers shows that the greatest effect can be achieved in the preparation of various mixtures in vibration mixers (Efremov 2011, Malakhov 2013), where blending components have the greatest number of impacts per unit time. The vast majority of constructions of vibrating mixers are forced mixer with balanced (inertial) vibration exciters, which relate to the vibration activator with kinematic excitation hesitation. Preference to use this type of vibrating actuators is caused by their simplicity in design concept and reliability (Hong 2012b, Hong 2012a, Paul 1999). Schematic diagrams of vibrating actuators are shown in Figure 1.

The housing of the eccentric vibratory activator makes a complex rotational motion, consisting of translational motion (crankshaft rotates around axis X) and relative rotation around the axis $X_1$. If the X and $X_1$-axes of the translational and relative rotation are parallel, the movement of the vibratory activator is called the plane parallel. If the axis $X_1$ is at a certain constant angle $\beta$ to the X-axis, the movement of the vibration activator is called processing. In such designs, called processing, the rotation of the crankshaft body of vibration activator makes movement called regular precession with the nutation angle $\beta$.

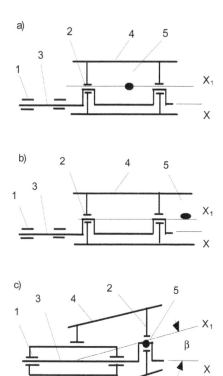

Figure 1.   Schematic diagrams of vibrating actuators. a) console, four-support; b) plane parallel, four-support; c) console, processing with a three-point support: 1—main bearings; 2—connecting rod bearings; 3—the drive shaft; 4—body of vibration exciter; 5—the center of the oscillating mass; 6—elastic element; X-axis of rotation of the exciter; $X_1$—the main central axis of inertia of the exciter housing.

Bearings whose centers coincide with the X-axis of the crankshaft rotation will be called main bearing 1; bearings installed along the axis of relative rotation of the vibrator $X_1$ will be called connecting rod bearings 2 (Fig. 1).

Vibration actuator is called the console if the oscillating weight is mounted on the connecting rod bearings, and located behind one of the last main supports (Fig. 1, a, b). Console vibration activator with precession is designed with a three-point support. In this case, the connecting rod bearing is located at the point of application of the resultant of the centrifugal forces of inertia resulting from vibration and a second bearing between the crankshaft bearings by an elastic element (Fig. 1c).

Plane-type vibration activators are characterized by a static unbalance, vibration activator for processing type—dynamic unbalance. Under balance (static or dynamic) is understood a state of vibration activator, in which the reaction in the main bearings of the centrifugal forces and moments is equal to zero, i.e., the degree of transmission of dynamic loads on the supporting pillars, foundation, and drive is low. Contemplated vibrating actuators can be brought into a state of dynamic or static equilibrium with the two (one) adjusting weights (counterweights), arranged in two (one) correction planes.

## 2   TASK OF CALCULATION

Task of the calculation is to determine the position and location of the resultant centrifugal forces generated during rotation of the unbalanced (oscillating) masses.

## 3   CALCULATION

Unbalanced mass of vibration exciter can be represented as a sum of linearly distributed and concentrated masses along the length. Typical schemes for the most common (possible) processing exciters are presented in Figure 2.

The resultant centrifugal forces arising from the circular vibrations of the body, shown in Figure 2, for typical schemes are determined from the following equations:

$$Q_i = \int_\ell dQ_i = \int_\ell \omega^2 q_i (tg\beta) x dx \qquad (1)$$

where $dQ_i$ is the elementary centrifugal force and $\omega$ is the angular velocity of rotation of the crankshaft of the vibration exciter.

After substituting the values of the distributed loads in Equation (1), the following integration can be obtained:

a. $q_i = const$; $Q_1 = \omega^2 q_1 tg\beta \int_0^l x dx = \dfrac{1}{2}\omega^2 q_1 l^2 tg\beta$

b. $q_2 = const$;

$Q_2 = \omega^2 q_2 tg\beta \int_{l_1}^l x dx = \dfrac{1}{2}\omega^2 q_2 (l^2 - l_1^2) tg\beta$

c. $q_3 = q^* \left(1 - \dfrac{x}{l}\right)$;

$Q_3 = \omega^2 q^* tg\beta \int_0^l \left(1 - \dfrac{x}{l}\right) x dx = \dfrac{1}{6}\omega^2 q^* l^2 tg\beta$

d. $q_4 = q^* \left(1 - \dfrac{x - l_1}{l - l_1}\right)$;

$Q_4 = \omega^2 q^* tg\beta \int_{l_1}^l \left(1 - \dfrac{x - l_1}{l - l_1}\right) x dx$

$\qquad - \dfrac{1}{6}\omega^2 q^* (l^2 + l l_1 - 2 l_1^2) tg\beta$

e. $q_5 = q^* \dfrac{x}{l}$; $Q_5 = \omega^2 q^* tg\beta \int_0^l \dfrac{x}{l} x dx = \dfrac{1}{3}\omega^2 q^* l^2 tg\beta$

f. $q_6 = q^* \dfrac{x - l_1}{l - l_1}$; $Q_6 = \omega^2 q^* tg\beta \int_{l_1}^l \dfrac{x - l_1}{l - l_1} x dx$

$\qquad = \dfrac{1}{6}\omega^2 q^* (2l^2 - l l_1 - l_1^2) tg\beta$

g. $Q_7 = \sum_{i=1}^n M_i l_i \omega^2 tg\beta$

The coordinate of points of the application of the resultant centrifugal forces is determined by the following equations with Expressions (1):

$$L_i = \dfrac{\int_\ell dM_i}{Q_i} = \dfrac{\int_\ell x dQ_i}{Q_i} = \dfrac{\int_\ell \omega^2 q_i (tg\beta) x^2 dx}{Q_i} \qquad (2)$$

where $dM_i$ is the elementary static moment of centrifugal force about the point O.

After substituting in Equation (2) values of distributed loads, corresponding to them centrifugal forces and after integration, we have:

a. $L_1 = \dfrac{2}{3}l$

b. $L_2 = \dfrac{2(l^3 - l_1^3)}{3(l^2 - l_1^2)}$

c. $L_3 = \dfrac{l}{2}$

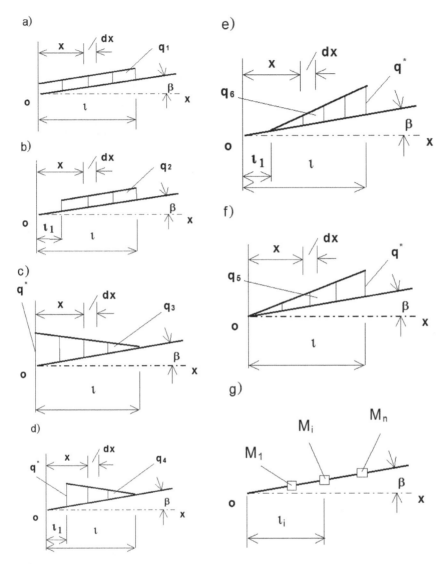

Figure 2. Typical schemes of distributed and concentrated masses: $x$-current coordinate; $q_i$—distributed load; $M_i$—concentrated mass; $l$—length.

d. $L_4 = \dfrac{1}{2} \dfrac{(l^3 + l^2 l_1 + l l_1^2 - 3l_1^3)}{(l^2 + l l_1 - 2l_1^2)}$

e. $L_5 = \dfrac{3}{4} l$

f. $L_6 = \dfrac{1}{2} \dfrac{(3l^3 - l l_1^2 - l^2 l_1 - l_1^3)}{(2l^2 - l l_1 - l_1^2)}$

g. $L_7 = \dfrac{\displaystyle\sum_{i=1}^{n} M_i l_i^2 \omega^2 tg\beta}{Q_7}$

## 3.1 Example

Let unbalanced elements of the exciter body be presented as a distributed masses $q_1$, ..., $q_2$ and a concentrated mass $M$, as shown in Figure 3.

Applying the method of superposition determines the resultant of centrifugal forces with the expressions (1 $a$, $b$, $g$):

$$Q = Q_1 + Q_2 + Q_7 =$$
$$= \omega^2 tg\beta \left( \frac{1}{2} q_1 l_1^2 + \frac{1}{2} q_2 (l^2 - l_1^2) + Ml \right) \quad (3)$$

301

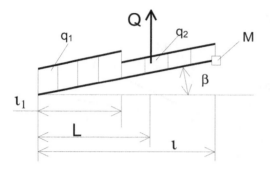

Figure 3. Design scheme.

And the point of its applications (distance $L$) with the expressions (2 $a$, $b$, $g$) can be represented as follows:

$$L = \frac{Q_1 L_1 + Q_2 L_2 + Q_7 L_7}{Q} =$$
$$= \frac{\omega^2 tg\beta \left(\frac{1}{3} q_1 l_1^3 + \frac{1}{3} q_2 (l^3 - l_1^3) + M l^2\right)}{Q} \quad (4)$$

Simultaneously with Expression (3), we get a valid expression as follows:

$$Q = M_p \omega^2 Ltg\beta \quad (5)$$

where $M_p$ is the reduced mass (mass of unbalanced exciter housing elements contained in the resultant point of application); $Ltg\beta$ is the amount of eccentricity in the point of application of the resultant (oscillation amplitude).

Comparing Expressions (3) and (5), we get:

$$M_p = \frac{1}{L}\left(\frac{1}{2} q_1 l_1^2 + \frac{1}{2} q_2 (l^2 - l_1^2) + Ml\right)$$

By using the concept of the reduced mass, the calculations are simplified, and as in (1) and (2), the factor $\omega^2 tg\beta$ can be neglected.

Schemes of distributed masses, as shown in Figure 2, practically cover all the possible variants of the unbalanced housing element of exciters.

In the case where the body of excited oscillates with constant amplitude (plane-type exciter), the calculation of centrifugal force and its point of application is given as follows:

1. By known methods, unbalanced mass $m$ of vibration generator and the position $L$ of its center of gravity can be defined—the distance from the end of the body;
2. The resultant centrifugal force is calculated by the formula: $Q = me\omega^2$, where $e$ is the magnitude of the eccentricity of the crank (oscillation amplitude); $\omega$ is the angular velocity of rotation of the crankshaft of vibration exciter.

## 4 CONCLUSIONS

By the above calculations, we can determine the magnitude and location of the application of the resultant of centrifugal forces generated during the rotation of the unbalanced mass. These calculations can help in determining the mass of the counterweight and the place of its application, which is a prerequisite for balancing.

## REFERENCES

Efremov, I.M. Lobanov, D.V. Figura, K.N. & Komarov, I.V. 2011. Patent-analytical survey and the extended classification of concrete mixing machines in the aspect of investigating the concrete mixtures vibration technologies. *Systems methods technologies*, 38–45.
Malakhov, K.V. & Lobanov, D.V. 2013. Experimental analysis of vibroactivation process of concrete mixing in gravity vibromixer. *Systems methods technologies*, 134–138.
Shen, H. Duan, Z.S. & Li, F. 2012a. The Evaluation about Vibratory Mixing Methods. *Applied Mechanics and Materials*, 217–219: 2678–2682.
Shen, H. Duan, Z.S. & Li, F. 2012b. The Significance of Vibration in Concrete Mixing. *Applied Mechanics and Materials*, 220–223, 509–512.
Wegman P.W. Vaynshteyn M. Abramov O.Y. Ryabov S.D. Yudin Y.A. Kashkarov A.G. Gerasimov A.N. & Kouzmitchev V.A. 1999. *Vibratory filler for powders*. USA patent application.

*Advanced Materials, Structures and Mechanical Engineering – Kaloop (Ed.)*
© 2016 Taylor & Francis Group, London, ISBN 978-1-138-02793-0

# Comparison between the effects of thoracoscopy and open thoracotomy on elderly patients with lung cancer

F.W. Lin & K.P. Cheng
*Department of Thoracic Surgery, The China-Japan Union Hospital of Jilin University, Changchun, China*

C. Zhang
*Department of Pediatric Surgery, The First Hospital of Jilin University, Changchun, China*

Y. Zhao
*Department of Endocrine, The Second Hospital of Jilin University, Changchun, China*

ABSTRACT: The objective of the present study was to investigate the effects of video-assisted thoracic surgery on the treatment of lung cancer in elderly patients. A total of 89 patients with lung cancer who underwent video-assisted thoracic surgery or open thoracotomy were enrolled in the present study, in which 44 cases underwent the video-assisted thoracic surgery while 45 cases were treated with open thoracic surgery. The mean operation time, the mean intraoperative blood loss, the average postoperative hospital stay, and the average postoperative chest tube duration were analyzed. The results showed that although no statistical difference existed in the mean operation time and the mean intraoperative blood loss, the average postoperative hospital stay and the average postoperative chest tube duration in the experimental group were significantly lower than those in the control group ($P < 0.05$). In conclusion, compared with the open thoracic surgery, the video-assisted thoracic surgery has better clinical effects and lower complications.

## 1 INTRODUCTION

Lung cancer is one of the common malignant tumors, and its incidence has been increasing in recent years (Zhang et al. 2011). Owing to longer operative incision and greater surgical injury in conventional open thoracotomy, Video-Assisted Thoracoscopic Surgery (VATS) has some advantages including smaller incision, less blood loss and ideal recovery in patients with lung cancer. However, there are only a few reports on VATS applied to elderly patients with lung cancer (Wang et al. 2014, Yang et al. 2013). This study compares the effects of VATS and open thoracotomy applied to elderly patients with lung cancer.

## 2 PATIENTS AND METHODS

### 2.1 Patients

From January 2012 to January of 2013, a total of 89 patients with lung cancer were enrolled in this study, and underwent VATS or open thoracotomy in China-Japan Union Hospital of Jilin University. All patients were divided into the control group (45 patients by open thoracotomy) and the experimental group (44 patients by VATS) according to the operational manner. All patients were aged in the range of 62~83 years, including 66 males and 23 females. All enrolled patients were free of hematological diseases, dysfunctions of heart, liver, spleen, kidney, stomach or intestine. Each patient signed an informed consent form. Approval was obtained from the institutional review committee of Jilin University. The two study groups had no significant difference in age, gender, pathologic type, and operational manner.

### 2.2 Study design

All patients in the two groups were anesthetized by the double lumen intubation method, and they were placed in the maximally flexed lateral decubitus position tilted slightly backward to prevent the hip from obstructing the downward movement. In the experimental group, an incision with a length of about 1.5 cm at the seventh intercostal space along the mid-axillary line was used as the camera port. An incision with a length of about 4~6 cm at the fourth or fifth intercostal space between the midclavicular line and the anterior axillary line was made and used as the operating port. The position of the lung tumor was detected carefully from the apex of the lung to the base of the lung during the operation. When the tumor position was confirmed, surgeons removed the tumor according to the screen and carried out lymph nodes dissection (Wang 2010). In the control group, a surgical incision of 25~30 cm was adopted at the fifth intercostal space to remove the neoplasm and lymph nodes. All the diseased tissues were sent for pathological analysis.

### 2.3 Evaluation criteria of treatment effects

Evaluation criteria included the time from ostomy to ostomy closure, the intraoperative bleeding, the mean operation time, the Average Length of Stay (ALS), and the incidence of postoperative wound infections.

### 2.4 Statistical analysis

All measured parameters including mean operation time, average postoperative LOS, intraoperative bleeding and average postoperative chest tube duration were weighted. Statistical analysis was performed using the statistical software program Statistical Product and Service Solutions (SPSS) 17.0 (SPSS Inc., Chicago, IL, USA) and the results are expressed as mean ± standard deviation ($\bar{x}$±s), and the t-test was used. Enumeration data including gender, prevalence frequency and postoperative pain score were analyzed by the $\chi 2$ test. $P < 0.05$ was considered as significant.

## 3 RESULTS

No significant differences were observed in the mean operation time and lymph nodes dissected between the two groups ($P > 0.05$). However, intraoperative blood loss, average postoperative chest tube duration and average postoperative LOS were significantly lower in the experimental group than in the control group ($P < 0.05$). All patients were followed up for 2 years after hospital discharge.

There were three patients lost to follow-up in the control group and two patients in the experimental group. The cancer re-emerged in seven patients of the control group and in five patients of the experimental group. Six patients died in the control group while seven in the experimental group. No apparent difference was observed between the two groups ($P > 0.05$), as can be observed from Table 1.

## 4 DISCUSSION

The incidence of lung cancer has increased in the past few years. As lung tissue has abundant blood supply and is prone to circulatory metastasis, it can cause great damage to patients with lung cancer. Operation has been the only method to radically cure lung cancer so far (Zhang et al. 2013). With the development of minimally invasive surgery, VATS has become a mature surgical technology of treating carcinomas of the lungs with a smaller incision and faster recovery than open thoracic surgery (Wang et al. 2013). Some studies have reported that VATS can lead to a higher recurrence and mortality as a result of limited visual field (Jones et al. 2008). However, there are only a few reports on the effects of VATS applied to elderly patients with lung cancer.

This study observed that intraoperative blood loss, average postoperative chest tube duration and average postoperative LOS were significantly lower in the experimental group than in the control group, which is consistent with previous reports (Scott et al. 2010, Whitson et al. 2008). Additionally, smaller operation injury and diseased blood loss contribute to rapid recovery. And lymph nodes dissected were similar in the two groups, which indicated that VATS can achieve the same effects with the open thoracic operation. Accordingly, there were no significant differences observed in lung cancer recurrence and mortality between the two groups (Liang et al. 2013). Furthermore, patients in the experimental group had significantly lower complications than those in the control group,

Table 1. Comparison of outcomes between the two groups.

| Item | Control group ($n = 45$) | Experimental group ($n = 44$) | P value |
|---|---|---|---|
| Mean operation time (min) | 156.74 ± 38.35 | 146.53 ± 30.42 | >0.05 |
| Blood loss (ml) | 284.53 ± 87.53 | 153.45 ± 34.85 | <0.05 |
| Lymph nodes dissected | 16.43 ± 4.35 | 16.08 ± 5.07 | <0.05 |
| Chest tube duration (day) | 7.43 ± 3.23 | 4.65 ± 1.32 | <0.05 |
| Postoperation LOS (day) | 13.57 ± 3.73 | 9.46 ± 2.85 | <0.05 |
| Recurrence | 7 | 5 | >0.05 |
| Dead | 6 | 7 | >0.05 |

which is consistent with earlier studies (Chen & Du 2015).

Conclusively, the VATS used in this study attained better postoperative effects and lower postoperative complications, and had practical implications for elderly patients. Undoubtedly, the VATS for lung cancer should draw the attention of thoracic surgeons as a mature operative technology and be worthy of popularizing.

## 5 CONCLUSIONS

The VATS used in this study attained better postoperative effects and lower postoperative complications, and had practical implications for elderly patients. Undoubtedly, the VATS for lung cancer should draw the attention of thoracic surgeons as a mature operative technology and be worthy of popularizing.

## REFERENCES

Chen, H.W. & Du, M. 2015. Video-assisted thoracoscopic pneumonectomy. *Journal of Thoracic Disease*, 7(4): 764–766.

Jones, R.O. Casali, G. & Walker, W.S. 2008. Does failed video-assisted lobectomy for lung cancer prejudice immediate and long-term outcomes. *The Annals of Thoracic Surgery*, 86(1): 235–239.

Liang, Z. Chen, J. & He, Z. et al. Video-assisted thoracoscopic pneumonectomy: the anterior approach. *Journal of Thoracic Disease*, 5(6): 855–861.

Scott, W.J. Allen, M.S. & Darling, G. 2010. Video-assisted thoracic surgery versus open lobectomy for lung cancer: a secondary analysis of data from the American College of Surgeons Oncology Group Z0030 randomized clinical trial. *Journal of Thoracic and Cardiovascular Surgery*, 139(4): 981–983.

Wang, J. 2010. Experience of removal of lung tumor in 32 patients by VATS. *Shangdong Medical Journal*, 50(27): 56–57.

Wang, T.K. Oh, T. & Ramanathan, T. 2013. Thoracoscopic lobectomy for synchronous intralobar pulmonary sequestration and lung cancer. *The Annals of Thoracic Surgery*, 96(2): 683–685.

Wang, W. Yin, W. & Shao, W. et al. 2014. Comparative study of systematic thoracoscopic lymphadenectomy and conventional thoracotomy in resectable non-small cell lung cancer. *Journal of Thoracic Disease*, 6(1): 45–51.

Whitson, B.A. Groth, S.S. & Duval, S.J. et al. Surgery for early-stage non-small cell lung cancer: a systematic review of the video assisted thoracoscopic surgery versus thoracotomy approaches to lobectomy. *The Annals of Thoracic Surgery*, 86(6): 2008–2016.

Yang, H.C. Cho, S. & Jheon, S. 2013. Single-incision thoracoscopic surgery for primary spontaneous pneumothorax using the SILS port compared with conventional three-port surgery. *Surgical Endoscopy*, 27: 139–145.

Zhang, C.M. Zhu. G.D. & Yan, Z.H. et al. 2011. Analysis of influencing factors of survival quality of elderly patients with lung cancer. *Chinese Journal of gerontology*, 31(1): 21–23.

Zhang, Y. Li, Y.B. & Liu, B.D. et al. 2013. Comparison of complete thoracoscope and small incision by VATS in lung cancers. *Chinese Medicine Journal*, 93(37): 2972–2975.

*Advanced Materials, Structures and Mechanical Engineering – Kaloop (Ed.)*
© *2016 Taylor & Francis Group, London, ISBN 978-1-138-02793-0*

# Territory of mining-industrial development as an object of recreation in the Far East federal region

L.T. Krupskaya
*Pacific National University, Khabarovsk, Russian Federation*
*Far East Scientific-Research Institute of Forestry, Khabarovsk, Russian Federation*

V.P. Zvereva
*Far East Federal University, Vladivostok, Russian Federation*
*Far East Geological Institute, FEB of RAS, Vladivostok, Russian Federation*

A.V. Leonenko
*Institute of Mining, Khabarovsk, Russian Federation*

N.G. Volobueva
*Northeastern State University, Magadan, Russian Federation*

ABSTRACT:  This article presents the results of a study of the assessment of the recreational-ecological potential of the mining-industrial technogenic territories of the former Kerbinsky mine as well as the contiguous recreation areas in the Amur River basin. We carried out a complex assessment on the nature-climatic specific features of the study district (Khabarovsky Krai, former Kerbinsky mine). The principles were worked out, and the recreation zoning was made using the ecological approach. The recreation capacity was calculated. Proposals were given for the organization of the industrial tourism with precious metal mining as a possible way to solve the social and environmental problems.

## 1 INTRODUCTION

Intensification of placer gold mining in the last century promoted the strengthening of technogenic action on the environment, and was responsible for the strained ecological situation on the territory of gold mining (Trubetskoy et al, 2009; Krupskaya et al, 2013). In recent years, the main contradiction of our epoch has become apparent between the nature possibilities to satisfy the human growing requirements and the necessity of life support and self-preservation of the society and the security of its ecological safety. Researchers have established the environment state worsening through the development of the gold placer deposit, the atmosphere alteration, disturbance of biodiversity, destruction of places of population rest of the adjacent settlements, and decreasing life quality (Khanchuk et al, 2012). The productive lands, withdrawn from the biological cycle, promote the pollution of surface and underground waters. Eventually, the recreation possibilities of the territories decrease. So of special importance are the investigations directed to the solution of the social-ecological problems. The task facing the science in the Far East Federal Region, and in Khabarovsky Krai in particular, is to search the most rational variants of reanimation of new technogenic landscapes, to reveal that their recreation potential is still weakly used, and to study the possibilities of creation of the zone of rest in the lands disturbed by mining works. At the mining enterprises of the South Far East, including Khabarovsky Krai, such problem was not even set practically, which defines and supports the urgency of investigations. So the aim of our investigation was to evaluate the recreation-ecological potential of the mining-industrial technogenic territories of the former Kerbinsky mine as well as the contiguous recreation areas in the Amur River basin to specify the possible ways of the solution of the social-ecological problems. The aim defined the following tasks: 1. to analyze, generalize, and systematize the home and foreign experience and the patent information on this problem; 2. to reveal the specificity of the nature-climatic conditions of the study region; 3. to carry out the territory zoning of the mining-industrial development with regard to the ecological situation; and 4. to work out the proposals on the recreation organization including the industrial tourism on this territory with precious metal mining.

## 2 OBJECTS AND METHODS OF RESEARCH

The objects of the present research are the mining-industrial systems formed in the last century by the mining-industrial operation of the former Kerbinsky mine, P. Osipenko district, Khabarovsky Krai, in the Amur River basin, which is of biosphere significance. The methodological base is the theory of V.I. Vernadsky (1967) about biosphere and noosphere. We used the methods of cartography, modeling, systematization, and forecasting.

## 3 RESULTS AND DISCUSSION

The literature analysis of the problem studied (Vedenin et al, 1973; Bol'shakov, 2006; Trubetskoy et al, 2009; Leonenko et al, 2011; Krupskaya et al, 2013) were carried out and others allowed the systematization of the available information, based on which the following conclusions can be made:

1. In Russia and in the Far East in particular, including Khabarovsky Krai, there is no experience in a complex solution of the problems of evaluation of the technogenic pollution of the territory of the mining-industrial development and elaboration of the measures for the improvement of the environment quality as well as the organization of the recreation zones on the mining dumps;

2. It remains to be worked out the question of the excursion demonstration of the memorials of the technogenic, industrial, and scientific-technical evolution of the society and mining-industrial territory, aimed at the cultural-historical

recreation that will allow a significant approach to the solution of social problems, and introduction of the principles of rational nature management through the gold mining process must help to solve the ecological problems. Many-year investigations allowed us to propose a new methodological approach to the evaluation of the disturbed ecosystem in the technogenic aspect and recreation use of the territory of the mining-industrial development under conditions of the deficient recreation resources and heightened demands in them regarding the structural changes in the economics of Russia in the transitional period. The definition of the concept is a development of the mining area.

In our opinion, the areas of development of mineral resources in which structural changes in industrial production are now accompanied by a depressive period, decreased quality of life, poor ecological environment and recyclable development, including in the direction of recreation, can be defined as the territory of mining development. For the first time on the example of the district with the strained ecological situation, we justified the system of indices of estimation of the environment quality as a non-traditional recreation resource with regard to its steadiness to the technogenic action. In the process of recreation evaluation, we first used the index of the land disturbance and established a directly proportional dependence between the landscape disturbance and the degree of territory pollution and the social-ecological problems. The regularities were revealed and the principles were worked out, as well as the ecological-recreation districting was proposed (Fig. 1). We made an effort

Figure 1. Ecological recreation zoning of the study district.

to make the recreation estimation of the little-attractive (in traditional understanding) territory of the mining-industrial development. The calculations of Climate Recreation Potential (CRP) showed that the study district is of a relatively favorable type of climate during a year: CRP = 58%. In June, August, and September, it was characterized by favorable comfortable weathers, and in July, it was characterized by subcomfortable weathers, because in this month, the days with sultry weathers predominate. The cold period of a year was excluded due to the strong action of the limiting factors—severe low temperatures. For the tourist trips, one should choose the "velvet" period (June-September) with comfortable climatic conditions. A set of new forms of recreation was established on the basis of the peculiar unique combinations of natural-climatic conditions reflecting the modern tendencies of change in the recreation demand for the possibilities of rest in the mine settlement of Khabarovsky Krai. The study region has a great natural-recreation potential necessary for organization of rest. All mentioned resources of P. Osipenko district allow fishing in the mountain river, water and walking marches, ascension to mountain tops, to high-altitude lakes, to "Dierovsky" water-falls, and to the "Radostnyi" mineral spring. It is quite possible to develop the tourism here. We proposed to organize the industrial tourism with the precious metal mining.

Development of diverse forms of the recreation service in scales answering the available requirements will make it possible to provide with work the people from the closed enterprises and will give a quick return and additional supplies into the budget. A new direction in the investigation of the recreation possibilities of the territory of The Mining-Industrial Development (TMID) is the concept of non-traditional recreation resources. The study of the recreation requirements of the territory was based on the ecological approach. We proposed a project of the recreation zone on the place of the mining dumps and routes that will allow the tourists to receive the aesthetic-emotional perception of the forest and mountain landscapes, the mountain river, as well as of the mining-industrial territories and possibility to participate in the gold mining process. The non-traditional use of the mining-industrial territories may stimulate the development of the ecological tourism in the region. The specificity of the study area and its natural resources (forest, useful minerals, and biological resources) creates the prerequisites for the intensive recreation development of the natural resources. Measures for a creation of the rest zone influence the landscape to one or another degree and change its inner relations and outer appearance. The tasks that are solved through the elaboration of a plan of organization of the recreation zone landscape on the territory of the mining-industrial development mainly concern its functional and aesthetic features. We proposed some organization-economic measures, the immediate one of which is the reclamation with the use of biotechnology (Golubev et al, 2015) for the quickest restoration of the land productivity. The worked-out measures on the creation of the recreation zone are directed to the development of cultural rest, management of the tourists' behavior, preservation of the natural resources, and solution of social problems (Leonenko 2006, Leonenko & Krupskaya 2011).

So, we worked out the proposals on the complex solution of the problem of the rational nature management at one of the former mining enterprises located in the Amur River basin. The results of the investigation allow the formulation of the ecologically balanced strategy of management of the recreation potential of the study area, providing the increase in the efficiency of its use. They may be of practical use in the creation of the territorial programs of the recreation nature management and territorial complex schemes of the nature protection. The original material obtained allows the revealing of the optimal functions of different territories of the mining-industrial development in the regions with a strained ecological situation.

## 4 CONCLUSIONS

For the first time, the ecological-recreational assessment of a traditionally unattractive territory of mining development was made. The studies allowed us to estimate (quantitatively and qualitatively) its impact on natural complex, including air, hydrosphere, and land objects in terms of their disturbance. Specifics of the natural environment, climate and recreation conditions for the studied area were determined. The study of the recreational needs of the territory was based on the environmental approach. The combined analysis of the environment and recreational potential of the territory was carried out and nonconventional recreational resources were identified, considering the level of development, the employment structure, health, and the overall tension of ecological situation. At the same time, the socio-environmental conditions of the ecological and recreational needs required to be met (recreation development, cultural and historical recreational resources, transport accessibility). On this basis, ecological-recreational zoning of the mining territory, based on the principles of the ecological-recreational situation assessment, was carried out.

## ACKNOWLEDGMENT

This research was performed at the expense of grant of the Russian Scientific Fund (Project 15-17-10016), FSBOU VPO "Pacific National University".

## REFERENCES

Bol'shakov P.M. 2006. Recreation forest management, *Syktyvkar: SLI*. 312.

Golubev, D.A. & Krupskaya, L.T. 2015. Advanced technologies of mined lands recultivation in the Far East federal region. *Izvestiya vuzov, Mining Journal.* 1: 79–85.

Khanchuk, A.I. Krupskaya, L.T. & Zvereva, V.P. 2012. Environmental problems of the tin resources in the Primorye and the Amur and solutions, *Geography and Natural Resources*, 33 (1): 46–50.

Krupskaya, L.T. Zvereva, V.P. Leonenko, A.V. & Babintseva, Y.N. 2013. Mining technological systems and their impact on the environment in the process of gold mining. *Vladivostok: Dal'nauka*, 142.

Leonenko, A.V. 2006. Social-ecological investigation of development zones of mining enterprises in the South Far East. *Mining Journal*. 9: 78–80.

Leonenko, A.V. & Krupskaya, L.T. 2011. Complex evaluation of natural-climatic and social-economic conditions of mining-industrial territory in Khabarovsky Krai for revealing its recreation possibilities and organization of rest (on the example of Kerbinsky mine). *Ecology of Mining Development.* 3: 5–9.

Leonenko, A.V. & Krupskaya, L.T. 2011. Assessment of recreation-ecological potential of mining territories in complex with adjacent areas in the Amur river basin (on the example of Kerbinsky mine). In Komarov A.P. (ed.) Forests and forest management under present conditions: Proc. intern. conf. 4–6 Oct, 2011. Khabarovsk: FGU DalNIILKh: 241–243.

Trubetskoy, K.N. Galchenko Y.P. Grehnev N.I. Krupskaya L.T. & Ionkin K.V. 2009. The main directions of solving environmental problems of the mineral—raw complex in the Far East. *Geoecology*. 6: 483–489.

Vedenin, Yu .A. & Zorin, I.V. 1973. Social aspects of study of territorial recreation systems. *Problems of Geography*. 93: 21–28.

Vernadsky, V.I. 1967. *Biosphere*. Moscow: Mysl, 287.

*Advanced Materials, Structures and Mechanical Engineering – Kaloop (Ed.)*
© *2016 Taylor & Francis Group, London, ISBN 978-1-138-02793-0*

# Validation of a temperature parameter in the expression for the efficiency of filtering magnetophoresis

A.A. Sandulyak & D.A. Sandulyak
*Moscow State University of Instrument Engineering and Computer Science (MGUPI), Moscow, Russian Federation*

M.V. Shitikova & Y.A. Rossikhin
*Voronezh State University of Architecture and Civil Engineering (VGASU), Voronezh, Russian Federation*

A.V. Sandulyak & V.S. Semenov
*Moscow State University of Civil Engineering (MGSU), Moscow, Russian Federation*

ABSTRACT: The expression for the efficiency of filtering magnetophoresis involving the temperature-dependent dynamic viscosity of a medium as one of its key parameters is analyzed in the present paper. Here the options for functional validation of the temperature parameter are demonstrated by obtaining phenomenological temperature dependences of logarithmic, exponential and power types for the viscosity of various liquids, among them: water, 25% ammonia aqua, ethanol, petrol, and liquid ammonia. The resulting modified expression for the efficiency of filtering magnetophoresis involving temperature in an explicit form fits well with the experimental data under the proper choice of special coordinates. It emphasizes the practicability of identifying a reasonable 'temperature point' of the technological line of some production unit for the most appropriate mounting location of magnetic equipment. It will enable one to increase the magnetophoresis efficiency, i.e., the removal of ferroparticles from a medium, and thus, to improve the quality of the product (as a rule, ferroparticles in the medium may appear due to the equipment corrosion and wear-out).

## 1 INTRODUCTION

The efficiency of ferroparticle magnetophoresis, in particular, during its utilization in filtering-type magnetic separators, depends on several primary parameters characterizing (besides the features of ferroparticles themselves) the operational regimes of a key working unit in separators of such a type (filter-matrices of ferromagnetic granules, grains, pins, wires), as well as the properties of a moving medium (the flow of liquid or gas). Among these parameters are the following: the filter-matrix length $L$ (the thickness of a layer), its average magnetic induction $B$ induced by the field intensity $H$, a characteristic size of its elements (for the granular matrix, it is an average diameter of granule-balls $d$), the speed of filtration $\upsilon$, and the dynamic viscosity of the medium $\eta$.

Theoretical and experimental studies of the ferroparticle magnetophoretic process in granular (or grain) filter matrices show that in order to determine the principle resulting parameter of the magnetophoresis, namely, its efficiency $\psi$, which characterizes the relative reduction of ferroparticles concentration due to magnetophoresis, the following calculating formula can be used:

$$\psi = \lambda \left[ 1 - \exp\left( -\frac{ABL}{\eta \upsilon d^2} \right) \right], \qquad (1)$$

where $\lambda$ is a fractional part of magneto-active particles, in particular, their segment in really diverse and most common iron-containing compounds (it could be defined by one of the special control methods, e.g., described in Sandulyak et al. 1985), and $A$ is a united parameter characterizing the susceptibility of ferroparticles to magnetophoresis. It primarily includes their magnetic susceptibility $\chi$ and size $\delta$; the role of these parameters has been discussed in Tsouris et al. 2006 and Nandy et al. 2008. The corresponding estimates based on the theoretical and experimental data reveal that $A \sim \chi \delta^2$.

Alongside with the resulting parameter $\psi$, the so-called $\xi$-parameter, which is equal to a module of the expression in round brackets of (1), i.e.,

$$\xi = \frac{ABL}{\eta \upsilon d^2}. \qquad (2)$$

is also important.

This parameter occurs to be especially useful for direct and also convenient and demonstrative testing, and it is also applicable for further analysis of experimental data.

As follows from (1) and (2), there exists the interconnection between parameters $\psi$ and $\xi$, which is given by

$$\psi = \lambda\left[1 - \exp(-\xi)\right], \ \xi = -\ln\left(1 - \frac{\psi}{\lambda}\right). \tag{3}$$

Apart from the primary parameters entering in (1) and (2), there is one more parameter that influences greatly the process of magnetophoresis, namely: the temperature $t$ of a medium that arrives at the magnetic separation zone, discussed in Sun et al. 2007, Wu et al. 2011, and Tarn et al. 2009. The absence of $t$ in the explicit form in relationships (1) and (2) may give a seeming impression of 'in-completeness' of the data provided by (1) and (2) on parameters of the magnetophoresis process. However, such superficial impression could be demolished by the obvious fact that the temperature parameter $t$ in its implicit form is included in the parameter of liquid and gas viscosity $\eta$, and the role of the latter in magnetophoresis is indisputable (Tsouris et al. 2006, Nandy et al. 2008, Sun et al. 2007, Wu et al. 2011, Tarn et al. 2009).

Therefore, when using (1) and (2) and/or other relative expressions, the time-dependent parameter $\eta$ can be classified much wider, e.g., $\eta$ can be considered as a viscous-temperature factor of the magnetophoresis process.

However, alongside with such expressions involving parameter $\eta$, e.g., (1) and (2), it is also interesting to obtain functional dependences of resulting parameters directly in terms of the temperature of the medium $t$. In other words, when using expressions (1) and (2) for the $\psi$- and $\xi$-parameters, there arises a necessity to 'legalize' parameter $t$, i.e., to get several modified expressions that would involve in the explicit form the parameter $t$, the temperature of the medium entering the magnetophoresis zone.

## 2   MAIN RESULTS AND THEIR DISCUSSION

It is possible to derive expressions that would involve the temperature t of the medium entering the magnetic separation zone as an argument (or one of the constituent arguments) in an explicit form. This could be done using the constitutive relationships (1) and (2) if the temperature t dependence of medium's dynamic viscosity η is found in a functional form.

For instance, the analytical (phenomenological) temperature-dependence of the dynamic viscosity $\eta$ for water can be obtained by providing the actual data of a nonlinear $t$-dependence of $\eta$ (Fig. 1a) that are represented in semi-logarithmic (logarithmic along the abscissa) coordinates (Fig. 1b). Then, it is easy to verify that in a quite wide and essentially working range of temperatures, from 5 ... 10°C to 70 ... 80°C, the approximating $t$-dependence of $\eta$ becomes quasi-linear in these coordinates (Fig. 1b), and thus could be described by a phenomenological expression of a logarithmic type:

$$\eta = \eta_* \ln(t_*/t), \tag{4}$$

with formal values of $\eta_* = 0.46 \cdot 10^{-3}$ Pa·s and $t_* = 170°C$.

The similar-type dependence as (4) is observed for 25% ammonia aqua in the temperature range from 5 ... 10°C to 70 ... 80°C with formal values of $\eta_* = 0.66 \cdot 10^{-3}$ Pa·s and $t_* = 150°C$. In order to assure, it is sufficient to represent the actual data of the nonlinear $t$-dependence of $\eta$ (Fig. 2a) in the same semi-logarithmic coordinates as in Figure 1b.

For other liquids, the nature of the approximating $t$-dependences of $\eta$ (also could be received using known, main reference, data) could be logically different and far from that of expression (4). As this takes place, different solutions of this problem are possible.

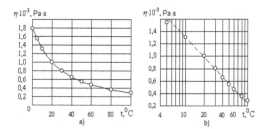

Figure 1.   Temperature dependence of dynamic viscosity for water: *a*) in conventional coordinates; *b*) in semi-logarithmic (abscissa) coordinates; the points represent factual data and the line is calculated according to Formula (4).

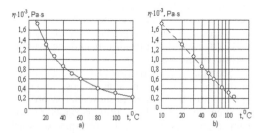

Figure 2.   The same as in Figure 1, but for 25% ammonia aqua solution.

312

Thus, for example, for 100% ethanol in the $t$-range from 0 … 5°C to 90 … 100°C and for motor petrol within the range of $t = (0–40)°C$, the factual data of $t$-dependence of $\eta$ (Figs. 3a and 4a, respectively) are also approximated by quasi-linear dependences (Figs. 3b and 4b, respectively), but in other semi-logarithmic coordinates (logarithmic along the ordinate axis). It means that the corresponding temperature dependences of $\eta$ are close to the exponential ones:

$$\eta = \eta_* \exp(-t/t_*), \qquad (5)$$

with formal values of $\eta_* = 1.7 \cdot 10^{-3}$ Pa·s and $t_* = 56°C$ for ethanol and $\eta_* = 0.6 \cdot 10^{-3}$ Pa·s and $t_* = 91°C$ for petrol, respectively.

As for 40% solution of ethanol in the narrowed interval of $t$ (in comparison with 100% ethanol) from 20°C to 50 … 60°C, the nonlinear $t$-dependence of $\eta$ (Fig. 5a) is closer to a power function, judging by the fact that it is approximated quite well by a quasi-linear dependence in logarithmic coordinates (Fig. 5b). In this case, it is practically inversely proportional dependence:

$$\eta = \eta_* t_* / t, \qquad (6)$$

with formal values of $\eta_* = 10^{-3}$ Pa·s and $t_* = 60°C$.

A functional form of the temperature dependence of viscosity $\eta$ for a liquid medium may be

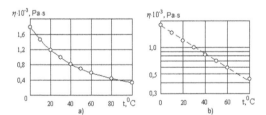

Figure 3.   Temperature dependence of dynamic viscosity for 100% ethanol: a) in conventional coordinates; b) in semi-logarithmic (ordinate axis) coordinates; the points represent factual data and the line is calculated according to Formula (5).

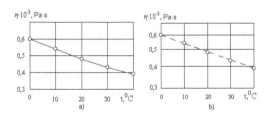

Figure 4.   The same as in Figure 3, but for motor petrol.

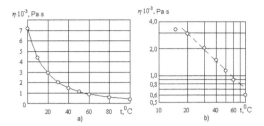

Figure 5.   Temperature dependence of dynamic viscosity for a 40% ethanol solution: a) in conventional coordinates; b) in logarithmic coordinates; the points represent factual data and the line is calculated according to Formula (6).

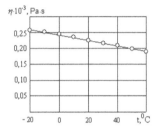

Figure 6.   Temperature dependence of dynamic viscosity for liquid ammonia; points are factual data and the line is calculated according to Formula (7).

the simplest as well, i.e., in a form of a linear dependence. Thus, it is valid for liquid ammonia (Fig. 6), since $t$-dependence of $\eta$ in the temperature interval from 40°C to 60°C in conventional coordinates obeys a linear dependence:

$$\eta = \eta_0 (1 - t/t_*), \qquad (7)$$

with the value of $\eta_0 = 0.244 \cdot 10^{-3}$ Pa·s and formal value of $t_* = 271°C$.

Employing functional relationships (4)–(7), together with other possible temperature dependences of dynamic viscosity, expressions (1) and (2) can be rewritten in a modified form involving the 'legalized' temperature.

In particular, for water systems, these expressions have the following form:

$$\psi = \lambda \left[ 1 - \exp\left( -\frac{ABL}{\upsilon d^2 \eta_* \ln(t_*/t)} \right) \right], \; \xi = \frac{ABL}{\upsilon d^2 \eta_* \ln(t_*/t)} \qquad (8)$$

Moreover, the established relationships of $\psi$ and $\xi$ with $t$ (which turned out to be somewhat unusual) are confirmed by the experimental data.

Figure 7*a* shows the temperature dependences of efficiency $\psi$ (as an outcome of filtering magnetophoresis) of the separation of ferroparticles from artificially prepared suspensions of a magnetite (lines 1–3) and an industrial condensate (line 4). For demonstrative testing of the experimental data (Fig. 7*a*) with the aim of establishing its agreement (or disagreement) with modified expressions (7), it is convenient to convert these data and present them in the appropriate coordinates.

In the given case, such coordinates could be easily elucidated if the expression for the $\xi$-parameter in (8) is written in the following form:

$$\frac{1}{\xi} = \frac{\upsilon d^2}{ABL} \eta_* \ln\frac{t_*}{t}, \qquad (9)$$

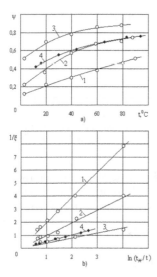

Figure 7.   Temperature impact on the efficiency of magnetophoresis (*a*), and illustration of a linear form of the same but converted data (*b*) in the coordinates according to (9); the data characteristics are given in Table 1.

In so doing, the actual data for $\psi$ (Fig. 7*a*) should be preliminary recalculated into the data for the $\xi$-parameter according to relation (3).

The latter expression points to the ordinate $1/\xi$ and abscissa $\ln(t_*/t)$ of the desired transformation, since precisely in these coordinates, the factual $t$-dependences of $\psi$ (Fig. 7*a*, lines 1–4) should be linearized as directly proportional relationships, and reference to Figure 7*b* shows that they really obey by such relations.

## 3   CONCLUSIONS

Usually, the expressions utilized for identifying the efficiency of ferroparticle magnetophoresis are independent of the temperature of the medium entering the zone of magnetic separation, at least in an explicit form. However, dynamic viscosity of the medium, which is a function of the temperature, is one of the active parameters in such relationships, which provides an opportunity to 'legalize' the temperature parameter by involving it in the constitutive relations.

By the example of the expression for the efficiency of filtering magnetophoresis, the options of functional validation of the temperature parameter are revealed by the determination and target utilization of phenomenological temperature dependences (of logarithmic, exponential and power types) of dynamic viscosity of different liquids, namely: water, 25% ammonia aqua, ethanol, petrol, and liquid ammonia.

The derived modified expression for the efficiency involving temperature is compared with the experimental data by adopting special coordinates graphically illustrating the agreement between the calculated values and the experimental data.

Considering that for liquid media, the increase in temperature results in the increase of the efficiency values due to the reduction of the value of the dynamic viscosity, a tentative opinion could be suggested concerning the preferable practical

Table 1.   Characteristics of the experimental data shown in Figure 7.

| № in Figure 7 | The medium under consideration, the part of active fraction $\lambda$ (filtration speed is 5.6 cm/s) | Matrix length, $L$, cm | Mean induction, $B$, T | Balls diameter, $d$, mm |
|---|---|---|---|---|
| 1 | Water suspension of magnetite with particle size $\delta < 1–2\ \mu m$, $\lambda = 1$ | 8.4 | 0.65 | 5.7 |
| 2 | Water suspension of magnetite with particle size $\delta < 2–4\ \mu m$ | 8.4 | 0.65 | 5.7 |
| 3 | Water suspension of magnetite with particle size $\delta = 10–15\ \mu m$ | 2 | 0.3 | 5.7 |
| 4 | Industrial condensate, $\lambda = 0.8$ | 100 | 0.55 | 5 |

realization of magnetophoresis for media with increased temperature. However, it scarcely allows us to make the conclusions recommending the increase in temperature (and thus the decrease in viscosity) of the agent entering a magnetophoresis zone (within the framework of any, as a rule, strictly standardized production process).

However, there is a limited capacity for the beneficial use of the considered 'viscosity-temperature' factor, since in a real production line, there are always alternatives for the magnetic equipment mounting location (in some point of a technological system). As this takes place, the information about the temperature of the medium can be crucial for the final decision in choosing the most optimal installation site.

Therefore, if there are no fundamental objections and limitations (technological, engineering, or locational ones), then developers and maintenance team are basically free in choosing the most convenient location of magnetic equipment, including its placement with regard to an acceptable 'temperature point' of the technological line (where the dynamic viscosity of the medium is minimal). As a result, the increase in the efficiency of magnetophoresis (removal of ferroparticles from the medium), and thus the improvement of the quality of the medium with respect to ferroparticle content (as a rule, caused by equipment wearout and corrosion) could be achieved.

## ACKNOWLEDGMENT

This work was supported by the Ministry of Education and Science of the Russian Federation.

## REFERENCES

Nandy, K. Chaudhuri, S. Ganguly, R. & Puri, I.K. 2008. Analytical model for the magnetophoretic capture of magnetic microspheres in microfluidic devices. *Journal of Magnetism and Magnetic Materials*. 320: 1398–1405.

Sandulyak, A.V. Garaschenko, V.I. & Korkhov, O.Y. 1985. *Method of Determining the Quantity of Solid Fraction of Ferromagnetic Matter in a Fluid*. Patent 4492921 US.

Sun, J. Xu, R. Zhang, Y. Ma, M. & Gu, N. 2007. Magnetic nanoparticles separation based on nanostructures. *Journal of Magnetism and Magnetic Materials*. 312: 354–358.

Tarn, M.D. Peyman, S.A. & Robert, D. 2009. The importance of particle type selection and temperature control for on-chip free-flow magnetophoresis. *Journal of Magnetism and Magnetic Materials*. 321(24): 4115–4122.

Tsouris, C. Noonan, J. Ying, T.yu. & Yiacoumi, S. 2006. Surfactant effects on the mechanism of particle capture in high-gradient magnetic filtration. *Separation and Purification Technology*. 51: 201–209.

Wu, X.Y. Wu, H.Y. & Hu, D.H. 2011. High-efficiency magnetophoretic separation based on a synergy of a magnetic force field and flow field in microchannels. *Science China Technological Sciences*. 54(12): 3311–3319.

*Advanced Materials, Structures and Mechanical Engineering – Kaloop (Ed.)*
© *2016 Taylor & Francis Group, London, ISBN 978-1-138-02793-0*

# Multi-focus fusion with a new sharpness parameter for textile testing

J.F. Zhou & R.W. Wang
*College of Textiles, Donghua University, Songjiang District, Shanghai, China*

ABSTRACT: Textile testing has played an important role in quality assurance and improvement, and is essential to textile production and trade. Recently, the testing has formed a developing tread with the image processing. However, in image processing, the captured image contains fuzzy parts and sharp parts—multi-focus image, because of the limited depth of the microscope system. In order to solve the multi-focus problem, the image fusion technology is applied. In this paper, a new judging function of the sharpness of image fusion is proposed and according to the function, the sharp parts from series of the multi-focus image are selected to combine a new fused image. Through the objective evaluations, it is confirmed that the fused image is more informative and appropriate for the purposes of human visual perception and further image processing. In this paper, the wool fiber is referred as the experimental sample. Through the image fusion, we can observe the wool scale and the vertical structures which can be beneficial to test textile materials.

## 1 INTRODUCTION

Textile testing has played an important role in quality assurance and improvement and is essential to textile production and trade. Therefore, developing textile measurement methods has been a significant research area. With the development and progress of the advanced science technology, modern textile measurement has formed a developing trend to realize the automation and intelligence, which improves the methods of testing and enhances the testing abilities.

Recently, based on the computer control system, many domestic and foreign scholars have used the digital image processing to measure the textile materials. For example, Bugao Xu (2009) measured the cotton maturity from the longitudinal view and Jingjing (2011) presented a novel wrinkle evaluation method to characterize and classify the wrinkling appearance of textile fabric. In the field of textile, the digital images are captured mainly through the automatic optical microscope in which the samples could be moved on the object stage. However, when observing the captured images, we can find that there exist fuzzy areas and sharp areas-namely the multi-focus problem. These are due to three dimensional states of textile materials and the finite depth of field of a microscopic system. When the samples are under the depth of the field, the captured images can be sharp. Out of the field, the images are fuzzy. In the microscope, a series of multi-focus images can be successfully captured and each image includes fuzzy and clear parts. For example, in Figure 1, the two wool fiber images were captured in the automatic microscope in the same view with different focal points. Observing the clear parts of the two

images, the wool scale, and longitudinal constructions could be gained for testing. However, the fuzzy parts could not realize those effects and could affect the accurate measurement of textile materials.

In order to overcome this problem-multi-focus problem and gain a sharp image to ensure the accuracy, multi-focus image fusion technology with a new judging function of sharpness is adopted in

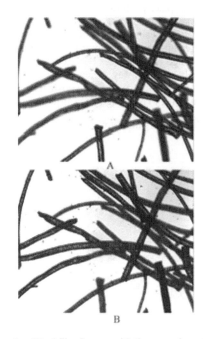

Figure 1. Wool fiber images with the same view at a different focus position.

the image processing through several images with different focus points. According to the existing achievements and problems, the novel method of multi-focus image fusion can solve successfully the multi-focus problem and gain a fused image for textile measurement.

## 2 EXPERIMENT

### 2.1 *Multi-focus images capturing*

Multi-focus images of wool fibers were captured on an automatic microscope in which the z-axis could allow the image to be focused at any depth. Figure 2 listed 20 multi-focus source images with the same view. The 20 images successfully contained all fuzzy parts of wool fibers.

### 2.2 *Multi-focus images preprocessing*

In order to gain the detail information of wool fiber, it was essential to separate the image object and image background. In the image processing, the Otsu's Method was commonly used to realize the image segmentation. However, if the images were directly processed in Otsu method, the fuzzy object parts could be wrongly recognized as background parts and 20 images could be able to gain 20 different segmentation images. In this way, it could affect the real results. In order to obtain the complete object information, in this paper, the fusion based on the pixel point was implemented in advance.

For the pixel position $(x, y)$, the images on 20 layers may have different gray-scale values-$G(x, y)$. The gray-scale values in 20 multi-focus images were compared and the minimum value was chosen. The pixel located on $(x, y)$ was replaced with the minimum value. In this way, a new matrix, named the minimum gray matrix can be constructed, and its element at $(x, y)$ is defined as in Equation (1)

$$MGM(x, y) = \arg\min_i G(x, y) \qquad (1)$$

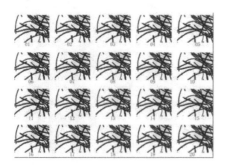

Figure 2. 20 multi-focus wool fiber images of the same way on different focus position.

In Figure 3(A), the image of the minimum gray matrix was shown. Afterward, the pixel point fusion image was processed in Otsu's Method to achieve image segmentation. Figure 3(B) was the separated image in which the black areas represented the object and the white stood for the background.

### 2.3 *Multi-focus image fusing*

Image fusion if defined as the process of combining substantial information from several sensors using mathematical techniques in order to create a single composite image. The resultant image acquired from image fusion technique is more informative and appropriate for the purposes of human visual perception and further image processing. Many image fusion techniques have been introduced to fuse multi-focus images. These methods can be categorized into two classes: spatial domain methods (Zhu et al. 2010, Kun et al. 2009, Li & Yang 2008) and transform domain methods (Phamila & Amutha 2014, Yang 2011, Li et al. 2011). Considering the merits and demerits of the two methods, in this paper, the spatial domain method based on pixels was employed and only the object parts were processed. In order to gain the clear parts and achieve the image fusion, domestic and foreign scholars have proposed many sharpness criterions to judge sharp parts for combing images, such as Spatial Frequency (SF), Energy of Laplacian (EL), Sum Modified Laplacian (SML) (Aslantas & Kurban 2010) and so on. When human

Figure 3. (A) The image of the minimum gray matrix (B) The image of segmentation.

318

observed many different images, we could recognize correctly the sharp parts and the fuzzy parts. Combing the characteristics of the human vision system, a new judging function was proposed to indicate image clarity to achieve image fusion—Gradient Variance Maximum Value (GVMV).

The new judging function contained two parts. The first part was to construct a new matrix named Gray Gradient Maximum Level (GGML). For the target areas, in the ith layer, the gradient between the pixel $(x, y)$ and its eight neighborhood pixels was defined as Equation (2)

$$Gradient_i(x,y) = \max\left(\|G_i(x,y) - G_i(m,n)\|\right) \quad (2)$$

$G_i(m,n)$ was the gray value of pixel $(x, y)$'s eight neighborhood. After the gradient values for 20 layers in position $(x, y)$ were calculated, the maximum value was chosen as GGML shown in Equation (3).

$$GGML(x,y) = \arg\max_{i=1}^{i=20}(Gradient_i(x,y)) \quad (3)$$

The second part was relied on GGML in the form of regions. The region was a circle of radius 5 pixels. At the junction of the target and background, the circle was changed according to the object parts. For each target pixel of 20 layers, in its range of regions, the variance of GGML was computed. Comparing the variances of 20 layers, the layer of the maximum was selected. The GVMV was defined as in Equation (4).

$$GVMV(x,y) = \arg\max_{\substack{i=1 \\ (m,n)\in R}}^{i=20}(GGML_i(m,n) - u_i)^2 \quad (4)$$

In Equation (4), R was the selected circle region, (m,n) was the pixels position in R.
$GGML_i(m,n)$ was the gray gradient maximum $(m,n)\in R$
level in the region of the ith layer, u was the averaged gray gradient maximum level in the selected region R.

With the same method, all pixels in target areas could correspond to a layer and the Layer Matrix (LM) was created. When the pixels were filled with the pixels in selected layers, a new image was built. Figure 4 was the newly built image.

However, when amplifying the selected red region in Figure 4, we could find that there were isolated noises, shown in Figure 5(A), this was because the layer of noisy pixel was different from those of its neighborhood and the gap between the layers was large. Though in this position of noisy pixel, GVMV of the selected layer was the maximum, it was essential to consider the continuity of the sharp parts and the

Figure 4. The newly built image with proposed method.

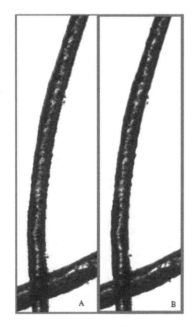

Figure 5. (A) The detailed image of LM and (B) The detailed image of NLM.

presence of noisy pixels seriously influenced the performance of the fused image. Therefore, in this paper, for the Layer Matrix, median filtering was implemented to eliminate the noisy pixels to form a New Layer Matrix (NLM). Figure 5(B) was the modified image of Figure 5(A). From the 5(B), the noisy pixels were removed and the community of sharp part was strengthened. The fusion performance of Figure 5(B) was superior to Figure 5(A).

Based on the New Layer Matrix, the pixels were filled with the pixels of modified layers to create a new fused image shown in Figure 6. Comparing Figure 6 with Figure 1, it could be witnessed the fused image combined all sharp areas of 20 multi-focus images to become a sharp image.

Figure 6. The fused image with proposed method.

Table 1. The objective assessment of fused image.

| Index | V | SF | Proposed method |
|---|---|---|---|
| E | 0.63 | 0.67 | **0.75** |
| AG | 12.28 | 12.08 | **10.89** |

From Figure 6, we could distinctly see the wool scale and longitudinal structure.

## 3 PERFORMANCE EVALUATION OF IMAGE FUSION

Generally, the performance evaluation of image fusion was divided into two categories: subjective assessments and objective assessments. In this paper, the objective assessments were adopted. The objective assessment relied on the mathematical algorithm to simulate the human visual perception of the image fusion to make a quantitative evaluation of the quality of the fused image. In order to confirm the superiority of the proposed fusion method, the fused image based on the function of the Variance (V) and Spatial Frequency (SF) were compared with the fused image in this paper.

Image Entropy (E) was an important index to evaluate the richness of image, and the larger the entropy is, the more information the fused image contains.

Average Gradient (AG) could reflect the image clarity.

From the Table 1, we could find there was more information of the fused image with proposed method. The proposed judging function could characterize the image clarity.

## 4 CONCLUSIONS

With the development and progress of the advanced science technology, modern textile measurement has formed a developing trend to realize the automation and intelligence. The method of image processing is applied in textile testing. However, when capturing images, there is a problem of multi-focus image. With the technology of image, fusion can solve the multi-focus problem. In this paper, a new method of image fusion is proposed with a novel judging function of image clarity. Through the experimental data, the proposed function is superior to the previous criterions such as variance and spatial frequency. From the fused image, the wool scale and longitudinal structure of the wool fiber can be easy to find. Based on the proposed method, the image of textile materials can be accurately captured and it is beneficial to analyze the textile materials in image processing system.

## ACKNOWLEDGEMENT

The authors would like to express our gratitude to the editors and anonymous reviewers for their comments and suggestions. The work is supported by the Natural Science Foundation of China (Grant No. 61172119), the Chinese Universities Scientific Fund (Grant No. 15D310101).

## REFERENCES

Aslantas, V. & Kurban, R. 2010. Fusion of multi-focus images using differential evolution algorithm. *Expert Systems with Applications*, 37(12): 8861–8870.

Kun, L. et al. 2009. Fusion of Infrared and Visible Light Images Based on Region Segmentation. *Chinese Journal of Aeronautics*, 22(1): 75–80.

Li, S. & Yang, B. 2008. Multi-focus image fusion using region segmentation and spatial frequency. *Image and Vision Computing*, 26(7): 971–979.

Li, S. Yang, B. & Hu, J. 2011. Performance comparison of different multi-resolution transforms for image fusion. *Information Fusion*, 12(2): 74–84.

Phamila, Y.A.V. & Amutha, R. 2014. Discrete Co-sine Transform based fusion of multi-focus images for visual sensor networks. *Signal Processing*, 95: 161–170.

Sun, J.J. Yao, M. Xu, B. & Bel, P. 2011. Fabric wrinkle characterization and classification using modified wavelet coefficients and support-vector-machine classifiers. *The textile Research Journal*, 81(9): 902–913.

Xu, B. Yao, X. Bel, P. Hequet, E.F. & Wyatt B. 2009. High volume measurements of cotton maturity by a customized microscopic system. *Textile research journal*, 79(10): 937–946.

Yang, Y. 2011. A Novel DWT Based Multi-focus Image Fusion Method. *Procedia Engineering*, 24: 177–181.

Zhu, X. et al. 2010. An enhanced spatial and temporal adaptive reflectance fusion model for complex heterogeneous regions. *Remote Sensing of Environment*, 114(11): 2610–2623.

*Advanced Materials, Structures and Mechanical Engineering – Kaloop (Ed.)*
© 2016 Taylor & Francis Group, London, ISBN 978-1-138-02793-0

# A novel three-dimensional reconstruction algorithms for nonwoven fabrics based on sequential two-dimensional images

L.J. Yu & R.W. Wang
*College of Textiles, Donghua University, Songjiang District, Shanghai, China*

ABSTRACT: This paper proposed a new method for Three-Dimensional Reconstruction based on sequential images. The method was realized by five steps: firstly, a series of images of different focal positions for the same view were obtained under an automatic optical microscope; Secondly, the sharpness of each pixel in the sequential images were calculated by a novel sharpness-judging function and recorded into a Three-Dimensional Matrix (DCM); thirdly, the trend of sharpness for the same $(x, y)$ position throughout the sequential images was drawn as a sharpness-focused position curve. The x-coordinate of each point on the curve was the layer number of the pixel and the y-coordinated was the sharpness. Multiple curves were drawn for all the $(x, y)$ positions in the view; fourthly, record the x-, y-positions and z-positions (the focused positions) of each peak in the curves to a dot matrix (DPM); lastly, the three-dimensional image of structure in the nonwoven fabrics was reconstructed by inputting the 3d-coordinates in DPM to the modeling software Geomagic Studio 12.

## 1 INTRODUCTION

Different from the traditional woven fabrics which consist of interlaced yarns in a regular structure, nonwoven structures are connected in unique ways, possessing different pore size and orientation distributions combined together to form an intricate microstructure (Chen et al. 2007). Fibers in nonwoven fabrics are forced to twist, bend and rotate around themselves and other fibers; therefore, a series of small, interlocking entanglement is formed (Ghassemich et al. 2001). The distribution of structural characteristics plays a very important role in determining the properties of the fabric, such as porosity, orientation, and mechanical properties.

To date, image analysis has been applied to measure the distribution of nonwoven fabrics. However, most studies used the two-dimensional images which cannot fully describe the behavior of a nonwoven because of the spatial structure of it. The degree to which fibers are redirected in the Z-direction (or called thickness direction) can certainly be responsible for the ultimate properties of the structure such as tensile and burst strength, but also can contribute to the three-dimensionality of the pore network and this influences such properties as liquid absorption and release (Lalith et al. 2012).

In this paper, a 3D surface reconstruction method was introduced to obtain a 3D reconstructed image derived from a sequence of 2D images. Because of the microstructure of nonwoven fabrics, the 2D images were captured under the microscope, Due to the limited depth of field; light microscope can ensure the clearness of the target partly clear while other parts are fuzzy (Wang et al. 2012). When the object is in the focus position, it exhibits clearly in the image. Therefore, for the same object, in a series of images captured at different focal positions, the object is clear in one certain image, while fuzzy in others. Consequently, according to the focal position of the image which declares the best clearness of the object, the best-focused position of the object can be determined. Since the focal length of the optical lens is constant, and the focused position of the object refers to the z-position of the platform when the object is in focus, the focused position can be considered as the z-position of the object. Once the focused positions of all the objects are acquired, the 3D surface reconstructed image can be built up.

The method will be completed by several steps. Firstly, sequential images of the same view at different focal positions were captured. Secondly, calculate the clearness of each pixel of each layer of the image and record the clearness into a 3-Dimensional Clearness Matrix (DCM). Thirdly, Mark the z-position of the image which declare the best sharpness of each pixel and record the x-position, y-position and z-position of each pixel into a Dot Matrix (DPM). Finally, build the 3D reconstructive image by using DPM with the software Geomagic Studio 12.

## 2 METHOD

### 2.1 *Image acquisition*

An automatic microscopic equipped with a motorized $x - y$ stage to transport the slide was used in this paper. 65 layers of images were captured at different focal positions which could cover the depth of field of the nonwoven fabric.

### 2.2 *Establishing DCM*

The 3-Dimensional Clearness Matrix consists of clearness of all the pixels throughout the sequential images, in which the third dimension is the layer where the pixel is. Assuming $S_i(x, y)$ is the sharpness of pixel $(x, y)$ in i-th layer of images, the Matrix can be expressed as:

$$DCM(i, x, y) = \arg S_i(x, y) \qquad (1)$$

The main and most difficult part of the establishment is finding an expression to describe the sharpness of each pixel. Generally, the sharpness of a pixel can be defined as its gradient, which is

$$S_i(x, y) = |G_i(x + 1, y) - G_i(x - 1, y)|$$
$$+ |G_i(x, y + 1) - G_i(x, y - 1)| \qquad (2)$$

However, due to the noisy in the images, there is a significant probability to select the wrong layer for the pixel. Previous studies have revealed that the focused position of s pixel has many relationships with pixels around it. Nearby pixels are probably focused at similar positions. The clearness of a pixel cannot be simply expressed by the gradient of itself and its neighboring points (that is how the nosy points come from), but an area covering it. When we talk about the sharpness of a pixel, we are talking about the sharpness of a region around the pixel. The expression to describe the sharpness of a section is a difficulty that this paper will discuss.

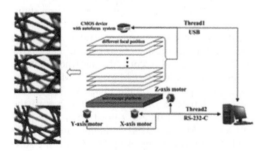

Figure 1. Microscope system for image acquisition.

This paper proposed a new algorithm to calculate the sharpness of a pixel based on circle regions around the pixel. To eliminate the computation burden and increase efficiency, the pixels in the background will not be considered. Therefore, an image division to extract the targets from images was carried out before the calculation of sharpness.

### 2.2.1 *Image segmentation*
The segmentation was produced on the dot-fusion image. The first step is calculating the sharpness of each pixel. The sharpness of a pixel in this part was calculated by using Equation 2.

The next step is to select the right layer number for each pixel and construct a matrix DM which consists of the layer number which declares the max sharpness of each pixel. The matrix DM can be expressed as follows:

$$DM(x, y) = \arg\max(S_i(x, y)) \qquad (3)$$

where, i is the layer number of the image.

The pixels in the dot-fused images were filled with the gray of the pixel in the corresponding layer of images according to DM.

The OTSU was used to divide the targets from the dot-fused image. All the pixels in the image were marked as targets or background according to the binary image after OTSU.

### 2.2.2 *Section selection*
The algorithm this paper introduced was based on the theoretical point that the sharpness of pixel can be expressed by the sharpness of an area covering the pixel. In this part, we select the area that was used to execute the following clearness evaluation.

Assuming the pixel we need to calculate the sharpness is the Pixel P. Pixels in the area were selected by the following two rules:

1. The distance between the pixel and P should be less than 5 pixel length.
2. The pixel marked as the targets were only taken into consideration.

### 2.2.3 *Sharpness-judging algorithm*
Many reaches have been studied several section-judging functions. Theses judge algorithms are based on the indicators such as Variance (Xin et al. 2012), max frequency (Kanjar & Masilamani 2013) and LBP-transformation (Xu 2014).

However, these algorithms did not work out well in our previous studies for complicated images such as the nonwoven fabrics in which fibers crossed with each other. By studying the differences of the gray-level matrix between the focused area and de-focused area and learn from human vision, we proposed a new sharpness-judging algorithm for an area.

Previous studies revealed that when humans observe an image, the borders which exhibit high gradient evoke their interests first. The gradients are high on the border of two targets and low in the body of one target. Therefore, the gradients are fluctuant when the image is in-focus. For the de-focused image or the de-focused area of an image, the border between two targets is not conspicuous; resulting in the gradients varies little among the image or the area.

In this paper, the sharpness of the area can be expressed by the variance of the gradient of each pixel in the area. $P_i(x, y)$ refers to the pixel in i-th layer of images at the pixel coordinate $(x, y)$. The new sharpness-judging algorithm was realized by two steps.

The first step was to record the gradient of each pixel in the selected region ($R_{(x, y)}$). The gradient was defined as Equation 2. The following three-dimensional matrix (GM) recorded the gradient of each pixel in the $P_i(x, y)$-centered region.

$$GM_i(m,n) = \arg S_i(m,n) \qquad (4)$$

In which, $P(m\ n) \in R_{(x, y)}$.

The second step was to calculate the sharpness of our selected region and construct the DCM with the sharpness.

Assuming $C_i(x, y)$, (i = 1, 2 ... 65) is the clearness of the $P(x, y)$-centered region, and $U_i(x, y)$ is the average gradient value in $R_{(x, y)}$, then

$$G_i(x,y) = \sum (GM_i(m,n) - U_i(x,y))^2 \qquad (5)$$

In which, $P(m\ n) \in R_{(x, y)}$. As discussed before, the sharpness of a pixel can be expressed by the sharpness of a circle region centralizing in the pixel. We can build the clearness matrix DCM in which the sharpness of a pixel can be calculated by Equation 4.

$$DCM(i,x,y) = \arg G_i(x,y). \qquad (6)$$

### 2.3 Establishing DPM

Once we obtain the DCM, we can draw a sharpness-focal position curve (Fig. 3). The curve clearly indicates the change of sharpness of a certain point in differently focused position. It can be seen from the curve that the clearness increases when the target is approaching the best-focused position and decreases when the target gets the best-focused position. The clearness declares its maximum at the in-focused position. If the point crossed only one object (Fig. 2), the position where the point exhibited the maximum was selected as the z-position of the point. However, if the point

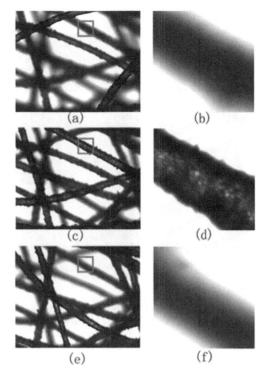

Figure 2. (a) The image in the 0-th layer. (b) Enlarged of window of Figure 2(a) in red. (c) The image in the 24-th layer. (d) Enlarged of window of Figure 2(c) in red. (e) The image in 58-th layer. (f). Enlarged of window of Figure 2(e) in red.

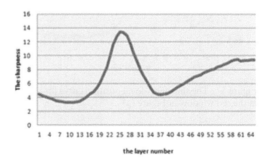

Figure 3. The sharpness-focal position curve for a point in the red window in Figure 2.

crossed two objects, reflecting in the curve that two peaks exists (Figs. 4 and 5), there are two target points for a corresponding position $(x, y)$. Two 3-d coordinates were recorded to the DPM matrix when the point crossed two objects. Therefore, in the sharpness-focal position curve for one certain point, the peaks of the curve were selected to record its z-positions. There is a sharpness-focal position

323

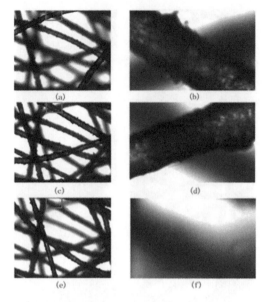

Figure 4. (a) The image in 0-th layer (b) Enlarged of window of Figure(a) in red. (c) The image in 25-th layer (c) Enlarged of window of Figure 4(c) in red. (e) The image in the 62-th layer. (f) Enlarged of window of Figure 4(e) in red.

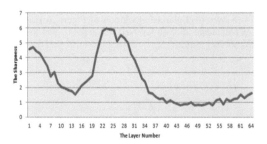

Figure 5. The sharpness-focal position curve for a point in the red window in Figure (a).

curve for each $(x, y)$—position. The x-coordinate of the peak in the curve conveyed the layer number at which the target point is focused. Since the focal positions of the sequential images were known, the focused position of the target point can be derived according to the layer number of the target point. Assuming Peak $(x, y)$ is the focal position of the peak in the curve of position $(x, y)$, the dot matrix DPM can be expressed as follows.

$$DPM(x) = x \qquad (7)$$

$$DPM(y) = y \qquad (8)$$

$$DPM(z) = Peak(x, y) \qquad (9)$$

It is ideal that there is only one peak in the curve for the pixel crossing one object, and two peaks in the curve when the pixel crosses two objects. However, invalid peaks exist in some curves. It can be seen from the curve that when the object deviates the best-focused position in a large scale, small peaks come up (such as the head and tail in the sharpness-focal position curve in Fig. 3). It is because the object was blurred too much.

The filtering algorithm was carried out to eliminating invalid peaks. Considering the invalid peaks were stroke by the situation that the object was blurred too much. It is common situation that in the blurred image the body of the object was the darkest, and the darkness spread from the center of the object to background (Fig. 2(f)) Figure 6 showed the direction of the gradient in the region in Figure 2. In order to show the macroscopic trend of varies of the gradient, the gray value of a pixel was replaced with windows with $(3 \times 3)$ pixels.

The direction of the gradient of the window was expressed by numbers from 0 to 7. As illustrated in Figure 6, if the gradient flows from the center window P to the pixel at the top-left corner of P, the gradient direction was expressed as 1. The sheet reflecting the gradient direction of all windows in regions in Figure 2 was displayed in Figure 7. It can be seen from Figure 7 that the gradient directions were consistent in the 0-th layer and the 58-th layer, while disorder in the 24-th layer. Therefore, the invalid peaks in Figure 3 (0-th layer and the 58-th layer) can be filtered based on the gradient directions.

### 2.4 Establishing the 3D reconstructive model

By inputting the Matrix-DPM in the fabrics-to the software Germanic Studio 12, the three-dimensional model was reconstructed with the x-, y- and z-positions of each target points. The result of the 3D image was showed in Figure 8.

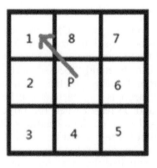

Figure 6. The expression of gradient direction.

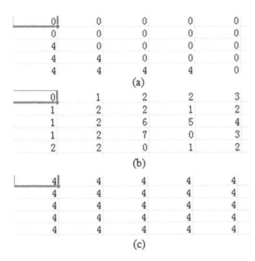

| | | | | |
|---|---|---|---|---|
| 0 | 0 | 0 | 0 | 0 |
| 0 | 0 | 0 | 0 | 0 |
| 4 | 0 | 0 | 0 | 0 |
| 4 | 4 | 0 | 0 | 0 |
| 4 | 4 | 4 | 4 | 0 |

(a)

| | | | | |
|---|---|---|---|---|
| 0 | 1 | 2 | 2 | 3 |
| 1 | 2 | 2 | 1 | 2 |
| 1 | 2 | 6 | 5 | 4 |
| 1 | 2 | 7 | 0 | 3 |
| 2 | 2 | 0 | 1 | 2 |

(b)

| | | | | |
|---|---|---|---|---|
| 4 | 4 | 4 | 4 | 4 |
| 4 | 4 | 4 | 4 | 4 |
| 4 | 4 | 4 | 4 | 4 |
| 4 | 4 | 4 | 4 | 4 |
| 4 | 4 | 4 | 4 | 4 |

(c)

Figure 7. (a) the Sheet of the gradient directions for a region in Figure 2(b); (b) The Sheet of the gradient directions for a region in Figure 2(d); (c) The Sheet of the gradient directions for a region in Figure 2(f);

Figure 8. The three-dimensional reconstructive model for nonwoven fabric.

## 3 CONCLUSIONS

This paper proposed a novel method for reconstruction of the three-dimensional image based on a series of images obtained at different focal positions. The main theoretical point in this paper is that the target observed under the microscope exhibits clear when it is in-focus, while blur when de-focus. By comparing the sharpness of one target-point at different focal positions, the in-focus position can be found. Considering the focus length is always constant for one lens, the in-focus position can be regarded as the z-position for a target-point in the 3D model. When there was one target point at a certain position, the focal position declaring the max clear-ness is the z-position for the target point. In the situation that objects crossed each other, there may be two or more target points at one $(x, y)$ position. In this case, the peaks on the sharpness-focal position curve for a $(x, y)$ position point conveyed the z-positions for each target points. To eliminate the interference of noisy point, the sharpness of a P-centered circle region was utilized to represent the sharpness of a pixel P. The sharpness judging algorithm introduced in this paper was based on the variance of gradients in the circle region. The final three-dimensional model can reflect the main structure of the nonwoven fabric, which is meaningful for further detection and analysis.

ACKNOWLEDGEMENT

The authors would like to thank the Natural Science Foundation of China (Grant No. 61172119) and the Program for New Century Excellent Talents in University (Grant No. NECT-12-0825) for the financial support given to this work.

REFERENCES

Chen, T. Li, L.Q. & Koehl, L. et al. 2007. A Soft Computing Approach to Model the Structure-Property Relations of Nonwoven Fabrics. *Journal of Applied Polymer Science*, 103(1): 442–450.

Ghassemich, E. Versteeg, H.K. & Acar, M. 2001. Microstructural Analysis of fiber Segments In Nonwoven Fabrics Using SEM and Image Processing. *International Nonwoven Journal*, 10(2): 26–31.

Kanjar, D. & Masilamani, V. 2013. Image Sharpness Measure for Blurred Images in Frequency Domain. *International Conference on Design and Manufacturing*, 64: 149–158.

Lalith, B.S.V. Eunkyoung, S. & Nagendra, A. et al. 2012. Three-Dimensional Structural Characterization of Nonwoven Fabrics. *Microscopy and Microanalysis*, 18(6): 1368–1379.

Wang, R. Xu, B. & Zeng, P. et al. 2012. Multi-focus image fusion for enhancing fiber microscopic images. *Textile Research Journal*, 82(4): 352–361.

Xin, X. Wang, Y.L. & Tang, J.S. et al. 2012. Adaptive Variance Based Sharpness Computation for Low Contrast Images. *Advanced Intelligent Computing Lecture Notes in Computer Science*, 6838: 335–341.

Xu, D. 2014. A Microscopic Image Sharpness Metric Based on the Local Binary Pattern (LBP). *Advanced Materials Research*, 902: 330–335.

*Advanced Materials, Structures and Mechanical Engineering – Kaloop (Ed.)*
© 2016 Taylor & Francis Group, London, ISBN 978-1-138-02793-0

# The effect of volume variation of silver nanoparticle solution towards the porosity and compressive strength of mortar

H.S.A. Tina, A. Andreyani & W.S.B. Dwandaru
*Yogyakarta State University, Yogyakarta, Indonesia*

ABSTRACT: As the world is growing rapidly, people need better building materials such as mortar. The aim of this research is to determine the effect of adding silver nanoparticle solution towards the porosity and compressive strength of mortar. This research was started by making silver nanoparticle solution from nitrate silver ($AgNO_3$). The solution is then characterized using Uv-Vis spectrophotometer. 5 mM silver nanoparticle is added in the process of mortar production with volume variation of the silver nanoparticle solution. The porosity, compressive strength, and the content of mortar were determined by digital scale, universal testing machine, and X-ray diffraction, respectively. For silver nanoparticle solution volumes of (in mL) 0, 5, 10, 15, 20, and 25 the porosity obtained are (in%) 20.38, 19.48, 19.42, 18.9, 17.8, and 17.5, respectively. The best increase in compressive strength is obtained for (in MPa) 29,068, 29,308, and 31,385, with nanoparticle solution volumes of (in mL) 5, 10, and 15.

## 1 INTRODUCTION

The growth and development of construction industry in Indonesia such as roads, buildings, and bridges are rapid. This development, certainly, demands a better quality of infrastructures. However, there are many buildings that are easily broken because of some problems such as natural disasters and low quality of the materials that are not strong enough to endure some pressure caused by environment interferences. The quality of a material can be determined from mixed-materials in which the building material is made from, treatments and the additive materials or additional materials that are added to the building material such as mortar. In addition, the amount of water also gives a big influence to the physical endurance such as the porosity and compressive strength.

Timuranto (2001) in Tjokrodimuljo examined the relation between mortar's strength and water reservation. The lower the water reservation means the less mortar's pores so that its strength is greater. The relation between mortar's strength, porosity, and density is influenced by the amount of cement (the ratio of cement and sand) and the factor of cement-water that are added to the mortar.

One of the ways to increase the strength of the mortar is by adding additional materials to the mortar in the mixing process. The functions of the additional material are to change some characters of the mortar to be more compatible for certain works and to lessen the expense so that the strength of the mortar can be increased and can be used for making a firm and strong building.

Mortar is a building material that is frequently used by people. The application of mortar is mostly in the building construction as an adhesive between the stones, walls, and concretes. Mortar is the mixing between cement, sand, and water with ratio of 1: 2.75: 0.5, respectively, in which the cement is the main adhesive. The correct portion or ratio of the materials is really important because it has a big influence toward the mortar's porosity and strength.

Porosity is a measure of empty space between materials. Moreover, it is also considered as the fraction of empty space (volume) that has a value between 0 and 1 or as the percentage between 0 and 100%. The aim of the porosity observation is to measure the empty space of the tested material. The bigger the porosity of the tested material means the less the strength and vice versa. Porosity examination is conducted to test the mortar's water absorbance by soaking the mortar for 7 days. The porosity can be determined using the following formula,

$$\% = \frac{m_b - m_k}{V_b} \times \frac{1}{\rho water} \times 100\% \qquad (1)$$

where, $m_b$ is the wet weight of the tested material (gram), $m_k$ is the dry weight of the tested material (gram), $V_b$ is the volume of the tested material ($cm^3$), and $\rho_{water}$ is the density of the water (1 gr/$cm^3$).

One of the ways to control the quality of mortar is by examining the sample or the tested material. The strength of the mortar can be defined as the comparison between the weight and the width of the mortar. One of the indicators of a good mortar is if it has high strength. Generally, the strength of a cement mortar is between 3–17 MPa. Wancik et al. (2008) researched the cement mortar in a styrofoam with the proportion of the mixed-mortar volume of 1: 15 and the number of cement-water factor of 0.4 which consists of 2.4 mm sifted sands and 1.5% viscocrete of the weight of the cement in order to make the production easier. The result finds that the 5 tested cube materials sized 5 cm × 5 cm × 5 cm produced an average strength of 79.01 MPa. Yulianingsih in Tjokrodimuljo (2004) examined the characteristics of cement mortar made from the rough sand with a different portion of the mixture volume. The result shows that the ratio of the mixture volume influences the pressing strength, pulling strength, water absorbance, and the density of the cement mortar. The mortar compressive strength can be found through the following formula,

$$\mathbf{f'_c} = \frac{F}{A} \qquad (2)$$

with, $\mathbf{f'_c}$ is the compressive strength (MPa), $F$ is the maximum weight (N), and $A$ is the surface area (m²).

Admixture is defined in standard deviation of terminology relating to concrete and concrete Aggregates (ASTM C.125-1999: 61) and in cement and concrete terminology (ACI SP-19) as a material beside water, aggregate, and hydraulic that are mixed with concrete and mortar. The additional materials are used to modify the characteristics of the concrete or mortar for example to make it more efficient to be used or for other purposes like energy saving (Mulyono, 2003: 117).

In order to increase the quality of the material, nanotechnology can be applied. Adding nanoparticle to mortar may decrease the porosity and increase the compressive strength of mortar. One application of nanotechnology in building material is specifically to strengthen mortar by using nanoparticle. According to Pacheco (2010), nanotechnology and nanomaterial can be used by the construction industry. It covers nanoscale analysis of Portland cement hydration products, the use of nanoparticles to increase the strength and durability of cement composites, and the photocatalytic capacity of nanomaterials. Tanvir (2010) said that the development of nanotechnology for cement and concrete has particular importance. The cement hydration chemistry and the physical behavior of hydration products may be potentially manipulated through nanotechnology. The main objective of the development of cement-related nanotechnology is to produce stronger, tougher, lighter, and more durable products.

The definitions of nanotechnology, including nanoscience and nano-engineering in concrete, are provided by Florence (2010). Recent progress in nano-engineering and nano-modification of cement based materials is presented. The development in nanoscience can also have a great impact on the field of construction materials. Portland cement, one of the largest commodities consumed by the public, is obviously a good product, but the potential is not completely explored. Better understanding and engineering of complex structure of cement-based materials at nano-level will definitely introduce a new generation of concrete, which is stronger and more durable, with desired stress strain behavior and, possibly, with the whole range of newly introduced 'smart' properties (Sobolev, 2008).

Mechanochemical activation was found to be an effective method to improve the strength of cement-based materials. It was proposed that this process is governed by the solid state interaction between the organic modifiers and cement. During this process, the surface of cement particles attaches the functional groups introduced from the modifiers; so the organo-mineral nano-layers are formed on the surface of the cement.

One of the products of nanotechnology is a silver nanoparticle. This research is conducted to determine the influence of silver nanoparticle solution toward mortar's porosity and compressive strength. Some benefits of using silver nanoparticle are the low cost and relatively easy to be made and to be applied. Nanoparticle is soluble with other materials and eco-friendly. According to Kim (2010) $Ag_2S$ crystal nanoparticles are in the size range of 5–20 nm with an ellipsoidal shape, and they form very small, loosely packed aggregates. Some of the $Ag_2S$ nanoparticles (NPs) have excess S on the surface of the sulfide minerals under S-rich environments. This study suggests that in a reduced, S-rich environment, such as the sedimentation processes during wastewater treatment, silver nanoparticle sulfides are being formed.

## 2 RESEARCH METHOD

### 2.1 The technique of data collection

#### 2.1.1 Research materials
The materials that were used in this research are i) Portland cement type I, ii) smooth aggregate (from Mount Merapi sand), iii) water, and iv) 5 mM silver nanoparticle solution.

#### 2.1.2 Research instruments
The research instruments used are given as follows: heater (1 unit), 500 ml chemical glass (3 units),

250 ml measuring cup (1 unit), pipette (3 units), solution stirrer (1 unit), thermometer (1 unit), test tube (12 units), test tube clamp (1 unit), hand gloves (1 unit), laser pointer (1 unit), digital scale (1 unit), scale (1 unit), stirring shovel (1 unit), tin plate (5 units), tin basin (5 units), cup for stirring cement (1 unit), mortar solidifying iron (1 unit), sand sifter 1.18 mm mesh on 16 (1 unit), mortar molding box 5 cm × 5 cm × 5 cm (10 units), pail (1 unit), oven (1 unit), ultimate testing machine (1 unit).

### 2.1.3 *Research procedure*
#### 2.1.3.1 Making 5 mM silver nanoparticle solution
5 mM nanoparticle solution is produced by the following procedure. The solution of $AgNO_3$ is made by dissolving $AgNO_3$ powder in water. In order to get $AgNO_3$ solution with concentration of 5 mM, a formula is applied which gives the ratio between $AgNO_3$ and water, viz.:

$$M = \frac{m}{mr} \times \frac{1000}{V} \qquad (3)$$

where, $M$ is molarity, $m$ is the weight, $mr$ is the relative weight, and $V$ is volume.

By applying the formula above, 0.85 gram of $AgNO_3$ powder is dissolved in 1000 mL of water. The process of dissolving $AgNO_3$ powder was done in some steps. Firstly, the $AgNO_3$ powder is gradually mixed into 100 mL of water and stirred. After they are mixed, $AgNO_3$ powder is continuously added into water and then stirred again.

Figure 2(a) shows the process of synthesizing silver nanoparticle $AgNO_3$ solution by adding 2 mL $AgNO_3$ solution of 5 mM concentration into the test tube and then heating it inside a tube filled with water of 100 °C for 10 minutes. Then the test tube was moved to the rack, and then added with 5 drips of sodium citrate ($Na_3C_6H_5O_7$) solution as much as 1%. Then, the solution is heated again until the solution is changed into a yellowish color.

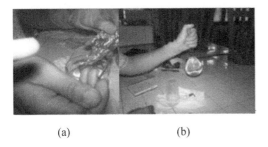

Figure 1. (a) dissolving $AgNO_3$ crystal with the aquatic liquid; (b) smoothing the $AgNO_3$ solution in the measuring cup.

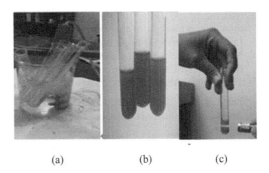

Figure 2. (a) the synthesis of silver nanoparticle inside boiling water; (b) the result of 5 mM of silver nanoparticle solution; (c) a beam of laser passing through the silver nanoparticle solution (Tyndall effect).

The yellowish color is considered as one of the indicators that the size of the particles in $AgNO_3$ solution has been changed. Figure 2(b) shows the picture of silver nanoparticle solution.

#### 2.1.3.2 The production of mortar
In this research, we used the handout of laboratory practical research and building material as guidance. In addition, we were also guided by the standard of American Concrete Institute (ACI). Here, 599 gram of Portland cement, 1375 gram of sand, and 242 ml of water are mixed with silver nanoparticle solution of 0 mL, 5 mL, 10 mL, 15 mL, 20 mL, and 25 mL in volume. There are three samples for each volume variation of the nanoparticle solution. The silver nanoparticle solution was mixed with water but the volume of water was decreased with varying amount of 0 mL, 5 mL, 10 mL, 15 mL, 20 mL, and 25 mL, respectively, for each sample. The decrease of water volume that was substituted with silver nanoparticle solution is conducted in order to make the volume of the mixture constant.

After all the materials are ready, we poured sand and cement to the stirring cup, and then stirred until all the materials were blended homogenously. Subsequently, the materials were blended until it became smooth; then the mixture was moved to the $5 \times 5 \times 5$ cm³ mortar molding box.

24 hours later, the mortar was moved from the molding box and soaked in water. This soaking process is needed to hydrate the cement and water. This was conducted for 7 days.

#### 2.1.3.3 Mortar porosity test
The procedure of testing the porosity of the mortar can be explained as follows:

1. The mortar sample was taken out from the soaking cup then dried with a fabric until there was no water dripping,

2. Measuring the weight of the mortar to determine the wet weight of the mortar,
3. Inserting the mortar to an oven for 24 hours,
4. Measuring the weight of the mortar to obtain the dry weight of the mortar,
5. The above steps were repeated to all samples.

The value of the porosity can be determined by applying the formula for porosity as given in Equation (1).

### 2.1.3.4 Mortar compressive strength test

The steps of testing the compressive strength of the mortar may be given as follows:

1. Taking out the mortar that has been soaked for a certain time (7 days), and then drying it with a dried fabric until there was no water dripping.
2. The mortar that would be tested was put in a pressing machine.
3. Giving a pressure to the mortar slowly by using UTM to test the compressive strength of the mortar. The UTM is completed with a plotter to draw the graph of the tension of the mortar. The pressing machine that has been arranged was linked to a controlling computer.
4. After the mortar was broken as required, the computer automatically stopped counting and the value of the compressive strength counted by the computer is the value of the maximum compressive strength.
5. The above steps were repeated for other mortar samples.

The value of the compressive strength can be obtained using Equation (2).

## 3   FINDINGS AND DISCUSSION

### 3.1   *The result of spectrophotometer Uv-Vis of silver nanoparticle solution*

Figure 3 above shows Uv-vis result of the silver nanoparticle solution sample. Silver nanoparticle

has an absorbance of 1.997 with the wavelength at maximum absorbance of 425 nm. The value of the wave length at maximum absorbance shows that silver nanoparticles are present in the solution.

### 3.2   *The result of porosity measurement*

Based on Table 1, we can obtain the results of wet mass and dry mass of the mortar, and then use the aforementioned results to find the porosity which is plotted in Figure 4.

Table 1.   The data result of porosity measurement.

| No | Varieti of mix volume | Wet mass (gr) | Dry mass (gr) | Porosity (%) | Average porosiy (%) |
|---|---|---|---|---|---|
| 1 | 0 | 303.1 | 276.3 | 21.44 | 20.38 |
|   |   | 287.8 | 269.3 | 14.80 |   |
|   |   | 291.4 | 261.0 | 24.32 |   |
|   |   | 291.1 | 264.9 | 20.96 |   |
| 2 | 5 | 297.2 | 273.8 | 18.72 | 19.48 |
|   |   | 275.7 | 250.2 | 20.40 |   |
|   |   | 287.4 | 264.2 | 18.56 |   |
|   |   | 286.0 | 260.7 | 20.24 |   |
| 3 | 10 | 296.4 | 275.2 | 16.96 | 19.42 |
|   |   | 290.3 | 264.7 | 20.48 |   |
|   |   | 291.6 | 266.9 | 19.76 |   |
|   |   | 283.8 | 258.2 | 20.48 |   |
| 4 | 15 | 296.3 | 274.1 | 17.76 | 18.9 |
|   |   | 286.8 | 262.3 | 19.60 |   |
|   |   | 291.8 | 267.8 | 19.20 |   |
|   |   | 291.1 | 267.3 | 19.04 |   |
| 5 | 20 | 294.5 | 272.3 | 17.76 | 17.8 |
|   |   | 287.6 | 265.3 | 17.84 |   |
|   |   | 287.6 | 265.2 | 17.92 |   |
|   |   | 288.0 | 265.9 | 17.68 |   |
| 6 | 25 | 292.6 | 273.8 | 15.04 | 17.5 |
|   |   | 293.5 | 270.8 | 18.16 |   |
|   |   | 295.7 | 273.3 | 17.92 |   |
|   |   | 289.4 | 265.8 | 18.88 |   |

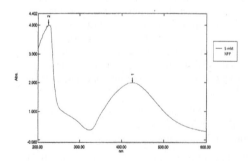

Figure 3.   The result of 5 mM silver nanoparticle test using spectroscopy Uv-Vis.

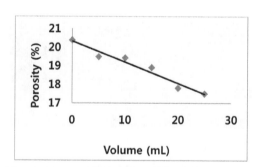

Figure 4.   The graphic of the porosity vs silver nanoparticle volume of the mortar samples.

Table 2. The data result of compressive strength.

| No | Varieti of mix volume | Surface area A (m²) | The max force (F) (N) | Compressive strength $f_c'$ (MPa) | Average compressive strength (MPa) |
|---|---|---|---|---|---|
| 1 | 0 | 0.0025 | 61220 | 24.488 | 24.276 |
|   |   |   | 61240 | 24.496 |   |
|   |   |   | 59840 | 23.936 |   |
|   |   |   | 60460 | 24.184 |   |
| 2 | 5 | 0.0025 | 56747 | 22.699 | 29.068 |
|   |   |   | 72980 | 29.192 |   |
|   |   |   | 69330 | 27.732 |   |
|   |   |   | 91620 | 36.648 |   |
| 3 | 10 | 0.0025 | 76850 | 30.740 | 29.308 |
|   |   |   | 73410 | 29.364 |   |
|   |   |   | 59050 | 23.620 |   |
|   |   |   | 83770 | 33.508 |   |
| 4 | 15 | 0.0025 | 66474 | 26.589 | 31.385 |
|   |   |   | 83580 | 33.432 |   |
|   |   |   | 85440 | 34.176 |   |
|   |   |   | 78360 | 31.344 |   |
| 5 | 20 | 0.0025 | 64390 | 25.756 | 28.926 |
|   |   |   | 83740 | 33.496 |   |
|   |   |   | 46870 | 18.748 |   |
|   |   |   | 94260 | 37.704 |   |
| 6 | 25 | 0.0025 | 61316 | 24.526 | 31.730 |
|   |   |   | 82130 | 32.852 |   |
|   |   |   | 86510 | 34.604 |   |
|   |   |   | 87340 | 34.936 |   |

From the data above, it can be seen that the value of the porosity without silver nanoparticle solution is 20.38%, while for the average porosity of the mortar with silver nanoparticle solution variation of 5 mL, 10 mL, 15 mL, 20 mL, and 25 mL, successively is 19.48%, 19.42%, 18.9%, 17.8%, and 17.5%. It can be seen that the more silver nanoparticle solution is added into the mixture, the less the mortar porosity becomes. This is caused by the increase of the number of the silver nanoparticles that occupy the pores inside the mortar such that it reduces the porosity of the mortar. The fitted model used in Figure 4 is the linear line. This means that the porosity decreases linearly as the volume of the nanoparticle solution is increased.

### 3.3 The result of compressive strength

Based on Table 2 we can obtain the results for surface area and maximum force, and then use these components to find the compressive strength as an input in Figure 5.

The graphic of the compressive strength of the mortar may be observed in Figure 5. From the graphic above, the compressive strength of the mortar without silver nanoparticle solution is 24.276 MPa, while for the average mortar compressive strength with variation of the silver

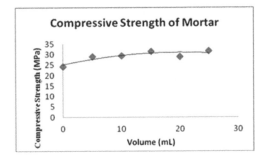

Figure 5. The graphic of the compressive strength vs the silver nanoparticle solution of the mortar.

nanoparticle volume of 5 ml, 10 ml, 15 ml, 20 ml, and 25 ml, successively, is 29.068 MPa, 29.308 MPa, 31.385 MPa, 28.926 MPa, and 31.370 MPa.

It is gained that the mortar compressive strength increases as the volume of the nanoparticle solution is increased from 5 ml to 15 ml that is 29.068 MPa, 29.308 MPa, and 31.385 MPa, while for 20 ml of nanoparticle solution the mortar compressive strength reduces to 28.926 MPa. However, for 25 ml nanoparticle solution the mortar compressive strength increases to 31.370 MPa.

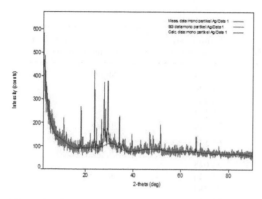

Figure 6. The graphic of the mortar's contents by XRD.

Table 3. The data of mortar's contents.

Weight ratio

| Phase name | Mono partikel Ag |
| --- | --- |
| | Content (%) |
| Quartz | 2.1 (7) |
| Anorthite, sodian, syn | 97 (6) |
| Silver | 0.80 (3) |

From the XRD test of the mortar which is already three months old, we can conclude that the content of the mortar consists of quartz (2.1%), anorthite, sodion, syn 97%, dan silver (0.8%). From the XRD test there is still a trace of silver content although the age of the mortar has been three months old, in solid condition, and drainage process.

## 4 CONCLUSIONS

From the discussion above we conclude as follows:

1. The compressive strengths of the mortar with the silver nanoparticle solution volume variation (in ml) of 5, 10, 15, 20, and 25 are obtained as 29,068 MPa, 29,308 MPa, 31,385 MPa, 28,926 MPa, and 31,730 MPa, respectively.
2. Based on the porosity test of the mortar, the more the silver nanoparticle solution is added into the mixture, the less the porosity becomes. The addition of 25 ml silver nanoparticle solution gives the lowest porosity of the mortar.
3. From XRD, it is obtained that the silver content in the mortar is around 0.8% although the mortar is already three months old.

## REFERENCES

ASTM, 2003. *Concrete and Aggregates: Annual Book of ASTM,* Vol. 04.02, Philadelphia.

Florence, S. & Konstantin, S. 2010. Nanotechnology in Concrete-A Review. *Construction and Building Material,* 24: 2060–2071.

Jain, P.K. & Jain, V. 2006. Impact of Nanotechnology on Healthcare-Application in Cell Therapy and Tissue Engineering. *Nanotechnology Law & Bussines,* 3(4): 411–418.

Kim, B.J. Park, C.S. Murayama, M. & Hochella, M.F. 2010. Discovery and Characterization of Silver Sulfide Nanoparticles in Final Sewage Sludge Product, *Environonmental Science & Technology,* 44(19): 7509–7514.

Mulyono, T. 2003. *Concrete Technology.* Yogyakarta: Andi Offset.

Pacheco-Torgal, F. & Jalali, S. 2010. Nanotechnology: Advantages and drawbacks in the field of construction and building materials. *Construction and Building Material, Elsevier. Elsevier.* 25(2): 582–590.

Ramsden, J.J. 2009. *Applied Nanotechnology.* Oxford: Elsevier.

Sobolev, K. Flores, I. Hermosillo, R. & Torres-Martinez, L.M. 2008. Nanomaterials and Nanotechnology for High-Performance Cement Composites. *International Concrete Abstracts Portal,* 254: 93–120.

Tanvir, M. & Nur, Y. 2010. Strength Enhancement of Cement Mortar with Carbon Nanotubes. *Transportation Research Board of the National Academies,* 2142: 102–108.

Tjokrodimuljo, K. 2004. *Concrete Technology.* Yogyakarta: UGM Press.

Wancik, A. Satyarno, I. & Tjokrodimuljo, K. 2008. Cement Mortar Composite Styrofoam Bricks. *Civil Engineering Forum,* XVIII (2): 780–787.

*Advanced Materials, Structures and Mechanical Engineering – Kaloop (Ed.)*
© *2016 Taylor & Francis Group, London, ISBN 978-1-138-02793-0*

# Design and implementation of reversible DC speed control system with logic non circulating current

L.N. Gao & S.Q. Yi
*School of Mechanical Engineering, Chengdu University, Chengdu, Sichuan, China*

ABSTRACT: The reversible DC speed control system with logic non circulating current can solve the problem of circulation in the thyristor-motor speed regulating the system. In this paper, the system has been designed and implemented. It includes the two sets of triggering devices controlling the positive and negative group of thyristors respectively. At any time, one group of thyristors works, and the other group would be blocked to avoid circulation. The experiment results show that the system has a good static characteristic.

## 1 INTRODUCTION

Due to processing and operational requirements, many production machineries driven by the electric motor would be in the transition process of starting, braking and reverse operating that is related to the motor motion control. The task of the motion control system is to change the torque, velocity, and displacement of mechanical working machine performed by controlling the motor voltage, current and frequency, so that a variety of machinery could operate according to the desirable requirements in order to meet the production technology and other applications. By establishing the coordinates to characterize the rotational speed and electromagnetic torque, the motion control system for four-quadrant operation on the coordinate system can be reversed running and called the reversible speed control system. The Direct-Current (DC) motor has a good starting and braking performance and is appropriate in a wide range of speed, which makes it widely used in the field of electric drive requiring speed regulation or fast forward and backward operation.

The reversible DC speed control system of high power often adopts thyristor-motor (V-M) system. The reversible V-M speed control system using two groups of anti-parallel thyristors can solve the problem of forward and reverse running and regenerative braking, but if the two groups of devices appear rectified voltage simultaneously, the short-circuit current called circulating current will flow directly between the two groups of thyristors and there will be no flow through the load. When the circulating current is too much, the thyristors can be damaged and therefore, the circulating current should be suppressed or eliminated. The reversible

DC speed control system with logic non circulating current by setting logic non circulating current control link (DLC) can avoid the emergence of circulating current. Based on understanding the principles of the reversible DC speed control system with logic non circulating current, this paper has designed and implemented the whole system, and thus analyzes the mechanical properties of the system.

## 2 SYSTEM PRINCIPLE AND DESIGN

### 2.1 *System principle*

According to the motor theory, the change of armature voltage polarity or the change of excitation flux direction can alter the rotation direction of the DC motor. The reversible circuit of the motor armature can take advantage of two groups of anti-parallel thyristors to form the reversible circuit. When the motor makes the positive rotation, the positive group of thyristors device VF will supply power; and when the motor makes the reverse rotation, the anti-group of thyristors device VR will supply power. The two groups of thyristors are controlled by two sets of the triggered device respectively that can flexibly control the starting, braking, decelerating and accelerating of the motor. However, the two groups of thyristors are not allowed to be in the state of rectification at the same time, otherwise it will cause the short circuit of power supply, therefore the reversible system puts forward strict requirements on the control circuit.

The state principle of rectifier and inverter by the two groups of anti-parallel thyristors devices forming the reversible circuit in V-M system is similar to the device powered by the single group thyristors. When the positive group of thyristors device VF makes power supply to the electric motor, it is in the state of rectification and the ideal no-load voltage polarity is shown in the Figure 1 among which $U_{d0f} > E$, $n > 0$. The motor is input energy from the circuit for the electric operation and the V-M system works in the first quadrant. When the electric motor needs braking, the counter electromotive force polarity of the motor is unchanged, the feedback of electric energy will produce the reverse current which is not possible through the VF circulation. At this time, the control circuit can be switched to a reverse group thyristor device VR, and make it work in the inverter state. The polarity of inverter voltage $U_{d0r}$ is shown in the figure, and when $E > |U_{d0r}|$, $n < 0$, the reverse current flow $-I_d$ can be passed through the VR circulation. To achieve regenerative braking, the motor outputs power and the V-M system works in the second quadrant. In the reversible speed control system, the VR thyristors can be used to realize feedback brake during the positive rotation and the VF thyristors can be used to realize feedback brake during the reverse rotation. In this way, the two groups of anti-parallel thyristors devices can realize the motor running on the four quadrants.

The main circuit of the reversible DC speed control system with logic non circulating current adopts two groups of reverse parallel thyristor devices, and because there is no circulating current,

the system do not set the loop reactor, but still remains the flat wave reactor $L_d$ to ensure the stable operation of continuous current waveform. In order to ensure that there is no circulation, the acyclic logic control link DLC is set up, which is the key link in the system. According to the working state of the system, DLC can control it and make the automatic switching of VR and VF, the output signal $U_{blf}$ is used to control the trigger pulse of positive group (VF) blocked or open, and $U_{blr}$ is used to control the trigger pulse of anti-group (VR) blocked or open. In any case, the two signals must be reversed, and the two groups of thyristor will never be allowed to open pulse at the same time to ensure the main circuit appears no circulation.

## 2.2 System design

The design scheme of the system is shown in Figure 2. The system is mainly composed of speed regulator (ASR), current regulator (ACR), reversed phase device (AR), torque polarity discrimination device (DPT), zero level detection device (DPZ), logic control device (DLC), speed conversion device (FBS) and other aspects. When the system has a positive starting, the value of given voltage $U_g$ is positive, the output $U_{lf}$ of DLC is "0" state and $U_{lr}$ is "1" state, that the trigger pulse of the positive bridge is opened and trigger pulse of the anti-bridge is blockaded. At this time, the three-phase full-controlled rectifier circuit on the positive bridge of the main circuit works and the motor is running forward. When the value of given voltage $U_g$ becomes negative, the rectifier on the positive bridge enters into the inverter state, the values of $U_{lf}$ and $U_{lr}$ are unchanged. When the main circuit current decreases to zero, the output $U_{lf}$ is shifted into "1" state and $U_{lr}$ is "0", the system will go into the braking state on the anti-bridge, and the motor will decelerate to the set speed and then switch to reverse electromotion. When the value of $U_g$ is 0, the motor will stop. When the motor makes

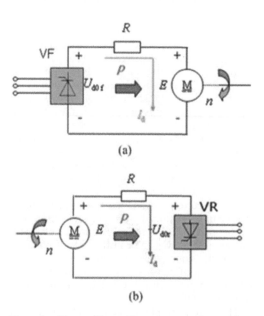

(a)

(b)

Figure 1. The rectifier and inverter state of two groups of anti-parallel thyristors devices forming the reversible circuit.

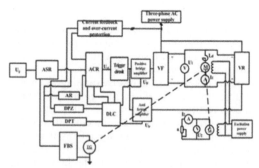

Figure 2. The design scheme of the reversible DC speed control system with logic non circulating current.

the reverse running, the three-phase full-controlled rectifier circuit on the anti-bridge of main circuit will work.

The output of DLC depends on the running state of the motor. When the motor is in the status of running forward or reverse braking, $U_{lf}$ is "0" state and $U_{lr}$ is "1" state that ensures the working of positive bridge and blockage of anti-bridge. When the motor is in the status of running backward or forward braking, $U_{lf}$ is "1" state and $U_{lr}$ is "0" state, that ensures the working of anti-bridge and blockage of a positive bridge. As the logic control role of DLC, the two rectifier bridges do not be triggered at the same time so that there is neither DC circulation nor pulsation circulation in the reversible speed control system with logic non circulating current.

## 3 SYSTEM IMPLEMENTATION

### 3.1 DLC

The logical control of DLC is the key link of logic non-circulation system, the mission is when it needs to switch to the working state of positive group thyristor VF, the anti-group trigger pulses are blocked and VF group trigger pulses are open and when it needs to switch to the working state of anti-group thyristor VR, the positive group trigger pulses are blocked and VR group trigger pulses are open. The principle of DLC circuit is shown in Figure 3. It is mainly composed of a logic judgement circuit, a delay circuit, a logic protection circuit, a pushing β circuit and a power amplifier circuit.

The task of the logic judgement circuit is to determine correctly whether the thyristor trigger pulses require to switch and switching conditions are in place based on the output $U_M$ of torque polarity discrimination and the output $U_I$ of zero level detected. When the sign of $U_M$ changes and the main circuit current reaches zero ($U_I$ = "1") by zero level detection, the logic judgement circuit will immediately flip over. The output $U_Z$ and $U_F$ of

the logic judgement circuit must be the opposite state at any time.

In order to make the positive and anti-group of rectification devices switch safely and reliably, when the switching instruction $U_Z$ or $U_F$ is issued by the logic judgement, it can be performed by the delay circuit after the waiting time $t_1$ (about 3 ms) and triggering waiting time $t_2$ (about 10 ms). VD1, VD2, C1 and C2 are used to achieve the prolonged time $t_1$. VD3, VD4, C3 and C4 are used to achieve the prolonged time $t_2$.

The Logical protection circuit is also known as "one more" protection link. When DLC occurs a fault and the outputs of $U_Z$ and $U_F$ are both "1" state, the two outputs $U_{lf}$ and $U_{lr}$ of DLC logic controller are both "0" state, which will result in two sets of rectification devices open simultaneously and cause a short circuit and circulation accident. After adding the logical protection link, when $U_Z$ and $U_F$ are both "1" state, the output point A of the protection link will shift into "0", so that $U_{lf}$ and $U_{lr}$ are high level and two groups of triggering pulses are blocked at the same time to avoid short circuit and circulation accident.

In the positive and anti-bridge switching, G8 of the logic controller DLC outputs "1" state signal, and this signal is sent into the input of the regulator ACR as pushing β pulse signal, thereby avoiding the impact of switching current.

Since the finite output power of NAND gate, in order to reliably push $U_{lf}$ and $U_{lr}$, the power amplifier consisted of V3 and V4 is increased.

### 3.2 System debugging and implementation

The debugging principles of speed control system with logic non circulating current include: firstly adjusting the parameters of each unit and then forming the system; firstly making the system run in open loop and then run in double closed loop comprised by the current and speed negative feedback loop; firstly making the double closed loop of the positive bridge and anti-bridge work correctly and then forming the system of logic non circulating current.

The specific debugging aspects include: debugging triggering circuit; determining the adjustment range of phase shift control voltage $U_{ct}$; adjusting the zero and amplitude limit of ASR and ACR; debugging the torque polarity discrimination; debugging the zero level detection; debugging reversed phase device (AR); debugging the logic control unit DLC; debugging speed feedback coefficient α and current feedback coefficient β and tuning the entire system.

When the system runs normally, by changing given voltage $U_g$, the mechanical properties of $n = f(I_d)$ are measured and recorded separately as

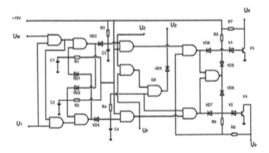

Figure 3. The design scheme of the DLC.

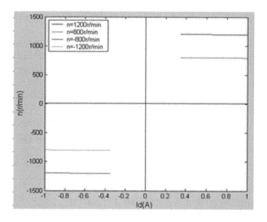

Figure 4.    The static mechanical property of the system.

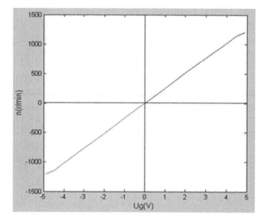

Figure 5.   The closed-loop control characteristic n = f(U_g) of the system.

n is ±1200 rpm and ±800 rpm. By decreasing the resistance of R on the generator circuit, the current on the motor will be increased to $I_d$ = 1.1 A. The experiment data have been recorded and drawn in Figure 4.

The closed-loop control characteristics n = f(U_g) can be measured by gradually increasing the positive given voltage. Similarly, the closed-loop control characteristics n = f(U_g) when reverse running can be measured by gradually increasing the negative given voltage. The related experiment data have been recorded and drawn in Figure 5.

As can be seen from the experiment data and figures, the static mechanical properties of reversible DC speed control system with logic non circulating current are good and the closed-loop control characteristics n = f(U_g) is basically symmetrical.

## 4   CONCLUSIONS

In order to solve the circulation problem of V-M reversible speed control system, the reversible DC speed control system with logic non circulating current can be used through the logic control. When a group of thyristors works, using the logic circuit (hardware) or the logic algorithm (software) to block another group of thyristors triggering pulses, it ensures that the two groups of thyristor do not work at the same time and cuts off the circulation channel, that is the reversible DC speed control system with logic non circulating current by logic control. In this paper, the system has been realized and the mechanical characteristics have been validated. The analysis of dynamic characteristics will be studied on the next step.

REFERENCES

Chen, Z. & Zhu, D.Z. 2009. Simulation of Logic Non-loop-current DC SR System Based on Matlab. *Micromotors*, 42(1): 95–97.
Liang, X.P. 2009. Design of a Logic Non-Loop-Current SR System Based on SCM. *Electronic Design Engineering*, 17(7): 44–46.
Ma, Q. 2010. Research of the Logic Speed Control System without Circulating. Current Based on Power system Module. *Mechanical & Electrical Engineering Technology*, 39(7): 24–26.
Ma, Q. 2011. DLC without Circulating Current Controller Design. *Computing Technology and Automation*, 30(1): 60–63.
Ward Leonard control. Information on http://en.wikipedia.org/ wiki/Ward_Leonard_control.

*Advanced Materials, Structures and Mechanical Engineering – Kaloop (Ed.)*
© *2016 Taylor & Francis Group, London, ISBN 978-1-138-02793-0*

# Effect of the plain-woven fabrics on mechanical properties of composites

S. Tanpichai
*Learning Institute, King Mongkut's University of Technology Thonburi, Bangkok, Thailand*

K. Kaew-in & P. Potiyaraj
*Department of Materials Science, Faculty of Science, Chulalongkorn University, Bangkok, Thailand*

ABSTRACT:  The plan-woven cotton fabric was used as reinforcement to prepare all-cellulose composites. Microcrystalline Crystalline (MCC) was firstly dissolved in a solution of lithium chloride and N, N-dimethylacetamide (LiCl/DMAc) to prepare a clear transparent solution. The plain-woven fabric was then immersed in the solution, and the all-cellulose composite was finally formed. Mechanical properties and morphology of the composites were investigated using tensile testing and scanning electron microscopy, respectively. Results of mechanical properties reveal the presence of the plain-woven fabrics improves mechanical properties of the cellulose films. This is because of the good interaction between the matrix and reinforcement.

## 1 INTRODUCTION

Natural fibers have been of significant interest due to low price, high mechanical properties, renewability and biodegradability (Eichhorn et al. 2010, Kalia et al. 2011, Klemm et al. 2005, Porras and Maranon 2012). Among forms of natural fibers such as continuous fibers, short fibers, nonwoven and woven fabrics used to prepare composites, the woven fabrics have been found to be the attractive reinforcement because the woven fabrics produced from sets of yarns by weaving have superior mechanical properties in two different directions (0° and 90°) as fibers are oriented as wasp and weft (Ramamoorthy et al. 2015, Alavudeen et al. 2015). Recent researches of the use of woven fabrics to enhance properties of polymers have been introduced. Porras and Maranon (2012) prepared the laminate composites of bamboo woven fabrics and poly (lactic acid). Tensile strength and elongation in the transverse direction of the composites with the woven fabrics were found to increase by 43 and 174%, respectively, compared to those of neat poly (lactic acid).

All-cellulose composites with better adhesion between the matrix and reinforcement have been firstly introduced by Nishino et al. (2004). The all-cellulose composites are derived from a wide variety of cellulose sources such as Microcrystalline Cellulose (MCC) (Gindl & Keckes 2005), ramie fibers (Nishino et al. 2004), regenerated cellulose fibers (Soykeabkaew et al. 2009), to name a few examples.

As both matrix and reinforcement in the all-cellulose composites are cellulose, the composite with the better interaction of the matrix and reinforcement can be produced, and superior mechanical properties are obtained from this material. This material has been termed fully bio and green composites (Nishino et al. 2004, Huber et al. 2012).

In this research, the plain-woven cotton fabric was impregnated in a matrix prepared from MCC. Mechanical properties and morphology of the composites were studied, compared to the cellulose films.

## 2 EXPERIMENTAL

### 2.1 Materials

Microcrystalline Cellulose (MCC) (Avivel PH-101) used in this work was purchased from Sigma Aldrich, and plain-woven cotton fabrics were supplied by Sumet Sampheng, Bangkok, thailand, as shown in Figure 1. Lithium chloride, *N,N*-dimethylacetamide, methanol, acetone, hydrogen peroxide, sodium silicate and sodium hydroxide were supplied by RCI Labscan. All chemical reagents were used without further purification.

### 2.2 Morphology

The morphology of the plain-woven cotton fabrics and composites was investigated using scanning electron microscopy (JSM-6480 LV) equipped

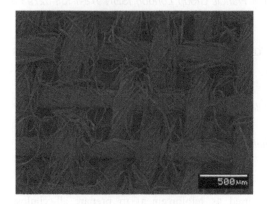

Figure 1.   SEM image of the plain-woven cotton fabric.

with a secondary electron detector under an accelerating voltage of 5 kV. The samples were coated with a thin layer of gold to avoid charging before analysis.

### 2.3   *Mechanical properties*

Mechanical properties of the composites and neat films were investigated using a universal testing machine (Lloyd LR100 K, West Sussex, UK) equipped with the 100 N load cell. The crosshead speed and the gauge length used in the test were 5 mm min$^{-1}$ and 25 mm, respectively. The width and length of samples were about 5 and 50 mm, respectively. At least five samples were tested for each material.

### 2.4   *Crystallinity*

X-ray diffraction photographs were taken by a flat camera having a length of 37.5 mm. The Cu KR radiation, generated with an RINT-2000 (Rigaku Co.) at 40 kV, 20 mA, was irradiated on the specimen perpendicular to the surface. The diffraction profile was detected using an X-ray goniometer with a symmetric reflection geometry. After subtracting the air scattering, the diffraction profile was curve resolved into noncrystalline scattering and crystalline reflections.

## 3   RESULTS AND DISCUSSION

### 3.1   *Morphology*

Figure 2 shows the fractured surface of the cellulose film, plain-woven cotton fabric and fabric reinforced composite. The cellulose matrix was observed to have the interaction with the top and bottom layer of the fabric. However, the cellulose matrix cannot penetrate into yarns in the fabric.

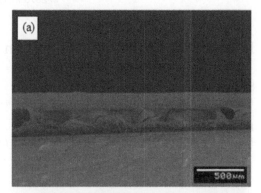
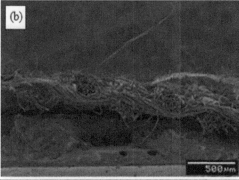
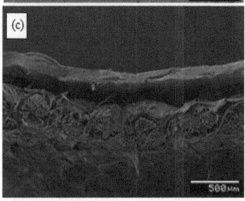

Figure 2.   SEM images of the fractured surface of (a) the cellulose film, (b) the plain-woven cotton fabric and (c) the cotton fabric reinforced composite.

Some cavities were seen between yarns in the fabric. This could be the weak point of the composites during deformation.

### 3.2   *Mechanical properties*

Table 1 shows values of tensile strength, strain at break and toughness of the cellulose film, the plain-woven cotton fabrics and composites reinforced with the fabrics. After impregnating the

338

Table 1. Mechanical properties of the cellulose films, composites and plain-woven fabrics.

| Material | Strength (MPa) | Strain at break (%) | Toughness (J) |
| --- | --- | --- | --- |
| Cellulose film | 21.9 ± 4.7 | 14.3 ± 2.6 | 0.03 ± 0.01 |
| Plain-woven cotton fabric (thin) | 11.9 ± 1 | 8.8 ± 1 | 0.02 ± 0.005 |
| Plain-woven cotton fabric | 16.3 ± 2.9 | 40.4 ± 8.1 | 0.05 ± 0.003 |
| Composite (Plain-woven thin) | 13.5 ± 1 | 28.1 ± 1 | 0.04 ± 0.004 |
| Composite (Plain-woven) | 24.4 ± 1.0 | 44.3 ± 2.8 | 0.16 ± 0.06 |

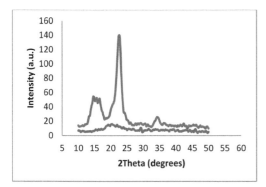

Figure 3. X-ray diffraction profiles of microcrystalline cellulose and film cellulose.

fabric into the cellulose matrix, tensile strength of the composites was increased by 11.4% from 21.9 to 24.4 MPa, and strain at break was significantly improved by 210% from 14.3 to 44.3%. Moreover, the toughness of composites was increased by 0.16 J from 0.03 J (433% increment). This is possibly due to the fact that fibers are able to absorb energy better than the film cellulose and good adhesion of the matrix and reinforcement. Although the tensile strength of the reinforcement used in this work (16.3 MPa) was lower than that of the matrix (21.9), when the fabric was in the composite, stress was transferred from the matrix to single yarns. This resulted in the higher strength (Tanpichai et al. 2012). This is also in agreement with an earlier work which bamboo woven fabrics were impregnated in poly(lactic acid) using compression molding (Porras and Maranon, 2012) and the higher mechanical properties of the composites with bamboo fabrics were reported.

## 3.3 *Crystallinity*

Comparison of the X-ray diffraction patterns of microcrystalline cellulose and film cellulose (Fig. 3) shows that the crystallinity of the composites decreased due to the dissolution of the crystals of the film cellulose in the LiCl/DMAc solution.

## 4 CONCLUSIONS

The all-cellulose composites with the plain-woven cotton fabrics were successfully prepared. Higher tensile strength and strain at break of the composites can be obtained, compared to those of the cellulose films. This was due to the good interaction between the matrix and cotton yarns in the fabric.

ACKNOWLEDGEMENT

This research was supported by the Ratchadaphiseksomphot Endowment Fund 2013 of Chulalongkorn University (CU-56-416-AM).

REFERENCES

Alavudeen, A. Rajini, N. Karthikeyan, S. Thiruchitrambalam, M. & Venkateshwaren, N. 2015. Mechanical properties of banana/kenaf fiber-reinforced hybrid polyester composites: Effect of woven fabric and random orientation. *Materials & Design*, 66: 246–257.
Eichhorn, S.J. Dufresne, A. Aranguren, M. Marcovich, N.E. Capadona, J.R. Rowan, S.J. Weder, C. Thielemans, W. Roman, M. Renneckar, S. Gindl, W. Veigel, S. Keckes, J. Yano, H. Abe, K. Nogi, M. Nakagaito, A.N. Mangalam, A. Simonsen, J. Benight, A.S. Bismarck, A. Berglund, L.A. & Peijs, T. 2010. Review: Current international research into cellulose nanofibres and nanocomposites. *Journal of Materials Science*, 45: 1–33.
Gindl, W. & Keckes, J. 2005. All-cellulose nanocomposite. *Polymer*, 46: 10221–10225.
Huber, T. Mussig, J. Curnow, O. Pang, S.S. Bickerton, S. & Staiger, M.P. 2012. A critical review of all-cellulose composites. *Journal of Materials Science*, 47: 1171–1186.
Kalia, S. Dufresne, A. Cherian, B.M. Kaith, B.S. Averous, L. Njuguna, J. & Nassiopoulos, E. 2011. Cellulose-based bio- and nanocomposites: A review. *International Journal of Polymer Science*. 2011(2011): 837875.
Klemm, D. Heublein, B. Fink, H.P. & Bohn, A. 2005. Cellulose: Fascinating biopolymer and sustainable raw material. *Angewandte Chemie International Edition*, 44: 3358–3393.

Nishino, T. Matsuda, I. & Hirao, K. 2004. All-cellulose composite. *Macromolecules*, 37: 7683–7687.

Porras, A. & Maranon, A. 2012. Development and characterization of a laminate composite material from Polylactic Acid (PLA) and woven bamboo fabric. *Composites Part B-Engineering*, 43: 2782–2788.

Ramamoorthy, S.K. Skrifvars, M. & Persson, A. 2015. A review of natural fibers used in biocomposites: Plant, animal and regenerated cellulose fibers. *Polymer Reviews*, 55: 107–162.

Soykeabkaew, N. Nishino, T. & Peijs, T. 2009. All-cellulose composites of regenerated cellulose fibres by surface selective dissolution. *Composites Part A-Applied Science and Manufacturing*, 40: 321–328.

Tanpichai, S. Sampson, W.W. & Eichhorn, S.J. 2012. Stress-transfer in microfibrillated cellulose reinforced poly (lactic acid) composites using Raman spectroscopy. *Composites Part A-Applied Science Manufacturing*, 43: 1145–1152.

*Advanced Materials, Structures and Mechanical Engineering – Kaloop (Ed.)*
© 2016 Taylor & Francis Group, London, ISBN 978-1-138-02793-0

# Reactively modified poly(butylene succinate) as a compatibilizer for PBS composites

V. Tansiri & P. Potiyaraj
*Department of Materials Science, Faculty of Science, Chulalongkorn University, Bangkok, Thailand*

ABSTRACT: A compatibilizer of PBS composites, PBS-*g*-GMA, was prepared by using the reactive extrusion technique with a twin-screw extruder. This compatibilizer was used to enhance the interaction between PBS and organoclay in organoclay/PBS composites. The mechanical properties showed that the addition of PBS-*g*-GMA had a good influence on the tensile strength, elongation at break, and impact strength. Moreover, PBS composites with the addition of PBS-*g*-GMA had slightly increased flexural strength. Furthermore, incorporation of 5 phr PBS-*g*-GMA into the 0.5 phr organoclay/PBS composites showed an optimum increase in tensile strength, elongation at break and impact strength compared with neat PBS. The decomposition temperature decreased after the addition of organoclay and the presence of the compatibilizer had no role in enhancing the thermal stability of the PBS composites. The changes in these properties were explained by the morphological properties, which were illustrated by Scanning Electron Microscopy (SEM) of a tensile fractured surface of the PBS composites.

## 1 INTRODUCTION

In recent years, studies on biodegradable polymeric materials have increasingly attracted the attention of both researchers and industrial scientists worldwide. This is because the consumption of polymeric materials from petrochemical feedstock such as Polyethylene (PE) and Polypropylene (PP) has increased the plastic waste-disposal problems. Many aliphatic types of polyester can be degraded in compost and moist soils. Poly(butylene succinate) or PBS is one of the most commercially available aliphatic polyesters, which has many advantageous properties such as biodegradability, processability as well as chemical resistance (Calabia et al. 2013, Phua et al. 2013). PBS has excellent processability, it can be processed in the field of textiles into melt blow, multifilament, monofilament, nonwoven, and split yarn and also in the field of plastics into injection-molded products, thus being a promising polymer for various potential applications (Makhatha et al. 2007). Moreover, its mechanical properties resemble those of PE or PP (Fujimaki 1998). PBS is chemically synthesized by the polycondensation of 1,4-butanediol with succinic acid. Now, some companies can produce succinic acid from natural resources, thus making PBS more eco-friendly (Xu & Guo 2010). However, some of its properties such as softness and thermal properties for further processing restrict its extensive applications. To improve the properties of PBS, the addition of organoclay as a filler in PBS composites is also expected to assist in enhancing the limitation of these drawbacks (Phua et al. 2011).

Nevertheless, the compatibility between the matrix and the filler is very poor due to incompatible hydrophobic-hydrophilic species, which leads to undesirable properties of PBS composites. Therefore, the compatibilizer is the most crucial way that is widely used for further enhancing the properties of polymer composites. For example, Aggarwal et al. (2013) investigated the effect of a functional group of compatibilizer on the mechanical properties of PE composites, and reported that PE-*g*-GMA was better than PE-*g*-MA. Moreover, Liu et al. (2012) used the PLA-*g*-GMA copolymer as a compatibilizer in PLA/starch blends. Their results showed that the starch granules can be clearly observed without the compatibilizer. Conversely, when using PLA-*g*-GMA as a compatibilizer, the starch granules were better dispersed and covered with PLA.

The main objective of this research is to focus on the preparation of modified PBS, namely, PBS-*g*-GMA, and to study the properties of organoclay/PBS composites using PBS-*g*-GMA as a compatibilizer. The compatibilizing effects of PBS-*g*-GMA on the mechanical and thermal properties of organoclay/PBS composites were studied accordingly.

## 2 EXPERIMENTAL PROCEDURE

### 2.1 Materials

Commercial PBS granules (AZ71TN) for injection molding grade used as the matrix polymer were purchased from Mitsubishi Chemical, Thailand.

Commercial organoclay (Cloisite® 30B, Southern Clay Products, USA) was used as the filler. Dicumyl Peroxide (DCP) and Glycidyl Methacrylate (GMA) were purchased from Sigma Aldrich. Chloroform and acetone obtained from RCI Lab Scan, Thailand were used as solvents. The chemicals and solvents were used without further purification.

## 2.2 Preparation of PBS-g-GMA

PBS granules were dried at 60°C for 24 h. Drying is necessary to avoid the hydrolytic degradation of the polymers during melt processing in the extruder. Then, 90 wt% of PBS, 10 wt% of GMA and 1.5 phr of DCP were physically premixed. The mixture of PBS, GMA and initiator (DCP) was reacted into a twin-screw extruder (Thermo Prism, DSR-28, Germany) at a temperature range of 120–130°C and a screw speed of about 35 rpm. After completion of the grafting reaction, the compound was dried at 60°C for 24 h. To confirm the successful grafting, the purification process was obtained. The compound was refluxed in chloroform under $N_2$ gas for 2 h. Then, the hot solution was precipitated by cold acetone and washed several times for removing any unreacted reagents. Finally, the precipitated polymer was dried at 60°C for 24 h. The purified PBS-g-GMA was acquired.

## 2.3 Preparation of PBS composites

PBS, PBS-g-GMA and organoclay were dried in a vacuum oven at 60°C for 24 hr. The mixture of PBS, PBS-g-GMA and organoclay, as shown in Table 1, was physically premixed and melt-mixed in a twin-screw extruder (Thermo Prism, DSR-28, Germany) at the temperature range of 120–130°C and the screw speed of about 35 rpm. Then, the compound was dried at 60°C for 24 h. It was then processed in an injection molding machine (Battenfeld BA 250/050 CDC, Austria) at 135°C into various shapes for mechanical testing.

Table 1. Composition of PBS and organoclay/PBS composites.

| Formula | PBS (phr) | PBS-g-GMA (phr) | Organoclay (phr) |
|---|---|---|---|
| PBS | 100 | – | – |
| PBS/O0.5 | 100 | – | 0.5 |
| PBS/PgG5/O0.5 | 100 | 5 | 0.5 |
| PBS/O1 | 100 | – | 1 |
| PBS/PgG5/O1 | 100 | 5 | 1 |
| PBS/O2 | 100 | – | 2 |
| PBS/PgG5/O2 | 100 | 5 | 2 |
| PBS/O6 | 100 | – | 6 |
| PBS/PgG5/O6 | 100 | 5 | 6 |

## 2.4 Characterization

The grafting of PBS-g-GMA was characterized by Fourier Transform Infrared Spectroscopy (FTIR) (PerkinElmer, USA) at the ambient temperature. The purified sample was determined in ranges of 4000 to 400 cm$^{-1}$ with 64 scanning times.

## 2.5 Mechanical testing

Mechanical properties of PBS and PBS composites including tensile strength, Young's modulus and elongation at break were measured according to ASTM D638 using a universal testing machine (LLOYD, LR100K, England) equipped with a 10 kN load cell. Flexural strength was measured according to ASTM D790 using a universal testing machine (LLOYD, LR500, England) with a 2500 N load cell. In addition, the impact strength was measured using an impact pendulum tester (GOTECH, GT-7045-MD, Taiwan), according to ASTM D256, with a pendulum of 2.75 J. The results of the mechanical properties were the average values of at least five measurements taken for each sample. Oneway Analysis of Variance (ANOVA) was used to compare the mechanical properties.

## 2.6 Thermal testing

The thermal behavior of PBS and PBS composites was determined by using a thermogravimetric analyzer (Mettler Toledo, TGA/SDTA851, Switzerland) from 30 to 1000°C at the heating rate of 10°C/min under nitrogen atmosphere.

## 2.7 Morphological testing

The tensile fractured surface of PBS composites was examined using a Scanning Electron Microscope (SEM, Jeol JSM 6400, Japan) at an accelerating voltage of 15 kV. Before observation, the samples were sputter-coated with a thin layer of gold to avoid electrical charging during the examination.

## 3 RESULTS AND DISCUSSION

### 3.1 Characterization of modified PBS

The FTIR spectra of PBS and PBS-g-GMA are shown in Figure 1. It can be seen from the figure that two spectra existed showing C-H stretching at 3000–2850 cm$^{-1}$, C=O stretching at 1711 cm$^{-1}$, C-H bending (in-plane) at 1470–1000 cm$^{-1}$ and C-H bending (out-of-plane) at 1000–400 cm$^{-1}$. However, the FTIR spectrum of PBS-g-GMA showed different peaks at 1730 and 1152 cm$^{-1}$ assigned to C=O stretching and C-O stretching from the acrylate group, respectively. Moreover, the peaks at

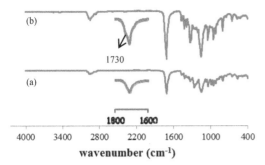

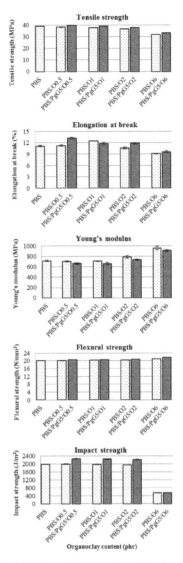

Figure 1. FTIR spectra of (a) PBS and (b) PBS-*g*-GMA.

1150 and 856 cm$^{-1}$ showed in the PBS-*g*-GMA spectrum indicating the presence of the epoxy group onto PBS (Sharif et al. 2013).

### 3.2 *Mechanical properties*

The epoxy groups in PBS-*g*-GMA interacted with the hydroxyl groups of organoclay, and the PBS chain of PBS-*g*-GMA had good compatibility with the PBS matrix. Therefore, PBS-*g*-GMA can be used as a compatibilizer to improve the interaction between PBS and organoclay (Xu et al. 2012). The mechanical properties of the PBS and PBS composites are shown in Figure 2. The tensile strength of PBS composites decreased with increasing organoclay content due to the agglomeration of organoclay in the PBS matrix, especially at 6 phr of organoclay. After using 5 phr of PBS-*g*-GMA as a compatibilizer, the tensile strength slightly increased (P < 0.05) due to the fact that PBS-*g*-GMA could improve the interaction between PBS and organoclay. The addition of PBS-*g*-GMA also increased the elongation at the break of PBS composites, except when the amount of organoclay was 1 phr. The impact strength increased significantly (P < 0.05) when using PBS-*g*-GMA compared with neat PBS. These results related to the better interaction between two phases; therefore, the composites can transmit stress from the PBS matrix to organoclay. However, the addition of 6 phr of organoclay in PBS composites still decreased the elongation at break and the impact strength along with the addition of PBS-*g*-GMA. This may be attributed to the agglomeration of organoclay. On the other hand, Young's modulus and flexural strength of PBS composites increased (P < 0.05) with increasing organoclay content because of the addition of the hard particles such as organoclay, which had a higher modulus than the PBS matrix (Chuai et al. 2011, Ray et al. 2005). The flexural strength of PBS composites also increased slightly (P < 0.05) when PBS-*g*-GMA was added.

Figure 2. Mechanical properties of PBS and PBS composites. (□ Without PBS-*g*-GMA, ▨ With PBS-*g*-GMA).

These results may be related to the good interaction between the two phases. Meanwhile, Young's modulus decreased (P < 0.05) when PBS-*g*-GMA was added to the PBS composites compared with those without the compatibilizer, which indicated that the addition of PBS-*g*-GMA may extend the movement of PBS into the PBS composites.

### 3.3 *Thermal property*

The thermal behavior of PBS and its composites is presented in Table 2. The decomposition

343

Table 2. Decomposition temperature of PBS and PBS composites.

| Formula | Td (onset) (°C) | Td (endset) (°C) | Residue (%) |
|---|---|---|---|
| PBS | 377.33 | 417.44 | 0.00 |
| PBS/O0.5 | 372.68 | 417.24 | 0.61 |
| PBS/PgG5/O0.5 | 372.69 | 416.81 | 0.45 |
| PBS/O1 | 367.35 | 413.52 | 1.30 |
| PBS/PgG5/O1 | 368.97 | 413.75 | 1.23 |
| PBS/O2 | 364.18 | 408.66 | 1.10 |
| PBS/PgG5/O2 | 364.25 | 409.70 | 1.35 |
| PBS/O6 | 351.82 | 390.89 | 3.90 |
| PBS/PgG5/O6 | 352.53 | 393.11 | 3.78 |

temperature can be considered through various points in the TGA curves. In this study, the onset and endset of the decomposition temperature were determined. The results from Table 2 indicated that the addition of organoclay into PBS lowered the $T_{onset}$ and $T_{endset}$ compared with neat PBS due to the thermal instability of the surfactant present in the organoclay. This means that the presence of organoclay has a significant effect on the thermal stability, as reported by several researchers (Cai et al. 2008, Cui et al. 2008). They reported that the tetraalkylammonium cations modified on the organoclay surface can be decomposed into an olefin and an amine, leaving an acidic proton on the organoclay surface. Therefore, this acid site is able to promote the degradation of polymers at high temperatures, resulting in the thermal instability of the composites. Furthermore, the incorporation of PBS-g-GMA as a compatibilizer into PBS composites did not present any significant change in the $T_{onset}$ and $T_{endset}$ compared with PBS composites. The residual mass of organoclay loading is larger than PBS. These results indicated that the addition of organoclay into PBS increased the residual mass of the PBS composites.

### 3.4 Morphological property

The tensile fractured surface of PBS composites is illustrated in Figure 3. The SEM micrograph of the 0.5 phr organoclay/PBS composite without the compatibilizer (Fig. 3(a)) shows that the organoclay in this composite agglomerated and was easily left-off from the PBS matrix due to the poor adhesion between the filler and the matrix. After the addition of PBS-g-GMA as a compatibilizer, a better adhesion between the two phases was acquired, as shown in Figure 3(b). This results implied that PBS-g-GMA can improve the compatibility between the organoclay and the PBS matrix.

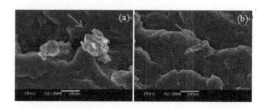

Figure 3. SEM micrographs of 0.5 phr organoclay/PBS composites (a) without PBS-g-GMA and (b) with PBS-g-GMA.

### 4 CONCLUSIONS

The modification of PBS by grafting GMA onto PBS was performed by using the reactive extrusion technique with a twin-screw extruder in order to utilize the obtained PBS-g-GMA as a compatibilizer. The grafting was confirmed by the FTIR analysis. The compatibilizer was then used for the preparation of organoclay/PBS composites. It was found that this compatibilizer slightly improved the tensile strength, elongation at break, and impact strength because of the better adhesion between the organoclay and the PBS matrix, as can be observed in the SEM micrographs. In this study, the optimum ratio of PBS-g-GMA and organoclay in PBS was 5 and 0.5 phr, respectively. In the case of thermal property, the decomposition temperature of the PBS composites was insignificantly enhanced by using PBS-g-GMA, but this property was mainly depended on the amount of the organoclay in the PBS composites.

ACKNOWLEDGMENT

This research was supported by the Ratchadaphiseksomphot Endowment Fund 2013 of Chulalongkorn University (CU-56-416-AM).

REFERENCES

Aggarwal, P.K. Chauhan, S. Raghu, N. Karmarkar, S. & Shashidhar, G. 2013. Mechanical properties of biofibers-reinforced high-density polyethylene composites: effect of coupling agents and bio-fillers. *Journal of Reinforced Plastics and Composites*, 32: 1722–1732.

Cai, Y. Wu, N. Wei, Q. Zhang, K. Xu, Q. Gao, W. Song, L. & Hu, Y. 2008. Structure, surface morphology, thermal and flammability characterizations of polyamide6/organic-modified Fe-montmorillonite nanocomposite fibers functionalized by sputter coating of silicon. *Surface and Coatings Technology*, 203: 264–270.

Calabia, B. Ninomiya, F. Yagi, H. Oishi, A. Taguhi, K. Kunioka, M. & Funabashi, M. 2013. Biodegradable Poly (butylene succinate) Composites Reinforced by Cotton Fiber with Silane Coupling Agent. *Polymers*, 5: 128–141.

Chuai, C.Z. Zhao, N. Li, S. & Sun, B.X. 2011. Study on PLA/PBS Blends. *Advanced Materials Research*, 197–198: 1149–1152.

Cui, L. Khramov, D.M. Bielawski, C.W. Hunter, D.L. Yoon, P.J. & Paul, D.R. 2008. Effect of organoclay purity and degradation on nanocomposite performance, Part 1: Surfactant degradation. *Polymer*, 49: 3751–3761.

Fujimaki, T. 1998. Processability and properties of aliphatic polyesters, 'BIONOLLE', synthesized by polycondensation reaction. *Polymer Degradation and Stability*, 59: 209–214.

Liu, J. Jiang, H. & Chen, L. 2012. Grafting of Glycidyl Methacrylate onto Poly (lactide) and Properties of PLA/Starch Blends Compatibilized by the Grafted Copolymer. *Journal of Polymers and the Environment*, 20: 810–816.

Makhatha, M.E. Ray, S.S. Hato, J. Luyt, A.S. & Bousmina, M. 2007. Thermal and Thermomechanical Properties of Poly (butylene succinate) Nanocomposites. *Journal of Nanoscience and Nanotechnology*, 8(5): 1–11.

Phua, Y.J. Chow, W.S. & Mohd Ishak, Z.A. 2011. Poly (butylene succinate) Organo-montmorillonite Nanocomposites: Effects of the Organoclay Content on Mechanical, Thermal, and Moisture Absorption Properties. *Journal of Thermoplastic Composite Materials*, 24: 133–151.

Phua, Y.J. Chow, W.S. & Mohd Ishak, Z.A. 2013. Reactive processing of maleic anhydride-grafted poly (butylene succinate) and the compatibilizing effect on poly (butylene succinate) nanocomposites. *Express Polymer Letters*, 7: 340–354.

Ray, S.S. Bousmina, M. & Okamoto, K. 2005. Structure and Properties of Nanocomposites Based on Poly (butylene succinate-co-adipate) and Organically Modified Montmorillonite. *Macromolecular Materials and Engineering*, 290: 759–768.

Sharif, J. Mohamad, S.F. Fatimah Othman, N.A. Bakaruddin, N.A. Osman, H.N. & Guven, O. 2013. Graft copolymerization of glycidyl methacrylate onto delignified kenaf fibers through pre-irradiation technique. *Radiation Physics and Chemistry*, 91: 125–131.

Xu, J. & Guo, B.-H. 2010. Microbial Succinic Acid, Its Polymer Poly (butylene succinate), and Applications. *Microbiology Monographs*, 14: 347–388.

Xu, T. Tang, Z. & Zhu, J. 2012. Synthesis of polylactide-graft-glycidyl methacrylate graft copolymer and its application as a coupling agent in polylactide/bamboo flour biocomposites. *Journal of Applied Polymer Science*, 125: E622–E627.

*Advanced Materials, Structures and Mechanical Engineering – Kaloop (Ed.)*
© 2016 Taylor & Francis Group, London, ISBN 978-1-138-02793-0

# Preparation and characterization of Ag–Bi$_{1.5}$Y$_{0.3}$Sm$_{0.2}$O$_3$ composite cathode

C.H. Hua & C.C. Chou
*Department of Mechanical Engineering, National Taiwan University of Science and Technology, Taipei, Taiwan*

ABSTRACT: The electrochemical performance of Ag mixed with Bi$_{1.5}$Y$_{0.3}$Sm$_{0.2}$O$_3$ or Y$_{0.5}$Bi$_{1.5}$O$_3$ composite cathodes on a Ce$_{0.78}$Gd$_{0.2}$Sr$_{0.02}$O$_{2-\delta}$ (GDCS) electrolyte was estimated through electrochemical impedance spectroscopy. The cathode mixtures consisting of Ag–Bi$_{1.5}$Y$_{0.3}$Sm$_{0.2}$O$_3$ or Ag–Y$_{0.5}$Bi$_{1.5}$O$_3$ exhibited lower overpotentials and higher exchange current densities than commercial Ag paste and perovskite cathodes. Favorable oxygen reduction reaction properties and low activation energies were observed when the cell was operated at temperatures near 600 °C, which is the ideal operating temperature for the energy integration of solid oxide fuel cells as part of in next-generation triple combined cycle power plants.

## 1 INTRODUCTION

### 1.1 Design of a triple combined cycle power plant

Solid Oxide Fuel Cells (SOFCs) are promising energy conversion devices for electricity generation because of their high efficiency and environmental friendliness. Recently, a leading advanced gas turbine manufacturer announced the development of a triple combined cycle power plant with a power generation efficiency over 70% (Choi et al. 2014). In this concept, the exhaust flue gas of the turbine was fed into the heat exchanger for increasing the temperature of compressed air, which promotes the kinetics of the SOFC electrochemical reactions (Jia et al. 2013). Therefore, the ideal operating temperature for the energy integration of SOFCs in next-generation triple combined cycle systems in the near future should be close to the exhaust temperature, which is currently ≤ 650 °C for the most advanced large-scale J- and H-class gas turbines.

### 1.2 Material selection for intermediate temperature SOFC

Significant research efforts have been a focus on the development of Intermediate-Temperature (IT; 500–800 °C) SOFCs. Operating SOFCs at an intermediate temperature not only reduces the degradation of fuel cell components but also prolongs the lifetime of the fuel cell systems. In addition, it broadens the selection of potential cathode materials. However, reducing the operating temperature reduces both electrolyte conductivity and cathode kinetics thus cell performance decreases dramatically at reduced temperature. Using an alternative high-ionic-conductivity material reduces the SOFC working temperature to

lower than 800 °C (Leng & Chan 2006). The ionic conductivity of gadolinia-doped ceria is about one order of magnitude larger than that of Yttria-Stabilized Zirconia (YSZ); it reaches 0.1 S cm$^{-1}$ at 800 °C (Inaba & Tagawa 1996).

Because of their excellent mechanical properties and toughness, doped ceria ceramics has been considered promising electrolytes for superseding YSZ (Reddy & Karan 2005). In the operation temperature range of IT-SOFCs, ceria reduction can be neglected at low temperatures of 600–700 °C. Moreover, with a Vickers hardness of 8.27 GPa and a fracture toughness of 2.49 MPa m$^{0.5}$, Ce$_{0.78}$Gd$_{0.2}$Sr$_{0.02}$O$_{2-\delta}$ (GDCS) is suitable for flat plate SOFC stack design (Dudek 2008). New cathode materials are commonly developed by adding an ionic-conducting second phase to electrically conducting cathode material. Using a mixed ionic and electronic conducting material appears to be an effective way in extending the Triple Phase Boundary (TPB) for increasing electronic and ionic conductivities as well as high catalytic activities for the Oxygen Reduction Reaction (ORR). The ceramic conducts the oxygen ions in the electrolyte and the cathode. Stabilized Bi$_2$O$_3$ is an ideal alternative ionic-conducting cathode material because the high oxygen ionic conductivity of δ–phase Bi$_2$O$_3$ is stable from 730 °C until Bi$_2$O$_3$ melts at 825 °C. Furthermore, silver has a high catalytic activity and high electrical conductivity and has been investigated as a low-temperature ionic-conducting cathode (Simner et al. 2005, Sarikaya et al. 2012, Muranaka et al. 2009). Ag–Y$_{0.5}$Bi$_{1.5}$O$_3$ (Ag–YB) cathodes have been reported to exhibit high performance on gadolinia-stabilized ceria electrolytes (Mosiałek et al. 2014) and long-term stability tests indicated that Ag–YB cathodes can be stably operated at temperatures lower than 700 °C

(Wang & Barnett 1992). Moreover, such a composite cathode has a low sintering temperature and thus the production cost can be further reduced (Xia et al. 2003).

### 1.3 *Structural modification of co-stabilized bismuth oxide*

This study examined the possibility of improving the ionic conductivity and electrochemical properties of a porous composite cathode. Among the numerous ionic–conducting materials, $Y_2O_3$ and $Sm_2O_3$ co–doped $Bi_2O_3$ (BYS) are highly chemically stable in highly reducing environments; in addition, they exhibit high oxygen ionic conductivity (Yang et al. 2011). Materials co–stabilized with $Y_2O_3$ trivalent cations, such as $Sm^{3+}$, exhibit higher ionic conductivity than does single-doped bismuth oxide (Kim & Lin 2000). By partial substituting $Y^{3+}$ with $Sm^{3+}$ in solid– solution $Bi_{1.5}Y_{0.3}Sm_{0.2}O_3$, co–stabilized bismuth cathodes in IT-SOFCs yielded high ORR and current densities.

## 2 EXPERIMENTAL

### 2.1 *Specimen preparation*

BYS and YB powders were prepared by calcinating the original powders of $Bi_2O_3$, $Sm_2O_3$, and $Y_2O_3$ (Nippon Yttrium Co., Ltd) at 750 °C for 3 h through oxide–mixing method. After wet ball–milling, the YB particle size decreased from several microns to the submicrometer range; 40 vol% of calcined $Bi_{1.5}Y_{0.3}Sm_{0.2}O_3$ and 40 vol% of calcined $Y_{0.5}Bi_{1.5}O_3$ were separately mixed with an appropriate amount 60 vol% of Ag powders (Gredmann, Taiwan); hereafter referred to as Ag–BYS and Ag–YB, respectively. Terpineol solvent and ethyl cellulose were added to fabricate Ag–BYS and Ag–YB composite cathode pastes; subsequently, the pastes were screen-printed on both sides of GDCS electrolyte pellets (approximately 0.6 mm) and fired at 670 °C for 1 h. Dense GDCS discs were obtained by sintering the pellets at 1500 °C for 10 h. Adding Sr promotes densification; therefore, the relative densities of the sintered electrolytes exceeded 98% of the theoretical density. The electrochemical properties of the fabricated Ag composite cathode (ca. 10 μm)/ GDCS (electrolyte)/Ag composite cathode (ca. 10 μm) symmetrical cells were observed through electrochemical impedance spectroscopy (EIS; Solartron 1260–1287 AC impedance analyzer). The total ionic conductivity of the cells was studied through two–probe AC impedance spectroscopy. Data collection for Cyclic Voltammetry (CV) and Tafel measurements was performed using Corrware software at a scanning rate of 25 mV/s to obtain exchange current density ($i_0$) data.

## 3 RESULTS AND DISCUSSION

### 3.1 *EIS analysis*

The performance of an SOFC cathode must be evaluated in terms of its conductivity, catalytic activity, adequate porosity for gas transport, compatibility with the electrolyte, and stability. The equivalent circuit used in this study and a typical example of the impedance spectrum obtained at 650 °C in the air are presented in Figure 1. The real axis ($Z' = 0$) intercept at high frequencies represents the resistance of the sandwich electrolyte ($R_\Omega$). The low-frequency real-axis intercept corresponds to the cathode polarization resistance $R_p$ and equals $R_H + R_L$ (Lu & Shen 2014). The low-frequency semicircle ($R_L$) in the impedance spectrum is attributed to oxygen adsorption and dissociation and its surface diffusion, whereas the high-frequency semicircle ($R_H$) corresponds to oxygen ion incorporation and transport in the cathode (Adler 2000). Cathode Area Specific Resistance (ASR) used in this study was calculated using Equation 1:

$$ASR = (R_p * A)/2 \qquad (1)$$

where, A is the cathode area; the factor '2' is applied because each cell contains two identical cathodes.

Adding BYS powder to silver created a higher concentration of oxygen vacancies and assisted electrochemical oxygen reduction through the catalytic promotion of the dissociation and surface diffusion of oxygen species on the cathode to the TPB, thereby significantly improving the performance of the Ag–BYS cathode.

At 600 °C, the ASR for the porous Ag–BYS and Ag–YB cathodes were 0.36 and 0.49 Ω cm², respectively, which were more than one order of magnitude lower than that of the traditional Lanthanum Strontium Manganite (LSM) cathodes on YSZ electrolytes (Huang et al. 2005).

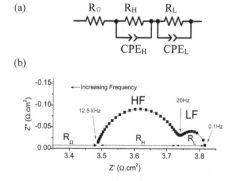

Figure 1. (a) Equivalent circuit used for fitting the EIS spectra. (b) Typical impedance spectrum for the Ag–BYS cathode at 650 °C.

Temperature dependences of the composite cathodes' polarization resistance were evaluated and are illustrated in Figure 2. The ASR of the Ag–BYS and Ag–YB cathodes at 650 °C were as low as 0.17 and 0.2 Ω cm², respectively; this performance is higher than that of perovskite cathode materials w/o expensive precious metals. For example, the overall cathode polarization resistances of LSCF and LSCF + Pd cathodes at 650 °C are 1.7 and 0.8 Ω cm², respectively (Chen et al. 2009).

## 3.2 Activation energy and kinetic for ORR

The composite cathode structure comprising micrometer-scale Ag and smaller BYS was expected to benefit from the catalytic activity and electrical conductivity of Ag, while the BYS ceramic phase prevented the silver from densifying, resulting in retention of the porosity required for superior cathode performance. Porosity is essential for effective oxygen transport to the electrochemical reaction sites throughout the entire cathode.

Furthermore, the ceramic phase of the BYS increased the stability of the Ag backbone structure and lengthened the TPB for improving the thermal compatibility and adhesion between the cathode and electrolyte. The lower ASR of the Ag–BYS cathode is likely due to the high chemical compatibility between the electrolyte and cathode. Cathodes generally exhibit low ASR when electrolytes with higher conductivity are employed. The apparent activation energies (Ea) for the Ag–BYS and Ag–YB cathode ionic conductivities are shown in Figure 3. The activation energies of the cathodes ranged from 1.11 to 1.31 eV for Ag–BYS and Ag–YB, which were lower than previously reported for pure Ag (1.42 eV; Wu & Liu 1998).

CV and Tafel analyses were performed for all samples. The exchange current density was obtained from the intercept on the logarithmic current density axis. A higher value of the slope of the

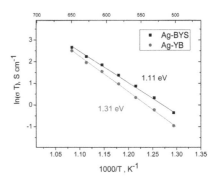

Figure 3. Arrhenius plots of the ionic conductivity as a function of temperatures for the porous Ag–BYS and Ag–YB composite cathodes, sintered at 670 °C for 1 h.

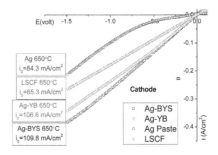

Figure 4. Typical cyclic voltammograms of the ORR and exchange current densities for various cathodes on GDCS electrolyte at 650 °C.

Tafel curve indicates a higher current density and thus a higher oxygen reduction rate; $i_0$ increased from 21 to 109.8 mA cm⁻² as the cell temperature of Ag–BYS cathode was increased from 500 °C to 650 °C. Ag–BYS exhibited the highest exchange current density at 650 °C (Fig. 4). BYS ceramics mixed with Ag enhanced the oxygen-ion conductivity and catalytic activity for oxygen reduction.

## 3.3 Composite cathode microstructures analysis

As shown in Figure 5, the average particle size was approximately 0.2 μm and 1–3 μm for bismuth oxide and silver powders, respectively. Adding bismuth oxide changed the pore profile and the porosity to approximately 25%. The contact between cathode and electrolyte was increased by adding the fine bismuth oxide powder. The continual bismuth oxide ceramic phase enhanced the bonding and TPB between cathode and electrolyte.

The percentage theoretical density and shrinkage of BYS and YB pellets are shown as a function of the sintering temperature in Figure 6. The bulk density of YB was higher than that of BYS at same sintering temperature, implying that a higher

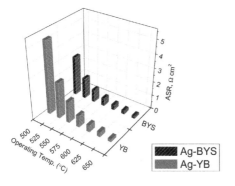

Figure 2. Composite cathode ASR of different composite cathodes at various operating temperatures (550–650 °C).

Figure 5. Surface (a) and Cross–sectional (b) scanning electron microscope images of the porous Ag–BYS cathode.

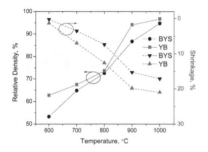

Figure 6. Relative density versus sintering temperature for bulk YB and BYS. All specimens were sintered for 1 h.

sintering temperature was necessary to densify BYS. Thus, BYS has a higher chemical stability at high temperatures.

## 4 CONCLUSIONS

The effect of adding bismuth oxides to silver for use as composite cathodes was studied. A porous microstructure was obtained through screen-printing at a low sintering temperature (670 °C). Incorporating co–stabilized bismuth oxide (BYS) particles in the silver paste prevented densification and led to porous structures with enhanced electrochemical properties. The ionic conductivity of the composite cathode was improved for IT-SOFCs. Ag–BYS exhibited a high exchange current density (109.8 mA cm$^{-2}$) and the ASR was as low as 0.17 Ω cm$^2$ at 650 °C, these characteristics benefit to the IT-SOFC stack design.

## REFERENCES

Adler, S.B. 2000. Limitations of charge-transfer models for mixed-conducting oxygen cathodes. *Solid State Ionics*, 135(1–4): 603–612.

Chen, J. Liang, F. Chi, B. Pu, J. Jiang, S.P. & Jian, L. 2009. Palladium and ceria infiltrated La$_{0.8}$Sr$_{0.2}$Co$_{0.5}$Fe$_{0.5}$O$_{3-\delta}$ cathodes of solid oxide fuel cells. *Journal of Power Sources*, 194(1): 275–280.

Choi, J.H. Ahn, J.H. & Kim, T.S. 2014. Performance of a triple power generation cycle combining gas/steam turbine combined cycle and solid oxide fuel cell and the influence of carbon capture. *Applied Thermal Engineering*, 71(1): 301–309.

Dudek, M. 2008. Ceramic oxide electrolytes based on CeO$_2$—Preparation, properties and possibility of application to electrochemical devices. *Journal of the European Ceramic Society*, 28(5): 965–971.

Huang, Y. Vohs, J.M. & Gorte, R.J. 2005. Characterization of LSM-YSZ Composites Prepared by Impregnation Methods. *Journal of The Electrochemical Society*, 152(7): A1347–A1353.

Inaba, H. & Tagawa, H. 1996. Ceria-based solid electrolytes. *Solid State Ionics*, 83(1–2): 1–16.

Jia, Z. Sun, J. Oh, S.R. Dobbs, H. & King, J. 2013. Control of the dual mode operation of generator/motor in SOFC/GT-based APU for extended dynamic capabilities. *Journal of Power Sources*, 235(0): 172–180.

Kim, J. & Lin, Y.S. 2000. Synthesis and oxygen permeation properties of ceramic-metal dual-phase membranes. *Journal of Membrane Science*, 167(1): 123–133.

Leng, Y.J. & Chan, S.H. 2006. Anode-Supported SOFCs with Y$_2$O$_3$-Doped Bi$_2$O$_3$/Gd$_2$O$_3$-Doped CeO$_2$ Composite Electrolyte Film. *Electrochemical and Solid-State Letters*, 9(2): A56–A59.

Lu, K. & Shen, F. 2014. Long term behaviors of La$_{0.8}$Sr$_{0.2}$MnO$_3$ and La$_{0.6}$Sr$_{0.4}$Co$_{0.2}$Fe$_{0.8}$O$_3$ as cathodes for solid oxide fuel cells. *International Journal of Hydrogen Energy*, 39(15): 7963–7971.

Mosiałek, M. Nowak, P. Dudek, M. & Mordarski, G. 2014. Oxygen reduction at the Ag|Gd$_{0.2}$Ce$_{0.8}$O$_{1.9}$ interface studied by electrochemical impedance spectroscopy and cyclic voltammetry at the silver point cathode. *Electrochimica Acta*, 120(0): 248–257.

Muranaka, M. Sasaki, K. Suzuki, A. & Terai, T. 2009. LSCF–Ag Cermet Cathode for Intermediate Temperature Solid Oxide Fuel Cells. *Journal of The Electrochemical Society*, 156(6): B743–B747.

Reddy, K. & Karan, K. 2005. Sinterability, Mechanical, Microstructural, and Electrical Properties of Gadolinium-Doped Ceria Electrolyte for Low-Temperature Solid Oxide Fuel Cells. *Journal of Electroceramics*, 15(1): 45–56.

Sarikaya, A. Petrovsky, V. & Dogan, F. 2012. Silver Based Perovskite Nanocomposites as Combined Cathode and Current Collector Layers for Solid Oxide Fuel Cells. *Journal of The Electrochemical Society*, 159(11): F665–F669.

Simner, S.P. Anderson, M.D. Pederson, L.R. & Stevenson, J.W. 2005. Performance Variability of La(Sr)FeO$_3$ SOFC Cathode with Pt, Ag, and Au Current Collectors. *Journal of The Electrochemical Society*, 152(9): A1851–A1859.

Wang, L.S. & Barnett, S.A. 1992. Lowering the Air-Cathode Interfacial Resistance in Medium-Temperature Solid Oxide Fuel Cells. *Journal of The Electrochemical Society*, 139(10): L89–L91.

Wu, Z. & Liu, M. 1998. Ag-Bi$_{1.5}$Y$_{0.5}$O$_3$ Composite Cathode Materials for BaCe$_{0.8}$Gd$_{0.2}$O$_3$-Based Solid Oxide Fuel Cells. *Journal of the American Ceramic Society*, 81(5): 1215–1220.

Xia, C. Zhang, Y. & Liu, M. 2003. Composite cathode based on yttria stabilized bismuth oxide for low-temperature solid oxide fuel cells. *Applied Physics Letters*, 82(6): 901–903.

Yang, C. Xu, Q. Liu, C. Liu, J. Chen, C. & Liu, W. 2011. Bi$_{1.5}$Y$_{0.3}$Sm$_{0.2}$O$_3$–La$_{0.8}$Sr$_{0.2}$MnO$_{3-\delta}$ dual-phase composite hollow fiber membrane for oxygen separation. *Materials Letters*, 65(23–24): 3365–3367.

*Advanced Materials, Structures and Mechanical Engineering – Kaloop (Ed.)*
© *2016 Taylor & Francis Group, London, ISBN 978-1-138-02793-0*

# Formation of silver nanowires in ethylene glycol for transparent conducting electrodes

N. De Guzman & M.D. Balela

*Department of Mining, Metallurgical and Materials Engineering, University of the Philippines, Diliman, Quezon City, Philippines*

ABSTRACT:   Silver Nanowires (AgNWs) with diameter of about 85 nm and lengths of up to 30 μm were successfully synthesized by electroless deposition in ethylene glycol at 160°C. Polyvinyl Pyrrolidone (PVP) was used as a surfactant and the structure-directing agent. The effect of increasing PVP concentration on the morphology of the nanowires was investigated. An increase in the yield and aspect ratio of Ag nanowires was observed at high PVP concentration. Ag ink in ethanol was spin coated on glass substrates at 2000 rpm for 40 s to fabricate transparent conductors. Annealing was performed to remove the PVP layer at a surface of the wires. Four-point probe and UV-Vis analysis of the AgNW electrodes indicated a sheet resistance of as low as 3 ohm/sq at a transmittance of 65%.

## 1 INTRODUCTION

To sustain the growth of high quality yet affordable electronic products, such as flat-screen displays, smart phones, tablets and other advanced electronic devices, various approaches are being explored to reduce their manufacturing cost (Wu et al. 2013, Leterrier et al. 2004, Song et al. 2013). One means is by replacing their films of Indium Tin Oxide (ITO) as the material of choice of the transparent conductors (Leterrier et al. 2004). Though ITO has the ability to be effective due to its good combination of electrical and optical transparency, its high production cost and brittleness limit its applicability in future devices (Song et al. 2013). Thus, alternative materials, including metal nanowires (Wu et al. 2013), graphene (Song et al. 2013), carbon nanotubes (Xu & Zhu 2012) and conducting polymers (Leterrier et al. 2004) are currently being investigated to determine their potential as replacement for ITO. Recent studies have shown that conductors made from Ag nanowires exhibit comparable properties to that of ITO. Though Ag is a precious metal, its noble properties make the production of Ag nanowires in solution relatively easy and economical. A number of methods have been developed to produce metal nanowires including template wetting, liquid phase synthesis and polyol synthesis (Mao et al. 2012). Among these different methods, polyol method is the most widely used because it is simple, low cost and large-scale uniform nanowires with high aspect ratios can be produced (Sun & Xia 2002). The structure-directing agent, plays a great role in the formation of metal nanowires. Varying the concentration and the parameters may result in the formation of different nanostructures such as nanoparticles, nanorods, nanocubes and nanowires (Lin et al. 2013).

In this study, Ag nanowires were successfully prepared by electroless deposition in ethylene glycol at 160°C. The effect of PVP concentration on the dimensions of the nanowires was investigated. The Ag nanowires dispersed in ethanol were then spin coated on glass substrates to fabricate the transparent electrodes.

## 2 EXPERIMENTAL

### 2.1 *Synthesis of Ag nanowires*

Silver Nitrate (AgNO3) was purchased from RTC. Ethylene glycol, PVP (Molecular weight = 55,000), and ethanol were purchased from Sigma-Aldrich. All the reagents were used as received.

In a typical synthesis, 10 mL of EG in a three-neck flask was heated to 160°C for 1h in an oil bath with reflux. Then, 3 mL of 0.1 M AgNO3-EG solution and 3 mL of 0.1 M PVP-EG solution were simultaneously injected drop-wise into the hot ethylene glycol solution for 10 min. The concentration of PVP added in the solution was varied. The total solution was then stirred at a rate of 260 rpm for 1 hour. When the reaction was finished, the Ag nanowires were washed with ethanol and centrifuged at 3000 rpm for 20 min. The Ag nanowires were washed three times to remove excess organic contents such as EG and PVP.

## 2.2 Fabrication of glass electrodes

For the transparent conductors, 1 cm x 1 cm glass slides were used as substrates. The glass substrates were sonicated in acetone to remove the grease and oil. Then, the slides were sonicated with ethanol to remove the excess acetone. The substrates were dipped in 30% nitric acid to etch the surface then washed with deionized water. Nitrogen gas ($N_2$) was used to dry the substrates. Ag nanowires dispersed in ethanol was diluted to a concentration of 3 mg/ml. 20 µl of the AgNW solution was spin-coated on a glass substrate at a rate of 2000 rpm for 40 s. Further, annealing of the coated substrates was done in a tube furnace at a temperature of 200°C for 1h to remove the organic material (PVP) surrounding the nanowires.

## 2.3 Characterization

The morphology of the AgNWs was examined using a Scanning Electron Microscope (SEM, JEOL 5300). The structure was determined by Shimadzu X-Ray Diffractometer (XRD-7000) with Cu Kα radiation in the range of 30–80°. The sheet resistance and visible transmittance of AgNWs coating on the glass substrates were estimated using the four-point probe and UV-Vis spectrophotometer respectively.

## 3 RESULTS AND DISCUSSION

At a PVP to Ag molar ratio of 1, mostly spherical Ag nanoparticles were present in the solution. The Ag nanoparticles have a mean diameter of about 260 nm as seen Figure 1a. When the molar ratio is increased to 2, both Ag nanoparticles and nanowires were formed in the solution. It can also be observed that the diameter of the nanoparticles became smaller. Further increase in the PVP to Ag molar ratio led to the growth of high aspect ratio nanowires. The thinnest and longest nanowires reached a diameter of about 105 nm and a length up to 18 µm. Interestingly, Ag nanoparticles were no longer present in the solution. At a PVP to Ag molar ratio of 4, the highest aspect ratio of Ag nanowires was achieved with an average length of 30 µm while having an average diameter of 85 nm. This can be attributed to the capping effect of PVP on Ag resulting to an anisotropic growth (Yang et al. 2011, Zhu et al. 2011, Sun et al. 2003).

At about 160 °C, ethylene glycol is oxidized to glycoldehyde and water. The aldehyde then acts as a reducing agent for Ag (I) ions to form minute Ag nanoparticles (Mao et al. 2012). These primary Ag nanoparticles serve as seeds for the growth of nanowires (Jun & Zhu 2011). PVP then acts as a

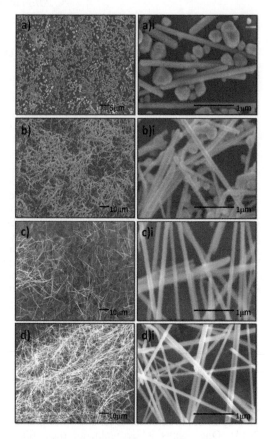

Figure 1. Shows the SEM images of the silver nanostructures synthesized at increasing molar concentrations of PVP (MW = 55,000).

surfactant and prevents the agglomeration of the primary nanoparticles. At the same time, it promotes the unidirectional growth of Ag nanowires. It may have capped the {100} facets of the Ag seeds preventing the deposition of Ag atoms on these higher energy planes and further allowed the growth on the lower energy {111} planes since it has the highest atomic density on FCC lattice (Sun et al. 2002, Kirchmeyer & Reuter 2005, Yang et al. 2011, Zhu et al. 2011, Sun et al. 2003). An increase in PVP concentration may have increased the coverage of PVP on {100} planes which promotes the anisotropic growth of the nanowires (Sun et al. 2003). Consequently, longer and thinner nanowires are formed.

Figure 2 shows the typical XRD pattern of the Ag nanowires prepared in the solution. The peaks at 38.32°, 43.96°, 64.65°, and 77.76° are peaks of the 111, 200, 220 and 311 face-centered cubic (fcc) Ag. (Mao et al. 2012) No peaks related to $AgNO_3$ are present in the diffraction pattern, suggesting

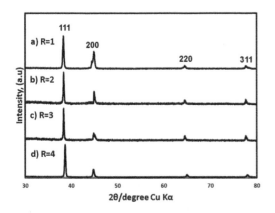

Figure 2. XRD pattern of as-synthesized Ag nanowires.

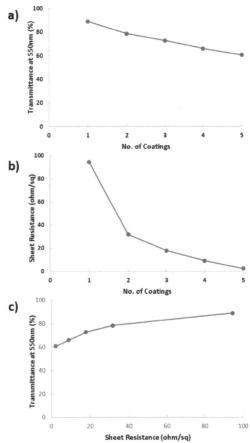

Figure 3. Plots of a) visible transmittance, b) sheet resistance and c) sheet resistance vs transmittance of glass electrodes after annealing.

that almost all Ag (I) are reduced to metallic Ag. As observed from the XRD patterns, there was a change in the intensity ratio between the (111) and (200) planes from 1.88 to 3.95 when the PVP to Ag molar ratio was increased from 1 to 4. This can be attributed to the increase in the yield of high aspect ratio silver nanowires (Lin et al. 2013, Sun & Xia 2002).

Ag nanowires with lengths of up to 30 μm at an average diameter of 85 nm were evenly deposited on glass substrates by spin coating. Spin coating parameters were optimized to produce the most even coating. Lower speeds produced coffee rings on the surface of the glass while fewer nanowires were deposited at higher speeds (Sun et al. 2003). The optimum spin-coating parameters were found to be at 2000 rpm for 40 s. Initial conductivity tests confirmed that Ag nanowires on glass without annealing exhibited a sheet resistance greater than $10^4$ ohm/sq even at a transmittance of less than 50% at 550 nm. The PVP organic layer on the surface of the wires possibly hinders the passage of current between the Ag nanowire junctions resulting to a high sheet resistance value (Dinh et al. 2013).

Shown in Figure 3 is the effect of nanowire density on the sheet resistance and optical transparency at 550 nm for annealed Ag nanowire electrodes. At 1 coating of Ag nanowire solution, the Ag electrode showed a transparency of about 90% with a sheet resistance of almost 100 ohm/sq. An increase in the amount of Ag nanowires on the electrodes led to a decrease in both the sheet resistance and transmittance as seen in Figure 3(a-b). Increasing the density of Ag nanowires in a random network possibly leads to more junctions and improves the connectivity between the wires. However, this also decreases the holes between the wires which lessens the amount of light that passes through the substrate (Dinh et al. 2013). As

a result, the optical transmission of the electrodes is reduced. The trend of the graph in Figure 3c showed the correlation between the sheet resistance and optical transmittance values at increasing nanowire density. The lowest sheet resistance of about 3 ohm/sq was achieved after 5 Ag nanowire coating layers. However, the transmittance was decreased to about 65%.

## 4 CONCLUSIONS

In summary, high aspect ratio Ag nanowires were successfully synthesized by electroless deposition in ethylene glycol with the aid of PVP as a structure-directing agent. Increasing the concentration of PVP in the solution produced longer and thinner nanowires at a high yield. XRD analysis showed that the ratio between the intensity of the 111 and 200 Ag peaks increases with increasing silver nanowire yield and aspect ratio. Transparent conducting electrodes were fabricated by spin coating Ag nanowire in ethanol solution on glass substrates. Annealing the electrodes at 200 °C for 1h drastically improved the conductivity. The results showed that the fabricated electrodes with 1 layer of Silver nanowire coating achieved a transmittance of 90% at 550 nm wavelength while having a sheet resistance of 95 ohms/sq. On the other hand, 5 layers of coating resulted to a large decrease in sheet resistance with a value of 3 ohms/sq at a transmittance of 65%. Indeed, silver nanowires have a great potential to replace ITO as the material of choice for electronic devices and flexible electrodes.

## ACKNOWLEDGEMENT

This study is supported by the Department of Science and Technology—Philippine Council for Industry, Energy and Emerging Technology Research and Development (DOST-PCIEERD).

## REFERENCES

Dinh, D. et al. 2013. Silver nanowires: a promising transparent conducting electrode material for optoelectronic and electronic applications. *Advance Science and Engineering,* 2: 1–22.

Kirchmeyer, S. & Reuter, K. 2005. Scientific Importance, properties and growing applications of poly (3,4 ethylenedioxythiophene). *Journal of Materials Chemistry,* 15: 2077–2088.

Leterrier, Y. et al. 2004. Mechanical integrity of transparent conductive oxide films for flexible polymer-based displays. *Thin Solid Films,* 460: 156–166.

Lin, J. Hsueh, Y. & Huang, J. 2013. The concentration effect of capping agent for synthesis of silver nanowire by using the polyol method, *Journal of Solid State Chemistry,* 214: 2–6.

Mao, H. Feng, J. & Ma, X. 2012. One dimensional silver nanowires synthesized by self-seeding polyol process, *Journal of Nanoparticles Research,* 14: 887–902.

Song, M. et al. 2013. Highly Efficient and Bendable Organic Solar Cells with Solution-Processed Silver Nanowire Electrodes. *Advance Functional Materials,* 23: 4177–4184.

Sun, Y. & Xia, Y. 2002. Large-scale synthesis of uniform silver nanowires through a soft, self-seeding, polyol process, *Advanced Materials,* 14: 833–837.

Sun, Y. Gates, B. Mayers, B. & Xia, Y. 2002. Crystalline Silver Nanowires by Soft Solution Processing, *Nano Letters,* 2(2): 165–168.

Sun, Y. Mayers, B. Herricks, T. & Xia, Y. 2003. Polyol synthesis of uniform silver nanowires: a plausible growth mechanism and the supporting evidence, *Nano Letters,* 3: 955–960.

Wu, H. et al. 2013. A transparent electrode based on a metal nanothrough network. *Nature Nanotech,* 8: 421–425.

Wu, Z. et al. 2004. Transparent, conductive carbon nanotube films. *Science,* 305: 1273–1276.

Xu, F. & Zhu, Y. 2012. Highly Conductive and Stretchable Silver Nanowire Conductors. *Advance Materials,* 24: 5117–5122.

Yang, C. Gu, H. Lin, W. Yuen, M. Wong, C.P. Xiong, M. & Gao, B. 2011. Silver nanowires: from scalable synthesis to recyclable foldable electronics. *Advanced materials,* 27: 3052–3056.

Zhu, J. Kan, C. Wan, J.G. Han, M. & Wang, G.H. 2011. High-yield synthesis of uniform Ag nanowires with high aspect ratios by introducing the long-chain PVP in an improved polyol process, *Journal of Nanomaterials,* 2011(2011): 982547.

*Advanced Materials, Structures and Mechanical Engineering – Kaloop (Ed.)*
© 2016 Taylor & Francis Group, London, ISBN 978-1-138-02793-0

# Study on the wave-truss composite bridge design

Y. Li
*Shenzhen Bridge Doctor Design and Research Institute Co. Ltd., Shenzhen, P.R. China*

C.L. Zhu
*Shenzhen Bridge Doctor Design and Research Institute Co. Ltd., Shenzhen, P.R. China*
*Shenzhen Graduate School, Harbin Institute of Technology, Shenzhen, P.R. China*

X.X. Zha
*Shenzhen Graduate School, Harbin Institute of Technology, Shenzhen, P.R. China*

ABSTRACT: In the engineering background of the Hengyang Qingjiang Bridge, by adjusting the arch axis, the bridge is a basket-type special-shaped composite structure arch bridge and the main girder is the corrugated steel web and truss composite structure, in which the corrugated steel web forms the longitudinal box structure and the concrete filled steel tubular truss forms the horizontal cantilever structure. The top of the truss is fixed on the cantilever plate, the bottom of the truss is fixed on the bottom slab of the box girder, the rigidity and load bearing capacities of CFST truss are improved, which is attributed to the benefit of filled concrete for steel tube. Corrugated steel-web and cantilever truss composite structure is a new type of bridge structure that wave-truss jointly work, which mainly solves the problem of longitudinal shear and lateral bending. The new-type structure not only fully utilizes the mechanical property of composite materials, but also improves the transverse stability of the bridge. Compared with steel beam, the wave-truss composite structure can reduce steel quantity, has good performance in vibration and can achieve good economic benefits. Compared with the pre-stressed concrete box girder, the structure can effectively reduce the self-weight of the bridge, has less construction formwork, and a beautiful shape. The new-type composite structure will provide the reference of "lightweight, high strength, long span" for a special-shaped combination arch bridge, which is a new trend in bridge development.

## 1 INTRODUCTION

France built the world's first corrugated steel web composite bridge (Cognac bridge) in 1986. Japan imported it from France in 1993. In recent years, the new style of structure has been already applied in the field of bridge engineering more and more extensively. The box girder with corrugated steel webs is a new type of composite steel-concrete structure, where the corrugated steel web shear stress distribution is more uniform, so it has an excellent shear capacity. Corrugated steel web does not resist the axial force, so it can effectively apply pre-stress on the roof and floor of concrete. Box-girder bridge with corrugated steel webs, compared with the ordinary pre-stressed concrete box girder, has many advantages for the use of a bridge. Steel truss structure has the advantages of less weight, fast built, beautiful shape, good anti-seismic properties and a high degree of industrialization. The project practice indicates that the wave-truss composite structure has been extensively applied in engineering recently because of its big rigidity,

economically applicable, good ductility, good stability, large bearing capacities and economic returns. With the development of pump irrigation techniques and the emergence of pre-stressed composite structure, composite structure has come into a new era.

## 2 GENERAL SITUATION OF ENGINEERING

Hengyang Qingjiang Bridge is located in Hunan, China. To the north of the bridge is the Zhengyang Avenue, and to its south runs across the Binjiang Road and the Binhe Road. The bridge is 405 m long, its main bridge is half-through basket-handle-tied arch bridge with the main span of 205 meters long, the bridge deck of 24.5 m wide, and the rise-span ratio of 1/3.6. The arch axis is composed of the catenary, arc and straight line, the arch axis coefficient is 1.756, and the pre-stressing camber is set to be L/600. The southern approach bridge is the pre-stressed concrete continuous box girder

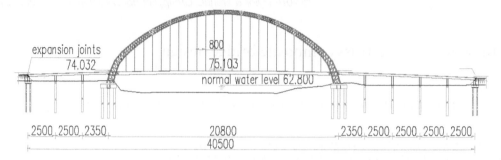

Figure 1. Elevation of the bridge type.

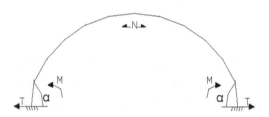

Figure 2. Optimization of the arch axis diagram.

bridge with a span of 5 × 25 m, and 16.5 m wide and 1.5 m high. The northern approach bridge is also the pre-stressed concrete continuous box girder bridge with a span of 3 × 25 m, and 16.5 m wide and 1.5 m high.

## 3 MAIN DESIGN TECHNICAL STANDARDS

1. Grade of road: urban roads;
2. The design speed: 50 km/h;
3. Design load: city—A, the first grade highway;
4. The width of cross section: 25 m;
5. The deck longitudinal slope: controlled by 5%;
6. The deck transverse slope: two-way 1.5%;
7. The clearance under the bridge: not less than 4.5 m;
8. Safety levels: one-level;
9. The design reference period: 100 years.

## 4 KEY POINTS OF DESIGN

### 4.1 *Main arch axis adjustment*

Arch bridge is a structure with thrust, bridge construction. We must first solve the horizontal thrust at both ends of the bridge, which is one of the most recognized problems worldwide. Qingjiang Bridge is a special-shaped concrete filled steel tube and basket-type arch. According to the piano-arch composite theory, we optimize the arch axis by a combination of catenary, arc and straight line, where the main arch axis is changed from divergent to convergent. After optimization, the angle α between the arch axis and the horizontal line is increased, which reduces the horizontal thrust caused by the rib axial pressure. The stiffness of the arch rib and lateral stability are also increased significantly. There are 7 wind braces that strengthen the horizontal connection between two arch ribs. Because half through the tied-arch bridge can rely on the horizontal tension of the tie bar to balance the horizontal thrust of the arch, each piece of the arch rib is, respectively, provided with 12 high strength low relaxation pre-stressed steel strands, which forms no thrust arch bridge.

### 4.2 *Main girder design*

The main girder is corrugated steel webs and cantilever truss composite structures. The beam is 2.1 m high, and the center spacing of the longitudinal beam is 18.5 m. The longitudinal beam is the box girder with corrugated steel webs, the box is 1.2 m wide, the web center spacing is 1.3 m, and the corrugated steel web is 12 mm high, 1000 mm long and 200 mm wide. The lower edge of the longitudinal beam is the pipe-concrete trusses, the lower chord adopts ¢ 600 × 12 mm, steel bottom chords using ¢ 600 × 12 mm, steel flat-linking uses ¢ 500 × 12 mm, the upper edge of the longitudinal beam uses full length steel plate, plate width is 2.0 m, and the thickness is 20 mm. The bridge has 39 beams with a triangular steel pipe, the beam is 16.7 m long, the beam longitudinal spacing is 8 m, the beam bottom chord material uses ¢ 500 × 12 mm, and the web members material is ¢ 299 × 10 mm. The sidewalk cantilever plate uses a concrete-filled steel tube cantilever supporting tube, the material is ¢ 299 × 12 mm, and the micro-expansion concrete is C60, as shown in Figure 3.

### 4.3 Diaphragm plate design

The composite beam with a corrugated steel web, compared with the previous concrete web, has a weak resistance toward the torque, and the torsion problem of the curved bridge is more prominent than the straight bridge. Therefore, the diaphragm plate should be sited in the proper position to resist the distortion. Usually, the concrete diaphragm is set in the fulcrum and midspan, in order to reduce the dead weight of the main girder and meet the requirements of anti-torsion, and the two bracings are set up as the diaphragm in the fulcrum and midspan. The lateral bracings are steel pipe with a diameter of 245 mm and a thickness of 12 mm, made by Q345qc. The whole bridge calculation results show that the diaphragm is arranged reasonably to meet the requirements of the bridge torsional strength.

### 4.4 Connector design

The connection between corrugated steel webs and the top-and-bottom concrete board is the most important binding sites, which ensures that bridge longitudinal horizontal shear can be efficiently transferred and also ensures that all parts of the box girder cross section can form the integrated bearing load. The connecting way of the project adopts the combination way of S-PBL keyboard and the shear stud. The connector mainly includes the upper and lower flange plate, the flange stiffening plate, the shear stud and penetrating reinforced (Fig. 4). The upper and lower flange plate widths are 50 cm, the thickness is 16 mm, the upper flange plate has four rows of welded nails, and the vertical and horizontal spacings are 15 cm and 10 cm, respectively; Because setting the welding nail too much will cause bleeding, which reduces the shear strength, the lower flange plate has only two rows of welded nails, with both vertical and horizontal spacings being 15 cm. The flange stiffening plate is 12 cm high and 12 mm thick. It has a radius of 2.5 cm reinforced holes on the board so that the node is connected through the steel reinforced hole.

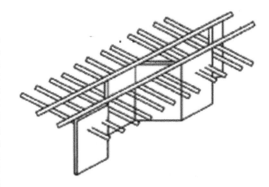

Figure 4. Embedded connection.

So the connection between the corrugated steel web and the upper and lower concrete slabs is more reliable.

## 5 STRUCTURE STABILITY ANALYSIS

### 5.1 Load combination

Combination I: 1.2D+1.4M; Combination II: 1.2D+1.4T;
Combination III: 1.2D+1.4M+1.12T; Combination IV: 1.0D+1.0M+1.0T;

where D denotes the dead load; M is a live load; T represents the temperature load; Combination III denotes the ultimate limit state load combination, and Combination IV is the serviceability limit state load combination.

The finite element model of the arch bridge is established using the finite element software Midas/civil 2013 with different element types for structure analysis and calculation. The whole finite element model has 4479 nodes and 9639 elements. Based on the stress analysis of the superstructure under different load cases, the bearing capacity calculation of least favorable component is computed.

The stability problems of the long span steel arch bridge often become a key factor in the design. The first stability problem has the equilibrium branch, and the second stability has a maximum point of 8. The engineering structures cannot be in the ideal compression state, so the first stability is widely used in engineering. But the critical load obtained from the first stability is the upper extreme point load of stability, so large stability coefficients should be used when employing the first stability. The stability factors of the main bridge are given in Table 1. The result shows that the stability of a structure is good and the design meets the requirement.

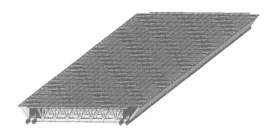

Figure 3. Main girder structure diagram (3d).

Table 1. Stability factors of the main bridge in using phase.

| Order | 1 | 2 | 3 | 4 | 5 |
|---|---|---|---|---|---|
| Eigenvalue | 21.85 | 25.25 | 39.55 | 44.52 | 46.79 |

## 6 CONCLUSIONS

1. The floor beam is a $\varphi 400 \times 14$ mm steel truss structure, which solves the problem of lateral bending; the stringer is a corrugated steel web. Vertical stiffness of the corrugated steel web is small. There is almost no resistance to the axial force and the imported longitudinal pre-stressed beam can concentrate the load on the roof and floor plate, which can effectively increase the efficiency of pre-stressed composite structure.
2. The stringer adopts a lighter corrugated steel web; its weight, compared with the ordinary pre-stressed concrete box girder, is lost by about 20%, which improves the seismic performance and avoids the web cracking problem.
3. Wave-truss composite bridge has a landscape impact effect. It is a good bridge type choice of the highway, the mountain and the scenic spot.

## REFERENCES

Ding. F.X. 2006. *The study of design method and mechanical behavior on CFST structure*. Central south university.

Elgaaly, M. & Seshadri, A. 1997. Girders with Corrugated Webs under Partial Compressive Edge loading. *Journal of the Structural Division ASCE*, 122(4): 783–791.

Elgaaly, M. Hamilton, R. & Seshadri, A. 1996. Shear Strength of Beams with Corrugated Webs. *Journal of Structural Division ASCE*. 122(4): 390–398.

Guan, L. & Wang, Y. 2007. Optimize the arch axial coefficient of CFST arch bridge. *Sichuan Architecture*, (03): 106–107.

He, W. Zhao, S.B. & Yang, J.Z. et al. 2010. Large-span wide double basket arc bridge dynamic characteristics and seismic performance. *Journal of Vibration, Measurement & Diagnosis*, 30(6): 32–35 + 109–110.

Li, Y. Chen, Y.Y. & Nie, J.G. et al. 2002. *Design and application of Steel-concrete composite bridge*. Beijing, Science Press, 2002: 113–116.

Li, Y. Fang, Q.H. & Zhu, H.P. et al. 2011. Study on prestressed transfer efficiency and moment amplitude modulation of steel-concrete composite bridge. *Journal of Harbin Institute of Technology*, 43(02): 357–361.

Li, Y. Nie, J.G. & Chen, Y.Y. et al. 2002. Design and Study on Shenzhen Rainbow Bridge. *China Civil Engineering Journal*, 35(05): 52–56.

Li, Y. Nie, J.G. Yu, Z.W. & Chen, Y.Y. 1998. The Stiffness study of Steel-concrete composite beam. *Journal of Tsinghua university*, 38(10): 38–41.

Luo. R. & Edlund, B. 1996. Shear Capacity of Plate Girders with Trapezoidally Cormgated Webs. *Thin-Walled structures*. 26(1): 19–44.

Sayed-Ahmed, E.Y. 2001. Behavior of Steel and Composite Girders with Corrugated Steel Webs. *Canadian Journal of Civil Engineering*. 28(4): 656–672.

Yun, D. 2007. Static and aseismic behavior of large span half-through CFST arch bridges. Harbin Institute of Technology.

*Advanced Materials, Structures and Mechanical Engineering – Kaloop (Ed.)*
© *2016 Taylor & Francis Group, London, ISBN 978-1-138-02793-0*

# The transient stability analysis based on WAMS and online admittance parameter identification

H.Z. Zhou, F. Tang, J. Jia & S.L. Ye
*School of Electrical Engineering, Wuhan University, Wuhan, China*

ABSTRACT: With the scale of interconnected power grid expanding, the topology of fault network and admittance parameters become unavailable, which makes it difficult for line personnel in the power system to do transient stability analyses. Aiming to solve the problem, this paper proposed the Multivariate Linear Regression algorithm (MLR) to identify the node admittance parameters of power system based on the datum of Wide Area Measurement System (WAMS). Firstly, the MLR algorithm was used to identify the pre-fault and after-fault admittance parameters; after that, the Transient Energy Function method (TEF) was applied to solve the energy function of the system, the Critical Clearing Time (CCT) could be obtained ultimately. Simulation results show that the presented algorithm can simultaneously consider specific failure type of the power system and uncertain network topological structures. Furthermore, when MLR algorithm applied in TEF method to analyze the transient stability, the calculation precision gets relatively higher and the admittance parameter calculating procedure gets simplified as well.

## 1 INTRODUCTION

The transient stability analysis and controlling is a core part of dynamic security analysis (Yu & Wang 1999). There are many blackouts happened in numerous countries caused by the power systems instabilities (Wang et al. 2012, Hu 2003, Zhao 2003, He 2004). Besides, as the demand for electricity is increasing quickly, it is common for the power systems to operate at their limits. It is a big challenge for power system to keep the system running correctly. Many actual cases show that the difficulty of obtaining the specific system network topological structure makes it hard to analyze the transient stability of power systems. Because some system loads or transmission line models are hard to acquire. However, the quick identification of system topologies and parameters from a global view is necessary for the operation personnel. Because it offers guidelines on taking the emergency control measures.

The previous papers about the transient stability analysis of power system can be classified to the three types: the Extend Equal Area Criterion method (Teng et al. 2003, Fang et al. 1999) and the Lyapunov transient energy function method (Liu & Thorp 2000, Li et al. 2007) and the intelligence artificial methods (Gu et al. 2013, Ye et al. 2012, Wang & Meng 2007). The emerging of Wide Area Measurement System (WAMS) offers a new opportunity for the transient stability analysis of power system. The data collected by WAMS is more accurate and dynamic, which makes the model more elaborate which will accelerate the computational time.

This paper proposed an identification method to establish the equivalent admittance parameter matrix of power system by using the Multiple Linear Regression algorithm (MLR) based on the real-time WAMS data. The identified admittance matrix is applied to solve the energy potential energy boundary surface of PEBS method for power system transient stability analysis.

## 2 THE NODE ADMITTANCE PARAMETER IDENTIFICATION BASED ON MLR

### 2.1 *The MLR application in power system*

When the classical model in power system, the generator rotor motion equation can be concluded as follows (Liu & Wang 1996):

$$\begin{cases} M_i \dfrac{d\omega_i}{dt} = P_{mi} - P_{ei} \\ \dfrac{d\delta_i}{dt} = \omega_i \end{cases} \quad (i = 1, 2, \ldots, n) \quad (1)$$

In Equation (1): $M_i$: the inertial time constant of the generator; $\omega_i$: rotor angular velocity and the synchronous speed deviation of the generator; $\delta_i$: the rotor Angle of generator; $P_{ei}$: the electromagnetic power of the generator; $P_{mi}$: mechanical power of the generator.

$$P_{ei} = E_i^2 G_{ii} + \sum_{j=1, j \neq i}^{n} (E_i E_j B_{ij} \sin \delta_{ij} + E_i E_j G_{ij} \cos \delta_{ij})$$

(2)

and $\delta_{ij} = \delta_i - \delta_j$ in Equation (2)

When considering the impact that a big disturbance such as error operation has on the mathematical model of power system, it is feasible to modify the node admittance matrix parameters of rotor motion Equation (2).

From the generator mathematical model, we can come to the conclusion that all kinds of energy exchange processes in the grid all via the changes of admittance matrix from the mathematical view. No matter these energy exchange processes occur between generators or between generators and the power grid. If there are imbalanced energy flow, there are variations in node admittance parameters. Therefore, if the system node admittance parameters can be accurately identified, we can accurately get the energy of the system before and after a fault. It means the transient stability of the system can be accurate analysis in both steady state and transient state.

As a consequence, $P_{ei}$ can be seen as the dependent variable, $\delta_i$ as the independent variable. The $B_{ij}$ and $G_{ij}$ is the overall regression parameters. Then the $B_{ij}$ and $G_{ij}$ can be obtained based on the multiple linear regression of the least squares parameter identification.

### 2.2 Multivariate regression algorithm steps and flow chart

The specific procedures are as follows (Chiang et al. 1988):

1. List the observation equations:

$$Y_{n \times 1} = X_{n \times (m+1)} \beta_{m+1}$$

(3)

2. Constructs the augmented matrix B:

$$B = [X\ Y]$$

(4)

3. Calculate the eigenvalue of the matrix $\mathbf{B}^T\mathbf{B}$ and the minimum eigenvalue $\lambda_{m+1}$ according to the theory of singular value decomposition.

$$\hat{\beta}_{m,1} = [X^T X - \lambda_{m+1} I_{m,m}]^{-1} X^T Y$$

(5)

4. Calculate the whole parameters of least squares

### 2.3 The TEF in multimachine power system model

Practical engineering calculation shows the COI coordinates has a better performance than the traditional synchronous coordinate on the accuracy of stability analysis (Kakimoto et al. 1978). Therefore, this paper use the COI coordinates to establish the model of the transient energy function for the multi-machine power system.

The transient energy function in COI coordinates consists of the transient kinetic energy and the transient potential energy:

$$V = V_k + V_p$$

$$= \frac{1}{2} \sum_{i=1}^{n} Mi\varpi_i^2 - \sum_{i=1}^{n} P_i(\theta_i - \theta_{si})$$

$$- \sum_{i=1}^{n-1} \sum_{j=i+1}^{n} C_{ij}(\cos\theta_{ij} - \cos\theta_{s,ij})$$

$$+ \sum_{i=1}^{n-1} \sum_{j=i+1}^{n} D_{ij} \frac{a}{b}(\sin\theta_{ij} - \sin\theta_{s,ij})$$

(6)

When calculating the transient energy after the fault is removed, it's only necessary replace the $\varpi$ and $\theta$ with $\varpi_c$ and $\theta_c$. Replace the $\varpi$ and $\theta$ with $\varpi_u$ and $\theta_u$ if the critical energy is required to calculate.

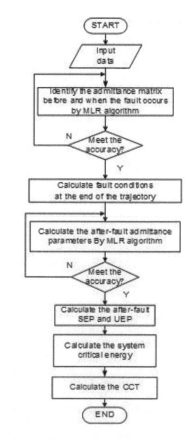

Figure 1. The flow diagram of solving CCT.

### 2.4 The modified potential energy boundary surface method

The physical meaning of the Potential Energy Boundary Surface (PEBS) is that when the system is in the transient state, it is easy to be affected by the imbalanced torque in the area covered by the PEBS. The imbalanced torque restricts the system in the potential energy boundary surface within the surface. As long as the generator trajectory across PEBS, the imbalanced torque will make the system leave the area as far as possible, which means the system is out of control.

### 2.5 The flow chart of the calculation of critical clearing time some common mistakes

The calculation of critical clearing time for the power system when the fault happens can be illustrated in Figure 1.

## 3 CASE STUDY

Ten transient stability cases is defined and calculated in New England 10-generator and 39-bus system in Table 1. It is assumed that the fault start time is 1 s and after 0.2 s the relay protection equipment works and the network is restored.

The simulation results are demonstrated by comparing the actual electromagnetic power and calculated power by identification. The case 2 and case 6 are as the representatives.

In the process of simulation, some following conclusions can be drawn by observing transient kinetic energy $V_k$, transient potential energy $V_p$ and total transient energy $V_{(t)}$ in both stable transient case and instable transient case:

1. The amplitude of system transient kinetic energy and system transient potential energy

Table 1. Fault event set in New England 10-machine 39-bus system.

| Case | The fault position Three-phase short circuit | Start time | End time |
|------|------------------------------|-----------|---------|
| CASE1 | (BUS26-29) BUS 26# | 1.0 s | 1.2 s |
| CASE2 | (BUS4-14) BUS 4# | 1.0 s | 1.2 s |
| CASE3 | (BUS6-11) BUS 6# | 1.0 s | 1.2 s |
| CASE4 | (BUS16-17) BUS 16# | 1.0 s | 1.2 s |
| CASE5 | (BUS25-26) BUS 25# | 1.0 s | 1.2 s |
| CASE6 | (BUS1-2) BUS 1# | 1.0 s | 1.2 s |
| CASE7 | (BUS1-39) BUS 1# | 1.0 s | 1.2 s |
| CASE8 | (BUS1-39) BUS 39# | 1.0 s | 1.2 s |
| CASE9 | (BUS23-24) BUS 23# | 1.0 s | 1.2 s |
| CASE10 | (BUS21-22) BUS 22# | 1.0 s | 1.2 s |

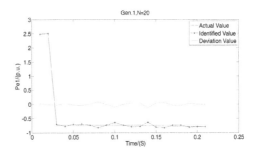

Figure 2. Calculated pre-fault electromagnetic power in case 2.

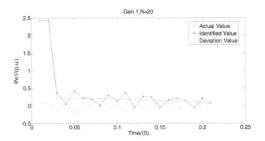

Figure 3. Calculated pre-fault electromagnetic power in case 6.

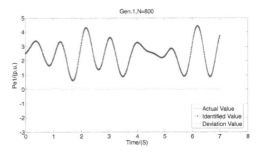

Figure 4. Calculated post-fault electromagnetic power in case 2.

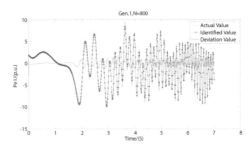

Figure 5. Calculated post-fault electromagnetic power in case 6.

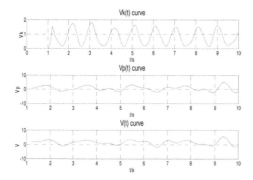

Figure 6. Transient kinetic energy $V_k$, transient potential energy $V_p$ and transient energy $V$ in case 2.

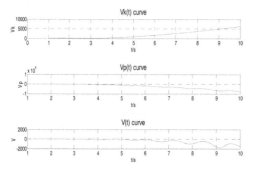

Figure 7. Transient kinetic energy $V_k$, transient potential energy $V_p$ and transient energy $V$ in case 2.

can go through a significant growth after the occurrence of a fault.

2. The transient kinetic energy curve generally oscillates with invariable frequency and constant amplitude after the fault is removed in the stable case. In instable cases, the transient kinetic energy curves also do oscillations with plus or minus amplitudes of fluctuations, while its frequency is not steady.

3. In instable case 2, the three-phase short-circuit fault just happens in 1.2 seconds and the transient kinetic energy $V_k$ reaches to zero at this moment. $V_k$ increases gradually, almost in an exponential growing speed and potential energy $V_p$ gradually turns into negative. However, $V_k$ does not reach extreme value or peak in the process of simulation. It can be concluded that the angle of the generator has been spreading out, which makes the kinetic energy increasing.

Therefore, the system has experienced the process that the kinetic energy is converted to potential energy. If the system is able to absorb the surplus kinetic energy, which means there is

Table 2. The CCT comparison between SBS and PEBS.

| Case | SBS $t_{cr}$ (S) | CCT $t_{cr}$ (S) | $V_{cr}$ (p.u.) | $V_{kc}$ (p.u.) | $V_{pc}$ (p.u.) |
|---|---|---|---|---|---|
| 5 | 0.13~0.14 | 0.1400 | −1.3632 | 7.0801 | −0.5773 |
| 6 | 0.19~0.20 | 0.0700 | 1.2474 | 7.0069 | 1.2899 |
| 7 | 0.19~0.20 | 0.2500 | 22.0708 | 7.1376 | 6.1033 |
| 8 | 0.14~0.15 | 0.1100 | 4.4190 | 11.3939 | 2.5740 |
| 9 | 0.15~0.16 | 0.1300 | −5.0523 | 8.4176 | −0.4040 |
| 10 | 0.39~0.40 | 0.5100 | 12.2602 | 1.5077 | 0.3075 |
| 11 | 0.39~0.40 | 0.5100 | 12.2602 | 1.5077 | 0.3075 |
| 12 | 0.39~0.40 | 0.1900 | 1.8515 | 1.8994 | 0.0346 |
| 13 | 0.20~0.21 | 0.2500 | 14.0431 | 5.8116 | 2.9296 |
| 14 | 0.20~0.21 | 0.2500 | 15.1421 | 5.7577 | 3.4494 |

no power preventing the generators form synchronous running and the system is stable. If the system can't absorb the residual kinetic energy and vice versa.

## 4 CONCLUSIONS

1. Comparing the CCT obtained by the modified PEBS potential energy boundary surface method with the CCT obtained by time domain simulation method, we can see that there are small differences between two of them in numerical value. But the situation of misjudgment also appeared in case 7 and case 12. Overall, due to the simplification of a transient energy function model for UEP, the CCT obtained by PEBS method could be more conservative.

2. This paper used the multiple linear regression algorithm to identify the pre-fault and post-fault admittance parameter matrixes, which can greatly reduce the computational burden for solving the node admittance matrix in the PEBS compared with traditional admittance parameter solution. As the system scale expanding, this advantage will be more apparent. The obtained admittance parameters can automatically correct with the transfer and change of the fault, making it consistent with the actual situation of the power system.

ACKNOWLEDGEMENT

Project Supported by State Grid Corporation of China, Research on the operation control technology of improving transmission capacity at the initial stage of Ultra-high Voltage AC/DC Power System and China.

# REFERENCES

Chiang, H.D. Wu, F.F. & Varaiya, P.P. 1988. Foundations of the potential energy boundary surface method for power system transient stability analysis. *IEEE Transactions on Circuits and Systems*, 35(6): 712–728.

Fang, Y.J. Fan, W.T. & Chen, Y.H. et al. 1999. An on-line transient stability control system of large power systems. *Automation of Electric Power Systems*, 23(1): 8–11.

Gu, X.P. Li, Y. & Wu, X.J. 2013. Transient Stability Assessment of Power Systems Based on Local Learning Machine and Bacterial Colony Chemotaxis Algorithm. *Transactions of China Electrotechnical Society*, 28(10): 271–279.

He, D.Y. 2004. Rethinking Over8.14'US-Canada Blackout after One Year. *Power System Technology*, 28(21): 1–5.

Hu, X.H. 2003. Rethingking and enlightenment of large scope blackout in interconnected North America Power Grid. *Power System Technology*, 27(9): T2–T2.

Kakimoto, N. Ohsawa, Y. & Hayashi, M. 1978. Transient stability analysis of electric power system via Lure-type Lyapunov functions, *Part I and II., IEEE Transactions of Japan*, 98(516): 566–604.

Li, Y. Zhou, X.X. & Zhou, J.Y. 2007. The perturbed trajectories prediction based on an additional virtual node. *Automation of Electric Power Systems*, 31(12): 19–22.

Liu, C.W. & Thorp, J.S. 2000. New methods for computing power system dynamic response for real-time transient stability prediction. *IEEE Transactions on Circuits and Systems I: Fundamental Theory and Applications*, 47(3): 324–337.

Liu, S. & Wang, J. 1996. *Power System Transient Stability Analysis on Energy Function*. Shanghai. Shanghai Jiao Tong University Press.

Teng, L. Liu, W.S. & Yun, Z.H. et al. 2003. Study of Real-time Power System Transient Stability Emergency Control. *Proceedings of the CSEE*, 23(1): 65–70.

Wang, H.W. & Meng, J. 2007. Multiple linear regression prediction modeling method. *Journal of Beijing University of Aeronautics and Astronautics*, 33(4): 500–504.

Wang, Y.Y. Luo, Y. & Tu, G.Y. et al. 2012. Correlation model of cascading failures in power system. *Transactions of China Electrotechnical Society*, 27(2): 204–209.

Ye, S.Y. Wang, X.R. & Zhou, S. et al. 2012. Power system probabilistic transient stability assessment based on Markov Chain Monte Carlo method. *Transactions of China Electrotechnical Society*, 27(6): 168–174.

Yu, Y.X. &Wang, C.S. 1999. *Theory and Method on the Stability of Electrical Power System*. Beijing. Beijing Science Press.

Zhao, X.Z. 2003. Strengthen power system security to ensure reliable power delivery. *Power System Technology*, 27(10): 1–7.

*Advanced Materials, Structures and Mechanical Engineering – Kaloop (Ed.)*
© *2016 Taylor & Francis Group, London, ISBN 978-1-138-02793-0*

# Polymer composites with ferrocene derivatives for fire-safe construction

A.A. Askadsky
*A.N. Nesmeyanov Institute of Organoelement Compounds of Russian Academy of Sciences,
Moscow, Russian Federation*

V.A. Ushkov & V.A. Smirnov
*Moscow State University of Civil Engineering, Moscow, Russian Federation*

ABSTRACT:  Wide use of polymers in the modern construction industry requires careful examination and optimization of operational properties related to fire safety. As an agent for smoke suppression, fire-retardant ferrocene was employed decades ago. In spite of this, to date, there are limited data related to the influence of ferrocene and its derivatives on the properties of several polymer composites important for construction. The aim of the present work is to investigate the suitability and effectiveness of ferrocene and different ferrocene derivatives as admixtures for two types of polymer composites. We discussed the experimental results of several laboratory tests. It is shown that ferrocene derivatives of relatively low volatility can be successfully used as efficient smoke suppressors and fire retardants. This is due to the barrier layer formation on the surface of the burning material. Such layer prevents the penetration of pyrolysis products and slows down the spread of the flame. At the same time, the results indicate that probably no universal chemical agent allows to achieve the preferred values of all operational properties related to fire safety at once: optimal values for weight loss, rate of decomposition, limiting oxygen index and smoke formation factor cannot be simultaneously reached in any single point of factor space.

## 1 INTRODUCTION AND PRIOR WORK

One objective of fire-safe construction involves the development of effective admixtures for flammable building materials, in particular, for polymer composites. Fire resistance of carbonizing polymers greatly depends on their chemical structure and the amount of synergists and smoke suppressors. It is known that ferrocene is an efficient smoke suppressor (Kulev et al. 1986, Ushkov et al. 1988). Its addition leads to a decrease in smoke density caused by sooting during the combustion of unsaturated (Zhang 1994) and saturated (Kasper 1999) hydrocarbons. Ferrocene structures in polymer chains lead to thermal stability and fire resistance (Kishore 1991); though at least for several ferrocene derivatives, there is a negative correlation between smoke suppression and flame retardancy (Carty 1996). As a methane flame inhibitor, ferrocene is as efficient as iron pentacarbonyl (Linteris 1991). The high efficiency of ferrocene is due to the fact that during oxidation, it acts as a source of iron-containing intermediates with high specific surface and catalytic activity. Ferrocene acts as a catalyst of the heterogeneous carbon oxidation process, promotes almost smoke-free burning of organic polymers and increases the completeness of combustion (Reshetova 1975).

Some aspects of the combustion catalysis of flammable materials by ferrocene have been examined in the literature (Sinditsky et al. 2014). Ferrocene influences the pyrolysis of polymers, inhibits oxidation and shows high reactivity to gaseous HCl, forming iron chloride during the reaction (Kulev et al. 1986).

Considering building materials for fire-safe construction, the drawback of pure ferrocene is due to its relatively high volatility at temperatures above 150 °C, despite the fact that ferrocene demonstrates thermal stability up to the temperature of 470 °C (Fomin 2007, Zhukov 2000). Ferrocene decomposes with a high rate only at temperatures higher than 550 °C (Fomin 2007). Because of this, there is a need to identify its derivative compounds that have similar fire and smoke suppression properties, but are less volatile at high temperatures. In the case of Polyvinyl Chloride (PVC), some results were obtained earlier (Ushkov et al. 1988, Carty 1996).

## 2 EXPERIMENTAL SETUP

We examined the thermal stability, combustion and smoke formation of composites based on several carbonizing polymers. Ferrocene according to RU TC 6-02-964-78 and its different derivatives

Table 1. Ferrocene and synthesized ferrocene derivatives.

| Compound/abbreviation | Chemical formula | Amount of iron, % by mass | Molecular mass |
|---|---|---|---|
| Ferrocene/FEC | $(C_5H_5)_2Fe$ | 30.12 | 186 |
| Oxyethylferrocene/OEF | $C_5H_5FeC_5H_4-C_2H_4-OH$ | 24.53 | 230 |
| Acetylferrocene/AF | $C_5H_5FeC_5H_4-CO-CH_3$ | 24.12 | 228 |
| Diacetylferrocene/DAF | $(C_5H_4-CO-CH_3)_2Fe$ | 20.28 | 271 |
| Ferrocenedicarboxylic acid/FDA | $(C_5H_4-COOH)_2Fe$ | 20.38 | 275 |
| Copolymer of acryloyl ferrocene and isoprene/PAIF | $-[C_5H_5FeC_5H_4-CO-CH-CH_2]_n-$ $[CH_2-C_3H_4-CH_2]_m-$ | 9.62 | $\sim10^5$ |
| Polydiisopropenyl ferrocene/PDPF | $-[(C_5H_4-C-(CH_3)_2)_2Fe]_n-$ | 20.98 | $\sim10^5$ |

(Table 1), which were synthesized and characterized in the laboratories of INEOS RAS, are used during an investigation.

Thermal stability, fire resistance and limiting oxygen index are studied for several types of polymer-based material that are widely used in the construction industry, which include: plasticized PVC (PPVC) and disperse-filled Epoxy Composites (EC). The latter type of composite is usually the subject of the so-called nanomodification process; the resulting nanocomposites with elevated operational properties can be successfully used for constructions operating in aggressive environments (Korolev 2012). Because of this, improvement of fire safety of EC is an important objective in construction. Epoxy oligomer ED-20, diabase and andesite powders are used as the matrix material and fillers for the investigated EC. Amounts of the filler and the admixture were 35% and 0.29%, respectively. Thermal stability of synthesized ferrocene derivatives and polymers was determined by means of thermogravimetry with an automated thermal-analytical complex DuPont-9900. Dynamic heating modes in air and nitrogen flows are used. Heating rates were 10 and 20 °C/min for air (during the determination of weight loss) and nitrogen (during thermal analysis), respectively. Several criteria of thermal stability, fire resistance and smoke formation are taken into account during the examination of fire rating, including: temperature $T_{bc}$ corresponding to the beginning of intensive decomposition, temperature $T_{10}$ of 10% weight loss and temperature $T_{max}$ of the maximal decomposition rate; rates $R_1$ and $R_2$ of decomposition at the first and second stages of the standard test; weight loss $\Delta W_{600}$ at 600 °C; thermal effect of decomposition $\Delta E$; Limiting Oxygen Index (LOI); ignition ($T_i$) and auto-ignition ($T_{ai}$) temperatures; smoke formation factor ($S_f$) in pyrolysis and combustion modes. The aforementioned parameters are determined according to RU GOST 12.1.044. Eligibility of piecewise-sigmoidal models for the statistical description of experimental data

(Korolev et al. 2009) is determined by the regression analysis of preliminary results.

3 RESULTS AND DISCUSSION

Obviously, combustion and smoke formation of polymer composites are related to the admixture properties. The results for the thermal stability of ferrocene derivatives are summarized in Table 2.

As can be observed from Tables 1 and 2, compounds with high molecular mass (PAIF and PDPF) are significantly less volatile at temperatures below 300 °C. As a rule, the low melting point corresponds to the low temperature of volatility; the dependence between the temperature of volatility and the rate of weight loss is not so clear. The matrix of correlation coefficients for columns of Table 2 is represented as follows:

$$r_{ij} = \begin{pmatrix} 1 & 0.95 & 0.46 \\ 0.95 & 1 & 0.18 \\ 0.46 & 0.18 & 1 \end{pmatrix}. \quad (1)$$

The tight linear dependence between a melting point and the temperature of volatility is reflected by the $r_{12}$ element, which is very close to the unit value.

On the contrary, $r_{13}$ and, especially, $r_{23}$ are considerably smaller; it can be shown that $r_{23}$ is statistically insignificant at the 95% level of confidence. Thus, a high temperature of volatility not always corresponds to a low rate of weight loss. In fact, while the OEF should be considered as the most volatile derivative (weight loss starts at 128 °C), it is also characterized by a low rate of weight loss.

It is revealed that the chemical structure and the contents of ferrocene and its derivatives have a considerable impact on the heat resistance and LOI of the examined EC at high temperatures (Table 3). At low temperatures, ferrocene derivatives do not significantly affect the decomposition of epoxy composites ($T_{bc}$ = 273–285 °C, $T_{max}$ = 300–306 °C).

366

Table 2.  Properties of synthesized ferrocene derivatives.

| Compound | Melting point, °C | Temperature of volatility, °C | Rate of weight loss, %/min |
|---|---|---|---|
| OEF | 75 | 128 | 12.5 |
| AF | 85 | 100 | 25.8 |
| DAF | 128 | 157 | 27.9 |
| FDA | 240 | 219 | 77.8 |
| PAIF | 250 | 314 | 26.5 |
| PDPF | 300 | 340 | 29.9 |

Table 4.  Smoke formation of EC with ferrocene derivatives.

| Type of admixture | $S_f$ during pyrolysis, $m^2/kg$ | $S_f$ during combustion, $m^2/kg$ |
|---|---|---|
| Ref. sample | 1030 | 890 |
| FEC | 720 | 480 |
| AF | 640 | 315 |
| DAF | 520 | 410 |
| OEF | 540 | 360 |

Table 3.  Properties of EC with ferrocene derivatives.

| Parameter | Values for different admixtures | | | | | |
|---|---|---|---|---|---|---|
| | Ref. | FEC | OEF | AF | FDA | PDPF |
| $T_{ai}$, °C | 515 | 490 | 470 | 480 | 480 | 480 |
| $R_1$, %/min | 19.9 | 19.9 | 20.0 | 21.2 | 18.4 | 21.4 |
| $R_2$, %/min | 18.9 | 24.2 | 24.5 | 20.8 | 18.8 | 16.0 |
| $\Delta W_{600}$, % | 65.7 | 67.7 | 59.1 | 64.3 | 59.4 | 67.7 |
| $\Delta E$, kJ/kg | 4070 | 4300 | 3960 | 4300 | 4300 | 3300 |
| LOI, % | 23.3 | 27.6 | 28.3 | 25.6 | 26.1 | 25.8 |

Table 5.  Correlation coefficients for the properties of EC.

| $r_{ij}$ | $T_{ai}$ | $R_1$ | $R_2$ | $\Delta W_{600}$ | $\Delta E$ | LOI |
|---|---|---|---|---|---|---|
| $T_{ai}$ | 1 | −0.11 | −0.23 | 0.48 | 0.15 | −0.77 |
| $R_1$ | −0.11 | 1 | −0.19 | 0.59 | −0.55 | −0.1 |
| $R_2$ | −0.23 | −0.19 | 1 | −0.27 | 0.55 | 0.69 |
| $\Delta W_{600}$ | 0.48 | 0.59 | −0.27 | 1 | −0.32 | −0.33 |
| $\Delta E$ | 0.15 | −0.55 | 0.55 | −0.32 | 1 | 0.07 |
| LOI | −0.77 | −0.1 | 0.69 | −0.33 | 0.07 | 1 |

The maximal rate of decomposition at the first stage is near to 20%/min.

The results of the examination of smoke formation for EC are summarized in Table 4.

The data presented in Tables 3 and 4 are in good agreement with the previously known fact that ferrocene and its derivatives can decrease only some of the parameters related to flammability, fire resistance and smoke formation, but not all of them.

In some cases, when the rate of decomposition is high, the high degree of carbonization (high weight loss) corresponds to high LOI (impeding candle-like combustion, FEC). In other cases, the high degree of carbonization corresponds to the low rate of weight loss and, at the same time, to the low LOI (PDPF).

As can be observed from Table 4, the ferrocene derivatives synthesized are superior smoke suppressors for epoxy composites. DAF is the most effective suppressor in the pyrolysis mode (smoke formation reduced to 51% of unmodified EC, 19% better than for FEC), and OEF is the best choice for the combustion mode (smoke formation reduced to 40%, 13% better than for FEC).

As before, the entire Table 3 can be represented by a matrix of correlation coefficients (Table 5).

As it follows from Table 5, for epoxy composites with ferrocene derivatives, there is a substantial positive linear correlation between:

– autoignition temperature and degree of carbonization ($r_{14} = 0.48$);

– decomposition rate at the first stage of the test and the degree of carbonization ($r_{24} = 0.59$);

– decomposition rate at the second stage of the test, the thermal effect of decomposition and the oxygen requirement for candle-like combustion.

At the same time, a high LOI corresponds to the low autoignition temperature ($r_{16} = -0.77$). The observed peculiarities can be attributed to the increased decomposition rate of the condensed phase, which, in turn, is caused by iron oxides formed during the decomposition of ferrocene derivatives. The latter is consistent with the results reported previously (Dyagileva 1979). In the presence of ferrocene derivatives, the LOI of EC increases from 23.3 to 25.6–26.1%. Higher values of LOI (28.3%) are observed for composites containing OEF. It is also found that while ferrocene derivatives lower $T_{ai}$ by 20–35 °C, they have a little effect on $T_i$ (which is nearly 225 °C for all studied EC). The results confirm that the effect of ferrocene derivatives is mostly on the high-temperature region of EC.

The increase in the concentration of ferrocene derivatives corresponds to the decrease in EC flammability. Significant improvement of LOI is observed for compositions containing OEF (Fig. 1): when the concentration is raised from 0.17 to 1.71%, the LOI increases from 23.4 to 28.9%.

In our opinion, ferrocene derivatives contribute to the formation of the carbonized barrier layer on the surface of the burning material. This layer

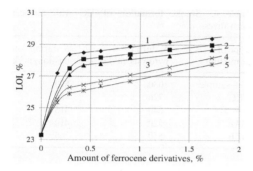

Figure 1. LOI of EC with different ferrocene derivatives: 1—OEF; 2—FEC; 3—DAF; 4—FDA; 5—AF.

Table 6. Properties of PPVC with ferrocene derivatives.

| Parameter | Values for different admixtures | | | |
| | Ref. | FEC | AF | OEF |
| --- | --- | --- | --- | --- |
| $T_{bc}$ | 240 | 240 | 245 | 247 |
| $T_{10}$ | 260 | 255 | 260 | 263 |
| $T_{max}$ | 285 | 280 | 285 | 290 |
| $T_{ai}$ | 420 | 420 | 420 | 435 |

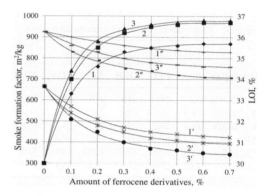

Figure 2. Dependencies of LOI (1, 2, 3) and smoke formation factor in pyrolysis (1′, 2′, 3′) and combustion mode (1″, 2″, 3″) for PPVC: 1—FEC; 2—OEF; 3—AF.

induces the penetration of volatile products from pyrolysis, and prevents heat and mass transfer and further spread of the flame. An essential increase of LOI is observed when the concentration of ferrocene derivatives is higher than 0.6%. At the same time, the concentration of ferrocene derivatives does not correlate with $T_i$ but negatively correlates with $T_{ai}$. The increase in DAF concentration ranging from 0.17 to 1.7% reduces $T_{ai}$ from 505 to 490 °C.

Due to revealed peculiarities, there can be complications during the selection of the appropriate admixture for fire-safe EC.

We also examined the smoke formation, thermal and fire resistance of PPVC with admixtures of ferrocene derivatives. The results are summarized in Table 6 and Figure 2.

The presence and type of ferrocene derivatives do not have a significant effect on either the characteristic temperatures for PVC materials or the dehydrochlorination processes and decomposition in the air. Weight loss during the standard test is 64.5–67%.

It is evident from Figure 2 that for higher concentrations of ferrocene derivatives (up to 0.7%), the LOI of the PVC increases from 30 to 36.8%. The smoke formation factor in the pyrolysis and combustion modes can be reduced from 925 and 660 to 710 and 350 m²/kg, respectively. As with EC, the ferrocene derivatives are more effective admixtures for PPVC than pure ferrocene. The higher efficiency of AF and OEF compared with pure ferrocene is due to their active decomposition that leads to the formation of ultrafine catalytically active iron oxides. Such oxides form a barrier layer on the surface and inhibit the formation of smoke. The optimum concentration of ferrocene derivatives in PVC materials is 0.3–0.5%.

## 4 SUMMARY AND CONCLUSION

In the present work, we summarized and discussed the results of several laboratory tests. The tests were carried out to answer the question concerning the suitability and effectiveness of ferrocene and its derivatives as admixtures for fire-safe construction.

It is discovered that ferrocene derivatives of relatively low volatility can be successfully used as efficient smoke suppressors and fire retardants during the production of polymer composites for fire-safe construction. For epoxy composites, the application of diacetylferrocene and oxyethylferrocene reduces smoke formation by 49% and 60% in the pyrolysis and combustion modes, respectively. However, it is again confirmed that probably no universal chemical agent allows increasing all operational properties related to fire safety at once. Examined admixtures provide low weight loss, slow rate of decomposition, high LOI and low smoke formation factor, though optimal values of these parameters cannot be simultaneously reached at any single point in factor space.

## ACKNOWLEDGMENTS

This work was supported by the Ministry of Science and Education of Russian Federation, Project # 7.2200.2014/K "Nanomodified polymer composites for fire safe construction".

# REFERENCES

Carty, P. Grant, J. & Metcalfe, E. 1996. Flame-retardancy and smoke-suppression studies of ferrocene derivatives in PVC. *Applied Organometallic Chemistry,* 10 (2): 101–111.

Dyagileva, L.M. Mar'in, V.P. Tsyganova, E.I. & Razuvaev G.A. 1979. Reactivity of the first transition row metallocenes in thermal decomposition reaction. *Journal of Organometallic Chemistry.* 175 (1): 63–72.

Fomin, V.M. 2007. Application of sandwich complexes of transition metals in electronics and catalysis. Oxidation reactions. Nizhny Novgorod: NSU.

Kasper, M. Sattler, K. Siegmann, K. Matter, U. & Siegmann, H.C. 1999. The influence of fuel additives on the formation of carbon during combustion. *Journal of Aerosol Science,* 30 (2): 217–225.

Kishore, K. Kannan, P. & Iyanar K. 1991. Synthesis, characterization and fire retardancy of ferrocene containing polyphosphate esters. *Journal of Polymer Science Part A: Polymer Chemistry,* 29 (7): 1039–1044.

Korolev, E.V. Samoshin, A.P. Smirnov, V.A. Koroleva, O.V. & Grishina, A.N. 2009. *Methods and synthesis algorithms of new-generation radiation-protective materials.* Penza: PGUAS.

Korolev, E.V., Smirnov, V.A. & Albakasov, A.I. 2012. Nanomodified epoxy composites. *Nanotechnologies in Construction,* 4(4): 81–87.

Kulev, D.H. Kitaygora, E.A. Golovnenko, N.I. & Mozzhukhin, V.B. 1986. *Reducing flammability and smoke forming ability of materials based on plasticized PVC.* Moscow: NIITEKHIM.

Linteris, G.T. & Rumminger M.D. 1999. Flame inhibition by ferrocene, carbon dioxide, and trifluoromethane blends: synergistic and antagonistic effects. NIST Interagency Report 6359.

Reshetova, M.D. 1975. *Ferrocene in the industry.* Moscow: NIITEKHIM.

Sinditsky, V.P. Chernuy A.N. & Marchenkov, D.A. 2014. Mechanism of combustion catalysis by ferrocene derivatives. Burning ammonium perchlorate and ferrocene. *Physics of combustion and explosion,* 50 (1): 59–68.

Ushkov, V.A. Kulev D.H. Lalayan V.M. Antipova B.M, Bulgakov B.I. & Naganovsky J.K. 1988. Ferrocene derivatives as inhibitors of smoke for plasticized PVC. *Plastics,* (7): 50–51.

Zhang, J. & Megaridis, C.M. 1994. Iron/soot interaction in a laminar ethylene nonpremixed flame. *Proc. of Twenty-Fifth Symposium on Combustion,* 25 (1): 593–600.

Zhukov, B.P. 2000. *Energy condensed systems: brief encyclopedic dictionary.* Moscow: Janus-K.

*Advanced Materials, Structures and Mechanical Engineering – Kaloop (Ed.)*
*© 2016 Taylor & Francis Group, London, ISBN 978-1-138-02793-0*

# Weight optimized main landing gears for UAV under impact loading for evaluation of explicit dynamics study

R.F. Swati, A.A. Khan & L.H. Wen
*Institute of Space Technology, Islamabad, Pakistan*
*School of Astronautics, Northwestern Polytechnical University, Xi'an, China*

ABSTRACT: Landing gears being the most critical part of an aircraft must be designed to withstand the worst scenario in a mission profile i.e., loading conditions. While in an unmanned aerial vehicles, weight/structural optimization is one of the important concerns for Analysis. The structure is designed considering the values of stresses, strains/deformations and stress intensities using computational tools for the maximum values of loads, with a reasonable & logical safety factor. Keeping in view all the parameters, weight is optimized in a way such that an optimized structure for the landing gear can withstand deformations. Commercially available computational tools are used for the evaluation of initial structure design in the original & modified model. Under the dynamic conditions during takeoff & landing, modes of vibration & frequencies have been computed to study their effects on the assembly. Explicit dynamics is a powerful and efficient analysis method used to model short duration, high energy dynamic events such as crash and impact, blast, drop testing, ballistics, and metal forming. The Explicit Dynamics tool performs analysis using computational tool package environment.

## 1 INTRODUCTION

Aircraft design is multidisciplinary subject. The design process comprises complex steps such as sizing of components from a conceptual sketch, airfoil and geometry selection, thrust-to-weight ratio and wing loading, configuration layout and lofts preparation etc (Swati & Khan 2014). It is an iterative process with compromises at each step to obtain optimum results that can fulfill the future mission and user requirements as per design. One of the critical parts to be developed during an aircraft design is its landing gears (Roskam 1985). These are designed for the maximum value of localized load. Although there are critical compressions & extensions during takeoff; but worst conditions in a mission profile appear during the landing phase of the aircraft (Currey 1988). In the impact phase of landing energy dissipates involving a wider spectrum of deformation in the elastic or plastic sphere thus making the landing gear design process challenging and demanding (Wu et al. 2005).

The design and positioning of the landing gear is determined by the unique characteristics associated with each aircraft, i.e., geometry, weight, and mission requirements. Given the weight and center of gravity range of the aircraft, suitable configurations are identified and reviewed to determine how well they match the airframe structure, flotation, and operational requirements. This necessitates a rigorous nonlinear Finite Element Analysis (FEA)

of the main landing gear to predict its behavior prior prototype manufacture. Energy absorbed by the main landing gear is stored in the form of elastic strain energy and hence the material used for making the main landing gear should have high elastic strain energy storage capacity. The desired characteristics of the main landing gear are high strength, lightweight, medium stiffness and high elastic strain energy storage capacity.

In this work, a functional specification of the part has to be specified in the beginning. General principles of composite design will be followed in arriving at suitable designs. In the design phase using the FEA method, starting form shape and wall construction, choosing a proper element type, loadings, constraints, materials and behavior modelling is performed. Various constants and lamination parameters are going to be used to define the element.

In terms of the design procedure, the landing gear is the last aircraft major component which is designed. In another word, all major components (such as wing, tail, fuselage, and propulsion system) must be designed prior to the design of landing gear. Furthermore, the aircraft most aft Center of Gravity (CG) and the most forward CG must be known for landing gear design. In some instances, the landing gear design may drive the aircraft designer to change the aircraft configuration to satisfy landing gear design requirements.

The primary functions of a landing gear are to keep the aircraft stable on the ground and during loading, unloading, towing and taxing allowing it to freely move and maneuver during taxing. It also provides a safe distance between other aircraft components such as wing and fuselage while the aircraft is on the ground position to maintain ground clearance. It absorbs the shocks during the landing operation. The structure must facilitate take-off by allowing aircraft acceleration and rotation with the lowest friction.

Designing of landing gear requires extreme care and accuracy for the severe load case scenarios. Literature reveals a number of accidents due to failure of landing gear assembly elements (Asi & Yesil 2013, Azevedo et al. 2002, Bagnoli et al. 2008, Azevedo & Hippert 2002, Franco et al. 2006, Lee et al. 2003, Ossa 2006). Therefore, the main objective of this work is redesigning the landing gear of an aerial vehicle under specified loadings & boundary conditions. Finite Element Method (FEM) has been used for the analysis. The used analysis tool provides a powerful and unified system for tackling the engineering problem. The selected tool effectively handled various parts of the work including static structural analysis, parametric study, optimization study and impact analysis (Pritchard 2001).

## 2 DEVELOPMENT OF MODEL-01

A detailed Computer Aided Design (CAD) is developed using modeling tool for further analysis. The dimensions from existing design were acquired through different analytical models. The design was modified for few dimensions to reflect true geometry while remaining in the design parameters limits.

The parts are modeled & integrated into assembly module of commercially available CAD software. The model further updated and

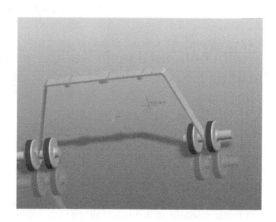

Figure 2. Fully constrained assembly.

Table 1. Material properties and other parameters.

| Parameters | Values |
| --- | --- |
| Distance between NLG and nose | 0.72 m |
| NLG & MLG (F) wheelbase distance | 4.45 m |
| Distance between MLG and aft CG (Lm) | 0.40 m |
| Distance between NLG and aft CG (Ln) | 3.75 m |
| Distance between MLG and fwd. CG (P) | 1.02 m |
| Distance between NLG and fwd. CG (L) | 4.5 m |
| Height (H) | 0.8 m |
| Design take of mass (Mt) | 560 kg |
| *MLG static loads* | |
| Static load on the MLG | 4629.38 N |
| Ultimate static load on the MLG | 6941.57 N |
| Static load on the MLG per strut | 2317.19 N |
| *Safety factor of 7%* | |
| Max static load on the MLG | 6944.57 N |
| Max static load per strut on the MLG | 2314.19 N |
| Design take-off weight (Mt) | 560 kg |
| Design max landing weight (WL) | 4629.38 N |
| Lift coefficient (CL) 0.8 | 4 |
| Number of MLG wheels (each strut) | 647 m² |
| MLG static load per tyre | (4629.38/4) N |

simplified at the major loading mounts for various loading conditions as shown in Figure 1. The integrated model for the analysis was fully constrained in the assembly module for all the parts & joints shown in Figure 2.

The engineering practice starts with the part modeling, linkages & generation of constraint assembly. This could be a solid component for a structural analysis or the air volume for a fluid or electromagnetic study. This geometry is produced as individual parts from scratch which were later assembled.

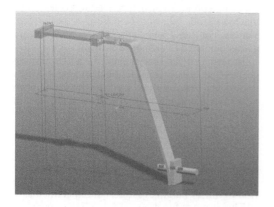

Figure 1. Simplified model of landing gear.

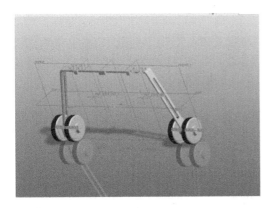

Figure 3. Modified model of landing gears—Model-02.

## 3 DEVELOPMENT OF MODEL-02

The geometric parameters are once again set for the optimized model, i.e. Model-02. The total volume has been reduced resulting in the weight reduction of the system as shown in Figure 3.

## 4 STATIC STRUCTURAL ANALYSIS FOR MODEL-02

Symmetric boundary conditions identical to Model-01 are set in computational tool workbench. Previously used engineering data is again applied for static loading, a similar mapped mesh and refinement is also generated for this step of analysis for Model-02. Static structural analysis of Model-02 after applying engineering data, geometry, model, setup & solution completed and all the results are analyzed and compared with the results acquired for Model-01.

## 5 RESULTS COMPARISON FOR BOTH MODELS

The result summary for both the cases is shown in Table 3. The static structural analysis for Model-01 & Model-02 is compared for Von Misses stresses, Von Misses strains, directional deformations, Maximum principle stresses & maximum shear elastic strains.

## 6 STATIC STRUCTURAL ANALYSIS FOR MODEL-02

After applying the Symmetric boundary conditions identically to Model-01 in Computational tool, Static structural analysis of Model-02 is analyzed and compared with the results of Model-01.

Table 2. Results comparison for static structural analysis.

| Object name | Minimum | Maximum |
|---|---|---|
| *Model-01* | | |
| Equivalent stress (MPa) | 0.00019772 MPa | 461.7 MPa |
| Equivalent elastic strain | 9.8862e-010 mm/mm | 5.3083e-003 mm/mm |
| Directional deformation | −63.082 mm | 66.648 mm |
| Max principal stress | −123.49 MPa | 1013.4 MPa |
| Max shear elastic strain | 1.4706e-009 mm/mm | 7.1826e-003 mm/mm |
| *Model-02* | | |
| Equivalent stress | 4.3721e-004 MPa | 542.2 MPa |
| Equivalent elastic strain | 2.1861e-009 mm/mm | 8.461e-003 mm/mm |
| Max principal stress | −226.04 MPa | 323.6 MPa |
| Directional deformation | −79.587 mm | 84.355 mm |
| Max shear elastic strain | 3.0308e-009 mm/mm | 1.1264e-002 mm/mm |
| Elastic strain intensity | 3.0308e-009 mm/mm | 1.1264e-002 mm/mm |

Table 3. Results comparison for both cases.

| Object name | Minimum | Maximum |
|---|---|---|
| *Model-01* | | |
| Equivalent stress | 9.5336e-010 MPa | 169.53 MPa |
| Equivalent elastic strain | 4.7668e-015 mm/mm | 8.4764e-004 mm/mm |
| Maximum principal stress | −15.887 MPa | 158.77 MPa |
| Maximum principal elastic strain | −3.4985e-009 mm/mm | 7.6463e-004 mm/mm |
| *Model-02* | | |
| Equivalent stress | 2.9115e-009 MPa | 142.86 MPa |
| Equivalent elastic strain | 1.4558e-014 mm/mm | 7.1432e-004 mm/mm |
| Directional deformation | −3.2234e-002 mm | 9.1212 mm |
| Maximum principal elastic strain | −2.1905e-007 mm/mm | 7.1486e-004 mm/mm |

Table 4. Comparison of parameters in weight reduction.

| Parameter | Model-01 | Model-02 |
|---|---|---|
| Height | 540 mm | 510 mm |
| Thickness | 320 mm | 290 mm |
| Width | 50 mm | 58 mm |
| Reduction | 0 mm³ | 200X 20X5 mm³ |
| Factor of safety | 1.2 | 1.12 |
| Weight | 20.089 kg | 13.174 kg |

Figure 4. Equivalent total stress (Von-Misses) contour plot.

Figure 5. Equivalent total strain (Von-Misses) contour plot.

Table 5. Comparison of parameters in weight reduction.

| Type | Total acceleration | Maximum principal elastic strain | Total velocity | Directional acceleration | Stress intensity |
|------|------|------|------|------|------|
| *Results* | | | | | |
| Min | 0. mm/s² | −8.877e-006 mm/mm | 0. mm/s | −2.3081e+009 mm/s² | 0. MPa |
| Max | 2.77E+09 | 1.1493e-003 mm/mm | 18370 | 2.0142e+009 mm/s² | 323.28 MPa |

Figure 6. Total deformation.

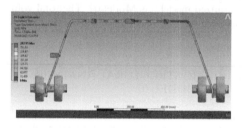

Figure 7. Equivalent stress.

## 7 OPTIMIZED PARAMETERS AND COMPARISON

The overall effect on the structure of landing gears keeping in mind the structural safety & integrity, the weight optimization overall summary is listed in the Table 4.

## 8 RESULTS CONCLUSION AND DISCUSSIONS

The maximum stresses from Figures 4 & 5 clearly depict that the directional deformations are a little increased but the equivalent stresses & strains lies in the safe region for the assembly. Similarly Table 4 clearly shows that the structure lies in reasonable safe limit with reduced/optimized parameters of its geometric dimensions.

## 9 RESULT PLOTS

The post processing plot results for nodal solutions have been obtained to identify the critical zone and the area of consideration for different values of impacts:

## 10 CONCLUSIONS

After completion of stress analysis for model-01 & model-02, it is clear from Table 5 that the structure is safe for the specified loading conditions. The maximum values computed as seen in Figures 6 & 7, clearly depict that the directional deformations are slightly increased but the equivalent stresses & strains have dropped while remaining in the safe region for landing gear assembly.

## REFERENCES

Asi, O. & Yeşil, Ö. 2013. Failure analysis of an aircraft nose landing gear piston rod end, *Engineering Failure Analysis,* 32: 283–291.

Azevedo, C.R.F. Hippert Jr, E. Spera, G. & Gerardi, P. 2002. Aircraft landing gear failure: fracture of the outer cylinder lug, *Engineering Failure Analysis,* 9: 1–15.

Bagnoli, F. Dolce, F. Colavita, M. & Bernabei, M. 2008. Fatigue fracture of a main landing gear swinging lever in a civil aircraft, *Engineering Failure Analysis*, 15: 755–765.

Borouchaki, H. & George, P.L. 2000. Quality mesh generation, *Comptes Rendus de l'Académie des Sciences—Series IIB—Mechanics*, 328: 505–518.

Currey, N.S. 1988. *Aircraft landing gear design: principles and practices*, Aiaa.

de Farias Azevedo, C.R. & Hippert Jr, E. 2002. Fracture of an aircraft's landing gear, *Engineering Failure Analysis*, 9: 265–275.

Franco, L.A.L. Lourenço, N.J. Graça, M.L.A. Silva, O.M.M. de Campos, P.P. & von Dollinger, C.F.A. 2006. Fatigue fracture of a nose landing gear in a military transport aircraft, *Engineering Failure Analysis*, 13: 474–479.

Lee, H.C. Hwang, Y.H. & Kim, T.G. 2003. Failure analysis of nose landing gear assembly, *Engineering Failure Analysis*, 10: 77–84.

Nicolai, L.M. & Dayton, O. 1975. *Fundamentals of aircraft design: School of Engineering*, University of Dayton Dayton, OH.

Ossa, E.A. 2006. Failure analysis of a civil aircraft landing gear, *Engineering Failure Analysis*, 13: 1177–1183.

Pritchard, J. 2001. Overview of landing gear dynamics, *Journal of aircraft*, 38: 130–137.

Roskam, J. 1985. *Airplane design*, DARcorporation.

Swati, R.F. & Khan, A.A. 2014. Design and Structural Analysis of Weight Optimized Main Landing Gears for UAV under Impact Loading, *Journal of Space Technology*, 4(1): 96–100.

Wu, C.Y. Li, L.Y. & Thornton, C. 2005. Energy dissipation during normal impact of elastic and elastic–plastic spheres, *International Journal of Impact Engineering*, 32: 593–604.

*Advanced Materials, Structures and Mechanical Engineering – Kaloop (Ed.)*
© 2016 Taylor & Francis Group, London, ISBN 978-1-138-02793-0

# Analysis of the influence factors of carbon emissions based on the LMDI approach—taking Jiangsu Province as an example

X.Y. Wang & S.Y. Jiang

*College of Economics and Management, Nanjing University of Aeronautics and Astronautics, Nanjing, China*

ABSTRACT: The rapid development of an economy is bound to cause the increase in carbon emissions. This article calculates the carbon emissions of 17 kinds of primary energy end-use from 2000 to 2012 of Jiangsu Province, and decomposes the carbon emissions into six factors by the LMDI method: population, per capita GDP, industrial structure, energy intensity, energy consumption structure and coefficient of carbon emissions (Log-Mean Divisia Index Method I). It finally provides a concrete analysis of each factor on carbon emissions.

## 1 INTRODUCTION

Jiangsu is one of the biggest provinces in China. In 2012, the primary energy consumption of Jiangsu Province was 290 million tons, accounting for 8% of the national energy consumption, and its terminal energy consumption was 56.029 million tons, which means there was a massive carbon emission. Controlling energy consumption and cutting carbon emissions is a serious question. In the 12th five-year plan, the Chinese government proposed that it will cut $CO_2$ emissions of per unit GDP by 40%-50% below the 2005 levels and increase the share of non-fossil fuels in primary energy consumption to around 15% by 2020. In this background, Jiangsu Province must actively explore the influencing factors of carbon emissions and formulate rational and effective energy conservation and emissions reduction policies.

## 2 LITERATURE REVIEW

The most commonly used analytical method of carbon emissions' influencing factors is Laplace's index method and the LMDI method. The former has residual, and its decomposed result may have a 0 value, but the latter is the preferred option. So scholars usually adopt the LMDI model to study the influencing factors of carbon emissions. Wang, C. Chen, J. & Zou, J (2005) used the LMDI method to decompose the Chinese carbon emissions, and analyzed four factors of carbon emissions including energy intensity, industrial structure, population, and GDP. Fan, Y. Liu, L.C. & Wu, G (2007) used the AWD method

to analyze the influence factors of carbon intensity, and found that the method showed a trend of gradual decline. D. Diakoulaki (2007) used Laplace's index model to analyze the influence factors of $CO_2$ emissions of European countries' industrial sectors. Regrettably, although European countries made efforts to cut $CO_2$ emissions, it appeared to have a little effect.

Domestic scholars also have done a lot of research from different angles in different ways. Song, D.Y. & Lu, Z.B. (2009) adopted a "dual stage" LMDI method to do research on the influence factors of carbon emissions. Guo, C.X (2010) decomposed carbon emissions into GDP, economic structure, energy efficiency, energy consumption structure and coefficient of carbon emissions from the perspective of industry and region. Dai, X.W (2013) studied the implicit carbon emissions, and argued that the progress of the mode of production and life style was the most important driving factor to change carbon emissions. Wang, Y. Jia, J.J. Zhao, P. Cheng, X. & Sun, T (2014) used the LMDI method to analyze the influencing factors of Tianjin Carbon emissions, and carried out a scenario analysis to seek the appropriate carbon reduction policies. All these above studies have performed a thorough analysis on the influencing factors of carbon emissions or $CO_2$ emissions at the national or regional level, and described the characteristics of carbon emissions and the main challenging problems well, which have an important reference value to the realization of the low-carbon economy. However, given that many regions of China have many significant differences, the existing literature about carbon emissions is not enough to reflect comprehensively the influencing factors.

## 3 RESEARCH METHODS AND DATA SOURCES

### 3.1 LMDI model introduction

According to the Kaya model, the influencing factors of Jiangsu's energy consumption presented in this paper can be decomposed into six components, namely population, per capita GDP, industrial structure, energy efficiency, energy structure and carbon emission coefficient, and then the contribution of each factor to carbon emissions changes can be analyzed, in order to provide references for carbon emission reduction policies. The formula for the Kaya model is represented as follows:

$$C = \sum_{ij} C_{ij}$$

$$= \sum_{ij} P * \frac{G}{P} * \frac{G_i}{G} * \frac{E_i}{G_i} * \frac{E_{ij}}{E_i} * \frac{C_{ij}}{E_{ij}}$$

$$= \sum_{ij} P * Y * S_i * I_i * M_{ij} * F_{ij} \qquad (1)$$

where $C$ is the carbon emissions of terminal energy consumption; $i$ refers to different industries, $i = 1,2,3$; $j$ indicates different kinds of energies, $j = 1,...,17$; $C_{ij}$ is carbon emissions of energy $j$ in industry $i$; $E_i$ is the energy consumption of industry $i$; $E_{ij}$ is the energy consumption of energy $j$ in industry $i$; $G$ is GDP; $G_i$ is the gross domestic product of industry $i$; $P$ is the population; $Y$ is per capita GDP; $S$ is the industrial structure; $I$ is the energy intensity; $M$ is the energy structure; and $F$ is the coefficient of carbon emissions.

The growth of carbon emissions in energy consumption is defined as the comprehensive effect and denoted as $\Delta C$. According to the additive decomposition of the LMDI model, the comprehensive effect is made up of six aspects: $\Delta C_p$, population effect; $\Delta C_Y$, effect of per capita GDP; $\Delta C_s$, industrial structure effect; $\Delta C_I$, energy intensity effect; $\Delta C_m$, effect of energy structure; $\Delta C_F$, carbon emission coefficient effect, which can be expressed as follows:

$$\Delta C = C^T - C^0 = \Delta C_p + \Delta C_Y + \Delta C_s$$
$$+ \Delta C_I + \Delta C_m + \Delta C_F \qquad (2)$$

where $C^T$ is carbon emissions in year t and $C^0$ is carbon emissions in year of benchmark. Because the energy coefficient of carbon emissions basically has no change, the effect of the coefficient of carbon emissions $\Delta C_F$ is equal to 0. The formula for the comprehensive effect can be represented as follows:

$$C^T - C^0 = \Delta C_p + \Delta C_Y + \Delta C_s + \Delta C_I + \Delta C_m \qquad (3)$$

In Formula (3),

$$\Delta C_p = \sum_{ij} \frac{C_{ij}^T - C_{ij}^0}{\ln C_{ij}^T - \ln C_{ij}^0} \ln \frac{P^T}{P^0} \qquad (4)$$

$$\Delta C_Y = \sum_{ij} \frac{C_{ij}^T - C_{ij}^0}{\ln C_{ij}^T - \ln C_{ij}^0} \ln \frac{Y^T}{Y^0} \qquad (5)$$

$$\Delta C_s = \sum_{ij} \frac{C_{ij}^T - C_{ij}^0}{\ln C_{ij}^T - \ln C_{ij}^0} \ln \frac{S_i^T}{S_i^0} \qquad (6)$$

$$\Delta C_I = \sum_{ij} \frac{C_{ij}^T - C_{ij}^0}{\ln C_{ij}^T - \ln C_{ij}^0} \ln \frac{I_i^T}{I_i^0} \qquad (7)$$

$$\Delta C_m = \sum_{ij} \frac{C_{ij}^T - C_{ij}^0}{\ln C_{ij}^T - \ln C_{ij}^0} \ln \frac{M_{ij}^T}{M_{ij}^0} \qquad (8)$$

When $C_{ij}^T = C_{ij}^0$, or other value is 0, the value is defined as $10^{-10}$.

### 3.2 Data sources

End-use energy consumption is regarded as an important part of carbon emissions, which is mainly fossil fuels. According to the data of the energy balance sheet, the fossil fuels are used in the primary industry (e.g. agriculture, forestry, animal husbandry and fishery and water conservation), the secondary industry (e.g. industry and construction), and the tertiary industry (e.g. transportation, warehousing and postal service, wholesale, retail and accommodation, and catering). In contrast to other studies, this paper argues that life energy consumption is also an important part, so it is included in the tertiary industry. In addition, this paper eliminates the energy consumption of electricity and heat.

Because China has not yet released carbon emissions data of every province directly, we need to calculate carbon emissions through the corresponding calculation method. The formula for the calculation of carbon emissions is given as follows:

$$C = \sum_{ij} e_{ij} * L_j * F_j,$$

where $i$ represents different kinds of industries and $j$ represents different kinds of energies. Accordingly, carbon emissions data can be calculated as follows: $E_{ij}$ (energy consumption of energy $j$ in industry $i$) times $L_j$ (conversion coefficient of standard coal) is standard coal and then multiplies $F_j$ (coefficient of carbon emissions). Carbon emissions coefficient has a low rate of change in different years, so this article assumes its value to be constant and it is calculated according to the (IPCC) 2006 carbon emissions calculation guide.

## 4 RESULTS AND ANALYSIS

Using the decomposition model LMDI, and regarding the year 2000 as the base period, carbon emissions of 17 kinds of energy end-use between 2001 and 2012 of Jiangsu Province are decomposed. The results of the cumulative effect of each decomposition factor are summarized in Table 1.

If the base year 2000 is replaced by T-1, the yearly effect of 5 factors can be calculated, the results of which are summarized in Table 2.

It can be seen from the decomposition results that the cumulative effect of the energy consumption intensity of Jiangsu Province in 2012 is negative, which indicates that the energy intensity's increase can inhibit the increase in carbon emissions. Industrial structure had a positive impact on the carbon emissions' increase from 2001 to 2010. However, in 2011 and 2012, the effect of industrial structure became negative, which indicates that the industrial structure adjustment can inhibit the increase in carbon emissions. In 2012, the population, per capita GDP and the energy structure all had a positive cumulative effect, which indicates that they can promote the increase in carbon emissions.

From the percentage cumulative effect, the cumulative effect of per capita GDP is the largest, and its contribution to promote the carbon emissions is 161.32%; the second is the energy intensity, whose contribution to curb carbon emissions is 79.87%. Population change for carbon emissions increase has a small drive, i.e. only 7.14%. The

Table 1. Cumulative effect factors of carbon emissions of Jiangsu Province.

| Year | Population | GDP per capita | Industrial structure | Energy intensity | Energy structure | Total effect |
|------|-----------|----------------|----------------------|------------------|------------------|--------------|
| 2001 | 12.9131 | 275.935 | 15.9560 | −430.42 | −18.5326 | −144.148 |
| 2002 | 32.3451 | 605.604 | 28.1868 | −664.925 | −30.7215 | −29.5103 |
| 2003 | 55.9914 | 1125.15 | 106.812 | −949.325 | −54.9476 | 283.676 |
| 2004 | 97.1690 | 2078.06 | 309.912 | −920.044 | −10.6194 | 1554.48 |
| 2005 | 145.079 | 3047.79 | 179.291 | −857.454 | 115.897 | 2630.60 |
| 2006 | 188.035 | 3831.68 | 236.467 | −1301.53 | 161.067 | 3115.72 |
| 2007 | 237.458 | 4774.87 | 172.822 | −1686.45 | 142.624 | 3641.33 |
| 2008 | 268.510 | 5650.01 | 94.9595 | −2091.27 | 144.043 | 4066.25 |
| 2009 | 376.379 | 7871.74 | 177.843 | −2992.15 | 1083.26 | 6517.07 |
| 2010 | 329.056 | 6923.07 | 48.6745 | −3382.37 | 486.555 | 4404.98 |
| 2011 | 361.467 | 8011.18 | −23.5462 | −3822.7 | 615.474 | 5141.87 |
| 2012 | 365.050 | 8249.85 | −91.8884 | −4084.55 | 675.539 | 5114.01 |

Table 2. Yearly effect of carbon emissions of Jiangsu Province.

| Year | Population | GDP per capita | Industrial structure | Energy intensity | Energy structure | Total effect |
|------|-----------|----------------|----------------------|------------------|------------------|--------------|
| 2001 | 12.91308 | 275.9354 | 15.956 | −430.421 | −18.5359 | −144.152 |
| 2002 | 18.85067 | 318.8105 | 11.605 | −220.754 | −14.2536 | 114.2586 |
| 2003 | 22.18469 | 492.4446 | 71.718 | −247.417 | −24.9641 | 313.9662 |
| 2004 | 33.55701 | 806.1248 | 179.477 | 212.2352 | 44.31807 | 1275.712 |
| 2005 | 43.32203 | 859.8072 | −211.064 | 222.0812 | 120.2283 | 1034.375 |
| 2006 | 52.64522 | 942.6186 | 71.275 | −573.788 | 21.56181 | 514.3126 |
| 2007 | 55.8228 | 1046.717 | −116.595 | −440.192 | −21.9588 | 523.794 |
| 2008 | 35.02571 | 1078.791 | −127.137 | −521.557 | −49.021 | 416.1017 |
| 2009 | 45.76086 | 902.4071 | 65.029 | −433.865 | −5.72711 | 573.6049 |
| 2010 | 56.08019 | 1239.268 | −122.478 | −1699.37 | 320.7367 | −205.763 |
| 2011 | 29.95687 | 1295.307 | −113.362 | −485.441 | 33.80045 | 760.2613 |
| 2012 | 21.88869 | 765.0363 | −114.668 | −633.215 | −61.2573 | −22.2154 |

contribution of energy consumption structure to increase carbon emissions is 13.21%. The contribution rate of the industrial structure changes to curb carbon emissions increase is very small, i.e. only 1.8%. From the viewpoint of the yearly effect, the demographic factor is positive, but smallest, which indicates that population growth has a small influence on carbon emissions. The effect of per capita GDP is positive, but largest, which means that it is the most important factor to increase carbon emissions.

Next, we will further study the effect of industrial structure and energy consumption structure effect.

### 4.1 Effect of industrial structure

According to the results of the above decomposition factors, it can be seen that the influence of the industrial structure change on carbon emissions is relatively small. In 2001–2010, the industrial structure played a positive role in promoting carbon emissions; in 2001–2004, the cumulative effect increased gradually, but showed an irregular decline after 2004, until it turned negative in 2011. Observing the change in the three industries of Jiangsu Province, the proportion of the secondary industry rose from 51.68% to 58.09% in 2001–2004, and ranged from 56.35% to 50.17% in 2006–2012, which are all in line with the cumulative effect of the industrial structure. The proportion of the tertiary industry increased from 36.3% in 2000 to 43.5% in 2012, and that of the primary industry ranged from 12.01% to 6.32%. In conclusion, reducing the proportion of the secondary industry can inhibit the growth of carbon emissions, so adjusting the industrial structure, promoting industrial structure optimization, and increasing the proportion of the tertiary industry particularly can significantly inhibit the growth of carbon emissions.

### 4.2 Energy consumption structure effect

Energy consumption structure that is regarded as the influencing factor of carbon emissions cannot be ignored. From 2001–2004, the cumulative contribution of the energy structure was negative, but after that, it turned into positive values, and it changed from 4.41% in 2005 to 13.21% in 2012. Raw coal, coke, diesel, gasoline, fuel oil and other petroleum products accounted for 80%–90% of the total energy consumption, which means that the total energy consumption structure has no change. While the proportion of coal consumption was decreased, the coke consumption increased, and the carbon emission coefficient of coke was higher,

up to 0.8542. Fall and rise leads to high carbon emissions all the time.

## 5 RESULTS AND ADVICES

Based on the 17 kinds of energy carbon emission coefficients, this article calculates the energy consumption carbon emissions of Jiangsu Province from 2000 to 2012, and then the carbon emissions' increment is decomposed into six factors by the LMDI method: population, per capita GDP, industrial structure, energy intensity and energy consumption structure and the coefficient of carbon emissions. Finally, it calculates the cumulative effect and the yearly effect of the carbon emissions. From the viewpoint of the decomposition results, we can make the following conclusions and advices:

1. Per capita GDP is the biggest influencing factor for carbon emissions' increase, which has a certain relationship with the high economic growth of Jiangsu Province. In the past 10 years, its average growth rate of per capita GDP is about 16.9%. Therefore, in order to reduce the effect of this factor, people must completely abandon the perception of focusing on GDP growth, but not ignoring the concept of energy consumption and the environment pollution. It also needs to adjust the industrial structure, and transform the pattern of economic development, which is a necessary step for the coordinated development of economy and environment.
2. Demographic factor promotes the increase in carbon emissions, but its effect is relatively small. It benefits from the family planning. Although the population growth rate has dropped a lot, policy should still strengthen the control of the population in the future, at the same time improve the quality of the population.
3. The influence of the industrial structure on carbon emissions changes from the start promoting effect to restraining. It shows that the adjustment of the industrial structure has significantly inhibited the increase in carbon emissions. It should eliminate the backward and unreasonable industries, especially the energy-intensive industries, and promote the upgrading of the industrial structure further.
4. To promote the progress of science and technology and innovation, improving energy intensity is a necessary step to curb carbon emissions. To improve the unreasonable energy consumption structure, the dependence on one-off energy such as coal and coke that have a higher carbon emission coefficient should be reduced, which also is an important way to reduce carbon emissions.

REFERENCES

Dai, X.W. 2013. China's implicit carbon factor decomposition research. *Science of finance and economics*, 2013: 101–109.

Diakoulaki, D. & Manddaraka. 2007. Decomposition Analysis for Assessing the Progress in Decoupling Industrial Growth from $CO_2$ Emissions in the EU Manufacturing Sector. *Energy Economics*, 29(4): 636–664.

Fan, Y. Liu, L.C. & Wu, G. et al. 2007. Changes in Carbon Intensity in China: Empirical Findings from 1980–2003. *Ecological Economics*, 62: 683–691.

Guo, C.X. 2010. China's emissions factorization: basing on LMDI decomposition technique. *China population resources and environment*, 20(12): 4–9.

Song, D.Y. & Lu, Z.B. 2006. The decomposition China's carbon emissions influence factor and cyclical fluctuations study. *China population resources and environment*, (6): 158–161.

The IPCC. 2006. *The IPCC in 2006 national greenhouse gas inventories guide*. The global environment institute for strategic studies in Japan.

Wang, C. Chen, J. & Zou, J. 2005. Decomposition of Energy-related $CO_2$ Emission in China: 1957–2000. *Energy*, 30(1): 73–83.

Wang, Y. Jia, J.J. Zhao, P. Cheng, X. & Sun, T. 2014. The analysis of LMDI method on the influence of the structure effect of carbon emissions in Tianjin and countermeasures. *Journal of Tianjin university (social science edition)*, (6): 509–514.

*Advanced Materials, Structures and Mechanical Engineering – Kaloop (Ed.)*
*© 2016 Taylor & Francis Group, London, ISBN 978-1-138-02793-0*

# Impedance characteristics of laboratory scale molten carbonate fuel cell fueled by ash free coal

S.W. Lee, T.K. Kim, Y.J. Kim & C.G. Lee
*Hanbat National University, Yuseong-gu, Daejeon, Republic of Korea*

M. Wołowicz
*Faculty of Power and Aeronautical Engineering, Warsaw University of Technology, Warsaw, Poland*

ABSTRACT: The paper presents results of experimental investigations of the molten carbonate fuel cell. Testing facility and components used for experiments are explained. Gasification of ash free coal process is described, as well as impedance characteristics of the molten carbonate fuel cell fueled by ash free coal. Comparison between AFC and hydrogen as a fuel for MCFC is presented. Then, simulation gases expected from gasified ash free coal were investigated. The simulations were provided for different composition of gases based on results of AFC gasification. Impedance characteristics for investigated case are presented as well as results are discussed.

## 1 INTRODUCTION

Nowadays, power is produced in the same way as it was generated almost 100 years ago, i.e. by utilization of heat cycles having their limitations (maximum efficiency on the level of <50%). Limited natural resources and increasing electricity consumptions force to research and develop more efficient devices. Fuel cells produce power as the consequence of electrochemical reactions, thus they are not bound by Carnot efficiency. This can result in very high efficiency.

Fuel cells are considered as the most perspective sources of energy for future generations (Kotowicz et al. 2011). Research and development associated with fuel cells are widely supported by national and international organizations (European Union, USA Department of Energy, Republic of Korea and others).

There are many types of fuel cells with various materials used as an electrolyte (Polymeric Membrane, Phosphoric Acid, Molten Carbonates, Solid Oxides, and others). All of them have advantages as well as disadvantages. Only very few (i.e. molten carbonate fuel cell) can utilize conventional fuels and provide the opportunity of producing additional energy for heating purposes. Fuel cells can also be fed by biogas (Budzianowski 2012) and can work as combined heat and power source for i.e. Distributed Generation (Milewski et al. 2012). Beside this advantages, molten carbonate fuel cells can be considered as a very efficient

way of separating CO2 from flue gases (Discepoli et al. 2012) which is competitive to other methods (Bartela et al. 2012, Janusz-Szymanska) that seek to reduce emissions from fossil fuel power plants (Bujalski 2012). Based on this statement, investigating molten carbonate fuel cells behavior seems to be justified. One of the ideas presented in this work is to use gasified ash free coal as a fuel for MCFC.

## 2 GASIFICATION OF ASH FREE COAL

Coal is relatively abundant energy source compared with other fossil fuels. Regardless of its abundance and a long history of usage, an inconvenience of its use reduces coal consumption at present. Coal as a solid fuel has a big problem of remaining ash in its use. Ash free coal has been attempted to overcome the problem. A trend of producing technologies of the ash free coal is described widely in (Kim & Lee 2012). Possibilities and effect of hydrogen production via steam gasification of ash free coals is presented in (Kim et al. 2013). In the same work also a comparison of steam gasification reactivity of AFCs and raw coals is described. Also more about gasification of AFC and using AFC as fuel for a direct carbon fuel cell can be found in (Lee et al. 2011, Kong et al. 2014).

To check the composition of syngas after gasification, experiment with AFC in the laboratory scale was performed. The gasification conditions were

850°C and Nitrogen flow at the level of 100 mL/min. Results changing in 60 min are presented in Table 1 for gas flow during gasification process of ash free coal and in Table 2 for gas composition during gasification of ash free coal. The data are in mL/min

Table 1. Gas flow during gasification process of ash free coal.

| Time [min] | Gas flow [mL/min] | | | | |
|---|---|---|---|---|---|
| | $H_2$ | $N_2$ | CO | $CH_4$ | $CO_2$ |
| 10 | 215.58 | 100 | 93.95 | 30.19 | 72.83 |
| 20 | 30.76 | 100 | 27.01 | 1.30 | 9.21 |
| 30 | 7.82 | 100 | 10.90 | 0.03 | 6.63 |
| 40 | 5.56 | 100 | 6.61 | 0 | 4.64 |
| 50 | 4.69 | 100 | 5.62 | 0 | 7.79 |
| 60 | 3.43 | 100 | 3.58 | 0 | 1.66 |

Table 2. Gas composition during gasification process of ash free coal.

| Time [min] | Gas composition [%] | | | | |
|---|---|---|---|---|---|
| | $H_2$ | $N_2$ | CO | $CH_4$ | $CO_2$ |
| 10 | 42.06 | 19.51 | 18.33 | 5.89 | 14.21 |
| 20 | 18.28 | 59.42 | 16.05 | 0.77 | 5.47 |
| 30 | 6.24 | 79.76 | 8.69 | 0.02 | 5.29 |
| 40 | 4.76 | 85.61 | 5.66 | 0 | 3.97 |
| 50 | 3.97 | 84.68 | 4.76 | 0 | 6.6 |
| 60 | 3.16 | 92.02 | 3.29 | 0 | 1.53 |

and percentage composition is also given. Percentage gas composition during gasification of ash free coal is presented also in Figure 1.

## 3 EXPERIMENTAL COMPARISON BETWEEN AFC AND HYDROGEN AS A FUEL FOR MCFC

To compare the behavior of molten carbonate fuel cell fueled by ash free coal and hydrogen, an adequate experiment was provided. A coin type molten carbonate fuel cell was investigated. The diameter of electrodes was about 3 cm. The anode was porous Ni-Al alloy, and the cathode was porous Ni. The matrix was made of $LiAlO_2$. The elements of the fuel cell testing facility were alumina tube, steel manifolds and current collectors which are shown in Figure 2. Fuel cell being installed in the manifold is shown in Figure 3. Before installation in the manifold, fuel cells (cathode, anode and matrix) have to be cut to their size (3 cm diameter). The long alumina tube at the upside of the cell was installed in the muffle furnace for the coal supply to the anode as shown in the Figure 4.

All experiments were provided in a muffle furnace. The operating temperature was 850°C. It has to be noticed, that providing experiments in the muffle furnace takes less time, than making them in the test bench. The difference is of course in the area of the cell (3 cm diameter in the case with furnace—so called "coin type" vs. $11 \times 11$ cm square shape in the test bench) and in

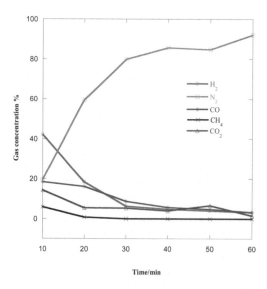

Figure 1. Percentage gas composition during gasification of ash free coal.

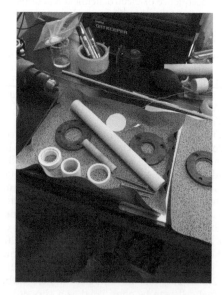

Figure 2. Components used for testing of coin type molten carbonate fuel cells fueled by ash free coal.

Figure 3.  Fuel cell being installed in the manifold.

Figure 4.  Coin type molten carbonate fuel cell with alu-mina tube at the upside of the cell used for experiment and installed in muffle furnace.

heating system. In the furnace, a fuel cell can be heat up faster because of its size and of the heating process. In the test bench scale experiments heating is realized by two heaters (located below and above manifold). It can cause thermal stresses that are why the start-up of fuel cell takes longer than in furnace. In an example, the start-up for a coin type fuel cell can be done in few hours, but the start up for test bench scale fuel cell should take 2 days to reach the reference point.

For the experiments, as an electrolyte a mixture of 62 mol% $Li_2CO_3$ and 38 mol% $K_2CO_3$ was used. The cathode gas was 70% air and 30% $CO_2$. The normal $H_2$ fuel for the anode was $H_2$—125 mL/min,

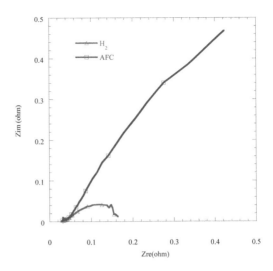

Figure 5.  Impedance characteristics of experimental comparison between molten carbonate fuel cell fueled by AFC (3 g of AFC and 3 g of carbonates) and $H_2$ (125 mL/min)+$CO_2$ (25 mL/min)+$H_2O$ (5 g) at 850°C.

$CO_2$—25 mL/min with ca. 5% of $H_2O$. The ash free coal was a mixture of 3 g coal and 3 g of Li-K carbonates. More details of the molten carbonate fuel cell operation with AFC as a fuel was described in a previous work (Lee et al. 2014).

In Figure 5 the impedance characteristics for both cases are presented. As described in (Kim et al. 2015) the ash free coal has much higher OCV than hydrogen. However, AFC has larger resistance than normal hydrogen fuel.

## 4  EXPERIMENTAL INVESTIGATION OF SUPPLYING GASIFIED AFC AS A FUEL FOR MCFC

To check the behavior of molten carbonate fuel cell fueled by gas generated from gasified ash free coal, following simulation of gas compositions were investigated: $H_2$+$CO$+$CO_2$+$H_2O$. Authors decided to take under consideration this composition, because it is adequate to composition of syngas after gasification, which is shown in Table 2. It can be noticed, that after 40 minutes of gasification process there is no methane in the composition. After 60 minutes, about 92% of gas composition is nitrogen and only about 3% is hydrogen.

The experiments were made in the same laboratory test equipment as a comparison between AFC and hydrogen. They were provided with the same manifold and in the same muffle furnace. Also parameters of the testing coin type molten carbonate fuel cell were similar. All of the

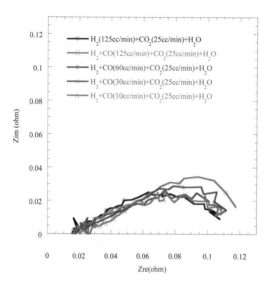

Figure 6. Experimental obtain impedance characteristics for $H_2$+CO+$CO_2$+$H_2O$ simulated gas at 850°C.

simulations were provided in temperature 850°C. For the simulated case, flow of the gases were 10, 30, 60 and 125 mL/min of $H_2$, 25 mL/min of $CO_2$, different amount of CO (10, 30, 60 and 125 mL/min) and also $H_2O$ (5%). The result is presented at Figure 6.

The OCV for an experiment with added CO at the flow of 125 mL/min is even higher than for reference one (Kim et al. 2015). However, CO raises the resistance at anode, so the slope is deeper than to hydrogen reference curve. For the experiments with amount of $H_2$ and CO at the level of 10, 30 and 60 mL/min OCV is lower than for reference fuel (Kim et al. 2015). It is caused by larger partial pressure of $CO_2$ at anode side ($CO_2$ flow is in constant level).

## 5 CONCLUSIONS

Experimental comparison between AFC and hydrogen as a fuel for MCFC was provided. The impedance characteristics were presented. The ash free coal has much higher OCV than hydrogen which authors described in their previous work (Kim et al. 2015). However, AFC has larger resistance than normal hydrogen fuel. For the experimental investigations of simulated gases delivering to the fuel cell, it can be noticed, that adding of CO is limiting hydrogen transport, so CO raises the resistance at the anode side. On the other hand, CO can reduce the resistance of hydrogen-only fueled fuel cell, so CO addition may not raise the resistance.

## ACKNOWLEDGEMENTS

This research was supported by the research fund of Hanbat National University in 2014. For one of the authors (Marcin Wołowicz) work has been supported by the European Union in the framework of European Social Fund through the "Didactic Development Program of the Faculty of Power and Aeronautical Engineering of the Warsaw University of Technology".

## REFERENCES

Bartela, L. Skorek-Osikowska, A. & Kotowicz, J. 2012. Integration of a supercritical coal-fired heat and power plant with carbon capture installation and gas turbine, *Rynek Energii*, 100(3): 56–62.

Budzianowski, W. 2012. Sustainable biogas energy in poland: Prospects and challenges, *Renewable and Sustainable Energy Reviews*, 16(1): 342–349.

Bujalski, W. 2012. Optimization of electricity and heat generation in large chp plant equipped with a heat accumulator, *Rynek Energii*, 101(4): 131–136.

Discepoli, G. Cinti, G. Desideri, U. Penchini, D. & Proietti, S. 2012. Carbon capture with molten carbonate fuel cells: Experimental tests and fuel cell performance assessment, *International Journal of Greenhouse Gas Control*, 9: 372–384.

Janusz-Szymanska, K. 2012. Economic efficiency of an IGCC system integrated with CCS installation, *Rynek Energii*, 102(5): 24–30.

Kim, J. Choi, H. Lim, J. Rhim, Y. Chun, D. Kim, S. Lee, S. & Yoo, J. 2013. Hydrogen production via steam gasification of ash free coals, *International Journal of Hydrogen Energy*, 38: 6014–6020.

Kim, S.H. & Lee, C.G. 2012. A Trend of Producing Technologies of the Ashless Hyper Coal as a Clean Energy Source, *Journal of Energy Engineering*, 21(4): 325–338.

Kim, T.K. Kim, Y.J. Lee, S.W. Lee, C.G. & Wołowicz, M. 2015. Operational characteristics of coin type Molten Carbonate Fuel Cell fueled by Ash Free Coal, *Applied Mechanics and Materials*, 752–753: 438–443.

Kong, Y. Kim, J. Chun, D. Lee, S. Rhim, Y. Lim, J. Choi, H. Kim, S. & Yoo, J. 2014. Comparative studies on steam gasification of ash-free coals and their original raw coals, *International Journal of Hydrogen Energy*, 39: 9212–9220.

Kotowicz, J. Skorek-Osikowska, A. & Bartela, L. 2011. Economic and environmental evaluation of selected advanced power generation technologies, *Journal of Power and Energy*, 225: 221–232.

Lee, C.G. Hur, H. & Song, M.B. 2011. Oxidation Behavior of Carbon in a Coin-Type Direct Carbon Fuel Cell, *Journal of Electrochemical Society*, 158(4): B410–B415.

Lee, I. Jin, S. Chun, D. Choi, H. Lee, S. Lee, K. & Yoo, J. 2014. Ash-free coal as fuel for direct carbon fuel cell, *Chemistry* 57(7): 1010–1018.

Milewski, J. Wołowicz, M. Badyda, K. & Misztal, Z. 2012. 36 kW Polymer exchange membrane fuel cell as combined heat and power unit, *ECS Transactions*, 42(1): 75–87.

*Advanced Materials, Structures and Mechanical Engineering – Kaloop (Ed.)*
© *2016 Taylor & Francis Group, London, ISBN 978-1-138-02793-0*

# Purification of plasmid DNA from *Escherichia coli* lysate using superparamagnetic nanoparticles modified with amino functional groups

X.L. Qiu, Y.P. Guan & K.K. Wang
*School of Materials Science and Engineering, University of Science and Technology Beijing, Beijing (USTB), China*

C. Guo
*Laboratory of Separation Science and Engineering, State Key Laboratory of Biochemical Engineering, Institute of Process Engineering, Chinese Academy of Sciences, Beijing, China*

ABSTRACT: Aiming at enhancing the efficiency and yield of purification, plasmid DNA purification by amine-functionalized $Fe_3O_4$ superparamagnetic nanoparticles was studied in this paper. Due to the large specific surface area, the adsorption capacity of these nanoparticles was found to be 447.3 mg/g, which have never been achieved with $Fe_3O_4$ particles whose average size is larger than 30 nm. The effects of ionic strength, pH and eluent mode during the elution process were investigated, and the yield was increased to 24.7 mg/g when an ultrasonic concussion was used.

## 1 INTRODUCTION

A plasmid is a small DNA molecule that can replicate independently, serving as an important tool in genetics and biotechnology laboratories (Ghanem et al. 2013, Rosati et al. 2005, Prather et al. 2003). The demand for plasmid DNA in medical applications shows a rapid growth rate with the development of medical research. Therefore, technology for the extraction and purification of plasmid DNA has become one of the most important factors. Methods for the purification of plasmid DNA rely mainly on chromatography methods. Chromatography is widely used for the purification of plasmid DNA because its product with high purity can meet the requirements of the gene pharmacy (Caramelo-Nunes et al. 2014, Diogo et al. 2005). However, limited by column media and pore structure, the productivity of plasmid DNA is low. Thus, the chromatography method is time-consuming and expensive (Paril et al. 2009, Shamlou 2003). In addition, non-chromatographic methods are also used, such as Polyethylene Glycol (PEG) precipitation or aqueous two-phase system separation, but these solutions are laborious and time-consuming in which a series of extraction and precipitation steps are needed, typically using toxic organic solvents such as phenol and chloroform. Due to such characteristics, these methods may not be suitable for the production of pharmaceutical grade plasmids (Humphreys et al.

1975, Ribeiro et al. 2002). The approaches of plasmid DNA purification within the past few years have indicate that there is an urgent need for high-capacity and high-productivity purification methods for plasmid DNA extraction.

As an alternative method, extraction of plasmid DNA with magnetic separation has been attractive because of their advantages of separation speed, easy use and affordability (Smith 2005). The principle of magnetic purification involves the use of magnetic particles modified with positively charged molecules to adsorb NDA via the electrostatic attraction, because the phosphate backbone of DNA is negatively charged (Tanaka et al. 2009), and then the complex particles binding with DNA are isolated under the action of an external magnetic field. Amino groups (−NH2) are widely used to provide cationic surface charge, and $Fe_3O_4$ magnetic particles are mostly preferred for DNA extraction due to their unique magnetic properties.

In 1994, T.L. Hawkins et al. (1994) used carboxyl-coated $Fe_3O_4$ magnetic particles (Cat No. # 8-4125B) obtained from PerSeptive Diagnostics (Cambridge, MA) to purify plasmid DNA, which they called SPRI (solid-phase reversible immobilization). The plasmid DNA purified by this method has a good structure, high biocompatibility and can be applied directly to the downstream application, such as Polymerase Chain Reaction (PCR) amplification. As $Fe_3O_4$ is superparamagnetic and can be uniformly dispersed in the solution, when a magnetic field is applied, $Fe_3O_4$ magnetic particles binding with plasmid DNA can be rapidly isolated from the suspension, and then separation can be achieved. Chiang et al. (2005, 2006) reported on the purification of plasmid DNA by using PEI-modified $Fe_3O_4$ magnetic nanobeads and silica-magnetite ($Fe_3O_4$) nanocomposites in 2005 and 2006. Due to the smaller size (31 nm) of particles, nearly 43 μg of high purity (A260/A280 ratio = 1.75), plasmid DNA was isolated from 3 ml of bacterial culture (Chiang et al. 2006). However, the nucleic acid extraction method based on hydroxyl also has disadvantages: the high concentration of chaotropic salt used during the binding step is harmful to the downstream enzyme digestion or sequencing analysis. Yoza (2002) purified DNA from *Escherichia coli* by using bacterial magnetic particles (size 50–100 nm) coated with 3-[2-(2-aminoethyl)-ethylamino]-propyltrimethoxysilane (A EEA). They claimed that the DNA binding efficiency increased with the presence of amino groups on the particle surface, and that the yield of purification DNA was 7.2 μg. But above all, the magnetic particles used today are all at micron and sub-micron levels, so the binding efficiency can be improved by reducing the particle size to increase the surface area.

In the present work, we aim at improving the efficiency and simplify the purification process by using the magnetic isolation method. Considering that the magnetic particles used before were all at the micron or sub-micron levels and in order to reduce the particle size for the purpose of increasing the specific surface area, nano-sized $Fe_3O_4$ particles modified with amino functional groups as the magnetic core were used in this research to enhance the throughput of plasmid DNA. The purity and the yield of plasmid DNA extracted from *Escherichia coli* lysate were detected by horizontal gel electrophoresis. The influences of other conditions such as ion strength, particles amount and eluting method on the purification of plasmid DNA were also investigated in this paper.

## 2 MATERIALS AND METHODS

### 2.1 Materials

Strains of Escherichia coli (pcDNA4) were supplied by the Institute of MateriaMedica, Chinese Academy of Medical Science. Superparamagnetic nanoparticles (primary amine group concentration 0.2 mmol/g) modified with amino functional groups were prepared in our laboratory (USTB). RNase A was purchased from Sigma. Supercoiled DNA Marker was prepared using the Trans2KTM Plus II DNA Marker. Deionized water and high purity water were prepared in our laboratory by using the ultrapure water purification system (Milli-Q SP, Nihon Millipore Ltd, Tokyo, Japan). (Hydroxymethyl)–aminomethane (Tris) were acquired from Aladdin (Shanghai). Glacial acetic acid (HAC), sodium hydroxide (NaOH), hydrochloric acid and disodium ethylenediaminetetraacetate (EDTA) were purchased from Sinopharm Chemical Reagent Co., Ltd (Beijing). All materials used were of analytical grade.

### 2.2 Zeta potential

Electrical property of modified particles depended on the pH condition. In order to confirm the suitable pH condition of the binding buffer and eluent, the zeta potential of the modified $Fe_3O_4$ nanoparticles in four different pH conditions was measured by a Delsa Nano C/Solid Surface Zeta-Potential-Analyzer. Measurements were replicated for three times at room temperature.

### 2.3 Purification of plasmid DNA

Crude samples of plasmid DNA were obtained from an *Escherichia coli* cell suspension by using a modification alkaline method described previously (H.C. Birnboim, J. Doly, Nucleic Acids Res. 7 (1979) 1513). The mixture of crude *Escherichia coli* lysates and the binding buffer containing 1% x(w/v) suspension of magnetic particles was placed in a 1.5 ml micro centrifuge tube, and then mixed gently for 2 minutes at room temperature. By using the permanent magnet, magnetic particles binding with plasmid DNA were immobilized. The addition of the magnet was continued for about 1 minute before removing the supernatant. The harvest particles were washed by the washing buffer after immobilization. Complex particles were placed in an elution buffer and mixed gently for about 1 minute. After removing the magnetic particles by using the permanent magnet, the supernatant was transferred to a fresh micro centrifuge tube, which is the pure plasmid DNA.

### 2.4 Absorption and elution strategy

In order to reveal the optimal amount of magnetic particles, two amounts of magnetic particles were studied at 0.1 mg and 0.5 mg. Moreover, during the purification, we found that the elution mode affects the purification result, and then two modes of elution were carried out, such as whirlpools vortex mixer or ultrasonic concussion.

### 2.5 Desorption of plasmid DNA

In the present study, the effect of the eluent NaCl concentration on the desorption of plasmid DNA was evaluated in the range of 0, 0.5 M and 1.0 M, respectively.

## 2.6 Agarose gel electrophoresis

Using the Bio-Rad (Gel Doc 2000, USA) horizontal gel electrophoresis unit, the purity of the plasmid DNA gained was detected. The running buffer was TAE (Tris-acetate-EDTA buffer), combined with a 1% agarose gel. Electrophoresis was then carried out at 80 V for 40 min. The concentration of plasmid DNA was measured by an ultraviolet spectrophotometer. By measuring the UV absorption wavelength value at 260 nm or 280 nm, respectively, the concentration, purification amount and purity were calculated.

## 3 RESULTS AND DISCUSSION

### 3.1 Binding buffer and eluent

DNA is a polyanionic molecule containing phosphate groups. It can be absorbed on the amine-functionalized magnetic nanoparticles. Meanwhile, the DNA molecule is conveniently eluted from the particles when the positive charge transforms to the negative one. The electrical property of modified magnetic nanoparticles depends on the pH condition of the binding buffer and eluent. The zeta potential of magnetic particles declines as the pH increases. It shows a positive electrical property on the left side of the isoelectric point, and an opposite effect on the other side. The zeta potential of nanomagnetic particles modified by amino groups was measured separately under different pH conditions. As shown in Figure 1, the isoelectric point is 7.2. Now we know that the modified magnetic particles will charge positive under the condition of pH 7.2, and they will also bind with plasmid DNA. Given that the DNA is acid

resistance, the binding buffer used here is HAc-NaAc at a condition of pH 5.0, and the eluent is Tris-HCl at a condition of pH 9.0.

### 3.2 Plasmid DNA absorption

By quantifying the amount of plasmid DNA in a crude solution (marked A), retained plasmid DNA in the supernatant after separation (marked B) and absorbed plasmid DNA (marked C) can be calculated. The results of ultraviolet spectrum detection are summarized in Table 1. The amount of plasmid DNA in crude *Escherichia coli* lysates is 450.90 μg. After the absorption by magnetic particles, the amount of plasmid DNA in the abandoned supernatant is 227.25 μg. Therefore, the amount of ideally harvested plasmid DNA should be 223.65 μg; however, the measurement value of the harvested plasmid DNA is 9.00 μg. In other words, there must be a large amount of plasmid DNA on the surface of magnetic particles that are not desorbed.

### 3.3 Effect of ionic strength on the desorption of plasmid DNA

NaCl was added to the eluent to increase the ionic strength. Excessive ionic concentrations intensify the competition for DNA adsorption on the surface of charged magnetic particles, thus accelerating the desorption of plasmid DNA. NaCl concentrations including 0 M, 0.5 M and 1.0 M combined with Tris-HCl (pH 9.0) elution were prepared for the subsequent experiments. Plasmid DNA absorbed by the $Fe_3O_4$ nanoparticles was eluted by using the prepared three reagents, and the results are summarized in Table 2.

As can be observed from Table 2, the elution yield of plasmid DNA increases with the increasing NaCl concentration. Therefore, the addition of

Table 1. Amount of plasmid DNA in different solutions.

| Group | A260 | A260/A280 | DNA (μg) |
| --- | --- | --- | --- |
| A | 20.04 | 2.06 | 450.90 |
| B | 6.11 | 1.96 | 227.25 |
| C | 0.90 | 1.84 | 9.00 |

Table 2. Amount of eluted plasmid DNA by elution at concentrations of 0 M, 0.5 M and 1.0 M NaCl.

| Group | A260 | A260/A280 | DNA (μg) |
| --- | --- | --- | --- |
| 0 M-NaCl | 0.592 | 1.83 | 4.44 |
| 0.5 M-NaCl | 0.690 | 1.78 | 5.18 |
| 1 M-NaCl | 0.738 | 1.82 | 5.54 |

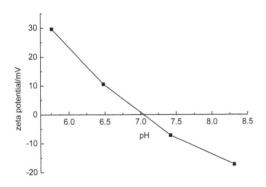

Figure 1. Zeta potential of amino-modified $Fe_3O_4$ nanoparticles.

NaCl facilitates the desorption of plasmid DNA from magnetic particles. Agarose gel electrophoresis detection was made to confirm the effect of NaCl concentration on the desorption of plasmid DNA directly. The results are shown in Figure 2.

### 3.4 The optimal amount of nanomagnetic particles and condition of desorption

Different amounts of $Fe_3O_4$ nanoparticles such as 0.1 mg and 0.5 mg were used for purification in order to confirm the optimal mass. It was found that the harvest amount increased from 2.56 μg to 9.00 μg when the particle mass reached up to 0.5 mg. The whirlpool blending or ultrasonic concussion was also conducted in the elution process for dispersing the aggregated particles, therefore increasing the production of plasmid DNA. The results indicated that whirlpool blending in the elution process increased the purification amount. The mass of the harvested plasmid DNA reached to 9.94 μg, increasing by 10 percent than the initial value. When using ultrasonic oscillation during purification, the maximum plasmid DNA elution was about 12.36 μg, increasing by 37.3 percent. All products had a very high purity and the ratio of A260/A280 was 1.84 and 1.79, respectively. The results are summarized in Table 3.

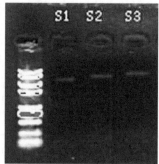

S1: 0M NaCl, S2: 0.5M NaCl and S3: 1M NaCl

Figure 2. Agarose gel electrophoresis of purified plasmid DNA at different ionic strength conditions.

Table 3. UV spectrophotometry results of yield and quality of eluted plasmid DNA from nanomagnetic particles under different conditions.

| Group | A260 | A260/A280 | DNA(μg) |
|---|---|---|---|
| 0.1 mg1E | 0.513 | 2.03 | 2.56 |
| 0.5 mg2E | 1.799 | 1.84 | 9.00 |
| 0.5 mg3E | 1.988 | 1.84 | 9.94 |
| 0.5 mg4E | 2.472 | 1.79 | 12.36 |

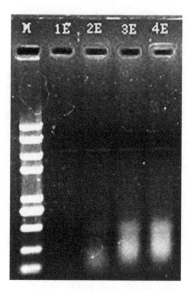

Figure 3. Agarose gel electrophoresis of purified plasmid DNA at different desorption methods.

Despite the increasing concentration of the product by using whirlpool blending or ultrasonic oscillation during purification, part of the plasmid DNA was destroyed. Electrophoresis results are shown in Figure 3. It should be noted that compared with group 1E, smeared bands at 0–500 pb observed in groups 2E, 3E and 4E are mostly due to the breakdown of supercoiled plasmid DNA during the desorption process. The above results suggests that whirlpool blending or ultrasonic oscillation is beneficial for desorption but may destroy the plasmid DNA. Therefore, these methods are unfavorable for practical applications.

## 4 DISCUSSION

During purification, we found that the mass of plasmid DNA binding on the surface of modified nanomagnetic particles was greatly large (approx. 300 μg), but no more than 13 μg was eluted. This is because the nanomagnetic particles showed excellent dispersibility during the binding step. A great amount of plasmid DNA was absorbed on the surface of modified magnetic particles through the interaction between the positive and negative charges. During the magnetic separation step, due to the agglomeration of nanomagnetic particles for its large specific surface area, part of the binding plasmid DNA was trapped in the aggregated particles. So the mass of desorbed plasmid DNA was less. Although it has been proved that the force or energy applied such as whirlpool blending and ultrasonic oscillation will improve the desorption amount that is affected by agglomeration, further research should be conducted to find a solution for avoiding nanoparticle agglomeration during the binding step.

## 5 CONCLUSION

In the present study, plasmid DNA purification was achieved by using the magnetic separation process with amino-modified $Fe_3O_4$ nanoparticles. The effect of purification was characterized by the application of separating pcDNA4. The results show that the desorption of plasmid DNA from modified magnetic particles was affected by ionic strength and pH condition of the elution buffer. Plasmid DNA was successfully eluted at the preferred ion concentrations (0.5 M) and pH (9.0) when the amount of the magnetic particles added was 0.5 mg. A product of high purity and free of impure molecules in the original bacteria can be obtained. The time periods of the whole process containing bacterial lysate for preparation and purification are less than 10 min. The mass of the binding plasmid DNA reached up to 300 µg while the amount of particles was 500 µg.

## ACKNOWLEDGMENTS

This study was financially supported by the National Natural Science Foundation of China (51274035) and the National Basic Research Program of China (2013CB632602).

## REFERENCES

Caramelo-Nunes, C. Almeida, P. & Marcos, J.C. et al. 2014. Aromatic ligands for plasmid deoxyribonucleic acid chromatographic analysis and purification: An overview. *Journal of Chromatography A*, 1327: 1–13.

Chiang, C.L. Sung, C.S. & Chen, C.Y. 2006. Application of silica–magnetite nanocomposites for the isolation of ultrapure plasmid DNA from bacterial cells. *Journal of magnetism and magnetic materials*, 305(2): 483–490.

Chiang, C.L. Sung, C.S. & Wu, T.F. et al. 2005. Application of superparamagnetic nanoparticles in a purification of plasmid DNA from bacterial cells. *Journal of Chromatography B*, 822(1): 54–60.

Diogo, M.M. Queiroz, J.A. & Prazeres, D.M.F. 2005. Chromatography of plasmid DNA. *Journal of Chromatography A*, 1069(1): 3–22.

Ghanem, A. Healey, R. & Adly, F.G. 2013. Current trends in separation of plasmid DNA vaccines: A review. *Analyticachimica acta*, 760: 1–15.

Hawkins, T.L. O'Connor-Morin, T. & Roy, A. et al. 1994. DNA purification and isolation using a solid-phase. *Nucleic acids research*, 22(21): 4543.

Humphreys, G.O. Willshaw, G.A. & Anderson, E.S. 1975. A simple method for the preparation of large quantities of pure plasmid DNA. *Biochimica et Biophysica Acta (BBA)-Nucleic Acids and Protein Synthesis*, 383(4): 457–463.

Paril, C. Horner, D. & Ganja, R. et al. 2009. Adsorption of pDNA on microparticulate charged surface. *Journal of biotechnology*, 141(1): 47–57.

Prather, K.J. Sagar, S. & Murphy, J. et al. 2003. Industrial scale production of plasmid DNA for vaccine and gene therapy: plasmid design, production, and purification. *Enzyme and microbial technology*, 33(7): 865–883.

Ribeiro, S.C. Monteiro, G.A. & Cabral, J.M.S. et al. 2002. Isolation of plasmid DNA from cell lysates by aqueous two phase systems. *Biotechnology and bioengineering*, 78(4): 376–384.

Rosati, M. von Gegerfelt, & A. Roth, P. et al. 2005. DNA vaccines expressing different forms of simian immunodeficiency virus antigens decrease viremia up-on SIVmac251 challenge. *Journal of virology*, 79(13): 8480–8492.

Shamlou, P.A. 2003. Scaleable processes for the manufacture of therapeutic quantities of plasmid DNA. *Biotechnology and applied biochemistry*, 37(3): 207–218.

Smith, C. 2005. Striving for purity: advances in protein purification, *Nature Methods*, 2(1): 71–77.

Tanaka, T. Sakai, R. Kobayashi, R. Hatakeyama, K. & Matsunaga, T. 2009. Contributions of Phosphate to DNA Adsorption/Desorption Behaviors on Aminosilane-Modified Magnetic Nanoparticles, *Langmuir*, 25(5): 2956–2961.

Yoza, B. Matsumoto, M. & Matsunaga, T. 2002. DNA extraction using modified bacterial magnetic particles in the presence of amino silane compound. *Journal of biotechnology*, 94(3): 217–224.

*Advanced Materials, Structures and Mechanical Engineering – Kaloop (Ed.)*
© 2016 Taylor & Francis Group, London, ISBN 978-1-138-02793-0

# Nanomodified bitumen composites: Solvation shells and rheology

E.V. Korolev, S.S. Inozemtcev & V.A. Smirnov
*Moscow State University of Civil Engineering, Moscow, Russian Federation*

ABSTRACT: Nanotechnology is now taking the worthy place in material science. One of the most effective nanomodification methods consists in formation of extra layers at the phase boundary either by means of chemical treatment or appropriate selection of preparation technology. The latter can lead to formation of solvation shells which stabilize dispersing system and improve operational properties of resulting building material. At present, despite the numerous models for a viscosity of disperse systems, there is no commonly accepted method which allows estimation of the thickness of solvation shells on the basis of rheological experiments. In the present work theoretical and experimental studies concerning origination and parameters of solvation shells on the surface of fine filler are performed. It is shown that proposed rheological method allows adequate estimation of the thickness of solvation shells.

## 1 INTRODUCTION AND PRIOR WORK

Current demands for advanced road construction materials lead to a shift of primary research direction towards a lower spatial level of constructional composites. Particular effects at micro- and nano-scale levels of bituminous and sulfur-bituminous mixtures are of most importance during structure formation of such disperse systems.

The state of nanotechnology in material science is adequately reflected in (Gupta et al. 2010): "nanomaterials are an important subset of nanotechnology". Strictly speaking, the nanotechnology was always employed in material science; it is only during last decades application of relevant methods becomes deliberate and systematic.

Concerning building materials for road construction, the modern trends of nanotechnology are mostly applications of high aspect ratio nano-scale filler platelets—layered silicates. Rheological properties of bitumen binder with an admixture of clays and organoclays were studied in numerous research works. It was shown that due to the addition of organomodified montmorillonite a number of physical properties can be enhanced (Jahromi et al. 2009), including resistance to rutting (Tao et al. 2010). It is discovered (Markanday et al. 2010) that bitumen systems with ethylene-vinyl acetate and organoclay are characterized by improved thermomechanical properties. Operational characteristics of bitumen-clay systems modified by styrene-butadiene-styrene copolymer are subject to an optimization in (Galooyak et al. 2010). Dedicated chapters about the application of organoclays are also in (Gupta et al. 2010).

Though positive prospective effect of organo-clay to bitumen binder is certain, there are also several drawbacks. The admixture of 2D nano-particles requires proper selection of homogenization methods (that are usually chemical surface treatments to make the clay more compatible with matrix material). Special measures for increasing the stability of mixes must also be taken (Galooyak et al. 2010).

This is why in our earlier research works we had proposed a different type of nano-modification. The proposed method is based on the assumption that surface treatment of fine filler is by itself can lead to significant enhancement of performance. Thus, while using 2D nanoplatelets is expedient, it is also not always necessary for some groups of fillers and treatment agents; in case of sulfur-based constructional composite, butadiene oligomers can successfully be used (Korolev et al. 2012). Moreover, since bitumen is a microheterogeneous system formed from primary (several nanometers in size) and complex (20–60 nm in size) structural units, for which the uniformity of system is disrupted when a mineral filler is introduced (Korolev et al. 1993), appropriate selection of preparation technology can lead to formation of bitumen absorption layer which stabilizes dispersing system and/or leads to improved operational properties of resulting asphalt concrete. In most cases, formation of oriented layers of bitumen is mostly due to physical sorption (Inozemtcev et al. 2014).

The aim of the present work is to perform theoretical and experimental studies concerning origination and parameters of solvation shells on the surface of fine filler. The shells can be formed either by fractions of bitumen (asphaltenes), or by nanomodifier with high molecular mass.

## 2 THEORETICAL STUDIES

The predominant influence of the fine filler to microstructure of asphalt concrete is mostly due to a large area of the phase boundary. Despite the relatively low amount of fine filler, total surface of its area is maximal among areas of other disperse phases. There is a simple way to demonstrate this. For the given specific surface $S_{s,i}$, density $\rho_i$ and volume $V_i$ of the $i$-th disperse phase, surface area can be expressed as:

$$S_{f,i} = S_{s,i}\rho_i V_i = \frac{1}{V} S_{s,i}\rho_i v_i, \quad i = \overline{1,N},$$

where, $V$—volume of the composition, $v_i$—volumetric fraction of the $i$-th phase, $N$—number of disperse phases.

Specific surface of the spherical particles:

$$S_s = \frac{6}{\rho d}.$$

Thus, for the area we get:

$$S_{f,i} = \frac{1}{V}\frac{6}{\rho_i d_i}\rho_i v_i = \frac{1}{V}\frac{6 v_i}{d_i}.$$

Total area of all phases is equal to sum:

$$S_f = \sum_{i=1}^{N} S_{f,i} = \frac{6}{V}\sum_{i=1}^{N}\frac{v_i}{d_i}.$$

Therefore, relative area of the $i$-th phase can be expressed in form:

$$\delta S_i = \frac{S_{f,i}}{S_f} = \frac{v_i}{d_i}\left(\sum_{j=1}^{N}\frac{v_j}{d_j}\right)^{-1}; \quad (1)$$

it can easily be verified that equality.

$$\sum_{i=1}^{N}\delta S_i = 1$$

holds for the obtained values.

Volumetric rates of the disperse phases can be determined a priori for the model of regular dense sphere packing with extra assumption.

$$d_{i-1} \gg d_i, \quad i = \overline{2,N}.$$

For such a model the volume occupied by the $i$-th phase equals to:

$$V_1 = V\eta_p, \quad V_i = \left(1 - \sum_{j=1}^{i-1}V_j\right)\eta_p, \quad i = \overline{2,N}.$$

In case of three disperse phases and hexagonal lattice the volumetric rates will be 60%, 24% and 9% (binder occupies about 6% of space); for the cubic lattice they will be 52%, 23% and 12% (binder occupies about 11%).

Due to several reasons model of regular dense sphere packing only roughly corresponds to asphalt concrete. The assumption about size ratio is incorrect for aggregate and coarse filler. The volumetric rate of aggregate is usually near 60%, and rates of coarse and fine fillers are near to 10% and 13%, respectively.

Relative areas of disperse phases obtained by (1) are presented in Tables 1 & 2.

As it follows from Tables 1 & 2, surface area of the fine filler dominates both for real asphalt concrete and for approximate model. The difference between Tables 1 & 2 is in the second row; such difference noticeably affects rheology of the composition.

A number of general considerations regarding interrelation between characteristic size and rheological properties of bituminous disperse systems with fine filler were former discussed in (Korolev et al. 2015), though obtained results allows neither determination, nor verification of shell size on the basis of rheological properties. It was also stated (Gladkikh et al. 2014) that dependence between mixture and rheology for a bituminous system with aggregates can only be subject of numerical investigation.

It is well known that viscosity of the bitumen in solvation shell (that is often called "structured

Table 1. Relative areas of disperse phases: cubic lattice.

| Phase | Average grain size, mm | Volumetric rate, % | Relative area, % |
|---|---|---|---|
| Aggregate | 5 | 52 | 0.8 |
| Coarse filler | 0.65 | 23 | 2.8 |
| Fine filler | 0.01 | 12 | 96 |

Table 2. Relative areas of disperse phases: real asphalt concrete.

| Phase | Average grain size, mm | Volumetric rate, % | Relative area, % |
|---|---|---|---|
| Aggregate | 5 | 60 | 0.9 |
| Coarse filler | 1.25 | 10 | 0.6 |
| Fine filler | 0.01 | 13 | 98 |

bitumen") is several orders of magnitude higher than viscosity of ordinary bitumen. One of the first methods of determination of solvation shell thickness was proposed in (Bakhrakh et al. 1969). The method is based on classical model (Einstein 1911) for viscosity of medium with non-interacting solid spheres. While such a model is only appropriate for diluted systems (Willenbacher et al. 2013), the obtained dependence

$$h = \frac{1}{\rho_{f,N} S_{s,N}} \left( \frac{k}{k_m} - 1 \right) \qquad (2)$$

(where, $\rho_{f,N}$ and $S_{s,N}$—density and specific surface of the fine filler, $k$ and $k_m$—empirically determined parameters in regression models which correspond to Einstein equation and were derived for bitumen and physical model system "filler-medium" without significant interaction on phase boundary) allowed adequate determination of the shell thickness (the results were in range 30–180 nm for temperatures 110–122 °C).

The drawback of the model presented in (Bakhrakh et al. 1969) is in assumption about identity in formation of layers for different dispersion mediums. In fact, the comparative analysis does not involve idealized systems that can be described by Einstein model. Experimental determination of parameters $k$ and $k_m$ is hindered by necessity to use dispersion medium which is:

a. Characterized by a thermal dependence of viscosity similar to the dependence for bitumen;
b. Inert to the surface of the fine filler (dispersion medium can be considered "inert" when contact angle is $\pi/2$: wetting does not change total energy of system).

The model (Bakhrakh et al. 1969) can be extended. Viscosity of dispersing system is affected by several factors. Among others, there are volumetric rate of filler and shape of particles of the fine filler. During preparation of the hot mix asphalt the formation of two surface layers takes place. First (absorptive) layer is mostly due to physical-chemical interaction between bitumen, nanomodifier and filler. Second (kinetic) layer depends on both shapes of the particles and viscosity of the dispersion medium. During experiment it is possible to directly measure the viscosity:

$$\eta_{ex} = \eta_0 \left( 1 + \alpha_0 \left( v_f + \Delta v_f \right) \right), \qquad (3)$$

where, $\eta_{ex}$ and $\eta_0$—viscosities of the dispersing system and dispersion medium, $\alpha_0$—form factor of the particle ($\alpha_0 = 2.5$ for spherical particles), $v_f$—volumetric rate of the filler, $\Delta v_f$—estimated

increment of the volumetric rate caused by absorption on the surface of filler:

$$\Delta v_f = \frac{\eta_{ex}(v_f) - \eta_0 \left( 1 + \alpha_0 v_f \right)}{v_f \alpha_0}. \qquad (4)$$

If we substitute,

$$v_f = N_f \frac{\pi}{6} d_f^3$$

into (4), we get,

$$\Delta v = v_f \left[ \left( \frac{d_f}{d_{f,0}} \right)^3 - 1 \right].$$

Taking into account,

$$d_f = d_{f,0} + 2h, \quad d_f = 6 / S_u \rho_f,$$

the thickness of the solvation shell will be:

$$h = \frac{3}{S_{f,N} \rho_{f,N}} \left( \left( \frac{\eta_{ex}(v_f) - \eta_0}{v_f \eta_0} + 1 \right)^{\frac{1}{3}} - 1 \right). \qquad (5)$$

If we take into consideration both shapes of the particles and presence of the absorptive layer, than:

$$\left( \frac{d_f}{d_{f,0}} \right)^3 = \frac{1}{\alpha v_f} \left( \frac{\eta_{ex}(v_f)}{\eta_0} - 1 \right),$$

and for thickness we get:

$$h = \frac{3}{S_{f,N} \rho_{f,N}} \left( \left( \frac{1}{\alpha v_f} \left( \frac{\eta_{ex}(v_f)}{\eta_0} - 1 \right) \right)^{\frac{1}{3}} - 1 \right). \qquad (6)$$

Since thickness of the solvation shell is the sum

$$h = h_a + h_k,$$

where, $h_a$ and $h_k$—thickness of the absorptive and kinetic layers, respectively, for the thickness of absorptive layer it can be derived from (5):

$$h_a = \frac{3}{S_{f,N} \rho_{f,N}} \left( \sqrt[3]{\frac{\eta_{ex}(v_f) - \eta_0}{v_f \alpha_0 \eta_0} + 1} - \sqrt[3]{\frac{\eta'_{ex}(v_f) - \eta'_0}{v_f \alpha_0 \eta'_0} + 1} \right), \qquad (7)$$

and (6):

$$h_a = \frac{3}{S_{f,N}\rho_{f,N}}\left(\sqrt[3]{\frac{\eta_{ex}(v_f)}{\alpha v_f \eta_0} - \frac{1}{\alpha v_f}} - \sqrt[3]{\frac{\eta'_{ex}(v_f)}{\alpha' v_f \eta'_0} - \frac{1}{\alpha' v_f}}\right),$$

(8)

where, apostrophe denotes model system. Viscosities and parameter $\alpha$ can be determined from the experimental studies.

## 3 EXPERIMENTAL SETUP

To determine the parameters required for calculations according to (5)-(8), we have performed series of experiments. Viscosity was measured for several bituminous and one model system. In the latter castor oil was used as a dispersion medium. The former were composed of BND 60/90 bitumen and four types of fine filler: fired diatomite (hereafter FDP), diatomite (DP), grinded quartz sand (QP) and grinded limestone (LP). For all types of fillers specific surface was 250 m²/kg. Volumetric rate of filling was varied in range 0–0.02. Viscosity was measured on MCR 101 rotation rheometer. The rotation rate was 20 s⁻¹. Temperatures of the bituminous systems were varied in range 120–150 °C.

## 4 EXPERIMENTAL RESULTS

Experimental dependencies between temperature and viscosity ($v_f = 0.01 = const$) and volumetric rate of filler and viscosity ($T = 135$ °C $= const$) are presented on Figures 1 & 2.

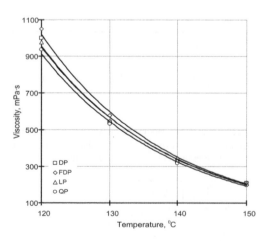

Figure 1. Dependencies between temperature and viscosity.

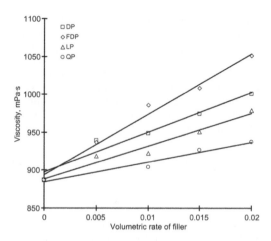

Figure 2. Dependencies between volumetric rate of filler and viscosity.

Table 3. Thickness of the solvation shell, nm.

| Type of filler | Temperature, °C | | | |
|---|---|---|---|---|
| | 120 | 130 | 140 | 150 |
| DP | 378 ± 40 | 355 ± 30 | 225 ± 20 | 41 ± 10 |
| FDP | 219 ± 40 | 178 ± 30 | 55 ± 20 | 27 ± 10 |
| LP | 448 ± 40 | 413 ± 30 | 272 ± 20 | 106 ± 10 |
| QP | 285 ± 40 | 185 ± 30 | 48 ± 20 | 25 ± 10 |

Series of experimental data similar to presented on Figure 2, upon completion of regression analysis, allowed to determine parameters for (5)-(8). Obtained values of the shell thickness are summarized in Table 3.

The data presented in Table 3 are in good correspondence with values obtained by means of different other methods (Zlotarev 1995). It is also evident from Table 3 that traditional limestone-based filler requires higher amount of binder.

## 5 SUMMARY AND CONCLUSIONS

Solvation shells formed on the particles of the fine filler considerably affect both rheological properties of the asphalt mixtures and operational properties of the asphalt concrete. Despite the numerous models for a viscosity of disperse systems (Willenbacher et al. 2013), there is no commonly accepted method which allows to estimate the thickness of such shells on the basis of rheological experiments.

In the present work we have proposed such a method. It is shown that obtained results correspond to values obtained by means of other methods.

## ACKNOWLEDGEMENTS

This work is supported by the Ministry of Science and Education of Russian Federation, Project # 7.11.2014/K "Theoretical and experimental study of the dynamics of constructions".

## REFERENCES

Bakhrakh, G.S. & Malinsky Yu.M. 1969. Estimation of the solvation shell thickness on the surface of mineral filler. *Colloid Journal*, (1): 21–24.

Einstein, A. 1906. Eine neue bestimmung der moleküldimensionen. *Annalen der Physik*, 19: 289–306.

Galooyak, S.S. Dabir, B. Nazarbeygi, A.E. & Moeini, A. 2010. Rheological properties and storage stability of bitumen/SBS/montmorillonite composites. *Construction and building materials*, 24(3): 300–307.

Gladkikh V.A., Korolev E.V. & Smirnov V.A. 2014. Modeling of the sulfur-bituminous concrete mix compaction. *Advanced Materials Research*, 1040: 525–528.

Gupta R.K. Kennel, E. & Kim, K.J. 2010. *Polymer nanocomposites handbook*. London: CRC Press.

Inozemtcev, S.S. & Korolev, E.V. 2014. Mineral carriers for nanoscale additives in bituminous concrete. *Advanced Materials Research*, 1040: 80–85.

Jahromi, S.G. & Khodaii, A. 2009. Effects of nanoclay on rheological properties of bitumen binder. *Construction and Building Materials*, 23(8): 2894–2904.

Korolev, E.V. Tarasov, R.V. Makarova L.V. & Smirnov, V.A. 2015. Model research of bitumen composition with nanoscale structural units. *Contemporary Engineering Sciences*, 8(9): 393–399.

Korolev, E.V., Smirnov, V.A. & Albakasov, A.I. 2012. Nanomodified composites with thermoplastic matrix. *Nanotechnologies in Construction*, (4): 81–87.

Korolev, I.V. & Solomentsev, A.B. 1993. Features of component interaction in asphalt-mineral systems. *Chemistry and Technology of Fuels and Oils*, 29(4): 192–195.

Markanday, S.S. Stastna, J. Polacco, G, Filippi, S. Kazatchkov I. & Zanzotto L. 2010. Rheology of bitumen modified by EVA-organoclay nanocomposites. *Journal of Applied Polymer Science*, 118(1): 557–565.

Tao, Y, Y. Yu, J.Y. Li, B. & Feng, P.C. 2010. Effect of different montmorillonites on rheological properties of bitumen/clay nanocomposites. *Journal of Central South University of Technology*, 15(s1): 172–175.

Willenbacher, N. & Georgieva, K. 2013. Rheology of disperse systems, in *Product Design and Engineering: Formulation of Gels and Pastes*. Weinheim: Wiley-VCH Verlag GmbH & Co. KGaA.

Zlotarev, V.A. 1995. Adhesion of bitumen on the surface of minerals: estimation. *Automotive roads*, (12): 13–15.

# Effects of the multi-step ausforming process on the microstructure evolution of nanobainite steel

C. Zhi, A. Zhao, J. He & H. Yang
*Metallurgical Engineering Research Institute, University of Science and Technology Beijing, Beijing, China*
*Beijing Laboratory for Modern Transportation Advanced Metal Materials and Processing Technology, Beijing, China*
*Collaborative Innovation Center of Universal Iron and Steel Technology, Beijing, China*

L. Qi
*Material Science and Engineering, Jiangxi University of Science and Technology, Jiangxi, China*

ABSTRACT:    The effect of the multi-step ausforming process on the microstructure and variants orientation relationship of nanobainitic steel was studied by using Scanning Electron Microscopy (SEM) and the electron backscatter diffraction (EBSD) technique. The relationship between bainite variants after multi-step ausforming is closer to the N-W orientation relationship. The volume fraction of variants V10 in nanostructured bainite variants is the highest (about 50%) and the morphology is continuously elongated needle, while variants V2 and V5 with an equiaxial island shape disperse in the matrix.

## 1 INTRODUCTION

Carbide-free bainite steels have been extensively studied because of their excellent tensile strength and toughness. Caballero and Bhadeshia (Bhadeshia 2013a, Bhadeshia 2013b, Lonardelli et al. 2012) designed a new kind of carbide-free bainite steel, which was called the "nanobainite steel" based on the theory of phase transformation strengthening. By isothermally treating at 200 °C for 5 days, the ultrafine microstructure with a thickness of 20–40 nm can be obtained. Nanobainite consists of a mixture of two phases, namely bainite ferrite and carbon-enriched retained of austenite. The ultimate strength of nanobainite steels is 2.5 GPa, and Vickers hardness is more than 600 HV. The best way of developing low-alloy high-strength steels is by using nanobainite because of its remarkable comprehensive mechanical properties, low cost and simple processing technique.

Although nanobainite steel has excellent comprehensive mechanical properties, it consumes long time to complete bainite transformation, which will be more than 72h when heat-treated in the range of 200~300 °C (Bhadeshia 2010, Podder et al. 2010, Bhadeshia et al. 2005, Huang et al. 2013). The speed of carbon diffusion becomes more slow and the strength of super-cooled austenite becomes stronger at low temperatures, so all these factors will extend the incubation period of bainite transformation and slow down the growth rate of bainite (Caballero et al. 2002).

Since 2010, Shanghai Jiao Tong University (Li H Y et al. 2010), Yanshan University (Long X Y et al. 2014) and Wuhan University of Science and Technology (Hu F et al. 2014, Bu C H et al. 2012) have begun to study ultra-fine bainite steel. Wuhan University of Science and Technology studied the effect of the ausforming process on super bainite steel, in which the carbon content was 0.393%. The results showed that the volume of bainite reduced when ausforming strain reached up to 50% (Bu et al. 2012). Recently, a new method called the "multi-step ausforming process" was found to accelerate high carbon (> 0.9%) nanobainite formation without decreasing the amount of bainite.

The aim of the present paper is to study the influence of the multi-step ausforming process on bainite microstructure evolution and crystallographic orientation relationship of nanobainite.

## 2 EXPERIMENTAL MATERIALS AND METHODS

The chemical composition of the investigated steel was Fe–0.91C–1.65Si–2.07 Mn–1.26 Cr–0.25Mo–1.56 Co–0.78Al (wt.%). After melting in the vacuum induction furnace, the cylinder casting ingot of Φ 50 mm × 600 mm was forged into 80 mm × 80 mm × 80 mm. Then, it was homogenized at 1552K for 24h, followed by hot rolling in the temperature range of 1173–1273 K to reduce the thickness of the ingot from 80 mm to 6 mm. A microstructure

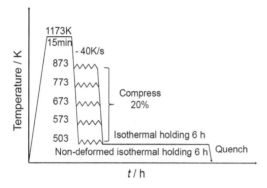

Figure 1. Schematic illustration of the ausforming process.

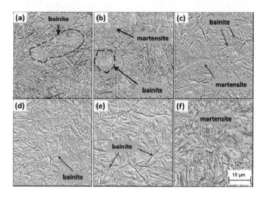

Figure 2. Microstructures deformed (20%) at different temperatures followed by 6 hours holding and quenching, (a) 230 °C, (b) 300 °C, (c) 400 °C, (d) 500 °C, (e) 600 °C and (f) undeformed.

study was carried out by using a Gleeble 3500 thermo-mechanical simulator machine, which has been described elsewhere (He et al. 2015). The specimens were austenitized at 1173 K for 15 min, and rapidly cooled to 230–600 °C, then compressed plastically to a 20% reduction, followed by isothermal treatment at 230 °C for 6h. The details are shown in Figure 1. Crystallographic orientation relationship analysis was carried out by the multi-step ausformed bainite process, which has been described elsewhere (He et al. 2015).

The specimens were cut from the middle of the warm rolling (multi-step ausformed) plate and were prepared by mechanical grinding followed by electrolytic polishing for metallographic and Electron Backscatter Diffraction (EBSD) examination. In this study, alcohol, perchlorate and glycerol (volume ratio 7:2:1) were used as electrolytes. Electrolytic polishing was carried out at room temperature at a voltage of 15 V, a current of 2 A and a time of 15 s. The EBSD measurement was performed at an accelerating voltage of 20 kV, a tilt angle of 70 ° and a step size of 0.06 μm using the ZEISS ULTRA 55 type field emission scanning electron microscope with the system of HKL.

## 3 RESULTS AND DISCUSSION

### 3.1 Effects of deformation temperature on nanobainite transformation

Figure 2 shows the SEM micrographs obtained by isothermal treatment at 230 °C for 6h after deformation at a different temperature. The microstructure deformed at 230 and 300 °C is illustrated in Figure 2(a) and (b). Bainite ferrite matrix with a small amount of block retained austenite and martensite is shown in Figure 2(a) and (b), which is highlighted by the black dotted line frame. The block microstructure in (a) is slightly less than that

in (b), but both of the bainite transformation are not complete (Jianguo He et al. 2015). Figure 2(c), (d) and (e) shows the microstructures obtained by deformation at 400, 500 and 600 °C. There is plenty of martensite, but a little of nanobainite, as shown in Figure 2(c). It is because the transformation of bainite is not complete and the retained austenite is transformed into martensite during quenching at room temperature.

It is also hard to find bainite ferrite sheaves in Figure 2(d) and (e), but for a little single bainite ferrite lath in martensite. The microstructure shown in Figure 2(e) indicates that bainite transformation has just onset after holding at 230 °C for 6h. The microstructure after isothermal treatment at 230 °C for 6h has no bainite formation, so martensite is completely obtained, as shown in Figure 2(f). Therefore, the incubation time for bainite transformation at 230 °C is longer than 6 hours. It can be concluded that deformation at 600 °C can accelerate bainite transformation. While the stored energy of deformation at higher temperatures is smaller than that at lower temperatures, the effect of accelerating bainite transformation is not obvious. However, the incubation period of transformation can also be shortened due to the plastic deformation before the isothermal treatment.

### 3.2 Effect of multi-step ausforming on nanobainite microstructure

The microstructure at all the stages consists of bainite ferrite (matrix) and retained austenite (block structure), and with the increase in the deformation, the block retained austenite decreases, as well as the directional nano-bainite lath increases. Figure 3 shows the SEM micrographs of

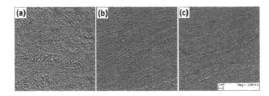

Figure 3. Thermal transformation microstructures of nanobainite steel after the multi-step ausforming process, with deformation strain values of (a) 25%, (b) 33% and (c) 50%.

nanobainite obtained by the multi-step ausforming process with deformation strain values of 25%, 33%, and 50%, respectively.

The increase in the deformation strain will lead to strong variation selection. Therefore, it is necessary to study the crystallographic of multi-step ausformed nanobainite. Studies have shown that ausformed at 800 °C leads to variant selection in bainite transformation (Gong et al. 2013, Gong et al. 2010). The deformation of super-cooled austenite will lead to a plastic flow of the grain, and the slip system of each grain is different. When the first slip system is predominant, the direction of dislocation will disperse along the {111} direction after deformation. The bainitic microstructure obtained is shown in Figure 3. When the Schmid factor of the second or third slip system is large enough, the dislocation can also slip and the bainite ferrite will form in the other direction. According to Takayama (Takayama et al. 2012), bainite variant selection will weaken with decreasing transformation temperature (non-deformed). At a lower transformation temperature, there is more resistance needed to overcome bainite transformation, because austenite becomes stronger; therefore, multi-variants will form at low temperatures. The stress and dislocation caused by bainite transformation is not sufficient to restrict the formation of multi-variants, so that almost all the 12 variants can be observed in a single austenite grain (Gong et al. 2010).

## 3.3 *The analysis of multi-step ausforming nanobainite variation orientation*

The nanobainite microstructure with multi-step ausforming presents great orientation selection on the mesoscopic scale. The angle between bainite films and the rolling direction is about 45 °. After multi-step ausforming, there are a lot of dislocations accumulating in super-cooled austenite, and the austenite grain is deformed and rotated. In order to investigate the influence of ausforming on bainite variants, we selected a single parent austenite grain for EBSD analysis. The Inverse

Pole Figure (IPF) of nanobainite crystal orientation with 50% deformation of multi-step ausforming is shown in Figure 4, where the different colors represent different crystallographic orientations and black indicates the retained austenite.

There is a strict orientation relationship when austenite (fcc) transforms to bainite (bcc). The crystallographic orientation relationship is found to include Kurdjumov-Sachs (K-S), Nishiyama-Wassermann (N-W) and Greninger-Troiano (G-T) (Jiang et al. 2008). The K-S and N-W relationship is presented in Table 1. Because different orientation relationships of bainite variants reflected in the location on the pole figure are unique, we can judge the relationship between the bainite variant orientation and the parent austenite phase.

Figure 5 shows the {100} polar diagram for all the bainite variant orientations illustrated in Figure 4. Figure 5(a) shows the {100} polar diagram of bainite, and the colors of the pole correspond to the colors of variants illustrated in Figure 4. The result of changing Euler Angle of the experiment polar diagram that rotated the original austenite crystal coordinate system to (001) [100] is shown as Figure 5(b). The pole density figure according to the extrusion of the high-frequency pole in order to compare with the polar diagram theory (Suikkanen et al. 2011) is shown in Figure 5(c). Obviously, the distribution of the pole in Figure 5(c) agrees with the N-W relationship polar diagram (Fig. 5(d)). One of the most obvious characteristics is that a pole of V10 variant is on the right side of the pole axis of the TD direction,

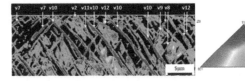

Figure 4. Inverse pole figure (IPF) map of multi-step ausforming nanobainite steel.

Table 1. Three typical orientation relationships between bainite (α) and austenite (γ) in steels.

| Orientation relationship | Plane | Direction |
|---|---|---|
| Kurdjumov-Sachs (K-S) | $\{111\}\gamma//\{011\}\alpha'$ | $<011>\gamma//<111>\alpha'$ |
| Nishiyama-Wassermann (N-W) | $\{111\}\gamma//\{011\}\alpha'$ | $<211>\gamma//<011>\alpha'$ |
| Greninger-Troiano (G-T) | $\{111\}\gamma\sim1°// \{011\}\alpha'$ | $<011>\gamma\sim2°// <111>\alpha'$ |

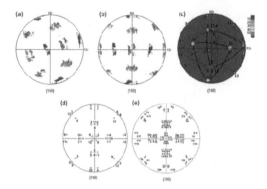

Figure 5. The {100} pole figure of bainite variants of multi-step ausforming nanobainite steel: (a) experimental pole figure, (b) pole figure after rotation, (c) pole figure contouring after rotation, (d) theoretical pole figure of N-W orientation relationship, (e) theoretical pole figure of K-S orientation relationship.

and a pole of V2 variant is on the lower part of the pole axis of the RD direction. There are only three poles of V2, V8, V5 and V11 on the RD pole axis on each half of the polar axis and arranged in a straight line, which disagrees with the K-S relationship polar diagram. Therefore, the relationship of multi-step ausforming nanobainite and original austenite agrees with the N-W relationship.

The experimental pole discretely distributes around the theoretical pole in Figure 5(b), and the V10 variant has the largest volume in all variants, which is more coherent than others, as indicated by the blue part in Figure 4, and agrees with the theoretical pole. The direction of V10 parallels to the maximum shear stress direction, which has a 45° angle with the rolling direction.

V2 and V5 variants in the microstructure have a 30–50% volume in all bainite variants and can be divided into small islands. The conclusions are different from the result of W Gong. The variants that are in agreement with the N-W relationship are V2, V3, V5 and V6, but there was no V10 in his study. In this paper, the super-cooled austenite was rolled at a higher temperature (starting rolling temperature 600 °C) with larger deformation (50%), so the plastic flow of grain and internal dislocation of the microstructure was more complex. The dislocation produced by ausforming at 600 °C can be easily removes and replaced, but the austenitic grain plastic flow produced by sliding cannot be replaced. Austenitic $\gamma(111)$ surface is the most prone to slip and turned to the direction of a 45° angle during ausforming. With the decreasing temperature during ausforming, the strength of super-cooled austenite increases, and the dislocation produced in the last several rolling passes is

along a certain direction on the $\gamma(111)$ surface. The direction is beneficial to the formation of bainite variant V10, thus the volume fraction of variant V10 by EBSD is the largest.

## 4 CONCLUSIONS

1. The multi-step ausforming process can accelerate the bainite transformation of carbon-rich silicon nanobainite steel. With decreasing ausforming temperature, the bainitic incubation period was shortened and the bainite transformation speed was accelerated.
2. Strong orientation selection can be found in the multi-step ausformed nanobainite microstructure. Large blocky retained austenite was eliminated as well.
3. The relationship between nanobainite (with multi-step ausforming of 50% strain) and the original austenite is close to the N–W relationship. The increasing multi-step ausforming deformation has a significant influence on the variant selection. The volume fraction of V10 variant is almost 50% in the whole nanobainite microstructure.

## ACKNOWLEDGMENT

The authors are grateful to the National Natural Science Foundation of China (Grant Nos. 51271035 and 51371032) and the Foundation of Jiangxi Education Department (GJJ14446) for funding this work.

## REFERENCES

Bhadeshia, H.K.D.H. 2010. Nanostructured bainite. *Proceedings of the Royal Society A: Mathematical, Physical and Engineering Science*. 466: 3–18.

Bhadeshia, H.K.D.H. 2013a. Computational design of advanced steels. *Scripta Materialia*. 70: 12–17.

Bhadeshia, H.K.D.H. 2013b. The first bulk nanostructured metal. *Science and Technology of Advanced Materials*. 14: 14202.

Bhadeshia, H.K.D.H. Caballero, F.G. & García-Mateo, C. 2005. Mechanical properties of low-temperature bainite. *Materials Science Forum*. 495–502.

Bu, C.H. Xu, G. & Liu, F. et al. 2012. Effect of low temperature deformation on microstructure of Fe-C-Mn-Si super bainite steel. *Heat treatment of metals*. 23–26.

Caballero, F.G. Bhadeshia, H.K.D.H. & Mawella, K.J.A. et al. 2002. Very strong low temperature bainite. *Materials science and technology*. 18: 279–284.

Gong, W. Tomota, Y. & Adachi, Y. et al. 2013. Effects of ausforming temperature on bainite transformation, microstructure and variant selection in nanobainite steel. *Acta Materialia*. 501: 65421.

Gong, W. Tomota, Y. & Koo, M.S. et al. 2010. Effect of ausforming on nanobainite steel. *Scripta Materialia.* 63: 819–822.

Hu, F. Wu, K. & Hodgson, P.D. et al. 2014. Refinement of Retained Austenite in Super-bainitic Steel by a Deep Cryogenic Treatment. *ISIJ International.* 222–226.

Huang, Y. Zhao, A.M. & He, J.G. et al. 2013. Microstructure, crystallography and nucleation mechanism of NANOBAIN steel. *International Journal of Minerals, Metallurgy, and Materials.* 1155–1163.

He, J.G. Zhao, A.M. Huang, Y. Zhi, C. & Zhao, F.Q. 2015. *Acceleration of bainite transformation at low temperature by warm rolling process.* Materials Today: Proceedings for the joint 3rd UK-China Steel Research Forum & 15th Chinese Materials Association-UK on Materials Science and Engineering.

He, J.G. Zhao, A.M. Huang, Y. Zhi, C. & Zhao, F.Q. 2015. Effect of Ausforming Temperature on Bainite Transformation of High Carbon Low Alloy Steel. *Materials Science Forum.* 817: 454–459.

Jiang, H. Wu, H. & Tang, D. et al. 2008. Influence of isothermal bainitic processing on the mechanical properties and microstructure characterization of TRIP steel. *Journal of University of Science and Technology Beijing, Mineral, Metallurgy, Material.* 15: 574–579.

Li, H.Y. & Jin, X.J. 2010. Determination of Dislocation Density in Nanostructured Bainitic Steels. *Journal of shanghai jiaotong university.* 613–615.

Lonardelli, I. Girardini, L. & Maines, L. et al. 2012. Nanostructured bainitic steel obtained by powder metallurgy approach: structure, transformation kinetics and mechanical properties. *Powder Metallurgy.* 55: 256–259.

Long, X.Y. Zhang, F.C. & Kang, J. et al. 2014. Low-temperature bainite in low-carbon steel. *Materials Science and Engineering: A.* 344–351.

Saha Podder, A. & Bhadeshia, H.K.D.H. 2010. Thermal stability of austenite retained in bainitic steels. *Materials Science and Engineering: A.* 527: 2121–2128.

Suikkanen, P.P. Cayron, C. & Deardo, A.J. et al. 2011. Crystallo-graphic Analysis of Martensite in 0.2 C-2.0 Mn-1.5 Si-0.6 Cr Steel using EBSD. *Journal of Materials Science & Technology.* 920–930.

Takayama, N. Miyamoto, G. & Furuhara, T. 2012. Effects of transformation temperature on variant pairing of bainitic ferrite in low carbon steel. *Acta Materialia.* 2387–2396.

# Author index